... *a must read for emergency managers, planners, first-line responders plus faculty and students involved in the study of emergency response, homeland security, and public health. Mike Fagel has a rare combination of both superb academic and hands-on, first responder credentials.*

Colonel Randall J. Larsen, USAF (ret)
Director
Institute for Homeland Security

Drs. Fagel and Hesterman provide an essential resource for security and emergency managers facing the daunting task of securing the most vulnerable of targets. Their comprehensive work covers all aspects of soft target hardening from planning and threat assessment through to emergency response. Soft Targets and Crisis Management *is an essential resource for every security and emergency manager's library.*

Tom Foley, D.Jur., CPP, PSP
Assistant Professor
College of Security and Intelligence
Embry-Riddle Aeronautical University–Prescott

In my 32-plus combined years in security, law enforcement, and the military, I am very impressed with the authors' extensive detail and expertise as it pertains to Soft Targets and Crisis Management: What Emergency Planners and Security Professionals Need to Know. *I had the opportunity to meet and discuss many subject areas of Drs. Fagel and Hesterman's book and would highly recommend this reading for law enforcement, emergency medical services, school administrators, risk management, and any others with responsibility for crisis management. Don't assume it can't happen to the people and/or property you are protecting; there is always a certain level of risk. What you are protecting could be next. As stated in the book, and I concur wholeheartedly, "Plan for the worst, hope for the best!"*

Michael D. Horak
Director and Chief of College Police
Dallas County Community College District

I had the privilege of meeting Mike Fagel a few years ago when he visited and taught at the London Fire Brigade to deliver his very personal account of his strategic command role at the scene of the 9/11 tragedy. I found his presentation compelling and moving and know that this latest book, with its many cross-collaborations, and coeditor and coauthors, will be of real value to anyone who has emergency management responsibilities. The world has witnessed some terrible natural and "man-made" disasters in recent years and the continuing threat from international terrorism has demonstrated yet again just how challenging responding to these types of incidents can be. I therefore know how important it is for people like Mike and his fellow authors to share their insight and experiences so that others can learn the important lessons and be better prepared for responding to the future major events and disasters that will inevitably happen.

Ron Dobson, CBE, QFSM, FIFireE
Commissioner
London Fire Brigade

I have had the pleasure to know and work alongside Dr. Michael Fagel for greater than 30 years. In that time, I have observed his commitment to emergency management response operations in small and large incidents in disaster situations in various parts of the United States, while having the ability to make extremely critical decisions in the provision of life and property. The decisions were delivered to the teams with grace based on his vast experience. Soft Targets and Crisis Management: What Emergency Planners and Security Professionals Need to Know *is presented as an understandable common-sense team-oriented approach to insure a successful outcome as based on prior experiences, planning, and execution of a plan. I completely enjoyed the opportunity to review and comment on this amazing resource publication, built for success in future events for all involved public safety agencies.*

Michael F. Spain
Chief of the District
Bensenville Fire Protection District
Bensenville, Illinois

The importance of effectively analyzing risks to soft targets, reducing vulnerabilities, and managing emergency situations is reinforced with astounding frequency on a global, national, regional, and local scale. Dr. Michael Fagel has drawn upon the expertise from a multidisciplinary team to produce a very accessible and useful tool to support decision making in times of crisis. Filling a massive and critical gap in homeland security education, the focus on soft targets and applicable guidance solidifies this publication as a must-read for first responders, planners, and policy makers alike. The authors bring both real-world experience and academic credentials to provide the rigor necessary to enhance the state of emergency management.

Dave Brannegan
Director
Risk and Infrastructure Science Center (RISC)
Argonne National Laboratory

SOFT TARGETS AND CRISIS MANAGEMENT

Uniting the best of Michael Fagel and Jennifer Hesterman's books in the fields of homeland security and emergency management, the editors of this volume present the prevailing issues affecting the homeland security community today. Many natural and man-made threats can impact our communities—but these well-known and highly respected authors create order from fear, guiding the reader through risk assessment, mitigation strategies, community EOC planning, and hardening measures based upon real-life examples, case studies, and current research in the practice.

SOFT TARGETS AND CRISIS MANAGEMENT

What Emergency Planners and Security Professionals Need to Know

Dr. Michael J. Fagel, CEM
Dr. Jennifer Hesterman
Colonel, US Air Force (Retired)

CRC Press
Taylor & Francis Group
Boca Raton London New York

CRC Press is an imprint of the
Taylor & Francis Group, an **informa** business

CRC Press
Taylor & Francis Group
6000 Broken Sound Parkway NW, Suite 300
Boca Raton, FL 33487-2742

© 2017 by Taylor & Francis Group, LLC
CRC Press is an imprint of Taylor & Francis Group, an Informa business

No claim to original U.S. Government works

Printed on acid-free paper
Version Date: 20160725

International Standard Book Number-13: 978-1-4987-5632-7 (Hardback)

This book contains information obtained from authentic and highly regarded sources. Reasonable efforts have been made to publish reliable data and information, but the author and publisher cannot assume responsibility for the validity of all materials or the consequences of their use. The authors and publishers have attempted to trace the copyright holders of all material reproduced in this publication and apologize to copyright holders if permission to publish in this form has not been obtained. If any copyright material has not been acknowledged please write and let us know so we may rectify in any future reprint.

Except as permitted under U.S. Copyright Law, no part of this book may be reprinted, reproduced, transmitted, or utilized in any form by any electronic, mechanical, or other means, now known or hereafter invented, including photocopying, microfilming, and recording, or in any information storage or retrieval system, without written permission from the publishers.

For permission to photocopy or use material electronically from this work, please access www.copyright.com (http://www.copyright.com/) or contact the Copyright Clearance Center, Inc. (CCC), 222 Rosewood Drive, Danvers, MA 01923, 978-750-8400. CCC is a not-for-profit organization that provides licenses and registration for a variety of users. For organizations that have been granted a photocopy license by the CCC, a separate system of payment has been arranged.

Trademark Notice: Product or corporate names may be trademarks or registered trademarks, and are used only for identification and explanation without intent to infringe.

Library of Congress Cataloging-in-Publication Data

Names: Fagel, Michael J., editor. | Hesterman, Jennifer L., editor.
Title: Soft targets and crisis management : what emergency planners and security professionals need to know / Michael J. Fagel, Jennifer L. Hesterman.
Description: Boca Raton, FL : CRC Press, 2017.
Identifiers: LCCN 2016018578| ISBN 9781498756327 (hardback) | ISBN 1498756328 (hardback) | ISBN 9781315451084 (web pdf) | ISBN 1315451085 (web pdf)
Subjects: LCSH: Terrorism--United States--Prevention. | Crisis management--United States. | Emergency management--United States. | Public safety--United States.
Classification: LCC HV6432 .S688 2016 | DDC 658.4/7--dc23
LC record available at https://lccn.loc.gov/2016018578

Visit the Taylor & Francis Web site at
http://www.taylorandfrancis.com

and the CRC Press Web site at
http://www.crcpress.com

Printed and bound in the United States of America by Publishers Graphics, LLC on sustainably sourced paper.

My coauthors and I have been practitioners and responders for many terrorism events. We were working in "Soft Targets and Crisis Management" before it was called Homeland Security.

We wish for you, our colleagues, to study, learn, and always maintain eternal vigilance as we look toward the future.

Sadly, all of us know "It's not a matter of if, but when again."

To the families of the victims, the responders, the planners, we humbly say thank you.

This work is designed to help those who may benefit from our experience and our collective thoughts as we help guide everyone to the future.

Our families have experienced as much turmoil as we have as responders, if not more.

To our families, we say thank you for supporting us and allowing us to do what we do.

Be prepared. Be vigilant. Be safe. Take care of one another.

Michael J. Fagel

Every day, there is a soft target attack in the world, carried out by the very same groups that threaten us here at home. We must fight complacency and wrap our minds around the idea that our citizens are the targets, where they work, study, worship, and play. Consequently, they deserve to understand the threats and how to protect themselves and their families. Arming them with this information will serve to lessen, not increase, fear and make them force multipliers for law enforcement and first responders.

My goal with this body of work is to ensure that all Americans are mentally and otherwise prepared for a scenario we hope never happens in our country. Living in the Middle East for two years opened my eyes to the threat and how we must engage now to protect people from attack. My experiences in the military securing installations and as an incident commander at airplane crashes, fires, hostage situations, and worse taught me much about preparedness and response. In short, not preparing for a soft target attack is naive and reckless. Burying our heads in the sand because it feels more comfortable than confronting the threat is something I am simply not willing to accept.

Thank you for caring about this important issue and being a person of action.

Jennifer Hesterman

CONTENTS

Foreword xiii
Preface xvii
Editors xix
Contributors xxiii

1 Soft Targets 1
   Jennifer Hesterman

2 The Psychology of Soft Targeting and Our Unique Vulnerability 9
   Jennifer Hesterman

3 Soft Target Hardening 101 27
   Jennifer Hesterman

4 The Common-Sense Guide for the CEO 51
   Michael J. Fagel

5 Planning for Terrorism 71
   Michael J. Fagel

6 Developing a Planning Team 97
   Michael J. Fagel

7 Developing an Emergency Operations Plan (EOP) 103
   Michael J. Fagel

8 Exercises: Testing Your Plans and Capabilities in a Controlled Environment 131
   James A. McGee

CONTENTS

9	ICS/EOC Interface *Michael J. Fagel*	157
10	EOC Management during Terrorist Incidents *Michael J. Fagel*	167
11	Emergency Management and the Media *Randall C. Duncan*	171
12	Deterring and Mitigating Attack *Jennifer Hesterman*	189
13	Soft Target Threat Assessment: Schools, Churches, and Hospitals *Jennifer Hesterman*	219
14	Soft Target Threat Assessment: Malls, Sporting Events, and Recreational Venues *Jennifer Hesterman*	253
15	Hospital Business Continuity *Linda Reissman and Jacob Neufeld*	281
16	Soft Targets, Active Shooters, and Workplace Violence *Lawrence J. Fennelly and Marianna A. Perry*	293
17	Sport Venue Emergency Planning *Stacey Hall*	307
18	Special Events *Patrick J. Jessee*	325
19	Coordinated Terrorist Attacks and the Public Health System *Raymond McPartland and Michael J. Fagel*	339

20	Hardening Tactics at Global Hotspots	353
	Jennifer Hesterman	
21	Developing Strategies for Emergency Management Programs	373
	S. Shane Stovall	
22	Soft Target Planning	391
	Michael J. Fagel and S. Shane Stovall	
23	Beyond the Response—The July 7 Bombings Inquest: A First-Person Account	403
	Gary Reason	
24	Infrastructure Protection: The Fusion Center's Role	421
	Vincent Noce	
25	Complex Coordinated Attacks	435
	J. Howard Murphy	
26	Violent Attacks and Soft Targets	457
	Rick C. Mathews	
27	Soft Target Cybersecurity: The Human Interface	465
	Michael J. Fagel, Erin Mersch, and Greg Benson	
Afterword		475
Index		479

FOREWORD

Soft Targets and Crisis Management: What Emergency Planners and Security Professionals Need to Know is an essential resource for law enforcement and security professionals, emergency managers, and students of homeland security and emergency preparedness. Dr. Michael Fagel and Dr. Jennifer Hesterman, both internationally recognized experts in the security and disaster management fields, and a superb team of very experienced, senior homeland security, intelligence, and emergency management professionals, provide information essential for those involved or preparing to become involved in such noble endeavors. I count it among my greatest privileges and highest honors to have worked with many of the contributors to this book within some of the most important national security and preparedness programs in the recent history of the United States, as well as on numerous and various local-, state-, national-, and international-level security, emergency preparedness, and disaster management operations.

The consummate professionals who invested knowledge, wisdom, effort, and time in developing the valuable tool that is this book clearly understand the challenges faced by members of the security and emergency preparedness community. The content they provide will not only address critical components and emerging trends and systems within homeland security and emergency management, but will also serve to fight the cognitive dissonance related to such topics within the minds of the readers and those with whom the readers and users of this text share its valuable and timely content. Of course, cognitive dissonance is a psychological process that prevents people from facing facts when the unthinkable becomes reality.

Warning! This book contains factual information about threats to people, places, systems, and freedoms that readers likely consider sacrosanct. Consider the recent attacks on soft targets in San Bernardino (2015), Paris (2015), Brussels (2016), and the less recent attacks on the Beslan primary school (2004), Mumbai (2008), Boston (2013), and Nairobi's Westgate Mall (2013)—not an all-inclusive listing of attacks, of course—as well as the consequences of those attacks. Consider the fact that the U.S. Department of Defense is now requiring Reserve Component facilities located within the United States, for example, Army Reserve Centers, Navy Reserve Centers, and so on, to develop and execute arming plans and operations reminiscent of the plans and operations required in areas characterized as semi-permissive environments around the globe. Semi-permissive environments are those with an operational environment in which the host nations' military and law enforcement agencies have limited control and capability to assist operations that a U.S. military unit intends to conduct. One could, therefore, logically assume that national-level decision makers using information about threats to U.S. military personnel operating within the United States have a reasonable belief that the operational environment within the United States is similar to that in nations like Qatar, the United Arab Emirates, Kuwait, Saudi Arabia, and others currently within a combat theater of operations.

If there was ever a time in recent U.S. history for security and emergency preparedness professionals to fight normalcy bias, now is that time! Normalcy bias is the mental state in which people underestimate the possibility of disaster and the possible effects of disaster. The consistent conflicts with cognitive dissonance and normalcy bias within the minds of security and emergency preparedness personnel should be like unto a truth stated by Ralph Waldo Emerson, "What goes on around you … compares little with what goes on inside you."* Grasping and embracing this truth is as essential for effective service as, to borrow a concept from Dr. Dave Grossman, a *sheepdog*. In his 2009 book, *On Killing: The Psychological Cost of Learning to Kill in War and Society*, Grossman provides descriptions of sheep, wolves, and sheepdogs within society. He describes sheep as kind, gentle, and productive creatures that only hurt one another by accident or under extreme provocation. Then, there are the wolves; Grossman describes the wolves as those who feed on the sheep without mercy. There are people in this world overcome by evil who are capable of evil deeds and, the moment you forget this fact or pretend it is not so, you become a sheep. There is no safety in normalcy bias, as there is no safety in denial.

So, what are members of the emergency management and security community to be? I firmly believe we are to possess both the Warrior Ethos and Service Ethos of what Grossman describes as sheepdogs. Sheepdogs live to protect the flock of sheep and confront the wolves. On average, out of every 100 people there are 98 sheep; people with no capacity for violence. Out of every 100 people there is on average, one wolf; a person with a capacity for violence and no empathy for the people around them. And, out of every 100 people there is, on average, one sheepdog; a person with a capacity for violence and a deep love for the people around you. Sheepdogs walk the hero's path.

As we enter what may be called a new era of hyperterrorism resulting from the interaction of complex geopolitical conflicts and a resurgence of apocalyptic religion, sheepdogs countering these and other threats will require an understanding of not only the tactical and operational aspects of security and emergency preparedness, but also strategic causes and effects. However, falling into the trap of narrowly focusing thoughts and efforts on countering terrorism is an error in judgment. Risks posed by major disasters are also projected to increase due to environmental changes, urbanization, mass population movement, and continuing technological and infrastructure innovation, or the lack of technological and infrastructure innovation in many communities. Owing to the interdependency of communities socially, politically, economically, and environmentally, whether regarding "community" as a national-level or an international community of nations, there are some clear realities associated with crisis management. These realities include the facts that the impacts of major disasters are more rapid, farther reaching, and interconnected, requiring a community approach to preventing, mitigating, preparing for, responding to, and recovering from such disasters, whether they be intentional or accidental, technological or natural.

As you consider the realities mentioned above and think critically to form judgments, and to strategically achieve success in your professional endeavors, use the invaluable information within this book. Use it as a tool, especially as you work to become more resilient as an individual while simultaneously facilitating and empowering resilience to those within your respective community, whether that community be local, state, national,

---

* Harrell, J. L. (2011). *The Peace Officer's Companion: 365 Days Worth of Wisdom of the Ages with Modern Commentary for Today's Peace Officer*. Bloomington, IN: AuthorHouse.

or international. In contemplating community resilience, I suggest that there are three steps. The first step is to anticipate the hazard(s) and/or threat(s) to a community and plan and prepare for them. The second step is to act in such a way during an incident that you mitigate cascading system failures: keep an emergency from becoming a disaster, and a disaster from becoming a catastrophe, much of which emergency managers accomplish through effective information and resource management at the operational to strategic levels, just as the incident commander(s) manage such at the tactical level. And the third step is to simultaneously bounce back both economically and culturally, and as rapidly as possible.

I commend you on your decision to use *Soft Targets and Crisis Management: What Emergency Planners and Security Professionals Need to Know* as a tool to prepare your respective community and build resilience. You will discover the invaluable information contained within this book fits well into the three steps of resilience, and will assist in your endeavors of the prudent but courageous as you seek to continuously grow within the emergency management and security professions.

J. Howard Murphy, MBA, MSS, CEM, FAcEM
Senior Homeland Security Program Manager

Assistant Professor and Emergency Management Degree Programs Coordinator
Anderson University (SC) School of Public Service and Administration

Former U.S. Strategic Command Senior Operations and Planning Officer
Syrian Chemical Weapons Demilitarization Support Operations
and
West Africa Ebola Response Operations

Former Commander of the U.S. Army's 1st CBRNE Incident Response Force

PREFACE

This textbook was assembled from the collective experiences of a dedicated team of practitioners who have worked to bring you this primer on Soft Target challenges and corresponding solutions.

As you explore each chapter, understand that these are guidelines to help determine the best plan to mitigate your vulnerabilities. Just like the complex threats we face, the solutions are also multifaceted and must be tailored to your situation—there is no "one-size-fits-all" solution.

The greatest benefits lie in planning, discussion, and research as you evaluate past incidents and move toward *your* solutions. Are there open discussions and teamwork, not just a paper plan? Do you consider and prepare for the worst-case scenario? Are you harvesting security best practices from other organizations and realms?

Please use our textbook and collective expertise as but one more tool in your tool box, and thank you for your efforts to keep our communities and citizens safe and secure.

Be Safe, Be Prepared

Michael J. Fagel
Jennifer Hesterman

EDITORS

Michael J. Fagel, PhD, CEM, CHS-V, has been involved in many phases of public safety service. His professional career spans more than four decades in fire, rescue, emergency medical services, law enforcement, public health, and emergency management, as well as corporate safety and security. Since 2003, he has supported many phases of Homeland Security operations in numerous capacities.

His latest textbook, *Crisis Management and Emergency Planning* (CRC Press/Taylor & Francis, 2014) was awarded the ASIS Inaugural Security Book of the Year in 2014.

Dr. Fagel is currently an adjunct professor at the Illinois Institute of Technology–Stuart School of Business, Masters in Public Affairs Program. Previously, he taught at Northwestern University in the Masters of Public Policy and Administration Program, delivering masters' level courses in biodefense, terrorism, and homeland security. In addition, Dr. Fagel teaches homeland security at Northern Illinois University, in Aurora University's Masters Program in the Criminal Justice Department, and is an instructor at Eastern Kentucky University, in the Safety, Security, and Emergency Management Master's Program. He supported the U.S. Army's SBCCOM at Aberdeen Proving Grounds in their Weapons of Mass Destruction (WMD) facility support operations for 48 months. He spent 32 months standing up the National Guard Bureau's CERIAC Fusion Center operations. He is a senior instructor at Louisiana State University's National Center for Biomedical Research and Training (NCBRT). He serves as a subject-matter expert (SME) for the National Center for Security & Preparedness, based in Albany, New York, supporting the New York State Division of Homeland Security and Emergency Services. He has been involved in numerous training classes for the Fusion Center and intelligence officials of the U.S. Department of Homeland Security (DHS).

Dr. Fagel has delivered several hundred lectures across the nation and written more than 250 articles on safety and disaster planning. He has served the National Domestic Preparedness Office (NDPO) SLAG team at the Federal Bureau of Investigation (FBI) in Washington, DC. Dr. Fagel spent 10 years at the Federal Emergency Management Agency (FEMA) in their Occupational Safety and Health Cadre in Washington, DC, responding to incidents and disasters such as the Oklahoma City Bombing, where he worked as a safety officer and a Critical Incident Stress Debriefer (CISD). He spent over 100 days at the World Trade Center for the New York City Fire Department (FDNY) at Ground Zero after the 9/11 attacks.

He was involved in numerous National Level Exercise efforts as well as Salt Lake City Emergency Operations Center (EOC) operations in 2002. He has been an exercise developer and lead for several regional operations as well as for specific federal partners.

Dr. Fagel has spent several deployments in the Middle East helping to create a national response plan and a new FEMA-type organization. He was a delegate to the European Conference on Emergency Management held in Budapest, Hungary, in 2007.

Along with other assignments, Dr. Fagel is a homeland security analyst at Argonne National Laboratories, where he is engaged in the protection of critical infrastructure.

He has served on numerous Occupational Safety and Health Administration (OSHA) Voluntary Protection Program (VPP) inspection teams as a Special Government Employee (SGE), with a background in safety, security, and disaster preparedness. He is a member of the Northern Illinois Critical Incident Stress Debriefing team, the International Association of Fire Chiefs Committee on Safety and Health, and served on their Terrorism Committee. He served on the Illinois Terrorism Task Force and was the Region V President for the International Association of Emergency Managers; he was also a Certified Emergency Manager Commissioner (CEM) for the International Association of Emergency Managers (IAEM). He spent 28 years at the North Aurora Fire Department as an Emergency Medical Services (EMS) coordinator and Emergency Manager, retiring as a Captain in 2003. Currently, he is a member of the board of trustees for the Sugar Grove (Illinois) Township Fire Protection District; he was a sheriff's deputy for 10 years.

Dr. Fagel has published several textbooks on emergency planning and crisis management:

> Fagel, Michael J., *Crisis Management and Emergency Planning: Preparing for Todays Challenges*, Boston, MA, CRC Press/Taylor & Francis, 2014. Earned the Security Textbook of the Year Award 2014.
> Fagel, Michael J., *Principles of Emergency Management: Hazard Specific Issues and Mitigation Strategies*, Boston, MA, Taylor & Francis, 2011.
> Fagel, Michael J., *Principles of Emergency Management and Emergency Operations Centers (EOC)*, Boston, MA, Taylor & Francis, 2010.
> Fagel, Michael J., *Emergency Operations: EOC Design*, Louisville, KY, Chicago Spectrum Press, 2008.
> Schuman, Michael S., Schneid, Thomas D., Schumann, B. R., and Fagel, Michael J., *Food Safety Law*, New York, Van Nostrand Reinhold, 1997.
> Fagel, Michael J., "Public Perception during the Oklahoma City Bombing," Dissertation, Columbia Southern University, 1996.

Dr. Fagel serves as a columnist for several national trade publications and has appeared on FOX, NBC, CBS, NPR, NY1, and local media outlets. He can be reached at Michael.Fagel@gmail.com.

Jennifer Hesterman, EdD, Colonel, U.S. Air Force (retired), served three Pentagon tours, and commanded in the field multiple times during her distinguished military career. Her last assignment was Vice Commander at Andrews Air Force Base, Maryland, where she was responsible for installation security, force support, and the 1st Helicopter Squadron, and regularly escorted the U.S. President and other heads of State on the ramp. Her decorations include the Legion of Merit, the Meritorious Service Medal with five oak leaf clusters, and the Global War on Terrorism medal.

She earned a doctorate from Benedictine University (Illinois), master of science degrees from Johns Hopkins University (Maryland) and Air University (Alabama), and a bachelor of science from Penn State University. In 2003, she was a National Defense Fellow at the Center for Strategic and International Studies in Washington, DC, where she immersed herself in a yearlong study of the nexus between organized crime and international terrorism. Her resulting book won the 2004 Air Force research prize and was published by AU Press. She is also a 2006 alumnus of the Harvard Senior Executive Fellows program.

Since her retirement from the military in 2007, Dr. Hesterman continues to serve as a cleared professional, studying international and domestic terrorist organizations, transnational threats, and organized crime. She is one of very few analysts specializing in the terror–crime nexus, and performs operational research on terrorist exploitation of the Internet through social networking, nonbank, Internet auction, and Dark Web sites. She is also a consultant on soft target vulnerabilities and hardening tactics for schools, churches, hospitals, and malls. Dr. Hesterman instructs and designs courses for federal government law enforcement and security organizations and serves on the ASIS Crime School Security Council. She is a senior fellow at the Center for Cyber and Homeland Security at George Washington University, serving on the Homeland Security and Emergent Threats panel.

Her recent book, *Soft Target Hardening: Protecting People from Attack*, was selected by ASIS as the Security Book of the Year for 2015. She also authored *The Terrorist–Criminal Nexus: An Alliance of International Drug Cartels, Organized Crime, and Terror Groups* (CRC Press/Taylor & Francis). Dr. Hesterman is a sought-after speaker, with thirty-five keynote and guest speaking events since 2007, including several venues abroad. She recently returned from a 2-year assignment in the Middle East. She can be reached at jenni@jennihesterman.org.

CONTRIBUTORS

Greg Benson has served in emergency services for over 35 years. He has worked on several Fire/EMS departments in capacities ranging from firefighter/paramedic to chief of department. He has also worked extensively in emergency management, assisting with development of an emergency management program in a Midwestern city of 100,000 people. Benson has designed, conducted, and evaluated regional training and exercises involving mass casualties and hazardous materials responses. Benson has extensive experience working with public- and private-sector entities in vulnerability assessments and development of emergency plans to address gaps. Benson teaches at Aurora University, College of DuPage, and is a field instructor with the Illinois Fire Service Institute, has an MPA from Northern Illinois University, and has been designated as a chief fire officer by the Center for Public Safety Excellence.

Roland Calia is the director of the Master in Public Administration program and the Master of Sustainability Management at the Illinois Institute of Technology–Stuart School of Business. Dr. Calia has extensive consulting experience in public financial management in the areas of budget, tax, and performance measurement for a wide variety of clients, including the State of Illinois, Cook County, and the Civic Federation of Chicago.

He previously served as a senior manager in the Research and Consulting Center of the Government Finance Officers Association (GFOA), where he managed GFOA's national training programs in financial management and staffed GFOA's Committee on Budgeting and Management. The Committee develops national recommended practices on state and local budgeting. He has served as a member of the Governmental Accounting Standards Advisory Council (GASAC), the standard-setting body for governmental accounting in the United States, and as a trustee of the Governmental Research Association (GRA). Dr. Calia earned a PhD in political science from the University of Chicago.

Randall C. Duncan, MS, MPA, CEM, recently concluded a 28-year career as a local government emergency manager in Kansas and Oklahoma. He served as the emergency management director for Wichita/Sedgwick County (Kansas) Emergency Management from 1998 to 2014. During his career, he administered a "baker's dozen" of presidential declarations of major disaster and emergency. In 2000, Duncan was the first local government emergency manager to accompany FEMA/EMI officials to the Republic of Turkey to assist Istanbul Technical University in setting up a center of excellence for emergency management. In 2001, Duncan provided support to the New York City Fire Department (FDNY) at the World Trade Center in conjunction with the U.S. Department of Justice. In 2012, Duncan provided assistance to the State of New York Department of Homeland Security and Emergency Services (DHSES) in their response to Superstorm Sandy. Duncan earned a master's in public affairs with an emphasis on disasters and emergency management from Park University and a bachelor of arts in political science from Southwestern

CONTRIBUTORS

College. He serves as an adjunct professor at the Hauptmann School of Public Affairs at Park University and Butler County (Kansas) Community College. Duncan is currently an instructor/lead exercise designer for Paradigm Liaison Services (http://www.pdigm.com). He can be reached at rduncan21@cox.net.

Lawrence J. Fennelly is currently the president of Litigation Consultants Inc. He is the author and editor of many security books for our industry. He is currently the Past Council Chairman of the Crime Prevention and Loss Prevention Council and of the School Safety and Security Council—an office that he held from 1985 to 1986. Mr. Fennelly retired from the Harvard University Police Department, where he was a staff sergeant in charge of Crime Prevention and Physical Security for the University. His recent books are titled *Handbook for School Safety and Security: Best Practices and Procedures*, *Crime Prevention Through Environmental Design*, 3rd Edition, *Security for Colleges and Universities*, *Security in the Year 2000 and Beyond*, *Security in 2025*, and *Physical Security: 150 Things You Should Know*.

Stacey Hall, PhD, is the associate director of the National Center for Spectator Sports Safety and Security (NCS4) and an associate professor of sport management at the University of Southern Mississippi (USM). Dr. Hall has been published in international sport management, homeland security, and emergency management journals. She has coauthored two textbooks—*Sport Facility Operations Management: A Global Perspective* and *Security Management for Sports and Special Events: An Interagency Approach to Creating Safe Facilities*. Additionally, she has been invited to publish in national magazines such as *Athletic Management*, *Athletic Administration*, and *Security Magazine*. Dr. Hall has been referred to as one of the nation's leading experts in sport security with interviews in *USA Today*, *ESPN the Magazine*, *CBS New York*, and *ESPN Outside the Lines*. Dr. Hall has presented papers at international and national conferences, and conducted invited presentations for U.S. federal and state agencies, college athletic conferences, and professional sport leagues. Dr. Hall has been the principal investigator on external grant awards in excess of $4M from the U.S. Department of Homeland Security to develop a sport event risk management curriculum, conduct risk assessments at college sport stadia, and develop training programs for sport venue staff.

Patrick J. Jessee, MS, NR-EMTP, CC-EMTP, is a firefighter/paramedic with the Chicago Fire Department. Since joining the department in 2000, he has worked in a wide variety of capacities within the department including as an officer for the Division of Emergency Medical Services, Hazardous Materials/Homeland Security instructor, program manager and policy advisor/author for Headquarters Staff, and firefighter/paramedic for the Division of Fire Suppression & Rescue. Currently, he is a member of Squad Company 2 of the Division of Special Operations.

Jessee earned undergraduate degrees in chemistry and biology from Southern Illinois University at Edwardsville and DePaul University, respectively.

He has also earned multiple master's, including a master's of science in biology from the University of Illinois at Urbana–Champaign, a master's of science in threat and response management from the University of Chicago (scientific discipline), and a master of arts in public policy and administration from Northwestern University.

CONTRIBUTORS

Rick C. Mathews, MS, is the Director of Simulations and Training for the College of Emergency Preparedness, Homeland Security, and Cybersecurity at the University at Albany State University of New York, at which he also serves on its affiliated faculty. He also hold an appointment as a Public Service Professor with the university's Rockefeller College of Public Service and Policy. Prior to his current appointment, Mathews served as the founding director of the National Center for Security & Preparedness at UAlbany. His professional experience spans over forty years with and includes emergency management, counter-terrorism, terrorism interdiction, risk management, and emergency medical services. He has provided technical assistance in his areas of expertise to state and federal governments, local emergency responder agencies, and major corporations.

James A. McGee, MS, earned a bachelor of science (BS) from California Polytechnic State University and a master of science (MS) from Virginia Commonwealth University. He has 21 years as a special agent with the Federal Bureau of Investigation (FBI). His experience includes 35 years addressing international security issues, counterterrorism investigations, crisis management, critical infrastructure protection, risk assessments, tactical operations, and homeland security initiatives. During his FBI career, McGee received numerous awards including the FBI Shield of Bravery, the FBI Medal of Merit, the FBI Medal of Valor, the Federal Law Enforcement Officers Association Medal of Valor, the U.S. Attorney General's Award for Exceptional Heroism, and the Department of Justice Certificate of Merit. McGee has been employed by the University of Southern Mississippi as a faculty member and as an adjunct professor at Tulane University. He is currently a faculty member and doctoral candidate with William Carey University's Department of Criminal Justice. He frequently teaches foreign law enforcement agencies on behalf of the U.S. Department of State Anti-Terrorism Assistance Program. McGee is designated as an expert witness in security issues regarding critical infrastructure protection, specifically venues of mass gatherings. McGee is an award-winning author, having written and published the book *Phase Line Green: The FCI Talladega Hostage Rescue* (2007). He also coauthored the textbook *Security Management for Sports and Special Events: An Interagency Approach to Creating Safe Facilities* (2011).

Raymond McPartland is the founder and chief executive officer of Tier One Associates, LLC. Tier One consults and develops training on matters of life safety, emergency preparedness, and homeland security for both the private and public sectors and specializes in creating unique, client-specific, realistic training, and preparedness programs. McPartland is a subject-matter expert with the National Center for Biomedical Research and Training (NCBRT) at Louisiana State University, and the Chemical, Biological, Radiological, and Nuclear Defense Information Analysis Center (CBRNIAC), as well as an active trainer and detective with the New York City Police Department (NYPD).

Currently assigned to the NYPD's Counterterrorism Division's Training Section as a lead instructor and curriculum development specialist, Detective McPartland is a leading instructor with his regional training team, responsible for instructing patrol, specialty personnel, and regional partners in various aspects of terrorism and counterterrorism. He is currently the Division's subject-matter expert on active shooter events and the primary author of the NYPD's published research work—*Active Shooter: Recommendations and Analysis for Risk Mitigation*. His other topics of instruction include active shooter preparedness and response, critical infrastructure protection, maritime terrorism, WMD and radiological awareness, hostile surveillance detection, and behavioral analysis and observation.

McPartland has attended numerous schools and training through the Department of Homeland Security's National Preparedness Consortium and FEMA's Emergency Management Institute.

Concurrent to his numerous duties as an NYPD detective and CEO of Tier One Associates, LLC, McPartland works part-time as an adjunct professor at Metropolitan College of New York in their MPA Program in Emergency and Disaster Management, and at Mercy College in their undergraduate program for Corporate and Homeland Security.

Erin Mersch has served as a coeditor of this textbook. She has earned a master's in public administration from the Illinois Institute of Technology in Chicago. She has taken numerous Homeland Security Masters Level Courses in bioterrorism, homeland security, and cybersecurity.

J. Howard Murphy, MBA, MSS, CEM, FAcEM, has contributed to the development of many of the existing United States' and allied nations' Emergency Management, National and Homeland Security, Complex Medical and Public Health, and Defense Support of Civil Authorities (DSCA) policies, programs, organizations, and operations. Murphy has mentored many professionals around the globe. Currently serving as the program coordinator and assistant professor for Anderson University's Emergency Services Management and Emergency Management degree programs, Murphy possesses 33 years of experience as an emergency responder, including tenures as a chief officer, director, and commissioner; 28 years in emergency management; 21 years in national and homeland security as a senior analyst and senior program manager; and 31 years of service in the U.S. Army and Army Reserve. In addition to his service as an Anderson University faculty member, he serves as the part-time Emergency Management Coordinator for Anderson County (South Carolina), and as a Colonel with the U.S. Army Reserve. Murphy is an accomplished writer, serving as author or coauthor of over thirty publications and courses for graduate and undergraduate emergency management, homeland security, and public health.

He can be contacted at jhmurphy@charter.net.

Jacob Neufeld is the emergency planning associate at Memorial Sloan-Kettering Cancer Center, in New York City. In addition to supporting the director with the hospital's preparedness program, Neufeld is directly responsible for enterprise-wide clinical and business continuity planning for the hospital, three research facilities, ten business locations, and seventeen regional care sites. He is a graduate of Metropolitan College of New York, where he earned an MPA in emergency and disaster management. He also attended the Disaster Recovery Institute International, where he achieved certification as Associate Business Continuity Planner. He also holds a certificate from the Israeli Military Industries School for Security and Anti-Terror Training. Neufeld is an active volunteer with his local American Red Cross chapter, serving as a coordinator of continuity of operations planner and disaster responder.

Vincent Noce has served in various law enforcement, investigative, and intelligence roles in government since 1997, including as the director of a U.S. Department of Homeland Security–designated intelligence fusion center. Noce earned his bachelor of arts from Marquette University and his master's at the Naval Postgraduate School.

Marianna A. Perry has over 35 years of progressive experience in law enforcement, physical security, safety, and loss control. Marianna earned her bachelor's from Bellarmine University and a MS from Eastern Kentucky University. She is a safety and security consultant with Loss Prevention and Safety Management, LLC and is an adjunct faculty member at Sullivan University in the Department of Justice and Public Safety Administration. Perry is a former detective with the Kentucky State Police and was previously the director of the National Crime Prevention Institute (NCPI) at the University of Louisville. She is a member of the School Safety and Security Council as well as the Women in Security Council. Her recent books are titled *The Handbook for School Safety and Security, Security for Colleges and Universities*, and *Physical Security: 150 Things You Should Know*.

Gary Reason, QFSM, former director of Operational Resilience and Training, London Fire Brigade, United Kingdom.

Reason served in the London Fire Brigade as an operational firefighter for over 30 years. During his long and distinguished career, he accrued a wealth of experience representing the Brigade at all levels of management, including the most senior leadership and operational command positions. Reason successfully managed operations at some of London's most challenging and complex emergencies and major incidents and regularly represented the Brigade at multiagency strategic coordination meetings and political briefings. In 2010, Reason was nominated the lead officer for managing the London Fire Brigade's preparations for the Inquests relating to the fifty-two civilians who were killed as a result of the terrorist attacks that occurred in London on July 7, 2005, and was also the most senior fire officer to give evidence during the Inquest proceedings.

Reason was awarded the Queen's Fire Service Medal (QFSM) in the 2013 New Year's Honours list for his exemplary commitment and contribution to the Brigade and to London as a whole, and in his final role as director of Operational Resilience and Training Reason managed the Brigade's functions covering operational and emergency planning, operational procedures, operational assurance, health and safety, human resources, staff training and development, and Inter-Agency Liaison, as well as the development of new response capabilities for emerging terrorist threats.

Linda Reissman has over 30 years of experience in the municipal, private, and governmental sectors of the emergency management, public health, and EMS disciplines and earned a master of science in emergency and disaster preparedness. Reissman is currently the director of emergency management for Memorial Sloan-Kettering Cancer Center (MSKCC) and Network. She is responsible for emergency preparedness and business continuity for the 450-bed hospital, including more than thirty clinical, research, and business sites in New York City (NYC) and regionally. From 1996 to 2010, Reissman served as training officer for HHS/NDMS Disaster Medical Assistance Team (DMAT), NY-4, and is currently assigned to NY-5, a developmental DMAT team. As a member of DMAT, Reissman has participated in the development and execution of integrated federal regional exercises and has deployed to disasters for team activations, including 9/11.

Prior to her position as MSKCC, Reissman served for 8 years as the senior EMS representative with the New York State Bureau of EMS, where she oversaw EMS agency compliance and coordinated hospital and EMS disaster-response plans in cooperation with local, state, and federal emergency management authorities. She has also served with the NYC Office of Emergency Management as a planner during its inception and assisted in the development of the NYC Bio-Terrorism Plan and Family Assistance Plan, and spearheaded a $500,000 Federal Terrorism Grant in 2010, in cooperation with Mount Sinai Medical Center. Reissman coauthored a $500,000 ASPR/NYCDOH (Office of the assistant Secretary for Preparedness and Response/New York City Department of Health) Consortium grant that will address disaster planning for patients with special health care needs.

Reissman is also a consultant/emergency management planner, with the Strategic Emergency Group (SEG), specializing in eight Emergency Support Functions (ESF) project aspects, including comprehensive emergency management planning, Homeland Security Exercise Evaluation Program (HSEEP) exercises, response and recovery, and critical infrastructure protection and resilience. She also serves as a core team member on a number of SEG projects including the statewide HSEEP exercise program for the State of Missouri Emergency Management (SEMA) and the Missouri Hospital Association (MHA), where she supports exercise design, operations, and analysis as an exercise planner, controller, and lead evaluator.

Reissman has been a national speaker on resiliency planning for health care and most recently presented at the International Association of Emergency Managers (IAEM) Annual Conference and the Baylor Emergency Management Symposium.

S. Shane Stovall, CEM, serves as the director of Emergency Services for Western Carolina University. Stovall has served over 20 years in the emergency management field, in both the public and private sectors. This includes serving as the director of emergency management for the City of Plano, Texas, and in various positions at the Charlotte County Office of Emergency Management in Punta Gorda, Florida. Stovall graduated in December 1995 from the University of North Texas with a bachelor's in emergency administration and planning. In addition to his education, Stovall holds certification as a certified emergency manager (CEM) from the International Association of Emergency Managers. Stovall is an active member of the emergency management community, and has served in many national roles promoting emergency management programs. Stovall has performed in various capacities in large-scale emergencies and disasters throughout his career. He has developed master's-level courses on terrorism, emergency management, homeland security, and cybersecurity for several universities. Stovall has also written several chapters in the CRC series on Emergency Planning and Crisis Management.

1

Soft Targets

Jennifer Hesterman

Contents

Introduction 1
The Difference in Threat, Vulnerability, and Risk 6
References 8

> I'm still at the Bataclan. 1st floor. Hurt bad! They are faster on the assault. There are survivors inside. They are cutting down all the world. One by one. 1st floor soon!!!!
>
> Facebook post
> *November 13, 2015*

INTRODUCTION

When hundreds of music lovers walked into the historic Bataclan concert hall in Paris, they were looking forward to an evening of good music, fun and fellowship. As American rock band *The Eagles of Death Metal* played its first set, laughing and joking with the crowd, ISIS terrorists were circling behind the Bataclan, preparing for their violent assault. At 9:42 pm, the gunmen sent a text to an unknown contact stating "We've left" and "We're starting." After throwing the mobile phone into a trash can, the group stormed the concert hall and began their assault with a barrage of AK-47 gunfire. People thought the action was part of the show, perhaps pyrotechnics or firecrackers, until the horrified band ran from the stage. What started as a night of fun in one of the world's most revered cities ended in tragedy, not only for those at the concert hall but at a stadium and several outdoor cafés. ISIS successfully carried out its first coordinated soft target attack against a western city, killing 130 people and injuring 350.

In the post-9/11 world, the United States has made great strides to further reinforce already hardened targets such as military installations, government buildings, and transportation systems. Those facilities now employ concentric rings of security, more cameras,

Photo 1.1 Man grieves for victims of the November 13, 2015 attacks in Paris for which the Islamic State in Iraq and Syria (ISIS) has claimed responsibility for the terrorist attacks in which multiple operatives attacked soft targets. (From the New Jersey Department of Homeland Security.)

and increased security manpower, serving to repel would-be terrorists and violent criminals. However, as these hard targets are further reinforced with new technology and tactics, soft civilian-centric targets such as those attacked in Paris are of increasing interest to terrorists. But this concept is not new; although lost in the news at the time, evidence collected following the 9/11 attack proved the aircraft hijackers also accomplished preliminary planning against soft targets, surveying and sketching at least five sites including Walt Disney World, Disneyland, the Mall of America, the Sears Tower, and unspecified sports facilities (Merzer, Savino, and Murphy 2001). Despite horrific terrorist operations against civilians such as the 2002 Beslan school massacre, the 2005 Moscow theater siege, the 2008 Mumbai attacks, and the 2013 Nairobi shopping mall assault, few resources have been applied toward hardening similar soft targets in the United States. A very small portion of our national security budget and effort is spent protecting civilian venues. Responsibility for security is often passed on to owners and operators, who have no training and few resources. In military terms, we are leaving our flank exposed.

The problem is far more complex than a simple lack of funding; the challenges are also psychological and tactical. First, contemporary terrorism has no moral boundaries. Who could have predicted, even 10 years ago, that schools, churches, and hospitals would be considered routine, legitimate targets for terrorist groups? Outrage and outcry at the beginning of this soft targeting era has given way to acceptance. Psychologically, it is more comfortable to pretend there is no threat here in the homeland, that these heinous attacks will always happen "somewhere else." Americans also have a very short memory—a blessing when it comes to resiliency and bouncing back from events like devastating civil and world wars, the Great Depression, and 9/11. However, in terms of security, this short-term memory can be our Achilles heel when faced with a determined, patient enemy. In fact, we almost have a revisionist history, downplaying and explaining away previous attacks as individual acts of violence by groups of madmen, not seeing the larger trends.

There is also a tendency to hide behind numbers; for instance, the number of violent crimes committed in our country is down, which makes us feel good about our security efforts. However, the scale of violence and deaths resulting from those crimes is rising, an underlying trend that is far more important. Some facility owners and managers may even choose to roll the dice, sacrificing a robust security posture to provide a more pleasant experience for students, worshippers, shoppers, or sporting or recreational event fans. We must remember the choice of target for terrorists is not random, particularly for radical religious groups seeking elevated body counts, press coverage, and a fearful populace in order to further their goals. Hardened targets repel bad actors, and unprotected soft targets invite.

Attacks against soft targets have a powerful effect on the psyche of the populace. Modern terrorist groups and actors have redrawn the battlefield lines, and places where civilians once felt secure have been pulled into the war zone. Persistent, lethal attacks by ISIS and al-Qaeda–affiliated terrorists against churches, hospitals, and schools in the Middle East and Africa have even successfully shifted the center of gravity in major conflicts. When places formerly considered "safe" become targets, frightened civilians lose the will to fight. They may flee and surrender the territory to the aggressor or rise up en masse, compelling the government to make concessions to insurgents to stop the brutality against noncombatants.

Due to our country's adherence to the Geneva Conventions, we will not attack soft targets without military necessity, even if the enemy takes shelter among the people. Civilians are not treated as combatants and therefore are not targets. Injured civilians are protected, not fired upon. Places of worship are never purposely hit. Schoolyards filled with children are not a target. We are intellectually unwilling to imagine an enemy who does not share what we believe to be universally accepted moral codes; therefore, we have a severe blind spot and are wholly unprepared to protect soft targets in our country. We must understand that our enemies see a busload of children and a church full of people as legitimate targets. Terrorists do not care about "collateral damage," a phrase Timothy McVeigh unremorsefully used to describe the daycare children killed in his attack at the Alfred Murrah Building. We also tend to forget that domestic terrorists have routinely hit soft targets in our country, from arson to shooting, and bombings. And, no matter your gender, race, religion, or sexual preference, there is a hate group in the United States actively or passively targeting you. Therefore, we must prepare both psychologically and tactically to harden our soft targets and lessen their vulnerability.

Protecting soft targets presents unique challenges for law enforcement because the buildings are privately owned and responsibility rests on the owners to secure the property and its occupants. Therefore, partnering is critical: educating the owners on the threat, assisting with vulnerability assessments, and helping to harden the venue or establishment. However, many soft target venue owners and operators we've spoken with about the possibility of a terrorist attack on their property convey a feeling of *hopelessness* (there is not much we can do to prevent or mitigate the threat), *infallibility* (it will never happen here), or *inescapability* (it is destiny or unavoidable, so why even try). Those who frequent soft target facilities—employees and patrons alike—typically believe "It can't happen to me," showing a sense of *invulnerability*. Even worse, others may believe that "If it is going to happen, there is nothing I can do about it anyway," expressing *inevitability*. People with these mindsets are a detriment in an emergency situation, with no awareness of the threat, mental preparation, or sense of determination to engage during the situation. In an emergency, those without a plan or resolve may wait for first responders and law enforcement to arrive and rescue them before taking steps to save their lives and those of others. The Sandy Hook shooting event was over in 6 minutes, with twenty-six dead: there is no time to wait for help when the attacker is determined and brings heavy firepower to the fight. You are the first responder.

Most experts agree that, with our newly robust intelligence capabilities, another coordinated mass casualty event in multiple locations such as the 9/11 attacks is unlikely. However, the threat of a Paris- or Mumbai-style event in a city with multiple avenues of approach (water and land) or a mass casualty bombing at a soft target location is more probable and will have an associated shock factor. There is a general hesitation for the government to share specific threat information with the public, perhaps because we do not want to cause panic; however, education is the best way to lower fear, as people will feel they can protect themselves and their loved ones. As witnessed in several natural disaster events in our country since 9/11, citizens are generally overly reliant on the government, lacking supplies at home as simple as flashlights, radios, batteries, nonperishable food, and water. Unfortunately, many police departments and hospitals have taken large funding cuts due to our country's financial crisis, meaning response time may be slower than anticipated. During mass casualty events such as shootings or a fire, victims routinely are unable to locate emergency exits and they have no plan to defend themselves and others. Furthermore, most people do not understand what it means to "shelter in place" or how to follow other orders given during a serious emergency. The combination of lack of education about the threat, a feeling of invulnerability regarding soft target attacks, overreliance on the government for help, and lack of first-response resources is potentially disastrous. Citizens must be educated on the threat and response and become valuable force multipliers instead of adding to challenges at the scene.

Security training and resources are typically the first to go during budget-cutting drills. When leaders are faced with a budgeting dilemma, a good question might be: "What is the cost for not protecting our people?" Certainly, most schools, churches, and hospitals are not flush with cash and find it difficult to spend valuable dollars on security. Often, decisions are arbitrarily made instead of using a system to assess vulnerability and threat, and then obtain the right mitigation tools to lower risk and protect the unique venue. With regard to profit-making soft targets such as malls and sporting and recreational venues, there is a desire for balance between security and convenience minimizing the impact to the customer. Business owners see customer backlash when other facilities add layers of active, hands-on security; it likely discourages them from pursuing similar activities. For example, the addition of backscatter technology at airports drew the ire and scrutiny of millions of people, many of whom did not even fly on a regular basis. Even news of the almost-catastrophes in flight with shoe and underwear bombs did little to persuade the public for the necessity for the systems. Security may not be popular and decisions should not be made by consent.

If businesses are concerned about garnering a reputation for long security lines and visitor frustration due to security measures, owners should consider the consequences should a mass casualty terrorist or violent criminal attack happen on their property. For instance, the movie industry as a whole was impacted by the shootings at the Cinemark Theater in Aurora, Colorado, on July 20, 2012, at the premiere of the movie *The Dark Knight*

Rises. In response to the violent attack, the film's producer, Warner Brothers Studios, pulled that movie and all of its violent movies from theaters. As attendance dropped dramatically worldwide, theater owners paid extra for increased security to reassure their customers. Cinemark not only paid for the burial expenses of the twelve victims, but also gave each family $220,000. The company was able to avoid paying millions of dollars of hospital bills for the seventy injured theatergoers, as they were forgiven and funded by the state. However, despite these actions, Cinemark was still sued by the families for not preventing the event, with decisions pending.

The lavish Westgate Mall, which was portrayed as a symbol of Kenya's future and cost hundreds of millions of dollars to build, had high-end stores and affluent customers who generated millions of dollars in weekly revenue. The mall was completely destroyed in the 4-day siege with al-Shabaab terrorists, and only half of the store owners had terrorism insurance (Vogt 2014). Stores were looted during and after the event by corrupt soldiers, adding to the financial ruin of shop owners. Rebuilding the mall was a must to show resilience and national pride; however, the cost is exorbitant for the country and insurers. Also, the inability of mall personnel to detect the planning stages of the attack, the ineffective response by armed mall security to put down the offensive, and the lack of communication with shoppers and store owners inside the mall about the unfolding events have been widely criticized. The downplaying of the severity of the situation by the government and its sluggish, uncoordinated response cast doubt on its ability to handle violent events in the country. The tourism industry, critical to Kenya's fragile economy, was hit hard, with 20 percent fewer visitors in the months following the attack, and hopes for hosting future Olympic Games and World Cup soccer events dashed. Lax security at one mall sent devastating ripples through an entire national economy and harmed future prospects for development.

Although international terror remains a viable threat to our country, domestic terrorism from right-wing, left-wing, and single-issue groups is perhaps a greater daily concern

Photo 1.2 Westgate Mall, Kenya, following the attack by al-Shabaab terrorists, September 21, 2013. (From the U.S. State Department.)

for our law enforcement agencies. The growing propensity of these organizations and their members to "act out" on soft targets and to step up and engage law enforcement is alarming. The radicalization of Americans continues, with several successful attacks by ISIS and al-Qaeda-inspired homegrown jihadists and more than sixty more thwarted since 9/11. Exacerbating the threat, a lack of a rehabilitation program means there is no way of ensuring those who serve their prison sentence and return to society will not go back to their old terroristic ways ... with a vengeance. Furthermore, the threat of the lone wolf, already embedded in society and acting alone with unyielding determination, is extremely worrisome. Factor in an unprecedented increase in hate groups and gangs in our country, and the domestic terrorism picture becomes quite grim, with resource-constrained law enforcement agencies struggling to juggle myriad challenges. Finally, brutal Mexican drug trafficking organizations are now operating in the United States; cartels are using gangs to move product and corrupting border patrol officers who open lanes and permitting people and drugs (and potentially worse) into our nation. As a paramilitary organization with the tactical knowledge and equipment of a small army, the Los Zetas cartel, now as far north as Chicago, provides the most vexing threat. Cartels use brutal tactics against soft targets in an effort to influence the political process and instill fear in the populace—factors elevating them beyond mere criminal groups.

This book explores the psychology of soft target attacks, our blind spots and vulnerabilities, and attributes that make civilian-centric venues appealing to bad actors. Next, looking through the lens of past, current, and emergent activities, the research yields an estimate of the motivations and capabilities of international and domestic terrorist groups, as well as drug trafficking organizations (arguably terrorists), to hit soft targets in our country. A current assessment of soft target attacks worldwide will give insight to trends and operational tactics. Studying the activities of terrorist groups that successfully and repeatedly strike soft targets, such as Chechen extremists, reveals security vulnerabilities and how poor government engagement and response can intensify the number of casualties. Training and tactics for psychological and infrastructure hardening, as well as planning and exercising strategies, provide a road map for those who own, operate, protect, and use soft target locations. Finally, the book discusses response during a soft target attack, as well as business continuity and reconstitution procedures.

THE DIFFERENCE IN THREAT, VULNERABILITY, AND RISK

A few working definitions will help frame the discussion of soft targeting. The concepts of threat, vulnerability, and risk are routinely intermingled and exchanged; the graphic in Figure 1.1 helps explain the concepts and their relationship.

Examining and understanding our vulnerabilities is the first step to address cultural and personal limitations keeping us from adequately comprehending and addressing soft target threats. Your ability to deter attack is amplified by understanding the threat/Case studies presented in the book about soft target attacks or attempted attacks give insight on how future attackers may emulate or change tactics to increase their likelihood of success. Typically, after a successful or failed terrorist attack, much of the focus is placed on the "how" of the operation—methodology and tactics used by the perpetrators. Certainly,

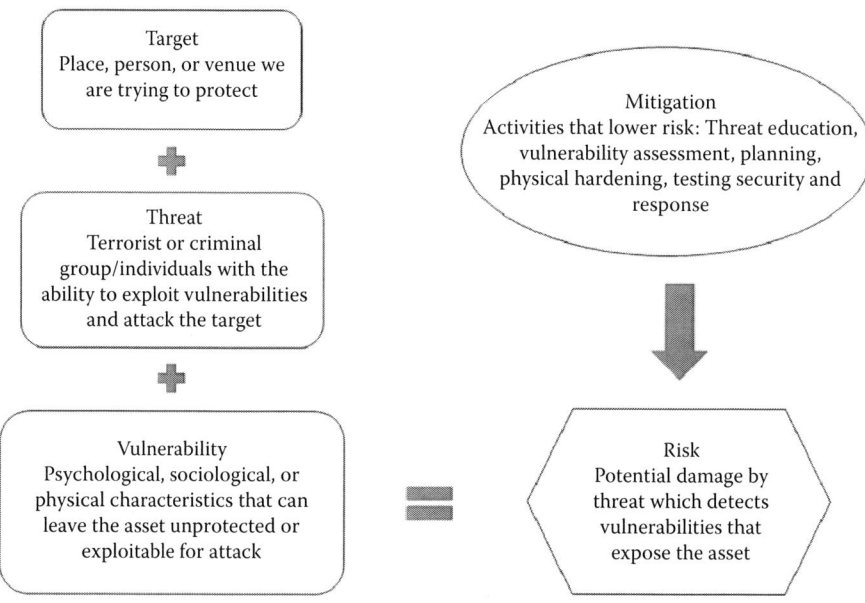

Figure 1.1 Threat, vulnerability, and risk model. (From Jennifer Hesterman.)

these data help us harden targets against copycats or other would-be terrorists and are included in the book. However, to harden psychologically, soft targeting science demands we also spend time on "why," contemplating why bad actors would strike a soft target and the possible gains. This question is addressed throughout the book to enhance understanding of how soft targeting evolved and how we might deter, discourage, and disincentivize actors from striking.

The use of imagination and intuition is critical when addressing soft target threats. Perhaps the most powerful statement in the 9/11 Commission report comes from its Chapter 11, "Foresight—and Hindsight." The investigators cite a lack of *imagination* as a root cause of the two worst attacks in our country's history: Pearl Harbor and 9/11. This book encourages critical thinking about soft targeting; and those entrusted with protecting venues should actively run "what if" scenarios through their minds while reading. Questions might include, "Is it possible my building and its users might be targeted in this manner? What is the one most vulnerable feature; is it access? Location? The users themselves? Our mission?" In addition to this type of brainstorming and soft targeting education, training, and planning activities, we also recommend in this book that facility owners accomplish military-style red teaming activities, as well as both tabletop and live exercises. Testing your security and response is the only way to reveal deficiencies in both planning and infrastructure and will build confidence in your ability to handle an attack.

This book is a broad brushstroke regarding the phenomenon of soft targeting and is meant to stimulate thought and start the conversation about vulnerability, threat, and response. It is not merely focused on response to an attack, but the critical front work of detection, prevention, deterrence, and mitigation. By reading this work, you are participating in mental hardening, making you a force multiplier no matter whether a citizen, community leader, principal, clergy, hospital administrator, sports or recreational venue or shopping mall operator, law enforcement, intelligence analyst, or first responder.

Let's get the national dialogue going.

REFERENCES

Merzer, Martin, Lenny Savino, and Kevin Murphy. "Terrorists reportedly cased other sites: Sports facilities, Disney World, Disneyland and the Sears Tower were surveyed." *Philadelphia Inquirer,* October 13, 2001.

Vogt, Heidi. "Shaken Kenya Aims to Rebuild Mall and Its Confidence." *Wall Street Journal,* February 28, 2014.

2

The Psychology of Soft Targeting and Our Unique Vulnerability

Jennifer Hesterman

Contents

Introduction 10
What Is Terrorism? 10
Who becomes a Terrorist? 11
Terrorist Motivations 11
Terrorist Behavior 12
Terrorism's Effectiveness 13
Terrorism's Target 16
Soft Target Violence: A Collective Twenty-First Century Black Swan 16
The Internet as a Tool 18
Soft Targeting Motivations 19
Our Unique Vulnerability 21
What Do We Most Fear? 23
How Should We Respond? 24
References 25

After all, a festive gathering of county health workers in San Bernardino would not seem likely to make the top million of a list of shooting targets. It was not an iconic symbol of American freedom or American muscle. It was not a target draped in ideological conflict.

New York Times covering the San Bernardino Terrorist Attack
Kleinfeld, 2015

INTRODUCTION

December 2, 2015, was a beautiful, sunny day in San Bernardino, California. Eighty workers from the county's Department of Public Health were holding a training event and holiday luncheon in the conference room at The Inland Regional Center, a state-run facility serving citizens with developmental disabilities. Syed Rizwan Farook, who worked in the department as a health inspector, attended the training and quietly posed with coworkers for photos. Farook left the training, and, as the group were transitioning for lunch, he returned to room with his wife, Tashfeen Malik. Both were wearing ski masks and black tactical gear; they entered the conference room and in just 4 minutes, fired at least sixty-five rounds from semi-automatic weapons into the trapped crowd, killing fourteen and wounding twenty-two. Fortunately, the San Bernardino SWAT team was conducting its monthly training exercise a few miles away, which allowed for quick response to the scene. But Farook and Malik were already gone, leaving behind a bomb constructed from instructions in al-Qaeda's *Inspire* magazine. The device was rigged in a way to target first responders, but thankfully did not detonate when the SWAT team entered the building. Both shooters were killed in a subsequent confrontation with police, but left a trail of evidence including earlier plans to attack a college campus and rush hour traffic. The San Bernardino attack was the fifth deadliest mass shooting in U.S. history and was a soft target attack trifecta: a radicalized U.S. citizen, lone wolf, insider. There is nothing more dangerous—or difficult to mitigate.

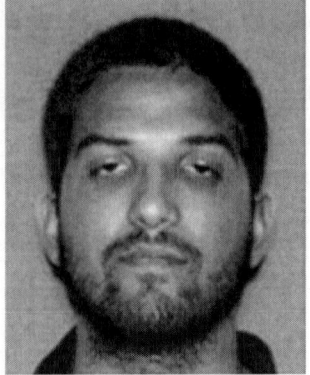

Photo 2.1 In December, Tashfeen Malik and Syed Farook, a married couple, killed fourteen people and injured twenty-two in a mass shooting in San Bernardino, California. The wife, Malik, pledged allegiance to ISIS's leader on Facebook. (From the New Jersey Office of Homeland Security.)

WHAT IS TERRORISM?

Terrorism is different from other crimes. Through acts of extreme violence, terrorists seek to advance an agenda, be it political change or religious domination. Terrorism is a crime, but does not lend itself to the same disincentives as normal offenses. The threat of prison or even death does not deter. We must also be careful in our approach to terrorists due to a dangerous paradox: the more brutal and oppressive we are, the more we make them martyrs to be emulated by others (Dershowitz 2002).

Terrorism is a complex phenomenon. There is no one root cause fueling terrorist activity; if so, we could simply fix the issue by addressing the aggravator and dissuade new group formation. Although the "hearts and minds" strategy is important to pursue, focusing solely on factors such as poverty and illiteracy fails to explain why groups with far greater grievances do not resort to terrorism (Dershowitz 2002). The approach also fails to explain why upper-class, educated individuals answer the jihadist call. Of course, we should continue to address perceived root causes of terrorism, but we must continue to study the ambiguity and myriad factors involved in the terrorism phenomenon.

WHO BECOMES A TERRORIST?

The terrorist believes his or her goals are just and actions taken to further the cause are justifiable. A very black-and-white picture is painted of the "enemy" or those who stand in the way of goal attainment. In the quest for social, religious, or political change, the terrorist lashes out directly either at the perceived opponent or at another target with the intent of bringing attention to the cause.

Terrorists are very rational actors. There is no conclusive evidence that terrorists are abnormal, psychotic, or have personality disorders. Consider that in the last few years, foiled terror plots included upstanding citizens who were educated, middle or upper class, and dedicated to their jobs and families. Statements from those closest to the terrorist suspect are usually very telling, and they typically indicate no suspicion the person they knew and loved for years would load a Jeep with explosives and park it in Times Square, or fly to an unfamiliar country and detonate a suicide vest. We routinely underestimate the terrorist's intelligence and sophistication—another blind spot.

TERRORIST MOTIVATIONS

What motivates an individual to join a terrorist group? Jessica Stern, a noted Harvard professor and terrorism expert, interviewed numerous members of terrorist groups. When discussing al-Qaeda, Stern (2003) notes:

> Militants have told me terrorism can become a career as much as a passion. Leaders harness humiliation and anomie and turn them into weapons. Jihad becomes addictive, militants report, and with some individuals or groups—the "professional" terrorists—grievances can evolve into greed: for money, political power, status, or attention.

So, the initial powerful "hook" for joining the jihadist movement may give way to other incentives over time.

The area of terrorist motivation is ripe for further exploration. Before taking up the cause, the terrorist group member was influenced somehow. What trigger flipped the switch from "off" to "on" and turned a father, son, daughter, student, or professor into a jihadist? By exploring this central question, we may develop a type of pre-emptive deprogramming operation preventing the spread of the radical Islamist ideology or supplanting a counternarrative, pushing the continuum toward our goals. There are three general schools of thought on what motivates an individual toward group membership, regardless of the type of organization:

a. *Instinct theory*: Group membership behavior is a function of a person's instinct, rather than activities that are conscious, purposeful, and rational.
b. *Reinforcement theory*: The decision to be part of the group is strongly affected by knowledge of rewards from its past behaviors.
c. *Cognitive theory*: Belonging to the group is influenced more by a person's belief and expectations for future rewards, not possible consequences.

Understanding the motivation to join a terrorist group is important for providing the counternarrative to recruiting efforts and to offering alternatives for those susceptible to the message of violence and hate. Looking at a mature group/corporation such as Hezbollah in total can be overwhelming when it comes to engagement and disruption. However, excavating the group down to the individual members and their motivations could be significant in developing deradicalization tools and dissecting the group from within. In postmodern society, microlevel changes are proving to be much more powerful than those taken at the macrolevel.

TERRORIST BEHAVIOR

Individuals are drawn to certain groups, feeling a unique connection at some level. Perhaps there is a demographic similarity among members such as religious belief or shared collective experience. Individual behavior and traits imprint on the entire group, so with the admission of a new member, the group changes in subtle but perhaps important ways. Group members bring their biographic background (including any "baggage"), abilities, intelligence, expectations, and personality into the organization. In terms of structure and culture, the group has norms, roles for members, and a level of cohesion. Size plays a role in a group's behavior, as well as its composition. In terms of performance, there are several possible outcomes; members are either satisfied or there could be turnover and even desertion.

We tend to act quite helpless when it comes to terrorist groups; however, they are quite easy to understand when viewed through the lens presented before, and we could take significant action to alter their course. For example, we could provide a counternarrative to group members and inject game changers. In fact, every variable and dynamic in a terrorist group could be specifically targeted in order to change the group's overall behavior and course. Viewing a terrorist group through a construct made popular by organizational behavior experts Szilagyi and Wallace is extremely valuable. If we adhere to the theory that terrorist groups have normative behaviors, despite their goals, adapting standard organizational behavior theory from Szilagyi and Wallace (1990) to terrorist behavior yields a new set of truths:

1. Behavior in terrorist groups is caused.
2. Behavior in terrorist groups is purposive and goal directed.

3. Behavior of the terrorist group is learned.
4. Individuals differ, but a terrorist group will have a shared set of values and characteristics.

Therefore, understanding individual and group dynamics could yield new solution sets for protection against soft target attack. We also might be able to understand the pull and influence a terrorist group like ISIS has on westerners, who leave comfortable homes to become fighters or even brides.

Photo 2.2 Douglas McAuthur McCain, the first known American to be killed while fighting alongside Islamic militants in Syria, was an undistinguished 33-year-old raised in Minnesota. (From the Sherriff's Office, Hennepin County, Minnesota.)

TERRORISM'S EFFECTIVENESS

Everything that makes democracies great makes them vulnerable to this enemy. Terrorism is much more effective against democracies than closed societies such as North Korea and China. For instance, consider the free press. Terrorism is an elaborate marketing campaign, and publicity adds to the legitimacy of the group and furthers its goals of creating fear. Democracies believe in freedom of information, providing the perfect tool for group member motivation, recruitment, and garnering public sympathy. Technology also plays a role, inasmuch as unfiltered news is often generated and distributed within seconds of an event taking place through Twitter, Facebook, and other social media outlets, in addition to traditional news channels. A closed society controlling the press and its "message" would not provide the terrorist organization this type of advantage.

For instance, how could a fundamental change in Western media's reporting of events impact the power of a terrorist group? Greatly, if you consider fear is a desired outcome of these groups' attacks. Consider social psychologist Kurt Lewin's "force field analysis," a system for predicting the motivated behavior of an individual in his or her "life space" (Lewin, 1943). Lewin defined the life space as an individual's perceived reality based on the many factors, real and imagined, at play. In terms of the threat of terrorism, it is important to emphasize the word "perceived" for this analysis. For instance, if a man believes he is being followed down a dark alley, this is his perceived reality and whether a pursuer actually exists is irrelevant. The man will suffer physiological changes associated with fear and panic, and his "fight-or-flight" instinct will involuntarily take over. Factors feed our perceived reality of the threat of terrorism such as press coverage; panic induced by quick spread of false information through social media; and, if the terrorists are savvy enough, a psy ops (psychological operations) campaign to make us believe something that is not true. Videos produced by ISIS depicting chilling execution scenes were choreographed, rehearsed, and edited to ensure maximum impact of horror and fear for viewers. The western press attempted to take the videos down, but horrific pictures were routinely printed on the front page of popular papers and magazines. This coverage seemed to feed ISIS, which kept raising the bar with ever more appalling execution tactics.

Also, closed societies would not exercise restraint when retaliating against terrorist groups and their members, instead taking swift, brutal countermeasures to suppress antigovernment activity. In a democracy, we believe one is innocent until proven guilty in a court of law. We provide due process and legal representation to even the most heinous bad actors in society and give them a chance to present their case to a jury of their peers for decision. For conviction, the law requires proof beyond reasonable doubt—something difficult to obtain when it comes to the arrest of sleeper cell members and would-be terrorists; if the trigger has not been pulled on an operation, would-be terrorists may receive light sentences and return to society quickly and without mandated rehabilitation.

Consider the case of Jose Padilla, former member of the Latin Kings street gang, al-Qaeda member, and so-called "dirty bomber." He was arrested in 2002 and it took 5 years for his case to go to trial and another year for sentencing. In non-Democratic countries, a decision would be quickly rendered with immediate and severe punishment. When a terrorist is identified and detained in our country, innocent family members are not a target for retaliation; however, in other societies, their homes may be razed and they are imprisoned and sometimes killed. The fear of collective punishment encourages families to engage if one of their own is turning to terrorism. Finally, the ease of moving about a democracy undetected is attractive to terrorists. Our country lacks internal checkpoints and is heterogeneous, with a broad demographic base. This allows bad actors to move more easily across our borders and then just disappear into society. In short, our democracy provides fertile ground for terrorists.

Alan Dershowitz, author of *Why Terrorism Works* (2002), proposes that amoral societies would engage in several (arguably successful) ways to stop the terrorist threat. For example, they would

- Completely control the media's reporting on terror incidents and, simultaneously, use the media for disinformation operations.
- Monitor all citizen communications.
- Criminalize "advocacy" of the terrorist group through inciting speech.
- Restrict movement within the country with layered identification checkpoints and even segregation of certain demographics within the country.
- Carry out collective punishment with family members and even entire villages to dissuade future activities.
- Initiate pre-emptive attacks against the group.

Naturally, this does not sound like a course of action taken by the United States due to our protections under the Constitution and innate sense of justice. However, consider the impact any of these activities, taken alone or in total, would have in our fight against terror. Is there a gray area? The events of 9/11 led to enhanced law enforcement activity, such as more extensive surveillance and wiretapping powers under the USA PATRIOT Act. The resulting skepticism was warranted, as we certainly must safeguard our constitutional protections and closely monitor any erosion of our civil liberties because this is a "slippery slope." However, in the asymmetric fight against an unpredictable, dangerous, and adaptable enemy, we may not be able to have it both ways when dealing with modern terrorist groups. We clearly need a corresponding modern and flexible solution set. To summarize, the very ideals constituting a democracy are leveraged and exploited by terrorist groups—and a primary reason why we are struggling to contain the threat.

A scientific approach to the phenomenon of terrorism is perhaps the purest way of viewing the topic. The very root of the word terrorism—"terror"—naturally provokes an emotional and personal response. Many Americans lost loved ones in acts of extreme violence perpetrated on 9/11 and the two major combat engagements fought in its aftermath. We also remember the horrific scenes broadcast live on the then-new CNN network of Marines being pulled out of the embassy rubble in Lebanon in 1983, and in 1985, we saw Navy diver Robert Stethem's lifeless body thrown from his aircraft onto the tarmac of the Beirut Airport by Hezbollah hijackers. The bombing of the World Trade Center in 1993 and the Oklahoma City attack in 1995 again brought the horror of terrorism, foreign and domestic, into our homes, minds, and hearts.

Photo 2.3 Nighttime shot of the Alfred P. Murrah building following the Oklahoma City Bombing. (From the Federal Emergency Management Agency.)

There are many persistent disagreements regarding the rise of modern terrorism, such as those revolving around democracy, capitalism, and Christianity, all of which may make the United States a target … or hegemon and invader. The specific characterization of terrorism is also very difficult to define: criminal act, holy obligation, reaction to oppression, or freedom fighting? Each government agency has its own definition and interpretation, but we intuitively know, whatever the reason or definition, one thing is crystal clear: terrorism is a threat to our national security. Investigating the terrorism phenomenon through a scientific lens eliminates both individual and institutional biases and removes emotions such as fear and anger. This approach also helps organizations move beyond sunk cost, "groupthink," and other unproductive and dangerous decision-making behaviors that stymie our efforts to eliminate terrorist groups.

TERRORISM'S TARGET

Unlike many criminal activities, terrorist attacks are not indiscriminate, as we explore in later chapters through case studies and analysis. Targets are carefully chosen to assure the advancement of political or religiously motivated objectives. In soft targeting, the most vulnerable place will be selected as a way to maximize casualties—often venues where people will not be armed and will be paralyzed by the element of surprise.

Perhaps it is helpful to think of terrorist targets in four categories: symbolic, functional, logistic, and expressive (Drake 1998). Symbolic targets are attacked to elicit a psychological reaction, whether the assassination of a key political figure or the physical penetration or destruction of a government building or monument. Functional targets are obstacles to the success of the terrorist or group, such as the military, security officers, or opposing groups. Logistical targets are hit for financial profit, to obtain money or goods such as weapons, fuel, or food. Kidnapping is a lucrative business for criminals and terrorists, and wealthy businessmen and their family members are logistical targets. Expressive targets are those hit as a response to emotion, for example, an attack on a lone American working for a nongovernmental agency in rural Africa. Expressive targets are not part of an overarching strategy and will not result in any political gain to the terrorist group. Usually, the victim was simply in the wrong place at the wrong time. Soft targets are a new addition to the playbook, complementing all of the preceding activities.

SOFT TARGET VIOLENCE: A COLLECTIVE TWENTY-FIRST CENTURY BLACK SWAN

The black swan theory is a metaphor describing a surprising event with major impact thought to be impossible, but then rationalized by observers in hindsight. The combination of low predictability and large impact of black swan events as related to nefarious groups and actors is central to this book's narrative concerning the asymmetric tactic of soft targeting.

In the first century AD, a Roman poet, Juvenal, called something presumed not to exist or deemed impossible a "black swan." Interestingly, Dutch explorer Willem de Vlamingh spotted a black swan in Australia in the 1600s while traveling through the area. The discovery of this creature, thought to be nonexistent, was deemed the "black swan problem" and introduced the concept of "falsifiability" to the scientific world. Scientists soon realized things which they thought were unequivocally true might be proven otherwise. Modern black swan events were characterized by Nassim Nicholas Taleb in his 2010 book, *The Black Swan: The Impact of the Highly Improbable*. Taleb regards almost all exceptional scientific discoveries, historical events, and artistic accomplishments as black swans, or undirected and unpredicted occurrences. The rise of Adolph Hitler, the personal computer, the Internet, and the September 11 attacks are Taleb's examples of black swans. Taleb's comments in a *New York Times* article are important to frame our discussion regarding the social networking phenomenon:

> What we call here a Black Swan (and capitalize it) is an event with the following three attributes. First, it is an outlier, as it lies outside the realm of regular expectations, because nothing in the past can convincingly point to its possibility. Second, it carries an extreme impact. Third, in spite of its outlier status, human nature makes us concoct explanations for its occurrence after the fact, making it explainable and predictable.
>
> I stop and summarize the triplet: rarity, extreme impact, and retrospective (though not prospective) predictability. A small number of Black Swans explains almost everything in our world, from the success of ideas and religions, to the dynamics of historical events, to elements of our own personal lives.

Major axioms of the black swan theory include:

1. The disproportionate role of high-impact, hard-to-predict, and rare events beyond the realm of normal expectations in history, science, finance, and technology
2. The noncomputability of the probability of the consequential rare events using scientific methods (owing to the very nature of small probabilities)
3. The psychological biases making people individually and collectively blind to uncertainty and unaware of the massive role of the rare event in historical affairs

The theory refers only to unexpected events of large magnitude and consequence and their dominant role in history. Such events, considered extreme outliers, collectively play vastly larger roles than regular, predictable occurrences. In short, these are game changers, such as the rise of modern terrorist groups, cartels, and transnational organized crime groups. The advent of the domestic terrorist, a citizen who turns on his or her own government and kills innocent civilians to further his or her cause, might be considered a black swan.

Black swan logic makes *what you do not know* far more relevant than what you do know. Therefore, in the counterterrorism business, we should be exploring "antiknowledge" or what we do not know and what we do not expect from the enemy. In the security realm, we typically use specific data to make strategic decisions, instead of stepping back and viewing the entire issue with all of its complexities and changing environmental factors. We expect groups to engage in a similar manner as they have in the past, and we harden facilities and screen people accordingly. Unfortunately, we expend an inordinate amount of resources to prevent history from repeating itself, while the groups and individuals with a violent ideology work to hit us from an unexpected or asymmetric angle. As we discuss in Chapter 3, this resource drain and diversion of manpower is exactly what enemies hope to elicit, and we often play right into their hands.

The story of the Maginot Line shows how we are conditioned in this manner. After WWI, the French built a wall along the previous German invasion route to prevent reinvasion. However, Hitler simply went around the wall and marched into France. Overreliance on past events as a predictor of future action, underestimating the creativity of the enemy, and the inability to think "outside the box" clearly leave us vulnerable. In black swan vernacular as related to terrorist events, history does not crawl, it jumps.

THE INTERNET AS A TOOL

Similarly, with the black swans of rapid technological and communication advances, we were wholly unprepared for how social networking would change our society, the definition of "community," and the manner in which people communicate. The Internet is a virtual world where people can be compelled (even unconsciously) to change their opinions, join movements, and even take up arms for a new, captivating cause. These new members are not forced to join; they are volunteers, who always make the best recruits in both licit and illicit groups. Sitting in their comfortable middle-class homes or at the local Wi-Fi–enabled coffeehouse, they do not have to be suffering, hungry, or uneducated to join in the dialogue. The impact of the Internet in the crime and terror realm was unforeseen and perhaps marginalized with the excitement of rapid growth, social networking sites, and virtual banking and commerce. Every piece of technology we enjoy—every application that makes our life easier or helps us communicate—is exploited by bad actors.

Photo 2.4 Abubakar Shekau, leader of Nigeria's Boko Haram, speaks in a video taken from the terrorist group's website on Monday, May 12, 2014. (From the U.S. Department of Defense.)

The main product of the terrorism marketing campaign is fear; however, by-products include recruitment, empathy seeking, and fund-raising. As with all marketing operations, terrorism is meant to shift the public center of gravity through use of symbolism or themes, and these techniques can be overt or covert. We must also remember that, for many groups, this is a long crusade; the enemy is patient and thinks in terms of millennia, not years. Our children's children may indeed struggle with the same terrorism and transnational crime issues with which we struggle today, and a quick fix is simply not possible.

Terrorist groups have changed how they use the Internet to recruit and spread their propaganda, with a shift from outright violent threats to a more subtle, understated threat. For instance, Facebook pages appearing friendly and inviting will pull in the curious and then redirect them elsewhere for further discussion on edgier topics. Of course, many groups do not sugarcoat the message or their hate for the United States and also use the open forums to blatantly spread false information and propaganda. With the Internet, they are literally "hiding in plain sight" and, indeed, why should groups go underground when their Internet activities are perfectly legal?

The perpetrators of recent shooting events have even signaled their intentions in advance on social networking websites. On the night before he killed fifteen students and teachers at his former high school in Germany, Tim Kretschmer chatted on the Internet about his intention to commit mass murder. He wrote (Davies and Pidd 2009):

> I've had enough. Everybody's laughing at me. No one sees my potential. I'm serious. I have weapons and I will go to my former school in the morning and have a proper barbecue. Maybe I'll get away. Listen out. You will hear of me tomorrow. Remember the place's name: Winnenden.

Closer to home, Elliott Rodger posted a sinister YouTube video entitled "Retribution" the day before stabbing his roommates to death and going on a shooting rampage at the University of California, Santa Barbara, on May 23, 2014 (CNN, 2014). The video gives rare insight into the mind of a killer:

> Tomorrow is the day of retribution, the day in which I will have my revenge against humanity, against all of you.
> I'll take great pleasure in slaughtering all of you.

Rodger killed six people and injured fourteen others before committing suicide.

Website administrators and users could become force multipliers if trained on what to look for, how to engage, and how, what, and where to report provocative comments such as those posted by Kretschmer and Rodger. Unfortunately, the companies hosting social media site often act as victims themselves. Lacking any government regulation compelling their engagement and partnering with law enforcement to *proactively* detect bad actors, or education on what to look for and report, it's easier and more comfortable to simply look the other way.

SOFT TARGETING MOTIVATIONS

We need to understand group or individual motives to attack a soft target and what gains are reaped from such an operation. Case studies in the book discuss specific attacks; however, there are similar motivations and goals crossing all brands of terrorist and criminal groups:

> *Easier, cheaper, and short planning cycle.* Consider that the Kenya Westgate Mall attack planning cycle was just 1 year and executed with a small team of shooters with basic weaponry,

whereas it took 7 to 8 years to plan the complex attacks of 9/11. The scope of the events differed, of course, with the events of 9/11 killing thousands and drawing the United States and its allies into protracted, expensive, and deadly wars in Afghanistan and Iraq. However, what if the target on 9/11 was a Paris- or Mumbai-like coordinated soft target attack in a major city, executed by al-Qaeda at the direction of Osama bin Laden? We likely would have followed the same military course of action into Afghanistan to penetrate the heart of the group. Therefore, a smaller-scale soft target hit in the United States may carry the same weight as a large-scale operation to hit symbolic or hardened targets.

Increased likelihood of success. The probability of a successful attack against an unprotected soft target is higher than against fortified hard targets.

Credibility. A successful attack garners instant status for the group or cause.

Recruiting value. Sympathizers are more likely to join the cause of a "winner."

Flexing muscle. In the past 7 years, U.S. government officials repeatedly stated al-Qaeda was drying up and dying. Many thought the splintering was a sign of decline, yet AQ-affiliated groups are successfully attacking and inspiring attacks worldwide. Government officials now say al-Qaeda is resurgent. Even small soft target attacks at hotels and restaurants prove the group is thriving and viable.

Compensating for weakness. If a group does not have the resources to hit a hard target, or ongoing planning is moving too slowly, they can quickly hit a soft target.

As a last gasp. If a group is declining, an easier soft target attack could be the last hope to recruit and gain credibility.

Backed into a corner. When a group is trapped, the situation for civilians becomes extraordinarily dangerous. Consider the LTTE terrorist group: when corralled and boxed into one area of Sri Lanka by military forces, the terrorists took a staggering 100,000 civilians hostage on a beach and used them as human shields; credible sources report nearly 20,000 were killed in the final showdown. This "final stand" scenario is especially worrisome if dealing with a desperate religious group with apocalyptic intentions; its members may field their final and best weapon, a weapon of mass destruction.

Test a new strategy, tactic, or weapon. Hitting a soft target could be a dry run for a group that is honing an operation. It also could be used to assess emergency response and crowd evacuation procedures and to gather points for a later, larger attack including secondary or tertiary devices and possibly chemical, biological, or radiation agents.

Fund-raising. Soft target attacks may result in kidnapping and hostage taking for ransom or for sale into slavery, such as the schoolgirls taken from their dormitory in Nigeria by Boko Haram. Piracy of a cruise ship or ferry could quickly raise tens of millions of dollars.

Quickly damage a market. As witnessed with the airline industry after 9/11, the movie theater industry after the shootings in Aurora, Colorado, and attacks against resorts and hotel chains in Africa and the Middle East, a soft target attack will immediately cause fear and avoidance of similar venues.

Delegitimize a government. A successful soft target attack casts doubt on the government's ability to protect its people, who will scrutinize intelligence operations, law enforcement, and first-response efforts. We already have a conspiracy theory–fueled public, with tens of thousands of "truthers" who not only distrust officials, but also believe events such as 9/11 and the Boston Marathon bombing were government "false flag" operations to cause fear and maintain tight control on citizens.

Cause political instability. If properly timed, a soft target attack could change the course of an election and history. The group propel a candidate more favorable to its cause into office or repel voters from polling places. This type of political pressure routinely happens in Mexico with the drug cartels, which violently engage prior to and during elections to shape outcomes by killing mayors, police chiefs, and political candidates.

Make a country look weak internationally. A soft target attack such as the Boston Marathon bombing makes the country appear weak and unable to protect its citizens. Terrorists appear to be able to attack at will. A country that cannot protect its own people has difficulty convincing it can protect others.

To attain global media coverage. A soft target attack receives coverage in the 24/7 news cycle such as the live television coverage of the Beslan massacre, the Mumbai attacks, and the events of 9/11. If the group wants immortality, press coverage is the best course as the images of destruction and suffering are shown over and over. Many groups will broadcast on social media, even using Twitter during an event to air their grievances and inject more fear with threats of further attacks. For example, Al-Shabaab terrorists were using Twitter throughout the attack on the Westgate Mall to shape the narrative.

A target-rich environment. Soft targets abound and groups can quickly bring a city to its knees, such as the Mumbai attack where a hospital, hotels, trains, and a train station were simultaneously targeted. In a small town, terrorists could hit a school, church, and mall all at once, or hit one and then the next in a campaign of terror. If the objective were to seize land and property, or even an entire small town, this type of timed multitarget operation may work in the United States as it has for the drug cartels in Mexico.

Psychological fear. A successful soft target attack will wreak havoc on the psyche of the average citizen who may think, "Not only can my government not protect me, but even worse, it cannot protect my family." Perhaps nothing is worse than fear for the safety of your family, especially children. Specifically targeting women and children fuels the terror multiplier effect (TME).

Make a domestic issue international. Hitting a soft target is a way to vault local or regional issues to the national scene, a hope of special-interest domestic terrorists such as those belonging to animal- and eco-rights groups.

So, why hit soft targets? After considering the potential benefits, the answer is obvious: why not?

OUR UNIQUE VULNERABILITY

Perhaps the softest, most vulnerable target is you—the human being. Due to our culture and history, Americans tend to cast a psychological blind eye toward soft targets. In the aftermath of World War II, which saw the mass slaughter of innocent civilians during combat operations by adversaries, the U.S. government helped establish the 1949 Geneva Conventions, which protect noncombatants and civilians in and around the war zone. Therefore, our military doctrine is to never purposely hit a soft target, even if the "most wanted" terrorist in the world is attending a religious service, visiting his child at school, or having a medical treatment. Instead, we will patiently wait for another opportunity to minimize collateral civilian casualties. We see the world through this lens, and believe (or hope) others will employ the same restraint.

Unfortunately, our enemies, foreign and domestic, do not play by the same rules. The new battlefields are wrapped around neighborhoods and communities, and civilians *are* the target. Terrorists have made this very clear in the United States with asymmetric

SOFT TARGETS AND CRISIS MANAGEMENT

attacks planned against civilian-rich, nonmilitary/nongovernment environments such as the Boston Marathon and Times Square. We routinely underestimate the sophistication of terrorists and criminals as they constantly exploit our vulnerabilities and naiveties to increase the chances of operational success. Also, for bad actors, there is no such thing as a "failed attack": every event is a learning experience. The free press typically overshares information about why the operation was successful or unsuccessful, and would-be attackers worldwide adjust their tactics. Professionals who fail to understand these emergent factors are at a disadvantage when protecting soft target venues, as we have transitioned to an entirely new security environment and paradigm in the last few years.

Photo 2.5 First responders respond to the aftermath of the Boston Marathon bombing on April 15, 2013. (From the National Counterterrorism Center.)

Hard targets including government buildings, military installations, and buildings with symbolic significance such as monuments have been further hardened over the years. Bad actors know hitting them not only will be more difficult, but also likely result in less shock factor to the American public. For instance, mass shootings on military installations such as the Navy Yard in Washington, DC, in 2013 and Fort Hood, Texas, in 2009 usually garner less public sympathy and outcry than those committed in schools, restaurants, and churches. Perhaps the occupants of these facilities are considered legitimate targets due to their occupation. This type of soft target rationalization pushes the would-be terrorist to strike elsewhere, to areas where the shock, outrage, and fear factors will be higher.

WHAT DO WE MOST FEAR?

Although typically avoided, we need to "go there" and desensitize when it comes to soft targets. What do we most fear? It seems the generations before us were much more adept at facing their collective fears and taking steps to educate the populace, including children. The employment of the atom bomb in 1945 and the Soviet Union's attainment of nuclear capability in 1949 transformed the meaning of civil defense in the United States. During the early days of the Cold War, the government produced a shocking video to teach children self-protection from a nuclear attack; the children practiced "duck-and-cover" exercises and were prepared to take care of themselves during attack if adults were not around (Mauer 1951). New York City issued 2.5 million identification bracelets, or dog tags, for students to wear at all times, with the unspoken purpose to help identify children who were lost or killed in a nuclear explosion. Over time and with the fall of Communism, the fear of nuclear attack waned along with civil defense exercises and preparation for WMD events. For decades, the government was quiet about threat and individual response.

After the attacks on 9/11, the Department of Homeland Security (DHS) attempted to re-engage in this realm and educate members of the public on how to protect themselves. For instance, on February 10, 2003, in response to intelligence information that terrorists were planning a WMD attack in the United States, DHS issued an advisory directing Americans to prepare for a biological, chemical, or radiological terrorist attack by assembling a disaster supply kit. Shelves were cleared of duct tape and plastic to seal homes and offices against nuclear, chemical, and biological contaminants. The DHS eventually faced ridicule and the advisory was jokingly referred to by comedians as "duct and cover." Certainly this type of response by the public should be factored into future advisories.

What types of psychological trauma are experienced after an attack like those perpetrated on 9/11? Typically, there are changes of behavior; for instance, a community or nation may pull together during a mass casualty situation, putting differences aside for the greater good. However, an attack may cause ruptures in relationships and lead to suspicions and poor treatment, similarly to what Japanese Americans experienced during World War II after the Pearl Harbor attacks and the American Muslim community suffers today. In Kenya, following the Westgate Mall attacks by the Somali-driven al-Shabaab, Somali citizens were rounded up in the country and many deported. Also, after an attack, anniversary dates take on importance. For example, April, with anniversaries of the Columbine school attack (April 15), Oklahoma City bombing (April 19), and the Branch Davidian siege at Waco (April 20), is particularly tough for Americans and nervous security experts who fear copycats. Naturally, the September 11 date will always be burned into the national psyche, similarly to December 7.

When a domestic terrorist attacks fellow citizens, a sense of denial typically sets in and we search for some type of foreign influence to explain the behavior. This reaction could hinder the investigative process, as it did immediately following the Oklahoma City bombing in 1995. The World Trade Center bombing executed by al-Qaeda was just 2 years before, so the immediate assessment was the Murrah Federal Building had undergone a jihadist attack. The national press fueled the fire with inflammatory, baseless statements:

"The betting here is on Middle East terrorists," stated CBS News's Jim Stewart just hours after the blast.

"The fact that it was such a powerful bomb in Oklahoma City immediately drew investigators to consider deadly parallels that all have roots in the Middle East," piled on ABC's John McWethy.

"Knowing that the car bomb indicates Middle Eastern terrorists at work, it's safe to assume that their goal is to promote free-floating fear and a measure of anarchy, thereby disrupting American life," the *New York Post* editorialized on April 20. "In due course, we'll learn which particular faction the terrorists identified with—Hamas? Hezbollah? the Islamic Jihad?—and whether or not the perpetrators leveled specific demands."

"It has every single earmark of the Islamic car-bombers of the Middle East," wrote syndicated columnist Georgie Anne Geyer in the *Chicago Tribune* on April 21.

Also on April 21, according to the *New York Times*'s A. M. Rosenthal, "Whatever we are doing to destroy Mideast terrorism, the chief terrorist threat against Americans, has not been working."

Even after the FBI released sketches of the suspects on April 24, the news media still would not let go of their assumptions. Thankfully, the VIN number from the vehicle used in the bombing was found at the scene and led to former soldier McVeigh, who was pulled over for speeding in a remarkable coincidence. Otherwise, the attack could have led us down the wrong path diplomatically and, possibly, militarily.

HOW SHOULD WE RESPOND?

A measured response to a terror event would be the best approach, although this is nearly impossible due to the proliferation of the Internet and 24/7 news broadcasts. As analysts and investigators, we need to avoid the "light pollution" obscuring the true threat as people gravitate and cluster around groundless theories as they did immediately following the Murrah Building explosion. Fighting this clustering activity and being a skeptical empiricist is critical, as well as valuing and encouraging out-of-the-box thinkers who may identify and prevent a looming black swan event that hits the "reset" button in history. With regard to data analysis, statisticians caution us to be careful of confirmation and distortion bias when looking at a terrorist attack situation. We also should not adopt a revisionist history of past attacks. Taleb calls this tendency the "triplet of opacity" and points out that how we understand history can be affected by three factors: the illusion of understanding, retrospective of distortion, and the overevaluation of facts.

Security fatigue is another phenomenon that is both dangerous and exploitable. Americans are weary of long airport security lines and, now, a loss of communications privacy. Security is often inconvenient and especially trying for those unaccustomed to being targets. Therefore, we must find the right balance in our world and our places of work, education, worship, and play between living vigilantly and normally. In the end, however, personal security is a personal choice: If citizens choose not to face their fears about the threat and play out a potentially horrific scenario in their minds, they will likely respond poorly during a soft target attack.

After we prepare mentally to engage the soft target threat, understanding the motivations and emergent tactics of those who might strike is key to planning efforts. Thinking through our response is critical. This book is a good first step toward those goals.

REFERENCES

CNN. Transcript of video linked to Santa Barbara mass shooting. May 27, 2014. http://www.cnn.com/2014/05/24/us/elliot-rodger-video-transcript/

Davies, Lizzy and Helen Pidd. Germany School Killer Gave Warning in Chatroom, *The Guardian*, March 12, 2009.

Dershowitz, Alan M. *Why Terrorism Works*. New Haven, CT: Yale University Press, 2002.

Drake, C. J. M. *Terrorists' Target Selection*. New York: St. Martin's Press, 1998.

The Geneva Conventions of 1949 and Their Additional Protocols. http://www.icrc.org/eng/war-and-law/treaties-customary-law/geneva-conventions/

Kleinfeld, N. R. Fear in the Air, Americans Look Over Their Shoulders, *The New York Times*, December 4, 2015.

Lewin, K. Defining the "Field at a Given Time." *Psychological Review*. 50: 292–310, 1943.

Mauer, Raymond J. Duck and Cover—Bert the Turtle Civil Defense Film, 1951. http://www.youtube.com/watch?v=IKqXu-5jw60

Stern, Jessica. The Protean Enemy. *Foreign Affairs* (Jul/Aug 2003).

Szilagyi, Marc J. and Andrew D. Wallace. *Organizational Behavior and Performance*. Glenview, IL: Scott, Foresman/Little, Brown Higher Education, 1990.

Taleb, Nassim N. *The Black Swan: The Impact of the Highly Improbable*. New York: Random House, 2010.

3

Soft Target Hardening 101

Jennifer Hesterman

Contents

Introduction 27
Steady-State and Crisis Leadership 29
Intuition 30
Insider Threat: A Fifth Column 31
    Good Hiring Practices: First Line of Defense 32
    Background Checks 33
Kidnapping and the Use of Human Shields in Terrorist Operations 36
Surrounded by Crises 38
First Response and the Threat of Secondary Attacks 39
Weapons of Mass Destruction 40
The SIRA Method 43
Suicide Terrorism 43
Face of Rage 45
Fifteen Soft Target Takeaways 48
References 49

> Security is always seen as too much until the day it's not enough.
>
> William Webster
> *Former FBI director (2002)*

INTRODUCTION

Mr. Webster made this statement during a debate entitled "National Security versus Personal Liberty" on March 3, 2002, at University of California, Santa Barbara. Little did he know that 12 years later, the idyllic, upscale campus would be the site of Elliot Rodger's stabbing and shooting rampage.

The balance of security and basic human rights such as privacy is important as we seek to secure our facilities and protect users. Terrorism, by its very nature, is a violation of human rights. However, when we become heavy handed in the face of a terrorist threat and overreact, stripping away elements of the very democracy our enemies seek to destroy, we have played into their hands.

Physical profiling is a counterproductive tool in postmodern terrorism and we must dismiss the notion a terrorist "looks" a certain way. After 9/11, the law enforcement community focused on foreign-born male Arabs; yet we soon learned radical jihadists can be American non-Arabs and, in some cases like Jihad Jane, a woman. Although hard for us to conceive, the use of women, children, and the disabled is another proven asymmetric tactic shared among terrorist groups worldwide. In 2008, two disabled women were unwillingly used by al-Qaeda in Iraq to attack an open market in Baghdad; the devices were strapped to their wheelchairs and remotely detonated. Also in Iraq, a female suicide bomber attacked a group of women and children at a playgroup gathering, killing fifty-four people. In 2011, a man dressed as a cleric and a small boy were walking toward a government building in Karachi; alert police approached the pair and found they were both wearing bomb vests. In 2012, a 14-year-old suicide bomber walked into a group of his friends playing outside the NATO building in Kabul, Afghanistan, killing six.

Profiling by age also does not work in the United States; radicalized Islamist terrorists in our country have ranged from 19 to 63 years old. Profiling can also lead to tunnel vision: while we are focused on one group or demographic, the real threat is on the periphery, taking advantage of our myopia. We must also take care when profiling based on a person's activities, which may be quite harmless, such as taking photographs or sketching locations. Our efforts can lead us down a wrong path, distracting and expending resources while infringing on a person's rights in the name of security. Certainly, calling law enforcement is prudent if the situation exceeds what you can handle; however, be aware—if the Joint Terrorism Task Force becomes involved, suspicious activity reports (SARs) filed with the FBI on an individual may be held for up to 20 years. Perhaps a popular saying in the security realm applies: If everyone is a terrorist, no one is a terrorist.

Furthermore, our hardening activities may unconsciously legitimize the terrorists or aggrandize them in a way where their power grows. For example, thanks to Richard Reid, the "shoe bomber," we all now have to take off our shoes at airport security. After Umar Farouk Abdulmutallab, the "underwear bomber," tried to detonate a device in his pants, the restriction on carry-on liquids was put into effect. The backscatter machines may deter the one bad actor who is thinking about an airliner attack, yet inconvenience and intrude on the privacy of millions daily. Certainly these knee-jerk overreactions and the understandable outrage by the flying public must delight the terrorists: our way of life drastically changed = they won.

Finally, our asymmetric enemies keep reinventing tactics and improving, so we must not only tighten security based on past events, but also be aware of what is happening worldwide and prepare for copycats here in the United States.

SOFT TARGET HARDENING 101

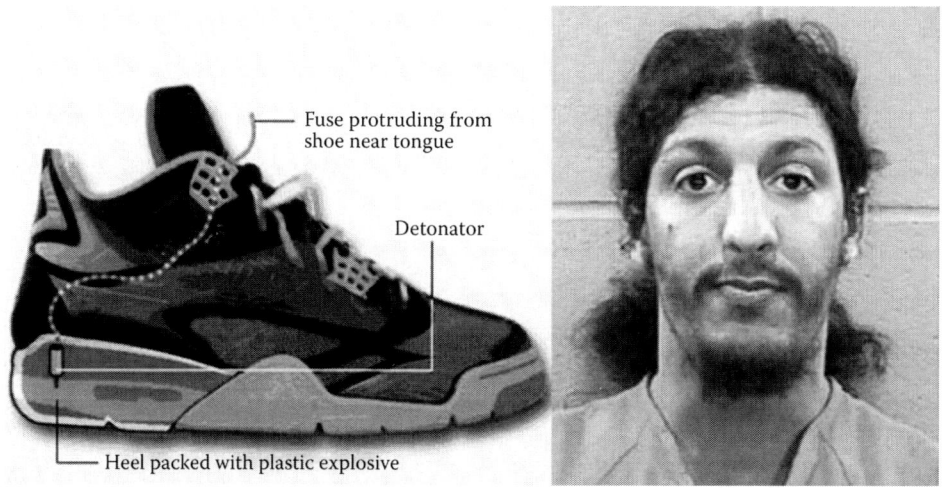

Photo 3.1 On December 22, 2001, Richard C. Reid was arrested after flight attendants on American Airlines Flight 63 observed him attempting to ignite an improvised explosive device in his sneakers. (From the U.S. Department of Energy, Office of Scientific and Technical Information.)

STEADY-STATE AND CRISIS LEADERSHIP

So now you may be thinking:

> Profiling will not work. I do not want to make my facility into a fortress. The security environment is dynamic and unlike a government building, I cannot merely barricade my buildings and occupants into a cocoon of safety. The threat keeps changing as groups shift tactics and learn based on past successes and mistakes and I do not have the resources for high-end equipment and a legion of guards. How should I proceed?

Security begins and ends with the decisions and actions of the organization's leadership—not just the front office, but everyone who has a key role on the leadership team. "Steady-state operations" is military vernacular describing periods of predictable activity, the day-in and day-out rhythm of an organization. During steady state, the leader ensures the organization trains, equips, plans, and practices. In terms of training, there are many fantastic, free resources regarding threat, vulnerability, risk management, emergency response, and hardening for owners, operators, leaders, staff, and building users. For example, Appendix B contains a list of free FEMA certificates and courses open to everyone, many of which also give credit hours for security certifications and college degree programs. Also, the Red Cross can always come to your facility and provide first aid, CPR, and emergency medical care training. The fire department would be happy to inspect your facility for hazards and law enforcement can assess the security of your property and building. During quiet periods, you may want to upgrade your security system, add a few cameras, and implement other suggestions you may find in this book. Dust off your emergency plans and checklists and gather key people to "what if" scenarios and talk through response. The point is that during steady state, you build and tend to the foundation which will hold you, your team, and the organization afloat during a crisis.

Crisis leadership is when you will rise to the occasion or fail; there truly is no in-between state. If danger crosses your doorstep, there will be a full investigation and history will judge *both* your actions and inactions. As leaders, we may get the glory for the wonderful things happening in the organization, but we are also held accountable for its failures, especially if they involve loss of life. Often, leaders fail during crises because they were not mentally or physically prepared for the acute stress of the situation. Perhaps they chose the easier path of invulnerability: "It will never happen to me," and "It will never happen here."

As previously discussed, the leader must identify vulnerabilities in the face of the threat and acknowledge and accept risk as needed. This book gives some ideas for how to work through these issues, but perhaps think about the one "Achilles heel" in your organization that, if leveraged by the adversary, will have the greatest impact on the safety and security of your facility and its occupants. In an organization that cannot completely control access, like a college campus, perhaps it is an active shooter on foot or in a vehicle. In situations with tight access, such as sporting events and concerts, open-air venues are vulnerable from the skies, and closed venues through their ventilation systems or food operations. A megachurch with thousands of worshippers and a few entrances/exit chokepoints may identify its weak spot as the central air conditioning system or the potential for an arson event leading to mass casualty. The hospital emergency room may be the most vulnerable place in the building; however, the main entrance must also be protected by an alert front desk staff looking for the first sign of a problem and the loading docks secured and monitored. Thus, some hardening activities can be broadly applied, whereas others must be tailored.

A resource-constrained operation should try to identify security tactics and equipment that will have the greatest impact in terms of saving lives. A "cafeteria" option—choosing from a wide slate of ideas and best practices—is also a possibility. This book provides a slate of hardening tactics and checklists as a point of departure for your organization. The ambiguity of modern terror dictates, however, that soft targeting can be random and unpredictable—an attack conducted by an opportunist who is enraged and pushed to act on his or her impulses at your facility. The attack in the parking lot of the Jewish Center in Kansas City, Kansas, by an enraged white supremacist provides the best recent example of little to no planning and acting on impulse. Also, remember history may not be the best predictor of soft target strikes; for example, prior to the shooting at Virginia Tech, there was no other event like it in our country. Prior to the Beslan school massacre or the Moscow theater siege, there were no similar prior events. Prior to the Paris attack, there was Mumbai, but the Paris attack differed in many ways. This is why soft targeting is a true black swan: nothing in the past could predict the situation we may find ourselves in today.

INTUITION

One of our most underappreciated natural gifts is intuition, or the ability to know something from instinctive rather than conscious reasoning or proof of evidence. Almost every case study in the book contains statements from either the families of the perpetrators, their friends, colleagues, or people at the scene saying "Something just didn't feel right," or, worse, "I knew something like this could happen" or "When I heard about the shooting, I just knew it was him." Trust your gut feeling, honed by years of experience and wisdom. If something looks or feels wrong, it likely is. The U.S. military has done a great deal of work on intuition, especially the Marine Corps, which includes honing and trusting the sixth sense as the "art" in the art and science of war discussed in their doctrine. Based on stories from combat veterans returning from Afghanistan and Iraq about situations where intuition saved countless numbers of lives on the battlefield, the Office of Naval Research is studying how military members can hone their intuition: "Research in human pattern recognition and decision-making suggest[s] that there is a 'sixth sense' through which humans can detect and act on unique patterns without consciously and intentionally analyzing them" (Office of Naval Research 2014).

Terrorists and criminals will often accomplish surveillance at the target prior to acting, sometimes returning to the scene more than once. In the Middle East, which has suffered decades of terrorist attacks, no apology is made for security actions. Yet in the United States, security guards apologize for asking questions. It is within your right as the facility operator, manager, or security specialist to *unapologetically* ask a stranger their purpose for being on the premises and if you can help. Engagement gives you a chance to gather information, also to assess verbal and nonverbal behavior. In fact, the entire staff can be force multipliers in this regard. Naturally, if the individual looks threatening, it is best to call security or the local police department instead of confronting the stranger. After any suspicious event, take notes on the person's physical appearance and what he or she was wearing and carrying. If the person drives away, note the make and model of the car, and even the license plate number. You will probably never need the information, but it will be of great help to investigators if your gut feeling was correct. Also, if the individual was an opportunist, the mere fact you engaged and saw him or her may deter further actions. Don't marginalize or ignore your intuition; as with all hardening activities, you may never know the impact of your actions.

INSIDER THREAT: A FIFTH COLUMN

As if the soft target hardening situation was not already complex enough with myriad actors and tactics, we need to factor in the growing insider threat to organizations. When you ask a security professional the question, "What keeps you up at night," the insider threat is likely the answer. As previously discussed with the San Bernardino shooting, radical Islamist Syed Farook targeted his co-workers at the time when he knew they would be together and most vulnerable: the annual training/holiday party. He was able to walk unopposed into a secure building with gear and weapons. He knew where to park and how to quickly escape.

In military terminology, the insider threat is a "fifth column"—a group of insiders who secretly sympathize with or support an enemy. They may engage in acts of espionage or data gathering for the larger group's planning efforts or subversion, destabilizing the organization from within and further "softening" the target. Consider the sabotage

Photo 3.2 Former intelligence contractor, Edward Snowden, is suspected of leaking the details of extensive and top-secret U.S. surveillance efforts. (From the Voice of America, www.voa.gov)

perpetrated by insiders Edward Snowden and Bradley Manning; they never pulled a trigger, but certainly put many others in the crosshairs with their sharing of extremely classified government information with the world, which includes state enemies and terrorist groups. The insider could also be a "lone wolf"—perhaps the most dangerous situation of all as planning activity will often go undetected.

Know that the insider is a force multiplier to an attack, with intimate knowledge of your facilities, operations, and vulnerabilities. This individual can also pre-position supplies for himself/herself or others, as well as assess the perfect time to strike and obtain the maximum results, whether casualty counts or damage to the physical plant or the organization's reputation. The military and nuclear realms spend a great deal of time and resources countering the insider threat which could do grave damage to our national security.

There is a persistent lack of research and data analysis regarding the insider threat, making it the least understood and least appreciated danger to an organization or venue. Compounding the problem, there are organizational and cognitive biases leading managers to downplay the insider threat. Indicators we use to detect threats with outsiders fall short and our countermeasures fall short; we inherently trust the insider and he or she has a legitimate job and role in the workplace. There is a myth these employees are disgruntled or have outward signs of hostility at work; although some workplace attacks occur by recently fired employees, many others come as a complete surprise, such as Sayed Farook. Risk assessments usually do not include insider threat, but they must to fully protect the organization or venue and its occupants. Add it now to the top of your list of risks.

Good Hiring Practices: First Line of Defense

Countering the insider threat takes a multifaceted approach. The body of academic work regarding the insider threat is fairly small, but we strongly recommend the book, *Managing the Insider Threat: No Dark Corners*, by Nick Catrantzos (2012). He applies the

Delphi methodology to this security challenge and arrives at the conclusion that infiltrators are more likely to be a threat than disgruntled insiders. In other words, the terrorist or group is more likely to try to plant one of their own rather than to leverage a disgruntled insider (although the threat of the latter must not be overlooked). First, the infiltrator needs to get through your screening process and then on the payroll and into your facility. Certainly, a thorough employee background check is necessary, and we recommend using an approach similar to that used in the government's security clearance process. In terms of past employment, absolutely call past employers, but do not ask only about the former employee's work ethic. Take the opportunity to probe a little. Perhaps, in the name of security and the workplace violence issues in our society, ask about the emotional state of the employee and whether there were any signs of anger or disloyalty. At the end of the discussion, ask an open-ended question: "Is there anything else you think I should know about this person?" When interviewing the potential employee, stay within the law but ask everything you can about background, family, and even financial status.

You should strive to apply psychometrics to your hiring process; there is plenty of information on the Internet about these practices, and you can hire an external company to do all of your pre-employment screening. But recognize that the screening process itself has many flaws; deception experts believe 40 to 50 percent of applicants lie on their resumes and job applications and 80 percent lie during the full screening interviews (Whetstone 2014). The potential employee is likely trying to impress the hiring official by inflating past work experience or abilities or, possibly, he or she does have something to hide that might preclude hiring. For those who try to read body language during an interview, untrained screeners are usually about 50 percent accurate, but body language deception experts are 80 to 90 percent accurate. Therefore, a pre-employment screening by experts is a good frontline of defense for your operation.

Background Checks

The National Employment Law Project (NELP 2016) estimates 65 million U.S. adults, or one in four, has an arrest or conviction record, though not all of these are felonies. The federal government has passed laws to allow businesses and organizations to conduct the background checks on potential employees who will come in contact with children in schools, churches, and hospitals. These checks are often performed by both state and federal law enforcement, often through simple fingerprinting and processing at the local police station, which will determine whether the person was ever convicted of a crime. Be advised that this system does not indicate other previous arrests or instances where the charges were dropped. As for other businesses, you may or may not be able to run background checks. However, the employment application (private and confidential) should include a questionnaire regarding the applicant's criminal history. If your state or church only requires a state-level check, then keep in mind—you will not receive any information from a crime committed outside that state. Also, a new movement is not in your favor: thirteen states have passed "ban the box" laws barring employers from asking potential employees about past convictions, believing the information creates an unfair barrier to employment. Also, more than one hundred city and county governments, including Atlanta, Indianapolis, and Kansas City, Missouri, do not allow the practice. Therefore, you may need to complement your screening process with professional help and through Internet searches and other legal avenues. The time and money involved with verifying the background of an employee are worth it.

We frontload a great deal of work and resource in the hiring process, but need to continue vetting new employees during onboarding and after assuming their workplace duties. After gaining employment, the mole, as we refer to the nefarious insider here, must gather information and determine how best to exploit your vulnerabilities. Detecting a radical jihadist or other religious or political ideology is more difficult than detecting outright anger. Mole hunters are used by the government to find spies in an organization; perhaps one of the best books ever written on this topic is *True Believer: Inside the Investigation and Capture of Ana Montes, Cuba's Master Spy*. The author, Scott Carpenter (2007), was a mole hunter at the Defense Intelligence Agency; his sole job was to identify insiders who were colluding with or spying for other governments (the "fifth column"). In the book, Mr. Carpenter is honest about how he failed to act upon early gut feelings regarding Montes, who, interestingly, was recruited while a night school student in a master's degree program at Johns Hopkins. Since Carpenter did not fully act on his instincts, Montes was able to spy for Cuba several more years, and the intelligence she gave them resulted in the deaths of several Americans at a classified outpost. Carpenter retrospectively looks at the case and the indicators Montes was not a "friendly," despite her ability to pass two polygraph tests.

The "no dark corners" approach offered by Catrantzos (2010) illuminates the work space. First, a cultural shift is needed: all of your employees should be seen as force multipliers in terms of security. One person in an organization does not "do" security; rather, everyone is responsible for securing the workplace. Therefore, hire your employees on a probationary basis and treat the situation like a pilot/copilot scenario; stay close or have one of your long-term, trusted employees act in the mentoring role. The "no dark corners" concept means new employees are never left alone during this "on-boarding" period, thus communicating to the insider that "someone" is watching at all times. The workplace should be very transparent in terms of physical work spaces and redundancy and accessibility. Other than the manager, employees should not be the sole source of information or access; for instance, one person has a combination to a safe or a person has an e-mail account or files to which only he or she has the password. In the spirit of transparency, have regular audits in the workplace, then if you suspect an employee is an insider threat, it is easier to approach and gather evidence prior to calling for outside help. The chart in Figure 3.1 shows the innovation of the "no dark corners" theory for securing against the insider threat.

The "no dark corners" approach might also be taken from a facility standpoint, with manager and security personnel regularly walking the perimeter looking for cut fences or other indications of stockpiling or operational planning. Every storage area and compartment should be routinely inspected—again, to look for stockpiling. Look up at ceiling tiles and down at floorboards, and at everything in between. Employees have the right to privacy, but construct an open area where the work space is visible, including the computer screens. Several cases discussed in this book involved insiders pre-positioning explosives and weapons in the workplace, sometimes over the course of months. Staying vigilant will not only deter the insider but also, it is hoped, detect their activity early in the planning cycle.

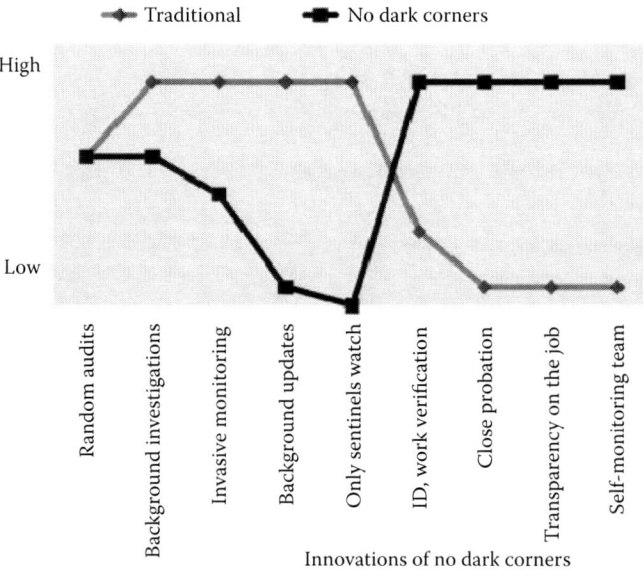

Figure 3.1 The "no dark corners" approach. (From Catrantzos, Nick. No dark corner: A different answer to insider threats. *Homeland Security Affairs Journal* VI, no. 2, May 2010. With permission.)

Much work has been done to mitigate the insider threat from the most critical, secretive, isolated workplace: nuclear power plants. Insider threats are the most serious challenge confronting nuclear facilities, and in every case of theft of nuclear materials the perpetrators were either insiders or had help from insiders. Sabotage is also a threat; in 1982, an employee detonated explosives directly on a nuclear reactor at South Africa's Koeberg nuclear power plant and in 2012 an insider sabotaged a diesel generator at the San Onofre nuclear facility in California. Matthew Bunn and Scott Sagan (2014) authored a very informative piece about worst practices and failures in the industry regarding the insider threat, with takeaways that may be cross-applied to soft targets. The authors encourage facility owners and security managers to fight their assumptions, stating:

- Don't assume serious insider threats are NIMO (not in my organization).
- Don't assume background checks will solve the insider problem.
- Don't assume red flags will be read properly.
- Don't assume insider conspiracies are impossible.
- Don't assume organizational culture and employee disgruntlement don't matter.
- Don't forget insiders may know about security measures and how to work around them.
- Don't assume security rules are followed.
- Don't assume only consciously malicious insider actions matter.
- Don't focus only on prevention and miss opportunities for mitigation.

Fighting these assumptions, use of psychometrics in the screening process, and the "no dark corners" approach will go a long way toward thwarting the insider threat.

Many soft target venues use volunteers, whether in school libraries, running church groups, staffing the reception desk at a hospital, or during sporting and recreational events. Due to budgetary woes, organizations may even replace paid staff and use volunteers in key front office positions. Someone with nefarious plans may think volunteering is one way to get inside an organization and its facilities by skipping the employment background checks. Apply the same rigor to checking the background of your volunteers and always have your staff do a face-to-face interview to pick up on any deception or gut feelings.

KIDNAPPING AND THE USE OF HUMAN SHIELDS IN TERRORIST OPERATIONS

The cases of Beslan and the Moscow theater siege have been the largest terrorist kidnapping and hostage-taking crises at soft target venues; however, there have been dozens of others and the tactic is escalating among terrorist groups. Not only al-Qaeda, but all of the international terrorist groups, as well as Mexican cartels, routinely kidnap tourists as a way either to raise ransom or to secure the release of imprisoned colleagues. With terrorist groups like ISIS, even if the ransom is paid, hostages are not released per the agreement or are killed anyway. Also on the rise are parental kidnappings during custody battles, a crime that schools and daycare providers must secure against.

Photo 3.3 A still shot from a video taken from Nigeria's Boko Haram terrorist group's website alleges to show dozens of abducted schoolgirls, covered in jihab and praying in Arabic. (From *Stars and Stripes*, U.S. Department of Defense.)

Managers and security personnel at soft target facilities and venues should also prepare for the possible use of children, women, and the elderly as human shields by terrorists (international or domestic) on our soil. Human shields tactics are the deliberate placement of civilians around targets or combatants to prevent the enemy from firing. The use of human shields is prohibited by the Fourth Geneva Convention, Protection of Civilian Persons in Time of War, which was passed in 1949 as a result of atrocities perpetrated by the Nazis during World War II. Specifically, the article states (Source: http://www.icrc.org/ihl.nsf/WebART/470-750065?OpenDocument):

> The presence or movements of the civilian population or individual civilians shall not be used to render certain points or areas immune from military operations, in particular in attempts to shield military objectives from attacks or to shield, favour or impede military operations.

However, as discussed in Chapter 2, terrorists do not adhere to Geneva Conventions or worry about worldwide condemnation of their activities and therefore routinely use this tactic when facing law enforcement or military response at the scene of their attack.

Several terrorist groups that previously attacked Americans and have operatives in the United States and/or Canada regularly use human shields. Independent reports give detailed evidence Hamas used hospitals, schools, homes, and mosques to hide weapons and soldiers during the Gaza War, an Israeli military initiative from December 2008 through January 2009. At 25 miles long and 6 miles wide, with a population of 1.5 million, Gaza is the sixth most densely populated place on earth, providing a very complex battleground situation. The UN report on the war mentions the possible use of children, women, and the elderly as human shields by Hamas; however, Malam, an Israeli intelligence think tank, produced a report using declassified material such as videotapes, maps, and operational plans recovered on the ground by Israel Defense Forces troops. The information indicates Hamas hid improvised explosive devices in and around civilian homes and hospitals, and video taken from a helicopter appears to show the use of children and the elderly as human shields for soldiers engaged in operations.

The Liberation Tigers of Tamil Eelam (LTTE) were defeated by the Sri Lankan government in May 2009 in a final, violent offensive on a northern beach in the Vanni region. During the months leading up to the conflict, the United States used satellites to monitor the situation, releasing photos to the public to show how LTTE herded hundreds of thousands of citizens on the beach for use as human shields in their final standoff with government forces. Many died from starvation, execution, or government shelling, and some escaped only to be captured and put into government internment camps. The remaining civilians, approximately 130,000, were forced to stay in a 1-square-mile area of beach and be part of the final battle. There has been no final accounting of civilians killed in the final offensive; the United Nations (2011) estimated at least 40,000. A BBC documentary entitled *Sri Lanka's Killing Fields* documents the atrocities performed by the government and the rebels in "no fire" and "safety" zones, as well as hospitals, schools, villages, and convoys of refugees deliberately pulled into the conflict (Snow 2011). In 2012, a Sri Lankan general admitted to war crimes, including the extrajudicial killing of civilians. According to intelligence agencies, LTTE members who fled the country are regrouping in Canada.

Although it seems inconceivable to us that a terrorist group, cartel, or lone wolf would use human shields, we must remember the power of ideology, especially radical religious dogma, which empowers believers with a sense of religious justification for their illegal and immoral activities. The Beslan example reminds us the terrorists will use civilians to absorb the line of fire. Naturally, it is difficult to "go there" mentally, but we should prepare for even the most horrific of scenarios, including the terrorist exploitation of innocent civilians to further goals.

SURROUNDED BY CRISES

Often, soft targets are in the area or even on the grounds of a hard target; they may not be part of the scenario, but get pulled into the disaster by virtue of location or services provided such as shelter, religious comfort, or medical care. In October 1997, LTTE terrorists arrived on the scene at the newly inaugurated World Trade Center in Colombo, Sri Lanka, with its twin thirty-nine-story towers as the target. However, the attackers could not get their truck bomb into the building's parking lot as planned. Instead, they parked next door, at the Galadari Hotel, and detonated the device by firing rocket-propelled grenades into the vehicle. The Murrah Federal Building in Oklahoma City had child daycare onsite, which unfortunately was on the bottom floor of the building and took the brunt of the blast from the 1995 domestic terrorist attack. On 9/11 there was a large daycare center located just a few hundred yards from the Pentagon. As previously discussed, there were a number of schools located near the Twin Towers, and nearby churches were pulled into the attack by being blanketed by toxic dust and used as mortuaries and sanctuaries for weary and devastated citizens and first responders. During the 2012 unrest in Libya, U.S. government officials identified a hotel and ordered all Americans to use it as an evacuation rally point even though the hotel was not consulted, putting its employees and guests directly in the line of fire. Even natural disasters can quickly turn into a period of unrest and violent crime impacting our facilities. For example, during Hurricane Katrina in August 2005, the Louisiana Superdome was used as a mass shelter of "last resort," becoming a cesspool of sewage, vandalism, a suicide, possible murders, and reported rapes and gang activity.

Several scenarios are ripe for discussion: What if there is a terrorist attack unfolding in your town or city during the school day or a church service? How do you shelter in place and protect your building occupants? Do you open your doors to those fleeing the event, possibly bringing the attacks to your property? Do you share your supplies? What if the event lasts 3 days? During Hurricane Katrina, Keesler Air Force Base, Biloxi, Mississippi, was the only location in the area that had some electricity (thanks to emergency generators and fuel) as well as water and food. After the crisis, people in the devastated town wandered to the base gates to see if they could get help. On 9/11, confused and upset people were congregating at churches and schools, which naturally opened their doors. These are difficult scenarios, but thinking through them is critical to response during crises. In the event of a criminal- or terrorist-related emergency in the community, you may want to turn off lit signage, decrease interior lighting, and lower profile—this is what properties in the Middle East due during attacks. Also, investigate "contingency contracts" with local water and food delivery companies in case you are sheltering a group of people and the crisis goes on for an extended period of time. Similar to preparations for natural disasters, your organization may want to have a significant amount of cash on hand to purchase water, ice, and other necessities from stores if the electricity and ATMs are out. Preparing for the worst will ensure success no matter what crisis you're facing.

FIRST RESPONSE AND THE THREAT OF SECONDARY ATTACKS

It is what people like us do: Without concern for our personal safety, we race to the scene of an attack to assist survivors, or we allow them entry into our secure sanctuary, be it our place of work or home. Law enforcement, chaplains, medical professionals, those with a military background—we are hardwired to respond this way, and our response is instinctive. Unfortunately, the instinct to run toward the fight, not away from it, could be deadly. My military colleagues often tell stories about the use of secondary and tertiary devices in Iraq and Afghanistan. The insurgents set off the first IED and wait until other soldiers rush in; then they set off a second device, sometimes a third. But this is not a new tactic; both international and domestic terrorists have employed it successfully for years. Facilities often serve as shelters, such as schools, churches, malls, and sporting and recreation venues, and can make themselves a target by assisting those fleeing a scene. As covered later in the book, terrorists will even strike a hospital providing care to victims of the first attack.

International terrorist and criminal groups have successfully employed this "double-bomb" tactic to target first responders, and evacuating civilians are often caught in the fray. For instance, secondaries targeting responders in other notable attacks include the Jemmah Islamiyah hotel bombing in Bali in 2002, an anti-American attack by Hezbollah at the McDonald's in Lebanon in 2003, and the 2004 police station bombing in Athens by the Revolutionary Struggle group. In March 2010 in Dagestan, a Chechen suicide bomber dressed in a police uniform approached investigators and residents who had gathered at the scene of a car bomb explosion near a school and detonated his explosive vest. In April 2010, the primary al-Qaeda–related bombing in Algeria was followed by a secondary, detonated 1 hour later and killing a soldier. The same month, Chechen rebels bombed a train in Dagestan and

Photo 3.4 FBI image of Eric Rudolph who was charged with detonating bombs at Atlanta's Centennial Olympic Park, as well as bombings at an Atlanta area health clinic and nightclub. (From the Federal Bureau of Investigation.)

remotely detonated a secondary device to target first responders. Most recently, Sayed Farook left incendiary devices behind to target responders to the shooting in San Bernardino.

Domestic terrorists also use this technique. Almost every attack perpetrated by Eric Rudolph included a secondary device specifically targeted toward emergency personnel. For example, prior to the Atlanta abortion clinic bombing in 1997, Rudolph called in a bomb threat and watched as office members evacuated and gathered in certain areas of the parking lot. He planned this into his operation months later; after he bombed the clinic, a second device went off an hour and a half after the first, injuring seven first responders. The Mexican cartels use secondary device tactics to ensnare and kill law enforcement. In fact, Louis R. Mizell, a terrorism expert and former U.S. intelligence agent for the State Department, has compiled a database of three hundred double-bomb attacks by more than fifty terrorist groups in the world over the last 10 years. His advice is "the reality of today's double-bomb tactic dictates that first responders have three primary jobs at a site: attending to the wounded, dispersing the crowd, and finding a second bomb" (Gips 2003).

The threat of secondary and tertiary bombing is real, and it must be factored into soft target hardening and emergency response procedures.

WEAPONS OF MASS DESTRUCTION

According to the FBI (2012), WMDs are defined in U.S. law (18 USC §2332a) as

(A) any destructive device as defined in section 921 of this title (i.e., explosive device); (B) any weapon designed or intended to cause death or serious bodily injury through the release, dissemination, or impact of toxic or poisonous chemicals, or their precursors; (C) any weapon involving a biological agent, toxin, or vector ... and (D) any weapon that is designed to release radiation or radioactivity at a level dangerous to human life. WMD is often referred to by the collection of modalities that make up the set of weapons: chemical, biological, radiological, nuclear, and explosive (CBRNE). These are weapons that have a relatively large-scale impact on people, property, and/or infrastructure.

Although we tend to worry about al-Qaeda groups using WMDs, chemical, biological, and radiological weapons are also on the table for use by domestic terror groups. A 2015 study conducted by Syracuse University's Maxwell School of Public Policy and the New America Foundation examined 182 cases of domestic terrorist acts and plots in the United States since 9/11, none involving al-Qaeda or groups/individuals motivated by Islamist radicalism. The ideologies studied span the spectrum from neo-Nazism and militant Christian fundamentalism to anarchism and violent environmentalism (New America Foundation 2016). Among the vast number of weapons, destruction of property, and conspiracy charges, the research also found five plots of domestic terrorism involving chemical, biological, and radiological agents that could have killed thousands of Americans if executed (New America Foundation 2016).

Naturally, al-Qaeda and ISIS are pursuing WMD components and technology. Al-Qaeda's ten-volume *Encyclopedia of Afghanistan Resistance* found in Jalalabad contained formulas for manufacturing toxins, botulinum, and ricin and provided methods for dissemination. At Tarnak Farms, the former Afghan training camp near Kandahar, Afghanistan, al-Qaeda not only provided firearms training, but also experimented with biological warfare in a special laboratory. Ahmed Ressum, the Algerian al-Qaeda member who was caught at the Canadian border before he could execute the 2000 "millennium attack" of the Los Angeles airport, testified al-Qaeda taught him to poison people by putting toxins on doorknobs, and he engaged in experiments in which dogs were injected with a mixture containing cyanide and sulfuric acid. Also, al-Qaeda members were seeking to fly crop dusters that analysts believe might have been used to disseminate anthrax and chemical or biological agents (Boureston 2002). Regarding al-Qaeda and biological warfare, in 2009, Algerian newspaper *Anahar al-Jadeed* reported forty AQIM (al-Qaeda in the Islamic Maghreb) terrorists died at a training camp in Algeria from their infection with bubonic plague (*Al Arabiya News* 2009). Speculation abounded: Was it a dead rat causing the deaths or an experiment gone bad? We will never know; however, al-Qaeda's *Inspire* magazine has encouraged its readers to manufacture ricin, botulinum, and sarin in their homes, even encouraging them to get a Muslim microbiologist to assist, if needed.

Finally, ISIS considered weaponizing *Yersinia pestis*, or the bubonic plague in 2014, as indicated on a laptop seized from a militant/engineer in Syria. ISIS was keen to exchange American hostages for Pakistani neuroscientist Aafia Siddiqui, who is serving 86 years in U.S. federal prison on terrorism charges. Also known as "lady al-Qaeda," Siddiqui, who received her PhD in the United States, was captured in Afghanistan with detailed plans to construct a variety of WMDs and employ them against U.S. targets. Therefore, it is very concerning that ISIS sought Siddiqui to become part of their campaign to expand their self-labeled "caliphate." ISIS also accessed chemical weapons stockpiles as they took cities in Iraq and Syria, and tested them at least twice on the battlefield.

Domestic terrorists may also use the al-Qaeda guides (widely available on the Internet) as a template for constructing their own weapons; in 2008, the FBI arrested Roger Bergendorff, who had ricin, a schematic for an injection pen, weapon silencers, and the jihadist *Anarchist Cookbook* in his Las Vegas hotel room. After recovering from ricin poisoning, he was sentenced to 3.5 years in prison. Bergendorff never gave an exact motive behind his activities, but clearly it is dangerous to have private citizens tinkering with toxic biological and chemical compounds. We need to continue to work to get materials off the Internet and out of the hands of would-be attackers.

Perhaps the most accessible WMD is chlorine, an easily obtained chemical that could sicken or kill hundreds of people, under the right conditions. AQIM used chlorine in several vehicle-borne IED (VBIED) attacks against coalition forces in Iraq in 2006; although more casualties arose from the bomb blasts, the terrorists kept trying different methodologies to perfect their technique and used it thirteen more times in the war. Chlorine can be a silent killer; in 2005, two trains collided in Graniteville, South Carolina, in the middle of the night, releasing 60 tons of chlorine gas. Unsuspecting residents who heard the collision drove through the cloud, stayed in their homes, or kept working their outdoor night shift jobs. In all, nine people died and over two hundred fifty were sickened by the gas, which causes nausea, dizziness, and vomiting. Certainly chlorine is easily obtained; therefore, we have to educate suppliers on how and what to report in terms of suspicious buys or patterns. Facilities with pools should keep chlorine products stored in a locked facility.

Do al-Qaeda or other foreign and domestic terrorist groups have WMDs? Those of us who live in the "open source world" do not have the definitive answer to the question; however, we can turn to government reports and Congressional testimony to provide an accurate barometer. For instance, take the December 2008 report entitled "The World at Risk," issued by the Graham/Talent Commission, a bipartisan group that spent 6 months examining the WMD issue. When Senator Graham briefed Congress, he ominously stated, "Terrorists could mount nuclear or biological attack within 5 years." Although this prediction has not yet come to fruition, statements from the report are very telling:

> The commission believes that unless the world community acts decisively and with great urgency, it is more likely than not that a weapon of mass destruction will be used in a terrorist attack somewhere in the world by the end of 2013. The Commission further believes that terrorists are more likely to be able to obtain and use a biological weapon than a nuclear weapon. (Commission on the Prevention of WMD Proliferation and Terrorism 2008)

Synthetic manufacturing is causing increased concerns, as chemically synthesized DNA can replicate DNA found in biological or chemical agents occurring in nature and, more worrisome, enhance the effects. According to Vahid Majidi, the assistant director of the FBI's Weapons of Mass Destruction Directorate, the agency is working to keep this emergent technology from falling into the wrong hands (Committee on Homeland Security and Governmental Affairs 2011).

What does this mean for soft targets? WMDs have been used against civilians, notably the sarin gas attack by terrorist group Aum Shinryko (now known as Aleph) in 1995, which killed thirteen people, severely injured fifty, and caused temporary vision problems for nearly one thousand others, including first responders who rushed to the scene without the proper equipment. In the United States, Bhagwan Shree Rajneesh led a religious group in the Dalles, Oregon, area in the early 1980s. In an attempt to keep voters away from the polls and sway a 1984 local election toward a candidate friendly to the group, he and his followers deliberately contaminated salad bars at ten local restaurants with salmonella. The incident was the first and single largest bioterrorist attack in U.S. history. Vulnerabilities of soft targets include contamination of the food

Photo 3.5 Bhagwan Shree Rajneesh, leader of religious cult that contaminated salad bars in Dalles, Oregon, with salmonella in order to influence votes in a local election. (From the Sandia National Laboratories, U.S. Department of Energy, National Nuclear Safety Administration.)

supply in cafeterias and food vendors, as well as use of central ventilation systems to release a chemical gas or biological toxins. Anthrax has been transported through the mail system to introduce it to the target area, so mail rooms should take extra precautions with suspicious letters or packages. A WMD agent might also be attached to a bomb, making it the dispersal method for radiation or chemical or biological agents. WMD is in the playbook for a variety of bad actors wishing to do us harm; therefore, we must prepare.

THE SIRA METHOD

As previously stated, physical profiling efforts will likely come up short when identifying suspicious actors. However, behavioral profiling is a powerful tool as humans are unable to suppress certain physical changes that occur due to stress and adrenaline surges. Michael Rozin, an Israeli security specialist who was later the security manager and captain of Special Operations at the Mall of America from 2005 to 2011, developed a methodology based on his experiences called the "Suspicion Indicators Recognition and Assessment" (SIRA)™. SIRA is an innovative behavior deterrence, detection, and security program, a proactive system to prevent acts of violence. Mr. Rozin believes there are some basic problems with the way organizations conduct their security program. The allocation of resources and effort is based on perceived threat level (e.g., "We've never had a shooting or a bombing here before, so the security posture is low"), there is a lack of comprehensive security assessment and design, and ineffective security objectives are in place. SIRA is based on human factors, in a complementary manner to physical security and security technology. Effective security objectives are to deter, detect, prevent, and respond; however, tangible activities to deter, detect, and prevent are often given less weight than the response element. Mr. Rozin asserts the right approach is to believe the threat is constant; and an incident can happen at any time ("Today might be the day"). The relationship between suspicion and threat should be that suspicion is a threat until it is refuted. The approach should be, "If there is a doubt, there is no doubt." A doubt is a threshold for action and sometimes a doubt is reason for action. "Suspicious" does not mean "guilty"; the violent act is definitive evidence of guilt. There is a gray area involved with suspicions, but a pre-incident indicator is enough to take investigative action.

We strongly recommend the SIRA program for those responsible for securing soft targets as it provides tools for engaging when something does not seem right with an individual, whether a stranger or an insider. It draws on a leader's intuition, mentioned previously as an important security tool to hone and trust, and it gives permission and a methodology for engagement (Rozin 2016).

SUICIDE TERRORISM

Suicide bombings typically result in higher casualty counts because the actor can access areas packed with people or a critical place in the facility and detonate the weapon at the time of his or her choosing. Suicide bombings are a relatively new phenomenon; Islamic Jihad Organization's attacks in 1983 during the Lebanese civil war are the first examples of modern suicide terrorism. They have been used extensively as a war-fighting and propaganda tool by nearly every radical Islamist terrorist group in the last 30 years. Although suicide attacks amounted to just 3 percent of all terrorist incidents from 1980 through 2003, they were responsible for 48 percent of all fatalities, not counting deaths resulting from the suicide attacks via airliner on 9/11. In addition to strategically placed IEDs, suicide bombings are the primary tactic used in conflicts in Southwest Asia and are typically used against soft targets. In the 10 years after September 11, 2001, there were 336 suicide attacks in Afghanistan and 303 in Pakistan, and there were a staggering 1,003 documented suicide attacks in Iraq between March 20, 2003, and December 31, 2010. Notable al-Qaeda suicide attacks since 9/11 include the July 7 attacks in London in 2005, which killed fifty-two people, and hotel bombings against U.S. properties in Bali, Casablanca, Jakarta, Istanbul, Jordan, and Islamabad. Failed attempts at suicide bombing on airliners include the shoe bomber and the underwear bomber. ISIS used suicide bombers at the Stade de France stadium in the Paris attack; fortunately, they were unable to access the arena and detonated their vests in the parking lots. As police closed in on the remaining terror cell members, a woman blew herself up in an attempt to kill officer. Finally, ISIS employed a suicide bomber to infiltrate and attack the Sultanahmet tourist district in Istanbul, Turkey, on January 2016, killing ten and injuring hundreds.

Suicide attacks can shift the center of gravity of a battle and break the will of fearful citizens. The effect is amplified when the suicide attack is against a soft target like an open market or a school. Rather than focus on the psychology or motivation of a suicide terrorist, efforts are better placed on how to detect, prevent, and prepare for such an attack if it presents on your doorstep.

According to Dr. Daniel Kennedy (2006), a forensic criminologist, the following are seven signs of suicide terrorist activity (along with my elaboration); these steps can also be applied to detecting any type of terrorist or criminal preoperational planning activities:

1. *Surveillance*: The actor may observe the target area to determine security strengths and weaknesses, and the number of security personnel that might respond to an incident. Taking pictures, videotaping, drawing maps of the area, and stepping off distances are signs of terrorist surveillance activities. The 9/11 hijackers performed such surveillance on Disney properties and other soft targets. The Mumbai terrorists had a front man, American David Headley, who accomplished extensive surveillance on the targets. However, surveillance is not always necessary, as in the case of insiders or angry opportunists.
2. *Elicitation*: The actor attempts to gain information about certain operations: for example, asking questions about building infrastructure, class or worship schedules and attendance, and the posting of security personnel. The terrorists in the Nairobi mall shooting asked for and obtained blueprints of the building when securing shop space in the months prior to the attack.
3. *Test of security*: The individual may try to access unauthorized areas of your facility to test physical impediments or response time. Exit areas will be tested and possibly altered to prevent escape, such as in the Virginia Tech shooting when the

shooter chained the classroom building escape doors closed; he was seen testing the chains the day before. David Headley walked the ground in Mumbai and knew every emergency exit at the hotels, later instructing the terrorists how to enter the building and block exits for high casualty numbers.
4. *Acquiring supplies*: The purchase or theft of explosives, weapons, ammunition, or security badges or uniforms in the local area is a red flag of possible impending attack. Insiders will also start stockpiling items in the facility such as in the case of the Westgate Mall attack.
5. *Suspicious people*: The actor may come to the future attack scene more than once. Looking out of place or wearing odd clothing, several of the perpetrators of mall shootings walked the same path they took the day of the shootings.
6. *Trial run*: Before the final attack, the terrorist may conduct a "dry run" to identify obstacles. A dry run might involve calling in a bomb threat to observe emergency response, evacuation plans, and rally points for those leaving the building, such as the tactic used by Eric Rudolph prior to the abortion clinic bombing. Terrorists may also try to assemble weapons or bombs in the facility.
7. *Deploying assets*: In this last stage, the attacker and any operational supplies are moving or being moved into position. This is the last chance to prevent the attack.

When the planning begins, the suicide bomber has already decided to give up his or her life for the cause—whether revenge or a religious or political ideology. When backed into a corner, this type of attacker will jump straight to the execution stage of the operation, even if the timing is off, as has been witnessed in several suicide bombings where the individual was challenged before reaching the target and detonated the explosives. Therefore, never attempt to investigate, to engage, or to unravel any type of plot or attack on your own; call local law enforcement at the first sign of any of the preceding activity.

FACE OF RAGE

Your security personnel are your first line of defense, and they should be taught by experts to visually scan crowds or individuals for the following (Whetstone 2014):

- "Faces of threat" such as rage, anger, or guilt of a premeditated criminal act
- Threatening mannerisms
- Unusual nervousness
- Body language that could indicate having a concealed weapon

If your facility has a main reception area or front office, you may consider tailored training for those staff members. Anger is the most dangerous emotion; in this stage, a human is most likely to hurt another. Extreme anger, or unbridled rage, is extremely dangerous, as the capacity for rational thought and reasoning is lessened and the person will likely act out in a violent manner against the source of the rage until it is destroyed or the person is restrained.

Recognizing the "face of rage" is important in order to react properly to an imminent act of violence. Rage is an extraordinarily strong emotion, a feeling of intense, violent, or

SOFT TARGETS AND CRISIS MANAGEMENT

Photo 3.6 Faces of rage. (From Jennifer Hesterman.)

growing anger. There is a high adrenaline rush associated with a "fight-or-flight" response that can give extra strength and endurance; senses are sharpened and physical pain sensations are dulled. The enraged person may experience tunnel vision as he or she approaches the object of the anger. Due to elevated blood pressure, the person's sense of hearing will be diminished and he or she may actually "see red" due to bursting blood vessels in the eyes. Time may seem to slow, as the temporal perspective is also affected.

Rage is distinguished from other emotions by certain distorted facial expressions that cannot be reproduced and are the same regardless of race, ethnicity, or gender. The skin above the bridge of the nose folds in a certain way (think of the shape of the letter A), the lips become thin, and upper and lower teeth are visible. Also, the face remains fixed in a state of rage for approximately 4 hours before relaxing back to its normal state. Based on expression, which one of the people depicted in Figure 3.2 (interpreted from top to bottom and left to right) is the biggest threat to your organization?

Picture 1: His face shows possible elements of rage, but because his hands are in the air, he better demonstrates "posturing," meaning he is mad and he wants you to know that he is mad.

Picture 2: This is a good example of "stage 1 of rage," where anger is turning into rage. The man's eyebrows are scrunched together and he is showing the whites of his eyes, but the nose bridge of his face is not unnaturally wrinkled and his lips are pressed, indicating anger.

Picture 3: This woman is showing "stage 2 of rage." The nose bridge is wrinkled, nostrils are flared, the white areas of the eyes are more profound, and the lips are beginning to separate.

Picture 4: The woman is posturing, as evidenced by the display of her hand and arm. Her nose is crinkled and her upper lip is raised, which indicates disgust, not anger or rage.

Picture 5: She is posturing and there is no scrunching of the nose bridge or the eyebrows being drawn together, which would indicate rage. The fact that her eyebrows are not lowered suggests she is not even truly angry.

Picture 6: This face reflects anger, but not rage. Anger is best defined by lowered eyebrows and lips pressed together.

Picture 7: His arm suggests posturing and there is no scrunching of the eyebrows or nose bridge, which would indicate rage. He is protesting and he wants you to know he is mad, but he is not in a state of rage and will probably not react violently.

Picture 8: While this man is displaying a scrunched nose bridge, his raised upper lip is more indicative of disgust, not rage.

Picture 9: The expression is not natural; the picture is staged. The man is posturing, and the lowered eyebrows might suggest anger, except that the lips are not pressed together.

Picture 10: This is a good example of "stage 4 of rage" due the scrunched nose bridge (with the letter A visible), showing the whites of the eyes, possible flared nostrils, and display of teeth with the lips forming a square. This is the look of animals (including humans) who are in unbridled rage and about to attack or are in the process of attacking.

For more information on reading facial expressions such as rage, we recommend *Unmasking the Face,* by Dr. Paul Eckman (2003), who is likely the best facial expression and lie detection expert in the world. Dr. Eckman consults to the popular television show *Lie to Me,* which uses his theories to explore criminal cases and deception.

As opposed to terrorist rage, which is aimed against those who represent an opposing political or religious system, criminal rage is more likely tied to low self-esteem where a person has been bullied or perceives ill treatment at the hands of others. A good example is Elliott Rodger, the University of California, Santa Barbara, campus shooter. Although he did not have the "face of rage" in the video he posted the day prior to the shooting (podcast audio, http://www.youtube.com/watch?v=5EVEq-NikxA), he clearly states his rationale for the murders he is about to commit. The day of the shootings, he was likely in a state of rage; for example, even though he had high-powered rifles, his first act was to stab his three roommates to death. He then went on a shooting spree throughout the campus, killing six more people.

Not all terrorists, even suicide terrorists, will be enraged. Some have been methodically planning and rehearsing the attack, and on the day of the event, they will be calm and cool as they approach the target. Others may even laugh or show joy once the event begins. Religious terrorists likely believe their reward in heaven is just seconds away, sometimes causing a state of euphoria. However, in the case of an enraged opportunist, such as the white supremacist shootings at the Holocaust museum in Washington, DC, and the Jewish center in Kansas City, Kansas, the individual is pushed over the brink by some environmental factor and moves quickly to the execution stage. Or maybe there is no exact plan, but a church, school, college campus, hotel, or concert event provides the venue for the rage attack. In any case, recognizing the face of rage is important for response. Engaging enraged persons is extremely dangerous and you are unlikely to be successful either talking them down or changing their intention.

As we say in the military, the human is the best weapon system. After reflective thought on steady-state and crisis leadership, honing intuition, training, and comprehending the general types of threats and possible avenues of response, you are ready to build your organization's plan to detect, deter, mitigate, and respond to the threat.

FIFTEEN SOFT TARGET TAKEAWAYS

There are fifteen main takeaways that should be revisited often and shared in your facilities:

1. Humans have a psychological blind eye to soft targeting. We cannot help it: our cultural instinct is that civilians are noncombatants and should be protected, not targeted. We do not want to think bad actors would hit our soft spot and engage where we are most vulnerable, but they will. Terrorism has no moral restraints.
2. Understand that soft targets are hit every day around the world. International terrorist groups are actively planning operations in the United States, but a homegrown soft target attack is more likely.
3. Know the threat in your world and in your community. Ask questions of local law enforcement; join groups such as the FBI's Infragard or ASIS and attend chapter meetings. There may be no evidence of a threat, but that is not the same as evidence of no threat. Remember, in the "black swan" realm, what we don't know is more important than what we know.
4. Intuitively understand your vulnerabilities and how much risk you are assuming. Identify your Achilles heel as the one vulnerability that exposes you to the greatest risk and mitigate it.
5. Before investing in or relying too heavily on technology, recall a popular military axiom: "The human is the best weapon system." By honing your intuition, understanding behavioral detection tactics, and unapologetically engaging people in situations that just "don't feel right," you are taking a significant step to protect your property and its occupants. Technology should complement your other efforts, not be the central focus of your security plan.
6. Practice good steady-state leadership for outstanding crisis leadership when it is required. Frontloaded planning, training, and exercising efforts will pay dividends in an emergency situation.
7. Take steps today to harden your facility. Consider lowering the "heat" of your building by decreasing its profile, removing signage and symbols that may be inflammatory. Installing fences and raising the height of current fencing are the easiest and fastest ways to protect your property from intruders and present a fortified appearance to outsiders.
8. Consider the deterrent effect of security officers and vehicles on your property.
9. Remember, as we harden facilities, the insider threat will grow. An insider is a serious threat. An insider lone wolf is dangerous. An insider lone wolf who is a radicalized jihadist is the gravest.

10. Fight the five emotional states that increase your vulnerability: *hopelessness* (there is not much we can do to prevent or mitigate the threat), *infallibility* (it will never happen here), *inescapability* (it is destiny or unavoidable, so why even try), *invulnerability* (it cannot happen to me), and the most dangerous, *inevitability* (if it is going to happen, there is nothing I can do about it anyway).
11. When faced with a budgeting dilemma, consider this question: "What is the cost of *not* protecting our people?"
12. Focus on vulnerability, not probability.
13. Invest in preparedness, not prediction.
14. Understand that we can paint an accurate picture of consequences even if we cannot predict how likely they are to occur.
15. There is desire for balance between normalcy and vigilance. The best approach is to "bake" security fully into the culture where it permeates decision making about infrastructure and equipment purchases, event planning, and hiring and staffing activities.

Understandably, making changes for the sake of security sometimes feels like we are conceding our way of life to adversaries. Sadly, the world has changed, and so must we.

REFERENCES

Al Arabiya News. Black Death Kills 40 Al-Qaeda Fighters in Algeria. http://www.alarabiya.net/articles/2009/01/20/64603.html, 2009.

Boureston, Jack. Assessing Al Qaeda's WMD capabilities. *Strategic Insights, Naval Postgraduate School* 1, no. 7, September 2002.

Bunn, Matthew and Scott D. Sagan. *A Worst Practices Guide to Insider Threats: Lessons from Past Mistakes.* Cambridge, MA: American Academy of Arts and Sciences, 2014.

Carpenter, Scott. *True Believer: Inside the Investigation and Capture of Ana Montes, Cuba's Master Spy.* Annapolis, MD: Naval Institute Press, 2007.

Catranzos, Nick. *Managing the Insider Threat. No Dark Corners.* Boca Raton, FL: CRC Press, 2012.

Catranzos, Nick. No dark corners: A different answer to insider threats. *Homeland Security Affairs Journal* VI, no. 2, May 2010.

Commission on the Prevention of WMD Proliferation and Terrorism. World at Risk: Report of the Commission on the Prevention of WMD Proliferation and Terrorism. 2008.

Committee on Homeland Security and Governmental Affairs. Ten Years after 9/11 and the Anthrax Attacks: Protecting against Biological Threats, October 18, 2011.

Eckman, Paul. *Unmasking the Face.* Los Altos, CA: Malor Books, 2003.

FBI (Federal Bureau of Investigation). Weapons of Mass Destruction. 2012.

Gips, Michael A. Secondary devices a primary concern. *Security Management* 47, no. 7, 2003.

International Humanitarian Law—Treaties & Documents. http://www.icrc.org/ihl.nsf/WebART/470-750065?OpenDocument, 2014.

Kennedy, Daniel B. A précis of suicide terrorism. *Journal of Homeland Security and Emergency Management* 3, no. 4, 2006.

NELP (National Employment Law Project). Ban the Box. 2016.

New America Foundation. Non-Jihadist Cases, 2001–2016. http://securitydata.newamerica.net/extremists/analysis.html, 2016.

Office of Naval Research. Basic Research Challenge—Enhancing Intuitive Decision Making through Implicit Learning. 2014.

Rodger, Elliott. Retribution. Video posted on YouTube day prior to massacre. Podcast audio. http://www.youtube.com/watch?v=5EVEq-NikxA

Rozin, Michael. Suspicion Indicators Recognition & Assessment (Sira) Training. http://rozinconsulting.com/, 2016.

Snow, Jon. Sri Lanka's Killing Fields. United Kingdom: BBC, 2011.

UN (United Nations). Report of the Secretary-General's Panel of Experts on Accountability in Sri Lanka. 2011.

Whetstone, Douglas. *Catch the Lie': Importance of Body Language Deception Detection for Security Officials*. Sterling, VA: Whetstone Security Group, Inc., 2014.

4

The Common-Sense Guide for the CEO

Michael J. Fagel

Contents

Introduction	52
The Public's Expectations and Your Responsibilities	52
Federal Expectations for Response to Domestic Incidents	53
What Are Your Goals?	54
Reducing Your Risk by Using This Kit	54
Conducting Self-Assessment	55
Scoring Your Self-Assessment	56
Building Your Own Survival Kit and Your Community's EOP	57
What to Put into Your Survival Kit	57
Kit Contents	57
Suggestions for Developing and Using Your Kit	58
The "Take-It-with-You" CEO Checklist	59
Background Information	60
Immediate Actions	61
Personal Actions	62
Legal Issues	63
Political Issues	63
Public Information	64
A Note about the Checklist	65
Appendix A: IEMS and the CEO	65
Appendix B: Camera-Ready Copy of the CEO Checklist	68
Appendix C: Action Steps to Reduce Your Community's Risk and Implement IEMS	69
Acknowledgments	70

INTRODUCTION

As one result of rapidly increasing technology, the fact that a greater percentage of the population lives and works in areas that are at risk of disaster, the increased mobility of the worldwide population, and other factors, the risk of catastrophic disasters that result in large numbers of casualties and damage to property and the environment has never been greater, and the risk is expected to grow. Additionally, foreign terrorist groups that once seemed to pose only a remote threat to the U.S. homeland have demonstrated that they *can* strike us where we live—and strike with devastating consequences to our population, our infrastructure, and our way of life.

The increased risk that a catastrophic event may occur has placed new responsibilities on government at all levels. As the first line of defense against all types of incidents, you, as the elected leader in your community, must work actively to prevent, prepare for, respond to, recover from, and mitigate the effects of emergencies and disasters—and do so with a new sense of urgency. Emergency managers and responders need your support for a comprehensive emergency management program more than ever before. And the public is looking to you for reassurance that, should an incident occur, public officials and employees will be able to control the situation quickly and effectively.

Contrast these new expectations with the reality that most chief elected officials (CEOs):

1. Are not typically involved in the emergency management process and are unaware of what is required to ensure that an effective Emergency Operations Plan (EOP) exists in their communities.
2. Have little experience in managing emergencies.
3. Do not have a clear perception of the key players' responsibilities in emergency preparedness or response.
4. Do not understand the legal requirements and constraints placed on public officials by enabling legislation and legal governing framework.
5. Do not play a significant role in public preparedness.

If the realities stated above apply to your situation, this kit can help you fill the gaps in your understanding and become a leader in ensuring that your community is as prepared as possible.

THE PUBLIC'S EXPECTATIONS AND YOUR RESPONSIBILITIES

A 2013 public opinion poll conducted for the Office of Citizen Corps surveyed 2,000 adults about their concerns regarding emergencies. When asked about the types of emergencies that worried them most:

- Thirty-three percent stated that they were most concerned about terrorism.
- Twenty-six percent stated that they were most concerned about man-made accidents.

- Eighteen percent stated that they were most concerned about hurricanes and/or violent storms.
- When asked who they would expect to rely on in these emergencies,
  - Sixty-two percent stated that they would expect to rely on first responders.
  - Thirty-four percent responded that they would expect to rely on government agencies.

Additionally, a high percentage of respondents indicated a desire to have a plan for emergencies. Seventy percent said that they would be more likely to develop plans if they had support from local government and community organizations.

What do these survey results mean for you as a CEO? Specifically, that means that today, with the influx of natural disasters that have struck the United States in the last decade, citizens are demanding that local government provide adequate and timely leadership in times of disasters at ALL levels. ALL disasters ARE LOCAL.

What type of leadership are you providing to your community? Ask yourself:

- *Have I taken an active role in support of the emergency operations planning process?* Have I made it clear to the emergency management agency, first responders, and other personnel who are involved in emergency preparedness and response that their activities are important to me personally? Have I supported the emergency management agency and response agencies by providing what they need to get the job done? Have I participated in "kick off" training and exercises to show my support?
- *Have I taken steps to ensure that the public is educated about the threats facing the community?* Do citizens know what they need to do to protect themselves? Do they know what they can do to help prevent emergencies from occurring or to reduce the damage they cause? Do they know what they can do to help first responders?
- *Have I supported measures to mitigate the risk of damage from high-risk hazards?* Are zoning requirements, building ordinances, and other measures in place to minimize the risk of physical and environmental damage from high-risk hazards?

If the answer to any of these questions is "No," you are not doing enough to protect your community from high-risk hazards.

FEDERAL EXPECTATIONS FOR RESPONSE TO DOMESTIC INCIDENTS

The Department of Homeland Security (DHS), which has overall responsibility for coordinating the federal response to disasters and acts of terror within the United States, has recognized the need to improve domestic incident response at all levels of government. As such, DHS has developed and implemented the National Incident Management System (NIMS). Among the initiatives promulgated by NIMS are

- The requirement for governments at all levels to adopt the Incident Command System (ICS) as the model for managing all types of incidents, regardless of type, size, or complexity.

- The establishment of systems to ensure coordination among government entities and agencies.
- The establishment of standards for training and exercises for response personnel.
- The requirement that equipment be typed according to capability.
- The establishment of interoperability requirements, to be facilitated by the NIMS Integration Center, which will publish lists of equipment meeting established requirements.

To be compliant with NIMS initially, governments at all levels were required to adopt ICS not later than October 1, 2004, or risk being ineligible for preparedness funding. NIMS will be undergoing a major change in the near future, with some enhancements underway. Time frames for compliance with other NIMS requirements will be determined at a later date.

WHAT ARE YOUR GOALS?

If, after reading the information presented above, you believe that you can and should do more to improve your community's preparedness for emergencies and disasters, you might begin by setting some goals for yourself. Some examples of goals that you might set are listed below.

- Understand the need to have a personal commitment to emergency management, to have the capability to judge its status, and develop the public policy to improve it.
- Understand how some emergencies differ from the management of normal operations and encourage the development of EOPs that all response agencies practice regularly and implement when responding to all major incidents.
- Ensure that ICS is adopted by all response agencies and implemented during exercises and in all responses.
- Understand the interaction among federal, state, and local governments during a disaster or severe emergency.
- Understand your responsibility to educate citizens about what they can do to prevent and prepare for emergencies that present a high risk to the community.
- Understand your responsibility to communicate to the community how different types of emergencies can affect their lives and property and how sound emergency management policy is a good investment in public safety.

REDUCING YOUR RISK BY USING THIS KIT

After reviewing and determining your personal emergency management goals, you need to follow up with several concrete actions. You should:

1. Review the contents of this kit.
2. Conduct the self-assessment.
3. Assign staff to prepare your survival kit.
4. Review the survival kit.

5. Take other actions, as necessary, including:
    a. Direct senior staff to develop similar kits for their own use.
    b. Chair monthly Emergency Management Council meetings to improve inter-agency working relationships for both day-to-day and disaster response.
6. Participate—and direct your senior staff to participate—in disaster exercises at least once each year.
7. Review and update the kit contents at least semiannually.
8. Review the action steps included in Appendix C for more ideas to strengthen your community's comprehensive emergency management program.

CONDUCTING SELF-ASSESSMENT

This section presents an informal quiz that you can take as a confidential personal assessment of you and your community's emergency management risk. The questions reflect some of the survival capabilities that past experience has associated with effective CEO participation in emergency management.

Following the questions is a scoring section. The scoring section is designed to help you decide whether action is needed to reduce your risk.

1. Has your community updated its EOP within the past 12 months?
    ☐ Yes   ☐ No
2. Have all response agencies adopted ICS as the single management system for all incidents?
    ☐ Yes   ☐ No
3. Does your community have mutual aid agreements with surrounding communities, standby contracts for needed resources, and other mechanisms to ensure an adequate response to a major incident?
    ☐ Yes   ☐ No
4. Has your community typed its response resources according to capability?
    ☐ Yes   ☐ No
5. Does your community work with surrounding communities to improve interoperability among response resources, communications, and technology?
    ☐ Yes   ☐ No
6. Have you personally reviewed your community's EOP within the past 12 months?
    ☐ Yes   ☐ No
7. Have you and your senior staff participated in a disaster exercise within the past 12 months?
    ☐ Yes   ☐ No
8. Has your community adopted a comprehensive program to educate the public on the risks faced by the community and how to prepare for them?
    ☐ Yes   ☐ No
9. Has your community involved the media in planning, training, and exercising the EOP?
    ☐ Yes   ☐ No

10. Does your community have adequate accounting and record-keeping procedures to document requests for reimbursement from state and federal assistance programs?
    ☐ Yes   ☐ No
11. Does your community ensure that the information needed to defend itself in a disaster-related lawsuit is maintained during an incident?
    ☐ Yes   ☐ No
12. Have you committed time in the past year for face-to-face discussion with your emergency manager discussing ways to improve disaster management?
    ☐ Yes   ☐ No
13. Have you supported zoning, building code, and other ordinances that would reduce your community's risk of damage from specific types of disasters?
    ☐ Yes   ☐ No
14. Do you understand state and local emergency management law, particularly as it relates to the CEO's powers during an emergency?
    ☐ Yes   ☐ No
15. Does your community's EOP include plans for continuity of operations in the event of a catastrophic disaster that renders critical facilities inoperable or incapacitates key personnel?
    ☐ Yes   ☐ No

SCORING YOUR SELF-ASSESSMENT

Obviously, the correct answer for all questions in the self-assessment is yes. To gauge how well you are doing, give yourself one point for each "yes" answer. Total your score and grade your risk according to the scale below.

14–15 points. Your community's risk is apparently well managed, but it can be managed better. Look over your "no" answers, and decide what you can do to reduce these areas of exposure.

12–13 points. Your community is making good progress, but there are a number of actions you can take to reduce your risk. You may wish to focus your attention on the areas to which you answered "no." Based on the results of reviews in these areas, you can decide what additional steps you need to take.

9–11 points. Your community may be at risk, but it is not too late. Scores in this range suggest that your community's emergency management responsibilities are being partially met, but there is room for (and need of) improvement. Start today by working with your emergency manager to improve your community's emergency management program.

Fewer than 9 points. Your community is at risk! Prompt action is indicated. You need to take immediate action to improve your community's ability to respond effectively to a major emergency or disaster. Begin by conducting a complete review of the emergency management organization, and take an active role in the review. *Review the remainder of this kit*, and take the steps necessary to improve your community's capability.

BUILDING YOUR OWN SURVIVAL KIT AND YOUR COMMUNITY'S EOP

Before building your own survival kit, it is important to understand the relationship between the kit and your community's EOP. Many progressive communities have developed a system of emergency response guides tailored to local needs. The basic EOP is the largest and most complete guide. It will contain the basic plan, functional annexes such as evacuation and mass care, and hazard-specific appendices such as flood and terrorism.

Because it is impossible for everyone to carry a copy of the complete EOP, some communities have developed a simplified guide to their EOPs. These guides may be called "Emergency Checklists," "Emergency Procedures," or another name but their purpose is the same—to provide a quick reference that spells out

1. Which agency has primary responsibility for the different types of emergencies facing the community.
2. The steps that are critical to take regardless of the emergency.

The guides are designed to provide department directors, response personnel, and other key personnel a short, easily readable, and readily available list of task assignments to be carried out as required for the categories of disasters faced by the community.

If your community has developed a similar guide, you should include it in your personal survival kit which, in turn, would be stored in your car or elsewhere. The CEO Checklist is designed for you to carry in your wallet or purse so that it will be available to you immediately in an emergency.

WHAT TO PUT INTO YOUR SURVIVAL KIT

Your survival kit should include the tools and equipment that you, as an elected official, need with you during the first hours following a disaster, when critical decisions often must be made. Use the list below as a guide to assembling your survival kit.

KIT CONTENTS

Information:

- A pamphlet-sized mini emergency plan or other checklists
- A CEO Checklist
- Maps of the community (e.g., by street, topography, satellite photo)

Supplies:

- A tape recorder
- A notebook or log book
- Spare pencils or pens
- Thumb drives
- Grease pencils
- Paper maps

- Markers
- Pens
- Pencils
- Tape

Identification:

- Your government-issued credential with photograph

Contact Lists:

- The call-down list for all key personnel
- Contact information for key state counterparts or liaisons

Clothing:

- Seasonable outerwear

Optional Items:

- A portable two-way radio with extra batteries (not to be used if a bomb is suspected)
- A flashlight with lithium batteries and a spare bulb
- A cellular phone
- Protective clothing (e.g., a hard hat, goggles, gloves, boots)
- A change of clothing
- Personal comfort items

SUGGESTIONS FOR DEVELOPING AND USING YOUR KIT

Use the guidelines below when putting your survival kit together:

- Keep it simple and small. Try to keep the kit compact—small enough to carry comfortably with you in your car.
- Direct all key subordinates to develop their own kits.
- Test the kits during preparedness exercises.
- Send enabling memos to all departments directing them to support plan development, revision, training, and exercises.
- Attend important training and exercises personally.
- Call for multiagency critiques after each significant incident.
- Use the checklist and other materials in the kit as a guide to reviewing your emergency management knowledge and the status of your community's preparedness.
- Review and direct the modification of the EOP and survival kits, as necessary—but not less often than once each year.
- Consider testing the response of your senior staff through unannounced drills.
- Establish a sequence for reporting critical information to you, and use this sequence as a guide for designing your own checklist.

THE "TAKE-IT-WITH-YOU" CEO CHECKLIST

No commercial pilot would think of taking even a routine flight without going over the preflight checklist carefully. If pilots with hundreds of hours of training and practice need them for safety reasons, shouldn't you, as a CEO, have a checklist available when you are "flying solo" during a major emergency?

You can use the CEO Checklist included in this kit as it is—but it will be better and more effective if you use it as a prototype for addressing your community's specific needs. The Checklist folds to credit card size, so you can take it anywhere with you. Essentially, the Checklist is a set of reminders for the following:

- Information you should gather
- Questions you should ask
- Immediate steps you should take
- Key points for you to remember as you begin to manage your community's response

The Checklist is designed to take you through the first hours following a disaster. Make your version of it a synthesis of your role in your community's EOP.

Remember that neither this nor any other tool will substitute for direct involvement in your community's emergency planning process. Rather, the Checklist should be a reflection of your involvement in planning, development, training, and exercising.

A sample of the Checklist is included in this kit for your use. The Checklist is organized into six panels:

- Background Information
- Immediate Actions
- Personal Actions
- Legal Requirements
- Political Requirements
- Public Information

Each panel lists key reminders in brief phrases. To help you understand the phrases, each panel is explained in more detail on the pages that follow. If the meaning of any phrase is not readily apparent to you, refer to the appropriate section in this kit for a reminder as to its meaning.

If you see the value of the Checklist and are convinced that it or one similar to it would be an important tool for you in meeting your emergency management responsibilities, you may want to skip to the final item in this section. That item is a draft memo that outlines points you might like to raise with your staff when you forward this material to them for action.

BACKGROUND INFORMATION

The Checklist's first panel gives you a quick list of background information you will need when the incident is first reported to you. You will need this information to assess the incident and to determine your immediate actions.

This information is organized in a logical sequence that fits most situations. Because there is always the chance that you could be cut off from the official alerting you, the Checklist has been designed to get the most critical information first. You can help yourself while helping the reporting official if you make it clear that you want the official to report the facts in the order presented on the Checklist.

A description of each item on the Checklist is included on the following pages.

Notified by—For legal reasons, it is important to document who first notified you about the incident. Obviously, it is important to validate the authenticity of the call by password or other means.

Time—Be sure to note exactly when you were first notified of the incident. Use the space on the Checklist to note the time of notification. Liability judgments may well rest on how you respond based on information you were given in this first call.

Type of Emergency—Understand the type of emergency scenario you are dealing with and the actual or, if unknown, potential scale of it.

Location—Get the best description of the location and extent of the incident as possible.

CEO Reporting Point—Clarify whether to go to the Emergency Operations Center (EOC), and alternate EOC, or another reporting point.

Open Routes—Determine the best available travel route and type of transportation required.

Available Communication—Confirm the availability of primary and backup communication channels and frequencies available to you. Remaining in communication at all times is critical to effective leadership during an emergency.

Damage—The scope of damage is often unreliable in the initial reports. Despite this reality, try to document what you can. The information may prove crucial in how soon a state or federal disaster declaration is made. Be sure to gather information about
- Injuries and deaths
- Property damage
- Other impacts (e.g., damage to critical facilities or the infrastructure)

Resources—You need to gather information about resources to get an immediate sense of the resources committed, the resources required, and the possibility of the need for outside resources. Often, mutual aid agreements require the CEO or a designee to formally request the assistance.

Even if no immediate action is required from you to summon additional resources, it is important to know what resource requirements exist or might be needed later. This information allows the CEO to give immediate direction to authorize the use of additional resources.

Gather the information below as a minimum:

- *Incident command status*—Under the requirements of NIMS, your community is required to adopt ICS as its incident management system. One of the key aspects of successful emergency management is knowing who is in overall command at the incident and at the EOC.
- *Internal/external resources committed*—At this point, you will find out what local agencies are involved in the response and their degree of commitment (i.e., partial or full). You will also want to know if outside agencies have responded and, if so, whether they are engaged in operations or serving as backup resources.
- *Internal/external resources required*—Ascertain if there already is, or is about to be, a shortage of internal or external resources. Such a shortage may require you to take some immediate action, such as calling the CEO of a nearby community to request assistance.
- *EOC status and location*—The EOC may require your authorization to be activated—or an alternate EOC site may need to be designated. By checking the status of the EOC, you will be certain that everyone is clear about where they should be.
- *Other authorities notified*—Gathering this information will help you determine whether all appropriate officials have been notified of the incident. By determining what officials have been notified, you can determine whether other personnel should be notified immediately.

IMMEDIATE ACTIONS

The Checklist's second panel will be your guide to actions you must take immediately to save lives and protect property.

A description of the Immediate Actions is included below and on the next page.

Begin a personal log—Keep a log of all key information, factors weighed, and decisions reached from the time you are notified that an incident has occurred. The log can be written or recorded using a mini-cassette recorder.

The log should include all information and orders given under background information and should be used continuously throughout the incident. This log will document the amount of information you had when making decisions and will protect you if liability issues are raised later. The log will also be essential when preparing your after action report.

Establish contact with the Emergency Management Agency—As the situation permits, and if you have not already done so, contact the emergency manager to ascertain the Emergency Management Agency's status.

Direct staff to assess and report on problems, resource shortfalls, policy needs, and options—Direct the emergency management staff to compile an initial assessment as it is reported from the incident site. The assessment should take the form of a situation report. It should give the status of dispatched resources, coordination efforts, and appropriate measures of problems, shortfalls, and options for the initial response.

Chair assessment meeting—If officials are conforming to local protocols—and if those protocols are sound—this initial meeting should begin to answer some of the questions including the following:
- Who is in charge? Of what? Where?
- Has there been a proper vesting of authority?
- Is continuity of government assured?
- What is the status of intergovernmental coordination?
- What options are open to deal with resource shortfalls?
- What financial issues are surfacing?
- What parameters should be established in contacting outside public officials?
- Is there a need to place other personnel on alert?
- How much media interest is there in the incident?

Based on the information provided at this meeting, you may decide to place additional resources on standby or to activate them.

Issue emergency declarations, as needed—Decide if an emergency declaration should be issued or remain in force. Issue emergency declarations as appropriate, and be sure official documentation is initiated and continued throughout the emergency. This documentation may be critical if liability issues arise and will be necessary if state and federal assistance is sought.

Establish reporting procedures—It is important to establish a regular schedule for bringing your senior staff together to hear from those planning the next tactical response steps. You and your key advisors will need this information to establish strategic priorities and assign resources.

Remind staff to keep complete logs and financial records—Each key official should maintain a log that records actions taken, information received, and any deviation from policy, together with the rationale for that decision.

Begin liaison with other officials—Consider this one of your primary responsibilities. Maintaining contact will foster cooperation.

PERSONAL ACTIONS

In the midst of thinking about the danger threatening your community, it may be difficult to recognize the need to take the time to consider your own and your family's needs. Experience has shown that public officials function better if they have made adequate provisions for their families' safety and to ensure that they have the personal items that they will need during the incident. The Checklist's third panel describes these actions.

A description of the personal actions you should take is listed below.

Provide family with contact information—Alert your family to the incident, and assure that they know how to respond if they are in any danger. Be sure that they know where you will be and how to contact you. Taking the time to leave word with your family about your whereabouts and how you plan to maintain contact will be reassuring to them and to you.

Take your personal survival kit—If you have planned ahead, your personal survival kit will be assembled and in your car or in another place where you can reach it easily. Be sure to include any personal items that you may need should the incident continue for an extended period.

Other things to remember—Jot down other personal reminders in the space provided.

LEGAL ISSUES

The fourth panel of the Checklist includes legal issues that you will want to check on early in the incident. Checking potential legal issues early may very well save you and the community time, effort, and money later.

A description of the legal issues you should address is listed below.

Contact legal advisors—Set up communications links with the organization or community's legal advisor(s). The earlier you do this, the better. Review legal delegations and legally binding authorities for the following:
- Declaring emergencies
- Delegating authorities
- Securing intergovernmental assistance through mutual aid
- Protecting the population within the appropriate legal safeguards

Other issues may arise based on the incident type, location, cause, and severity.

Monitor equity of service based on needs and risks—Defend against charges of discrimination by establishing and following criteria for treating all sectors of the community equitably. This means keeping the public informed of what is being done to restore the community's essential services and monitoring service restoration to see that all neighborhoods receive equal treatment.

Have status of contracts reviewed—Have your community's legal advisors review any current contract with suppliers of emergency goods or services as necessary.

POLITICAL ISSUES

Regardless of how politically savvy you are, some issues will invariably arise. Staying aware of what issues are likely will save you political capital and allow you to focus your energy on the incident at hand.

Panel five of the Checklist includes some of the political issues you should consider. A description of these issues is provided below and on the next page.

Recognize accountability—Ultimately, the public will hold elected officials responsible for perceived response and recovery failures. Therefore, the response strategy must be carefully analyzed, without interfering with operational tactics. Recognize that accountability is a continuing issue and keep your senior staff and those at higher governmental levels informed of the ongoing situation.

Check provisions for public officials—Ensure that public officials have space at the EOC. Also be sure that they receive periodic updates and that senior staff are kept up to date on politically sensitive issues.

Establish and evaluate policy decisions frequently—Identify and consider the political aspects of declaring an emergency and other policy decisions.

Confer with other elected officials when problems arise—Get advice in anticipation of or as problems arise. A good rule of thumb here is that a good time to seek counsel from city council members or their equivalent is when things start to fall apart. By having these officials assigned in one place, these consultations can be undertaken promptly.

Consider also contacting any peer or advisor who has handled a similar incident for their advice and guidance.

Use elected officials to request assistance from public and private organizations—Often, a key to cutting through red tape and obtaining a quick response from other public and private resources is to contact the authority that controls the resource directly. This is a particularly appropriate role for public officials if normal channels are not responsive enough. Political connections can often expedite a special request.

PUBLIC INFORMATION

The impact of the media on response operations is sometimes referred to as the "disaster after the disaster." Disasters often have no greater challenge than that of attempting to use the media to inform the public accurately and in a timely manner. To succeed, the CEO must be prepared to deal with both the local and national media.

The sixth panel of the Checklist presents key points for dealing with the media. A description of each of these points is covered below.

Designate a single Public Information Officer (PIO)—Appoint one PIO to avoid conflicts in official statements that could result in confusion, panic, or misdirected public outcry about the handling of the incident. Appointing a single PIO may mean reaffirming your day-to-day spokesperson or it may mean appointing someone with experience in the type of incident that is occurring. Whomever you appoint, be sure that this individual coordinates with the PIOs from response agencies and other communities that may be involved in the incident—and with you—before releasing information to the public.

Ensure that the PIO establishes a Joint Information Center—A JIC is a single location for disseminating incident-related information to the media. Operating through a JIC helps to ensure a coordinated message and reduces "freelancing" by journalists who are looking for a story.

Approve all media releases—One of the worst situations that can occur during a response is for you to be blindsided by questions about media releases that you did not know about. Establish an approval process for media releases so that they are routed through you and coordinated with other agencies as they are developed. Take particular care with evacuation and return announcements.

Establish a media update and access policy—Consider the type of incident and the hazards at the scene when determining media update and access policy. For some incidents, it may be necessary to designate a pool of a few reporters and camera crews to allow reporting from the scene rather than granting blanket access. In any event, update the media regularly, especially early in the response when things are moving quickly. You can gradually reduce the frequency of media updates as the incident site is brought under control and especially during recovery.

Enlist the help of experts—It is not possible for you to be an expert in every type of incident response. Enlist the help of experts and let them answer technical questions about the incident. Experts may be available from agencies within the local or state government, from local colleges and universities, or from private industry.

A NOTE ABOUT THE CHECKLIST

Before using this or a modified checklist, review local procedures and your responsibilities from the basic EOP. You may find one of the Checklist's greatest benefits is that it facilitates your thinking about your and your community's readiness to handle a major incident.

Finally, to make this kit easy for you to use, a sample action memo is included here (Figure 4.1) for you to consider in taking your next steps to reduce your risk.

APPENDIX A: IEMS AND THE CEO

Background: Nature of the Problem

We live with a wide range of potential hazards—floods, hurricanes, tornadoes, earthquakes, fires, chemical and toxic material spills, civil disruption, public service strikes, major accidents, hostage situations, and terrorist attacks—that could affect any community no matter how large or small. Seventy percent of our nation's population resides in cities, providing high-density targets for disasters and, thus, greatly increasing the likelihood that these disasters will cause extensive loss of life and damage to property. One of the major problems in dealing with these hazards is the duplication of emergency response efforts at all levels of government. Currently, these emergencies are most often addressed by type and independently by the individual agencies that might respond. Responding to emergencies using this strategy has proven inefficient and ineffective, usually resulting in unnecessary and unacceptable loss of lives and damage to property and the environment.

Your Role in Comprehensive Emergency Management (CEM) and IEMS

When a disaster strikes, the CEO immediately becomes the focal point for leadership, and the community's success in responding to an emergency situation will depend on the effectiveness of the local emergency management system. The foundation of such a system

SOFT TARGETS AND CRISIS MANAGEMENT

CEO's Disaster Survival Kit **Immediate Actions**	CEO's Disaster Survival Kit **Immediate Actions**	CEO's Disaster Survival Kit **Personal Actions**
• Begin personal log • Establish contact with Office of Emergency Management • Direct staff to assess and report on problems, resources, shortfalls, policy needs, and options • Chair assessment meeting • Issue emergency declarations, as needed • Set reporting procedures • Remind staff to keep complete logs of actions and financial records • Begin liaison with other officials	• Begin personal log • Establish contact with Office of Emergency Management • Direct staff to assess and report on problems, resources, shortfalls, policy needs, and options • Chair assessment meeting • Issue emergency declarations, as needed • Set reporting procedures • Remind staff to keep complete logs of actions and financial records • Begin liaison with other officials	• Provide family with contact information • Take personal survival kit • Other things to remember: **Remember! Your role is not operational.**
CEO's Disaster Survival Kit **Legal Issues**	CEO's Disaster Survival Kit **Political Issues**	CEO's Disaster Survival Kit **Public Information**

• Contact legal advisors. Review legal responsibilities and authorities:
  • Making emergency declarations
  • Line of succession
  • Mutual aid agreements
  • Other legal issues:
• Monitor equity of service based on needs and risks. Maintain balance between public welfare and citizens' rights
• Have status of contracts reviewed
	• Recognize accountability. Check provisions for public officials:
	  • Space at the EOC
	  • Periodic updates
	  • Staff updates on politically sensitive issues, such as life and property losses and service interruptions
	• Establish and evaluate policy decisions frequently
	• Confer with other selected officials when problems arise
	• Use elected officials to request assistance from public and private organizations
		• Designate a single PIO
		• Ensure that the PIO establishes a Joint Information Center (JIC)
		• Approve all media releases
		• Establish media update and access policies
		• Enlist the help of experts

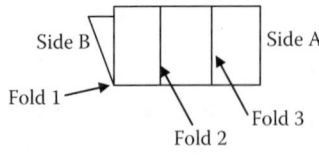

Trim away page except for the six panels. Fold (fold 1) and glue Side B to the back of Side A. Fold Side A panels (folds 2 and 3) as an accordion fold.

Figure 4.1 Camera-ready copy of a foldable CEO checklist. (Courtesy of the IAFC and Michael Fagel.)

should be an integrated emergency management plan and the capability of implementing the plan successfully.

One of the major roles of the CEO is to provide leadership in the development, training, and exercising of an effective integrated emergency management system (IEMS).

Often, CEOs take office with little or no experience in managing emergencies and without knowing what is involved in a comprehensive IEMS. The CEO may have little understanding of the laws that relate to emergency management or of liability issues revolving around action, or inaction, during an emergency. Community policies and systems that affect community risks and the potential response requirements may be unclear.

It is imperative to understand one's responsibilities to communicate to the public about the impact of emergencies on their lives and property and the need for emergency policy.

FEMA's Role in IEMS

The Federal Emergency Management Agency (FEMA)/DHS developed the IEMS to assist CEOs in establishing an effective system. IEMS has proven effective in communities throughout the country.

What Is IEMS?

IEMS is a comprehensive system that integrates and coordinates vital agencies and resources into a program of disaster prevention, preparedness, response, recovery, and mitigation. The establishment of an IEMS requires a systematic process to

- Identify risks and potential vulnerabilities.
- Inventory community resources.
- Outline roles and responsibilities of key agencies.
- Ensure strict coordination and communication among agencies, businesses, and nongovernmental organizations (NGOs).

What Is the Cost of an IEMS?

A comprehensive IEMS does not require extraordinary allocations of community funds. Because it is a management process, the program strives to use existing resources in a more efficient and effective manner rather than seek additional resources and generate new programs.

Phases of IEMS

When fully developed, IEMS is all hazard and all phase. IEMS is all hazard in that it addresses natural, man-made, and technological emergencies. IEMS is all phase in that it addresses prevention, preparedness, response, recovery, and mitigation. Each of the phases of IEMS are listed below:

- *Prevention* includes all activities initiated to prevent a disaster from occurring. Prevention includes measures such as ensuring the safe handling, transport, and storage of hazardous materials; safe storage of flammable materials; and enforcing building codes directed toward rendering structures safe from identified hazards.
- *Preparedness* readies governments to respond to emergencies. A response plan cannot be developed during a disaster and, if an emergency cannot be avoided, government must be prepared to cope with it. Planning, training, and exercises are essential elements of preparedness, as are proper and adequate supplies, equipment, facilities, and personnel.
- *Response* to a disaster depends on effective implementation of emergency operations that provide for an immediate, coordinated effort involving decisive actions that will eliminate or reduce the severity of the incident or will prevent it from intensifying. Response operations may include warning, evacuation, fire suppression, search and rescue, apprehension, treatment, and, in extreme cases, withdrawal for safety reasons.
- *Recovery* involves returning the community to its predisaster state. Recovery applies to individuals and organizations and includes physical, mental, and financial aspects such as repairing, replacing, or rebuilding property; regaining health and state of mind; and reopening damaged businesses.
- *Mitigation* includes all activities designed to reduce the recurring damage from known hazards. Mitigation and prevention are the best forms of emergency management but are often the most neglected.

IEMS: A System to Provide for Coordinated, Effective Response

IEMS is based on the recognition that there are common elements that form the foundation for responding to any emergency and increasing capabilities in these areas to improve the ability to deal with any type of emergency. The most important of these elements is the central management of all emergency activities under the ICS and Multiagency Coordination Systems.

Who is in charge during an emergency? *You*, as the CEO of your community, have the responsibility to provide for the protection of the lives and the property of your community's residents. Are *you* prepared to meet this responsibility? To carry out your responsibilities to the citizens of your community, you must ensure that local agencies are prepared to respond to any type of emergency.

APPENDIX B: CAMERA-READY COPY OF THE CEO CHECKLIST

Your printer can print these six panels back to front. Fold the Checklist as shown in Figure 4.1, and laminate it for longer use.

APPENDIX C: ACTION STEPS TO REDUCE YOUR COMMUNITY'S RISK AND IMPLEMENT IEMS

Getting Organized

1. Participate in monthly emergency management meetings.
2. Schedule presentations for elected and senior management officials.
3. Appoint representatives of other organizations to create an emergency management council or task force.
4. Hold a local IEMS workshop.
5. Assign an individual to be responsible for emergency management.
6. Provide line item budgets for emergency management activities.

Planning

1. Conduct or participate in community-wide planning.
2. Conduct or participate in multijurisdictional planning.
3. Conduct facility planning.
4. Conduct a hazard analysis and capability assessment.
5. Become involved in planning.
6. Involve the media, NGOs, and private entities in planning.
7. Evaluate and upgrade the EOP.
8. Designate a lead agency for each type of disaster.
9. Implement ICS for all responses.

Training and Exercises

1. Conduct or participate in community-wide training.
2. Conduct or participate in multijurisdictional training.
3. Conduct facility training.
4. Involve the media, NGOs, and private entities in training.
5. Conduct tabletop exercises every 6 months.
6. Conduct all-agency, multijurisdictional exercises at least annually.
7. Conduct ICS training.

Preparedness

1. Develop warning systems for the entire population.
2. Improve mutual aid agreements.
3. Give a customized Checklist to all senior officials.
4. Review legal responsibilities and authorities.
5. Develop procedures, media releases, and other emergency authorities before an emergency occurs.

Capability Development

1. Improve prevention and mitigation measures for all types of disasters.
2. Develop hazard-specific appendices to your plan for high-risk hazards.
3. Develop and maintain disaster resource lists.
4. Review liability and risk exposure.

Public Information

1. Conduct presentations on IEMS for senior officials.
2. Hold one-on-one meetings with senior officials to discuss emergency needs and procedures.
3. Conduct public and employee education programs on evacuation, sheltering in place, and disaster preparedness and response measures.
4. Improve relations with the media, NGOs, and the private sector.
5. Conduct seasonal public education campaigns on specific hazards.

ACKNOWLEDGMENTS

This form was largely developed in the 1980s by the Federal Emergency Management Agency and the International Association of Fire Chiefs under contract to the U.S. Government. The author gratefully acknowledges their role in providing the initial format (that has been out of print since 1982).

MAKE this document YOUR own by using it as a TEMPLATE for your community.

5

Planning for Terrorism

Michael J. Fagel

Contents

The Planning Process	72
Terrorism Hazards	73
WMD Hazard Agents	73
Chemical Agents	73
Biological Agents	74
Nuclear/Radiological Agents	75
Conventional Explosives and Secondary Devices	76
Combined Hazards	76
Other Terrorism Hazards	76
Low-Technology Devices and Delivery	77
Infrastructure Attacks	77
Cyberterrorism	77
Emergency Operations Plan: Situation	78
Potential Targets	79
Initial Warning	81
Initial Detection	81
Release Area	82
Investigation and Containment of Hazards	82
Emergency Operations Plan: Assumptions	83
Emergency Operations Plan: Concept of Operations	83
Direction and Control	83
Communications	86
Warning	86
Pre-Event Readiness	87
Emergency Public Information	87
Protective Actions	88
Mass Care	88

Health and Medical 89
Resources Management 90
Recovery 91
Urban Search and Rescue 91
Organization and Assignment of Responsibilities 91
    Local Emergency Responders 92
    Interjurisdictional Responsibilities 92
    State Emergency Responders 92
    State and Local Public Health Authorities 92
    Medical Service Providers 93
    Local Emergency Planning Committees and State Emergency Response
    Commissions 93
    Federal Emergency Responders 93
Administration and Logistics 93

Given the creativity of those committed to carrying out acts of terrorism, it is more important than ever that jurisdictions are equipped to respond to terrorist events. Planners must consider a broad range of incidents, including assaults on infrastructure and electronic information systems that could result in consequence affecting human, life, health, and safety.

State and local governments have the primary responsibility in planning for and managing the consequences of a terrorist incident using available resources in the critical hours before federal assistance can arrive.

THE PLANNING PROCESS

The process of developing a terrorism annex to your Emergency Operations Plan (EOP) is similar to that used to develop other EOPs. As is the case for these other plans, the terrorism planning process must begin before an emergency and before any planned special event that could be subject to a terrorist attack.

Traditionally, the planning process has consisted of six phases:

1. Initiation
2. Concept development
3. Plan development
4. Plan review
5. Development of supporting plans, procedures, and materials
6. Validation of plans using tabletop, functional and full-scale exercises

However, given the unusually intense and multifaceted nature of terrorist attacks, a seventh phase is recommended: thorough coordination of plans, internally and externally.

You should carefully compare plans for the various response functions within that agency and revise the plans, if necessary, to remove any discrepancies. This will help prevent disconnects between vital functions that support one another and help ensure that each does what the others expect on a timely basis. Similarly, the various departments and agencies within a local jurisdiction should also compare their plans, focusing on issues of consistency and coordination. Again, this review will ensure that each organization does what the other expects, when it is expected.

Such reviews are especially important in planning for response to a major terrorist incident, since local jurisdictions are likely to be aided during the response by neighboring communities, its own and neighboring counties, and its own and possible neighboring states.

TERRORISM HAZARDS

The terrorism annex to your EOP should identify and discuss the nature of the terrorist hazard. The hazard may be a weapon of mass destruction (WMD), including conventional explosives; secondary devices; and combined hazards, or it may be another means of attack, including low-tech devices; attacks on infrastructure; and cyberterrorism.

WMD HAZARD AGENTS

A WMD is defined as any weapon designed or intended to cause death or serious bodily injury through the release, dissemination, or impact of toxic or poisonous chemicals, disease organisms, radiation or radioactivity, or explosion or fire.

Two important considerations distinguish these hazards from other types of terrorist tools. First, in the case of chemical, biological, and radioactive agents, their presence may not be immediately obvious, making it difficult to determine when and where they have been released, who has been exposed, and what danger is present for first responders and medical technicians. Second, although there is a sizable body of research on battlefield exposures to WMD agents, there is limited scientific understanding of how these agents affect civilian populations.

Chemical Agents

Chemical agents are intended to kill, seriously injure, or incapacitate people through physiological effects. A terrorist incident involving a chemical agent will demand immediate reactions from emergency responders—fire, police, hazardous materials (HazMat) teams, emergency medical services (EMS), and emergency room staff—who will need adequate training and equipment.

Hazardous chemicals, including industrial chemicals and agents, can be introduced via aerosol devices (e.g., munitions, sprayers, or aerosol generators), breaking containers, or covert dissemination. Such an attack might involve the release of a chemical warfare agent, such as nerve or blister agent or an industrial chemical, which may have serious consequences.

Table 5.1 lists some indicators of the possible use of chemical agents.

Early in an investigation, it may not be obvious whether an outbreak was caused by an infectious agent or a hazardous chemical; however, most chemical attacks will be

Table 5.1 Indicators of Possible Use of Chemical Agents

Indicator	Examples of Indicator
• Stated threat to release a chemical agent	– Printed or verbal threat
• Unusual occurrence of dead or dying Animals	– Lack of insects – Dead birds – Sick or deceased livestock
• Unexplained casualties	– Multiple victims – Surge of similar 911 calls – Serious illnesses – Nausea, disorientation, difficulty breathing, or convulsions – Definite casualty patterns
• Unusual liquid, spray, vapor, or powder	– Droplets, oily film – Unexplained odor – Low-lying clouds/fog unrelated to weather
• Suspicious devices, packages, or letters	– Unusual metal debris – Abandoned spray devices – Unexplained munitions

Source: Adapted from FEMA. Emergency Response to Terrorism Job Aid. https://bookstore.gpo.gov/products/sku/064-000-00041-1

localized, and their effects will be evident within a few minutes. There are both persistent and nonpersistent chemical agents. Persistent agents remain in the affected area for hours, days, or weeks. Nonpersistent agents have high evaporation rates, are lighter than air, and disperse rapidly, thereby losing their ability to cause casualties after 10 to 15 minutes, although they may be more persistent in small, unventilated areas.

Biological Agents

Recognition of a biological hazard can occur through several methods, including identification of a credible threat, discovery of bioterrorism evidence, diagnosis, and detection.

When people are exposed to a pathogen, such as anthrax or smallpox, they may not know that they have been exposed, and those who are infected, or subsequently become infected, may not feel sick for some time. This delay between exposure and onset of illness, the incubation period, is characteristic of infectious diseases. The incubation period may range from several hours to a few weeks, depending on the exposure and pathogen.

Unlike acute incidents involving explosives or some hazardous chemicals, the initial detection and response to a biological attack on civilians is likely to be made by direct patient care providers and the public health community.

Terrorists could also use a biological agent that would affect agricultural commodities over a large area (e.g., wheat rust or a virus affecting livestock), potentially devastating the local or even national economy. The response to agricultural bioterrorism should also be considered during the planning process.

Responders should be familiar with the characteristics of biological agents. Table 5.2 lists some indicators of the possible use of biological agents.

Table 5.2 Indicators of Possible Use of Biological Agents

Indicator	Examples of Indicator
• Stated threat to release a biological agent	– Printed or verbal threat
• Unusual occurrence of dead or dying animals	– Lack of insects – Dead birds – Sick or deceased livestock
• Unusual pattern of casualties	– Unusual illness for region/area – Definite pattern inconsistent with natural disease
• Unusual liquid, spray, vapor, or powder	– Spraying – Suspicious devices, packages, or letters

Source: Adapted from FEMA. Emergency Response to Terrorism Job Aid. https://bookstore.gpo.gov/products/sku/064-000-00041-1

Nuclear/Radiological Agents

The difficulty of responding to a nuclear or radiological incident is compounded by the nature of radiation itself. In an explosion, the fact that radioactive material was involved may or may not be obvious, depending on the nature of the explosive device used.

The presence of a radiation hazard is difficult to ascertain, unless the responders have the proper detection equipment and have been trained to use it properly. Although many detection devices exist, most are designed to detect specific types and levels of radiation and may not be appropriate for measuring or ruling out the presence of radiological hazards.

Table 5.3 lists some indicators of the possible uses of nuclear or radiological agents.

The scenarios constituting an intentional nuclear/radiological emergency include the following:

- Use of an improvised nuclear device includes any explosive device designed to cause a nuclear yield. Depending on the type of trigger device used, either uranium or plutonium isotopes can fuel these devices. Although "weapons-grade" material increases the efficiency of a given device, material of less than weapons grade can still be used.

Table 5.3 Indicators of Possible Use of Nuclear or Radiological Agents

Indicator	Examples of Indicator
• Stated threat to release a nuclear agent	– Printed or verbal threat
• Presence of nuclear or radiological equipment	– Spent fuel canisters – Nuclear transport vehicles
• Unusual pattern of casualties	– Unusual illness for region/area – Definite pattern inconsistent with natural disease
• Nuclear placards/warning materials	– Placards indicating nuclear or radiological material

Source: Adapted from FEMA. Emergency Response to Terrorism Job Aid. https://bookstore.gpo.gov/products/sku/064-000-00041-1

- Use of radiological dispersal device (RDD) includes any explosive device utilized to spread radioactive material upon detonation. Any improvised explosive device (IED) could be using by placing it in close proximity to radioactive material.
- Use of a simple RDD spreads radiological material without the use of an explosive. Any nuclear material (including medical isotopes or waste) can be used in this manner.

CONVENTIONAL EXPLOSIVES AND SECONDARY DEVICES

The easiest to obtain and use of all weapons is still a conventional explosive device, or improvised bomb, which may be used to cause massive local destruction or to disperse a chemical, biological, or radiological agent. The components are readily available, as are detailed instructions on constructing such a device.

IEDs are categorized as being explosive or incendiary, using high or low filler explosive materials to explode and/or cause fires. Explosions and fires also can be caused by projectiles and missiles, including aircraft used against high-profile targets, such as buildings, as was the case for the September 11 attacks.

Bombs and firebombs are cheap and easily constructed, involve low technology, and are the terrorist weapon most likely to be encountered. Large, powerful devices can be outfitted with timed or remotely triggered detonators and can be designed to be activated by light, pressure, movement, or radio transmission. The potential exists for single or multiple bombing incidents in single or multiple municipalities.

Historically, less than 5 percent of actual or attempted bombings were preceded by a threat.

COMBINED HAZARDS

WMD agents can be combined to achieve a synergistic effect—greater in total effect than the sum of their individual effects. They may be combined to achieve both immediate and delayed consequences. Mixed infections or toxic exposures may occur, thereby complicating or delaying diagnosis. Casualties of multiple agents may exist; casualties may also suffer from multiple effects, such as trauma and burns from an explosion, which may exacerbate the likelihood of agent contamination.

Attacks may be planned and executed so as to take advantage of the reduced effectiveness of protective measures produced by employment of an initial WMD agent. Finally, the potential exists for multiple incidents in single or multiple municipalities.

OTHER TERRORISM HAZARDS

Planners also need to consider the possibility of unusual or unique types of terrorist attacks previously not considered likely. For example, prior to the attacks of September 11, 2001, the use of multiple commercial airliners with full fuel loads as explosive incendiary devices in well-coordinated attacks on public and private targets, was not considered a likely terrorist scenario.

Although it is not realistically possible to plan for and prevent every conceivable type of terrorist attack, planners should anticipate that future terrorism attempts could range from simple, isolated attacks to complex, sophisticated, highly coordinated acts of destruction using multiple agents aimed at one or multiple targets. Therefore, the plans developed for terrorist incidents must be broad in scope, yet flexible enough to deal with the unexpected.

These considerations are particularly important in planning to handle the consequences of attacks using low-technology devices and delivery, assaults on public infrastructure, and cyberterrorism. In these cases, the training and experience of the responders may be more important than detailed procedures.

LOW-TECHNOLOGY DEVICES AND DELIVERY

Planning for possibility of terrorist attacks must consider the fact that explosives can be delivered by a variety of methods. Most explosive and incendiary devices used by terrorists would be expected to fall outside the definition of a WMD. Small explosive devices can be left in packages or bags in public areas for later detonation at a time and place when and where the terrorist feels that the maximum damage can be inflicted.

The relatively small size of these explosive devices can also be brought onto planes, trains, ships, or buses, within checked bags or hand carried. Although present airline security procedures minimize the possibility of explosives being brought on board airliners, planners will need to consider the level of security presently employed on ships, trains, and buses within their jurisdictions.

Larger quantities of explosive materials can be delivered to their intended target area via car or truck bombs. Planners need to consider the possible need to restrict or prohibit vehicular traffic within certain distances of key facilities identified as potential terrorist targets. Planners may also need to consider the possible use of concrete barriers to prevent the forced entry of vehicles into restricted areas.

INFRASTRUCTURE ATTACKS

Potential attacks on elements of the nation's infrastructure require protective considerations. Infrastructure protection involves risk management actions taken to prevent destruction of or incapacitating damage to networks and systems that serve society.

Infrastructure protection is more often focused on security, deterrence, and law enforcement, than on emergency consequence management preparedness and response. Nevertheless, planners must develop contingencies and plans in the event critical infrastructures are brought down as the result of a terrorist incident.

CYBERTERRORISM

Cyberterrorism involves the malicious use of electronic information technology to commit or threaten to commit acts dangerous to human life, or against a nation's critical infrastructures in order to intimidate or coerce a government or civilian population to further political objectives. As with other infrastructure guidance, most cyber protection guidance focuses on security measures to protect computer systems against intruders, denial of service attacks, and other forms of attack, rather than addressing issues related to contingency and consequence management planning.

EMERGENCY OPERATIONS PLAN: SITUATION

The situation section of a terrorism annex to your EOP should discuss what constitutes a potential or actual terrorist incident. It needs to present a clear, concise, and accurate overview of potential events and discuss a general concept of operation for response. Any information already included in the EOP need not be duplicated in the terrorism annex, but should be referenced. The situation overview should include as much information as possible that is unique to terrorism response actions, including the suggested elements listed in Table 5.4.

Table 5.4 Situational Report Elements

Maps	• Use detailed, current maps or charts
	• Include demographic information
	• Use natural and man-made boundaries and structures to identify risk areas
	• Annotate evacuation routes and alternatives
	• Annotate in-place sheltering locations
	• Use geographic information and analytical tools, as appropriate
Environment	• Determine response routes and times
	• Include bodies of water with dams or levees (these could become contaminated)
	• Specify special weather and climate features that could alter the effects of a WMD
Population	• Identify those most susceptible to WMD effects or otherwise hindered or unable to care for themselves
	• Identify areas where large concentrations of the population might be located, such as sports arenas and major transportations centers
	• List areas that may include retirement communities
	• Note locations of correctional facilities
	• Note locations of hospitals, medical centers, schools, daycare centers, or any other locations where multiple evacuees may require assistance
	• Identify non-English-speaking populations
Regional	• Identify multijurisdictional perimeters and boundaries
	• Identify potentially overlapping areas for response
	• Identify rural, urban, suburban, and city mutual-risk areas
	• Identify terrorism-specific or unique characteristics, such as interchanges, choke points, traffic lights, traffic schemes and patterns, access roads, tunnels, bridges, railroad crossings, and overpasses or cloverleafs
Resources	• Identify mutual aid resources
	• Identify terrorism-specific resources

Source: Adapted from FEMA. Emergency Response to Terrorism Job Aid. https://bookstore.gpo.gov/products/sku/064-000-00041-1

As a state or local emergency manager or planner, you need to consider the possibility of unusual or unique types of terrorist attacks in addition to those that have occurred in the past. You need to think creatively about possible scenarios and response needs. The plans developed for terrorist incidents must be comprehensive in scope, yet flexible enough to deal with the unexpected.

Terrorism emergency response planning should include provisions for working with federal crisis and consequence management agencies. The key to successful emergency response involves smooth coordination among multiple agencies and officials from various jurisdictions regarding all aspects of the response.

Because of the need to interact with a wide range of organizations and individuals within these organizations, up-to-date directories of the points of contact must be maintained in the course of the planning process. It is important that these directories be updated to reflect changes in personnel and telephone numbers.

While assistance from federal agencies will be needed in the event of a terrorist incident, planning by state and local jurisdictions should take into account the difficulty that can be experienced in incorporating the federal resources into the initial local response. Coordination among state, local, and federal officials should take place well in advance of events that could be targeted so that all response organizations clearly understand the responsibilities of each organization, and how they will be integrated.

POTENTIAL TARGETS

In determining the risk areas within your jurisdiction, the vulnerabilities of potential targets should be identified, and the targets themselves should be prepared to respond to a WMD incident. In-depth vulnerability assessments are needed for determining a response to such an incident.

Areas of vulnerability may be determined by several factors:

- Population
- Accessibility
- Criticality
- Economic impact

It may be beneficial to coordinate vulnerable areas with the Federal Bureau of Investigation (FBI).

Table 5.5 gives you an idea of vulnerable areas in your jurisdiction that should be included in the planning process.

Various criteria may be used in determining the vulnerability of facilities to terrorist attack. In evaluating the vulnerability of facilities, state and local planners need to consider the existing security measures in place and the need, if any, to upgrade security.

In addition, the FBI has a standard vulnerability assessment paradigm that can be used for evaluating the vulnerabilities of potential targets. As a planner, you should also be aware that once target lists and vulnerability information are developed, careful decisions must be made regarding security considerations for handling this information based on applicable state and federal law regarding confidentiality and public information. Even

Table 5.5 Potential Areas of Vulnerability

Traffic	• Roads, tunnels, or bridges carry large volumes of traffic • Points of congestion that could impede response or place citizens in a vulnerable area. Note time of day and day of week that this activity occurs
Trucking and transport activity	• Hazardous materials (HazMat) cargo loading and unloading facilities • Vulnerable areas such as weigh stations and rest areas this cargo may transit
Waterways	• Pipelines and process or treatment facilities (in addition to dams) • Berths and ports for cruise ships, roll-on/roll-off cargo vessels, and container ships • International flagged vessels that conduct business in the area
Airports	• Carriers, flight paths, airport layout, and types of aircraft that use this facility • Air traffic control towers, runways, passenger terminals, and parking areas
Trains/subways	• Rails and lines, interchanges, terminals, tunnels, and cargo/passenger terminals • HazMat that may be transported via rail • Subway stations and ventilation control systems
Government facilities	• Federal, state, and local government offices • Post office, law enforcement stations, fire/rescue, town/city hall, and local mayor/governor residences • Judicial offices and courts • Monuments, memorial structures, and prominent governmental symbols
Recreational facilities	• Sports arenas, theatres, malls, special interest group facilities, and locations of special events
Symbolic buildings and locations	• National monuments, internationally well-known facilities and locations • Potential areas of congestion connected with such buildings and locations
Other facilities	• Financial institutions and business districts • Shopping centers or heavily populated downtown areas are congested at certain periods • Special events facilities that may have national importance. • Prominent high-rise buildings
Military installations	• Any type of military installation
HazMat facilities, utilities, and nuclear facilities	• Facilities, such as electricity generating stations, oil refineries, spent nuclear fuel storage facilities, etc.

(*Continued*)

Table 5.5 (*Continued*) Potential Areas of Vulnerability

Water supply facilities	• Water supply intakes from lakes or rivers • Water supply pipelines and holding areas, such as reservoirs and tanks • Water supply treatment plants
Food and agriculture	• Key agricultural facilities, such as large grain elevators and livestock concentrations • Food processing and packaging facilities
Computer systems	• Governmental and business-related computer systems located within the jurisdiction and ascertain their level of protection against terrorist cyberattack.

Source: Adapted from FEMA. Emergency Response to Terrorism Job Aid. https://bookstore.gpo.gov/products/sku/064-000-00041-1

when laws do not require strict confidentiality, you should use common sense regarding whether information that could be useful to terrorists should be made available.

INITIAL WARNING

While specific events may vary, the emergency response and protocol followed should remain consistent. When an overt WMD incident has occurred, the initial call for help will likely come through the local 9-1-1 center. This caller may or may not identify the incident as a terrorist incident, but may state only that there was an explosion, a major accident, or a mass casualty event. Information relayed through the dispatcher prior to the arrival of first responders on scene, as well as the initial assessment, will provide first responders with the basic information necessary to begin responding to the incident.

With increased awareness and training about terrorist incidents, first responders should recognize that a terrorist incident has occurred. The information provided in this chapter applies where it becomes obvious or strongly suspected that an incident has been intentionally perpetrated to harm people, compromise the public's safety and well-being, disrupt essential government services, or damage the area's economy or environment.

You need to be aware of the likely occurrence of false warnings. Since these cannot be ignored, they must be investigated, resulting in wasted resources and psychological stress. You should develop procedures and training to deal with such threats.

INITIAL DETECTION

The initial detection of a WMD terrorist attack will likely occur at the local level by either first responders or private entities (e.g., hospitals, corporations). Consequently, first responders, the business community—both public and private—should be trained to identify hazardous agents and to take appropriate actions. State and local health departments, as well as local emergency first responders, will be relied upon to identify unusual symptom occurrence, and any additional cases of symptoms as the effects spread throughout the community and beyond.

The detection of a terrorism incident involving covert biological agents, as well as some chemical agents, will most likely occur through the recognition of similar symptoms or syndromes by clinicians in hospital or clinical settings. Detection of biological agents could occur days or weeks after exposed individuals have left the site of the release. Detection will occur at public health facilities receiving unusual numbers of patients, the majority of whom will self-transport. Similarly, a biological attack aimed at agricultural assets might first be detected by veterinarians or agricultural inspectors.

First responders must be protected from the hazard prior to treating victims. Planning for response to terrorist acts must include provisions for appropriate personal protective equipment (PPE) for emergency responders, specifically first responders. This equipment should include protective clothing and respirators, with high-efficiency particulate air filters. Detection equipment for chemical, biological, nuclear, or explosive materials will assist in identifying the nature of a potential hazard.

You need to determine the present availability of this protective and detection equipment within your jurisdiction, determine if additional resources would be needed to adequately protect your first responders, and identify the funding needs to upgrade your resources, if needed.

RELEASE AREA

Standard models are available for estimating the effects of a nuclear, chemical, or biological release, including the area affected and consequences to population, resources, and infrastructure. Some of these models include databases on infrastructure that may prove useful when developing your terrorism annex.

Analogous to the area affected by a nuclear, biological, or chemical release is the area impacted by an explosive device. Models are also available for estimating the blast effects at various distances for various quantities of explosive materials.

These models can be useful in preparing your terrorism annex, especially in regard to determining minimum setback distances from a potential vehicle bomb to a vulnerable facility or structure. If a specific minimum distance cannot be maintained, then the planning effort may need to consider the cost and effectiveness of facility hardening to mitigate the effects of an assumed blast impact. You may also want to consider the removal or modification of window areas.

INVESTIGATION AND CONTAINMENT OF HAZARDS

Local first responders will provide initial assessment or scene surveillance of a hazard caused by an act of WMD terrorism. The proper local, state, and federal authorities capable of dealing with and containing the hazard should be alerted to a suspected WMD attack as soon as first responders recognize the occurrence of symptoms that are highly unusual or of an unknown cause. Consequently, state and local emergency responders must be able to assess the situation and request assistance as quickly as possible.

EMERGENCY OPERATIONS PLAN: ASSUMPTIONS

Although situations may vary, planning assumptions remain the same:

1. The first responder or health or medical personnel will, in most cases, initially detect and evaluate the potential or actual incident, assess casualties, and determine whether assistance is required. If so, state support will be requested and provided. This assessment will be based on warning or notification of a WMD incident that may be received from law enforcement, emergency response agencies, or the public.
2. The incident may require federal support. To ensure that there is only one overall lead Federal Agency, the US Department of Homeland Security (DHS) was formed in 2003. DHS retains the authority and responsibility to integrate fully with interagency partners throughout the Federal Response. In this capacity, DHS will coordinate federal assistance through local, state, and federal authorities under comprehensive guidelines and procedures as promulgated.
3. Federal response will include experts in the identification, containment, and recovery of WMD (chemical, biological, nuclear/radiological, or explosive).
4. Jurisdictional areas of responsibility and working perimeters defined by local, state, and federal departments and agencies may overlap. Perimeters may be used to control access to the affected area, target public information messages, assign operational sectors among responding organizations, and assess potential effects on the population and the environment. Control of these perimeters may be enforced by different authorities, which will impede the overall response if adequate coordination is not established.
5. Response activities may continue for an extended period of days or weeks. Early emergency responders may be pushed beyond their capabilities, and regional and federal resources may be needed. The incident will be covered extensively by the media. There may be many volunteer responders and donations of food and material that will require management.

EMERGENCY OPERATIONS PLAN: CONCEPT OF OPERATIONS

Your terrorism annex should include a concept of operations section to explain the jurisdiction's overall concept for responding to a WMD incident. Topics should include division of local, state, federal, and any intermediate interjurisdictional responsibilities; activation of the EOP; and other elements set forth in current guidance.

DIRECTION AND CONTROL

Local government response organizations will respond to the incident scene(s) and make appropriate and rapid notifications to local and state authorities. Control of the incident scene(s) most likely will be established by local first responders from either fire or police.

To ensure continuity of operations, it is important that the Incident Command Post be established at a safe location and at a distance appropriate for response to a suspected or known terrorist incident. In addition, in severe terrorist attacks, response operations may last for very long periods, and there may be more leadership casualties because of secondary or tertiary attacks or events. Therefore, planning should provide for staffing key leadership positions in depth.

The Incident Command System (ICS) should be used by all responding local fire, police, and emergency management organizations, and all relevant responder personnel should be trained in ICS use to prevent security and coordination problems in a multi-organizational response.

The ICS that was initially established likely will transition into a Unified Command System as mutual-aid partners and state and federal responders arrive to augment the local responders. It is recommended that local, state, and federal regional law enforcement officials develop consensus "rules of engagement" early in the planning process to smooth the transition from ICS to Unified Command. The Unified Command structure will facilitate both crisis management and consequence management activities.

The Unified Command Structure used at the scene will expand as support units and agency representatives arrive to support crisis and consequence management operations. The site of a terrorist incident is a crime scene as well as a disaster scene, although lives, health, and safety remain the top priority. Because of these considerations, as well as logistical control concerns, it is extremely important that this incident site and its perimeter be tightly controlled as soon as possible.

State and local planners must realize that the integration of the federal response into the local response efforts can be a difficult and awkward process. Whenever possible, each entity should involve the others in its planning process so as to facilitate a better understanding by all parties of the anticipated actions and responsibilities of each organization.

Planners should understand that integration of the federal response into an urban setting would be different from that into a rural setting. In an urban area, there will be substantial manpower and equipment resources, and the control of the emergency response. The rapid influx of federal resources can be a sensitive issue unless properly coordinated. The federal response should not overwhelm the local emergency response organization but should provide resources, as needed (Figure 5.1).

It is assumed that normal disaster coordination accomplished at state and local Emergency Operations Centers (EOCs) and other locations away from the scene would be addressed in the EOP. Any special concerns relating to state and local coordination with federal organization should be addressed in the terrorism annex.

Response to any terrorist incident requires direction and control. The planner must consider the unique characteristics of the event, identify the likely stage at which coordinated resources will be required, and tailor the direction and control process to merge these resources into an ongoing public health response.

With many organizations involved, there is the danger of key decisions being slowed down by too many layers of decision making. Planners should be aware of the need to streamline the decision-making process so that key decisions or authorizations regarding public health and safety can be obtained quickly.

PLANNING FOR TERRORISM

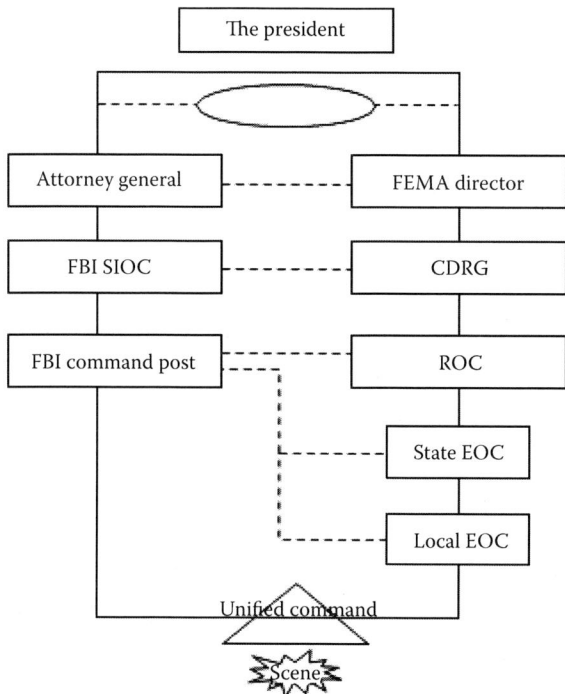

Figure 5.1 Incident Command from the federal level down. (From FEMA. Emergency Response to Terrorism Job Aid. https://bookstore.gpo.gov/products/sku/064-000-00041-1)

A primary EOC is necessary to properly coordinate response actions within the jurisdiction and to liaise with other jurisdictions and federal agencies. The EOC of the City of New York's Office of Emergency Management, a new state-of-the-art facility, was located at the World Trade Center, and was destroyed during the attack of September 11, 2001. This necessitated the establishment of an alternate EOC. Therefore, planning should address the possibility that operations might have to be shifted to an alternative EOC or even a secondary alternative location.

When considering direction and control, as well as continuity of operations, planners must determine the availability of usable alternate EOC facility locations that can be brought up to operational levels within a reasonable period of time. In a large-scale terrorist attack, the local EOC might become uninhabitable, especially if it is not a hardened facility. In identifying and evaluating alternative EOC locations, planners will need to consider the availability of communications systems, space to accommodate all key staff, materials and supplies, backup power, kitchen, bathrooms, and the overall capability to maintain around-the-clock operations for an extended period of time.

Local, state, and federal interface with the FBI On-Scene Commander is coordinated through the Joint Operation Center (JOC). FEMA will recommend joint operational priorities to the FBI on the basis of consultation with the FEMA-led consequence management group in the JOC. The FBI, working with local and state officials in the command group at the JOC, will establish operational priorities.

COMMUNICATIONS

In the event of a WMD incident, rapid and secure communications is important to ensure a prompt and coordinated response. Strengthening communications among first responders, clinicians, emergency rooms, hospitals, mass care providers, and emergency management personnel must be given top priority when planning. Planning should include adding 9-1-1 resources when an event requires extraordinary response.

In addition, terrorist attacks have been shown to overload non-dedicated telephone line and cellular telephones. In these instances, the Internet has proven more reliable for making necessary communications connections, although it should be recognized that computers may be vulnerable to cyberattacks in the form of viruses. It is recommended that response organizations both establish relevant Internet connections with all coordinating emergency response organizations and have the use of these connections formalized in plans and practiced during training, drills, and exercises.

Responders with different functions within a jurisdiction or from different jurisdictions may use different radio frequencies, hindering communications. Use of 800 MHz radios alleviates this problem. Therefore, a backbone communications system to interconnect local, state, and federal responders is recommended. Establishment of mutually agreed upon communications protocols are also needed during the planning and exercise stages so that all responding organizations will understand each other's codes and terminology during responses to real events.

Planning should also consider the need for an integrated communications system for all key state agencies and local emergency response organizations. The interoperability of such a system would facilitate the integrated response to a terrorist incident. Planning also needs to consider the importance of reliable backup communications systems for emergency responders. Terrorist incidents may include the loss of radio transmission capabilities, and telephone landlines and cellular phone connections will be overwhelmed. Satellite telephones, which can operate when cellular and non-dedicated landlines are overloaded, are another option for backup telephone systems.

State and local planners should consider the distribution of priority emergency access telephone cards to their emergency workers. Planners should determine the existing status of their emergency communications systems and identify the funding needed to upgrade them.

WARNING

Every incident is different. There may or may not be warning of a potential WMD incident. Factors involved range from intelligence gathered from various law enforcement or intelligence agency sources to an actual notification from the terrorist organization or individual. The EOP should have HazMat facilities and transportation routes already mapped, along with emergency procedures necessary to respond.

The warning or notification of a potential WMD terrorist incident could come from many sources; therefore, open but secure communication among local, state, and federal law enforcement agencies and emergency response officials is essential. The local FBI Field Office must be notified of any suspected terrorist threats or incidents.

Similarly, the FBI informs state and local law enforcement officials regarding potential threats. An integrated communications system would be an aid in maintaining these communications channels and would expedite the dissemination of warnings about suspected terrorist threats. The interoperability of such a system would eliminate the need to switch back and forth between different communications systems for different organizations.

PRE-EVENT READINESS

The Department of Homeland Security has developed the National Terrorism Advisory System (NTAS), using available information, the advisories will provide a concise summary of the potential threat, information about actions being taken to ensure public safety, and recommend steps that individuals, communities, businesses, and governments can take to help prevent, mitigate, or respond to the threat (dhs.gov/topic/ntas).

EMERGENCY PUBLIC INFORMATION

Terrorism is designed to be catastrophic. The intent of a terrorist attack is to cause maximum destruction of lives and property; create chaos, confusion, and public panic; and stress local, state, and federal response resources. Accurate and timely information, disseminated to the public and media immediately and often over the course of the response, is vital to minimize the accomplishment of these terrorist objectives.

Crisis research and case studies show that accurate, consistent, and expedited information calms anxieties and reduces problematic public responses, such as panic and spontaneous evacuations that terrorist hope will hamper response efforts.

The news media will be the public's primary source of information, from both official sources and other nonofficial sources, during the course of the incident. Ensuring that the media will receive accurate, consistent, and expedited official information from the outset and over what may be rapidly changing and lengthy response requires careful planning and considerable advance planning and considerable advance preparations. It is important to build and maintain a strong working relationship with the media. This relationship should include a clear commitment that government representatives will be immediately available to provide information over the course of the emergency.

Local plans should reflect responsibility for emergency information operations during the crucial initial response until state and federal personnel and resources can arrive to provide support. Planning should also reflect the following:

- A mechanism for sharing and coordinating information among all responder agencies and organizations
- Development and production of information materials

- Dissemination of information through various methods
- Monitoring and analysis of news media coverage with rapid response capabilities to address identified problems

A strong and ongoing public education program for terrorism response, built upon outreach and awareness programs for other types of emergencies, can enhance the response organization's credibility and benefit both members of the public and first responder efforts in the event of a terrorist attack.

PROTECTIVE ACTIONS

Evacuation may be required from inside the perimeter of the scene to guard against further casualties from contamination by primary release of a WMD agent, the possible release of additional WMDs, secondary devices, or additional attacks targeting emergency responders.

Temporary in-place sheltering may be appropriate if there is a short-duration release of hazardous materials or if it is determined to be safer for individuals to remain in place. Protection from biological threats may involve coercive or noncoercive protective actions, including isolation of individuals who pose an infection hazard, quarantine of affected locations, vaccinations, use of masks by the public, closing of public transportation, limiting public gatherings, and limiting intercity travel.

As with any emergency, state and local officials are primarily responsible for making protective action decisions affecting the public. Protocols should be established to ensure that important decisions are made by persons with the proper decision-making authority. The terrorism annex should include provisions for coordinating protective actions with other affected jurisdictions. Planning should also address ways of countering irrational public behavior that can hinder protective actions.

However, planning for evacuation should be flexible to account for difficult situations. After the attack on the Pentagon, federal buildings in Washington, DC, were evacuated as a precautionary measure. Evacuation was hampered by roads in the federal area being under the jurisdiction of several federal agencies and the roads in the rest of the District being under the authority of the District government. On the other hand, the Pentagon is located in Arlington County, Virginia, near a route persons from Washington, DC, were using to evacuate. Therefore, there was a situation where some evacuation was toward the incident site rather than away from it. Because of heavy traffic in the vicinity of the Pentagon, fire vehicles had difficulty reaching the incident site.

MASS CARE

The location of mass care facilities will be based partly on the hazardous agent. Decontamination, if necessary, may need to precede sheltering and other needs of the victims to prevent further damage from the hazardous agent to either the victims themselves or the care providers.

The American Red Cross (the primary agency for mass care), the Department of Health and Human Services (HHS), and the Department of Veterans Affairs (VA) should be actively involved with the planning process to determine both in-place sheltering and mobile mass care systems for the terrorism annex.

A midpoint or intermediary station may be needed to move victims out of the way of immediate harm. This action would allow responders to provide critical attention (e.g., decontamination and medical services) and general lifesaving support, then evacuate victims to a mass location for further attention. Some general issues to consider for inclusion in your terrorism annex:

- Location, setup, and equipment for decontamination stations, if any
- Mobile triage support and qualified personnel
- Supplies and personnel to support in-place sheltering
- Evacuation to intermediary location to provide decontamination and medical attention
- Determination of safety perimeters (based on agent)
- Patient tracking/record keeping for augmentation of epidemiological services and support

HEALTH AND MEDICAL

The basic EOP should already contain a Health and Medical Annex. Issues that may be different during a terrorist incident and that should be addressed in the terrorism annex include

- Decontamination
- Safety of victims and responders
- In-place sheltering and quarantine versus evacuation
- Multihazard and multiagent trage

Planning should anticipate the need to handle large numbers of people who may or may not be contaminated but who are fearful about their medical well-being. In addition, the terrorism annex should identify the locations and capacities of medical care facilities within the jurisdiction and in surrounding jurisdictions. The terrorism annex should also include a description of the capabilities of these medical care facilities, especially with regard to trauma care. Depending on the nature and extent of the terrorist attack, the most appropriate medical care facility may not necessarily be the closest facility.

In addition, first responders may be entering an environment rife with biological or chemical agents, radioactive materials, or hazardous pollutants from collapsed buildings, or collapsed buildings might be imminent. Other incidents may post environmental or physical risks to responders. Examples may be a structurally damaged and potentially deadly pipeline, tank car, tank truck, bridge, or tunnel.

In planning, you address the need for first responders to perform a risk assessment and to modify standard protocols (e.g., establish plans for inoculating first responders) if the risk assessment indicates.

The planning should also address how such assessments are made and what resources they may indicate are needed. The assessment may indicate monitoring and sampling resources before federal resources can arrive.

Responders will also need appropriate PPE, including respirators. The planning process needs to address the availability of regional monitoring and sampling capability and PPE.

A bioterrorism incident raises several other special issues. Such an incident may generate an influx of patients requiring specialized care. If an infectious agent is involved, it may be necessary to isolate the patients and use special precautions to avoid transmission of the disease to staff and other patients.

State planning should also consider the need to obtain and integrate supplementary medical professionals and technicians who may be needed to respond to a terrorist incident. In addition to physiological health consideration, planners should take into account the need for mental health considerations in the consequence management planning. Support must be provided not only to those individuals directly affected by a terrorist attack, but also to those surviving family members experiencing emotional stress.

Planning issues to consider include

- Immunization and prophylaxis for biological agents
- Notification to and receipt of information from doctors and clinics
- Augmentation of medical facilities and personnel
- Management of medical supplies and equipment
- Patient tracking and record keeping for augmentation of epidemiological services and support
- Analytical laboratory support, including memorandums of understanding (MOUs), specifying special considerations, as appropriate
- Mental health support services, including clinical psychologists, psychiatrists, social workers, etc.

RESOURCES MANAGEMENT

The following considerations are highly relevant to WMD incidents and should be addressed, if appropriate, in one or more appendices to your resource management annex:

- Nuclear, biological, and chemical response resources that are available through interjurisdictional agreements
- Unique resources that are available through state authorities (e.g., National Guard units)
- Unique resources that are available to state and local jurisdictions through federal authorities (e.g., the National Pharmaceutical Stockpile (is also known as the Strategic National Stockpile [SNS]; cdc.gov/phpr/stockpile/stockpile.htm), a national asset providing delivery of antibiotics, antidotes, and medical supplies to the scene of a WMD incident)

- Unique expertise that may be available through academic, research, or private organizations
- Trained and untrained volunteer resources and unsolicited donated goods that arrive at the incident site

RECOVERY

The basic EOP should already contain a Recovery Annex. However, different issues may arise during a WMD incident that should be addressed in the terrorism annex. A WMD incident is a criminal act, and its victims and their families may be eligible for assistance under your state's crime victim's assistance law, if one exists. Therefore, state crime victim's assistance staff should be included in the planning process. In addition, injured victims of a terrorist attack, those put at risk for injury, and the families of these persons may have suffered psychological trauma as a result of the attack and may be in need of crisis counseling.

In the event of an incident involving chemical or biological agents or radioactive materials, large areas or multiple locations may become contaminated. Decontamination may be required before buildings can be safely reoccupied. While decontamination is taking place, or until damaged areas are repaired or replaced, persons must be relocated from residences and offices, and office equipment must be relocated from office buildings.

Relocation after a terrorist incident tends to be of longer duration and entail greater costs than relocation following a natural disaster. You need to take these factors into consideration and make appropriate arrangements.

URBAN SEARCH AND RESCUE

Urban Search and Rescue (US&R) involves a rapid deployment of US&R task forces to provide specialized lifesaving support to state and local authorities, including locating and extricating, and providing on-site medical treatment to those trapped in collapsed structures.

There are several US&R task forces throughout the country. They have the ability to deploy within 6 hours and to sustain themselves for 72 hours. Currently, deployment plans rely on commercial air transport. Consequently, in incidents where air traffic is curtailed, arrival of remote US&R task forces may be delayed.

The capabilities of these US&R task forces are being enhanced to operate in a collapsed building environment contaminated with biological or chemical agents or radioactive materials. These enhanced task forces will have additional HazMat specialists and medical personnel and more monitoring and detection equipment.

ORGANIZATION AND ASSIGNMENT OF RESPONSIBILITIES

As with hazard-specific emergency, the organization for management of local response will probably need to be tailored to address the special issues involved in managing the consequences of a terrorist incident. This organization should be defined in the terrorism annex or your EOP.

The consequences of a terrorist attack have the potential to overwhelm local resources, which may require assistance from state or federal governments. The response by state and local governments, as well as the types of support and assistance from the federal government may be different than the response and support received for a natural disaster. Because of this, not only must the plans be upgraded to include response to a terrorist incident, but training and exercising must also be expanded to ensure that the unique aspects of response to a terrorist incident can be carried out in a coordinated, effective manner.

Training needs to be planned for local, state, and federal staff involved in the response. You should identify their training needs, establish budgets for the training, and determine what funding resources will be required to implement the training. Periodic integrated exercises must be conducted to ensure that the emergency response at the local, state, and federal levels can be adequately coordinated. The following response roles should be articulated in your terrorism annex.

Local Emergency Responders

Local fire-rescue departments, law enforcement personnel, HazMat teams, and EMS will be among the first to respond to terrorist incidents, especially those involving WMDs. In incidents associated with public transportation, workers and officials from these transportation organizations may be among the first responders. As response efforts escalate, the local emergency management agency and health department will help coordinate needed services.

Interjurisdictional Responsibilities

The formal arrangements and agreements for emergency response to a terrorist incident among neighboring jurisdictions, state, local, and neighboring states (and those jurisdiction physically located in those states) should be made prior to an incident. When coordinating and planning the Risk Assessment and Risk Area sections of the terrorism annex, interjurisdictional responsibilities should be readily identifiable.

State Emergency Responders

If requested by the local officials, the state emergency management agency has the capabilities to support local emergency management authorities and the Incident Commander.

State and Local Public Health Authorities

State laws grant the state and local public health authorities emergency powers to combat communicable disease. The powers available, diseases that trigger them, and procedures for enforcement vary from state to state. Typical powers include the power to isolate or quarantine persons and places and the power to compel vaccination and other preventive measure, such as wearing masks. In some states, these measures may be taken whenever there is a threat of communicable disease; in others, the powers apply to only one or more specific, named diseases.

Medical Service Providers

Hospitals generally perform emergency planning, both to protect their own facilities and patients and to respond to disasters in the community. State licensing and accreditation standards require hospitals to meet certain criteria for emergency preparedness, which often include participation in local or regional medical planning for disasters. Hospitals accredited by the Joint Commission on Accreditation of Healthcare Organization must be prepared for a variety of disaster scenarios, including facilities for biological, radioactive, or chemical isolation and decontamination, where appropriate.

Local Emergency Planning Committees and State Emergency Response Commissions

These entities are established under the Superfund Amendment and Reauthorization Act of 1986 (SARA) Title III, and the implementing regulation of the Environmental Protection Agency (EPA). Local Emergency Planning Committees (LEPCs) develop and maintain local hazardous materials emergency plans and receive notifications of releases of hazardous materials. State Emergency Response Commissions (SERCs) supervise the operation of the LEPCs and administer the community right-to-know provisions of SARA Title III, including collection and distribution of information about facility inventories of hazardous materials, chemicals, and toxins. LEPCs will have detailed information about industrial chemicals within the community. It may be advisable for LEPCs and SERCs to establish MOUs with agencies and organizations to provide specialized resources and capabilities for response to WMD incidents.

Federal Emergency Responders

Upon determination of a credible terrorist threat, or if such an incident actually occurs, the federal government may respond through the appropriate departments and agencies. These departments and agencies may include DHS, FEMA, the DOJ and the FBI, the Department of Defense, the Department of Energy, the HHS, the EPA, the Department of Agriculture, the Nuclear Regulatory Commission (NRC), and possible the American Red Cross and Department of Veterans Affairs.

ADMINISTRATION AND LOGISTICS

There are many factors that make consequence management response to a terrorist incident unique. Unlike some natural disasters, the administration and logistics for response to a terrorist incident require special considerations. For example, there may be little or no warning and because the release of a WMD may not be immediately apparent, caregivers, emergency response personnel, and first responders are in imminent danger of becoming casualties before the actual identification of the crime can be made. Incidents could quickly escalate quickly from one scene to multiple locations and jurisdictions.

The types of supplies needed to respond to a terrorist incident may differ from those needed for a natural disaster or other type of technological emergency. For example, the responders to the September 11 attacks incident needed compliant hard hats, steel-toed shoes, respirators that were appropriate for the hazards, and other PPE (Personal Protective Equipment as prescribed by Occupational Safety & Health Administration [OSHA] 1910.132(d)). These were not stockpiled and had to be purchased.

Your planning should address administrative protocols to ensure that proper purchasing procedures are followed and that duplicate purchases are avoided.

On September 11, and the days that immediately followed, commercial airlines were not allowed to operate. The shutdown delayed the arrival of supplies and federal responders from distant locations. To avoid the inefficiencies of ad hoc purchasing of supplies and of delays in the arrival of supplies related to air traffic curtailments, you need to consider regional warehousing of supplies and equipment for emergency responders, including equipment for use of US&R task forces.

One of the key logistical problems in the initial stages of emergency response to a terrorist incident is the establishment of an Incident Command Post from which direction of response activities can be made.

In a "routine" emergency, such as fire, small hazardous materials releases, or police actions, the Incident Command Post is established at a point that is close enough to observe the incident, but far enough away to maintain an overview perspective and a safe distance from the immediate hazards.

Because of the unique nature of terrorist activity and the inherent unpredictability of the incident, planners and emergency responders may need to rethink the protocol for locating the Incident Command Post.

One of the key administrative and logistical challenges in managing the emergency response to a terrorist incident is the successful integration of the federal response into the initial response by local and state emergency response organizations.

The very nature of a terrorist attack assumes federal response. Depending on the extent of the terrorist of the terrorist incident, the federal response could be swift and massive. The application, integration, and coordination of the federal resources into the existing local command and control structure can be a sensitive operation. Federal resources should not overwhelm the local response but should be made available as needed and requested.

You should involve federal agencies in your planning, to the extent possible, in order to develop a better understanding among all parties regarding the nature and extent of the federal response, including the logistical support needs of the federal agencies.

You should be aware that nonagricultural terrorist incidents are more likely to occur in urban areas than rural setting. Urban centers have significantly higher numbers of emergency personnel and material resources, and they routinely deal with emergency response. Local emergency response organizations will likely want to maintain the direction and control of the emergency response to a terrorist incident.

As a planner, you should be aware of the potential logistical problems that may be caused by the unsolicited influx of volunteers and donated goods, as experienced after September 11. Site and perimeter control is extremely important to avoid responder casualties and to prevent emergency operations from being disrupted by uncontrolled movement of such volunteers.

In developing plans for urban centers, you will need to identify potential staging areas for personnel and equipment and warehouses for materials, equipment, and supplies. Although these may not be needed for small-scale incidents, an inventory of available warehouse space and potential staging areas would assist in the response to a large-scale incident and/or a prolonged consequence management response and recovery effort.

6

Developing a Planning Team

Michael J. Fagel

Contents

The Emergency Planning Team 97
Who Should Be on the Planning Team? 97
Getting the Team Together 98
Team Operation 99
Stages of Team Formation 99
Team Roles 100
Characteristics of an Effective Team 100
Summary 100

THE EMERGENCY PLANNING TEAM

Emergency planning should be a team effort because disaster response nearly always requires coordination among every department within an organization. Planning as a team helps to ensure that everyone having the expertise to contribute in a response is a part of the process. Team planning also helps to ensure the buy-in of all key players or stakeholders.

WHO SHOULD BE ON THE PLANNING TEAM?

Different types of emergencies require different expertise and response capabilities. The specific individuals and organizations involved in a response will vary by the type of emergency or disaster. Some emergencies may involve only an internal response, whereas others may require the response of external agencies, such as law enforcement, fire services, and emergency medical services (EMS). These external agencies should be involved in large-scale planning, as well as other agencies, such as those responsible for hazardous materials (hazmat) response, should be included as well, as they may be needed in a response to, for example, an emergency involving a refrigeration leak or a different hazardous material.

Table 6.1 Potential Planning Team Members

Attorney	Information Technologies
Budget	Library
Building Inspections	Parks and Recreation
Chambers of Commerce	Planning
City/County Manager	Police/Sherriff
Emergency Management	Private Sector (businesses)
Facilities	Public Information
Finance	Public Works
Accounting	*Engineering*
Inventory	*Road and Bridge*
Purchasing	*Traffic*
Fire Department	Schools (public, private, colleges)
GIS	State Agencies
Health Department	Transportation Agencies
Health Care Facilities	Utilities (power, gas, sewer)
Human Resources	Volunteer Agencies

Source: From Shane Stovall. With permission.

Table 6.1 presents a breakdown of potential team members. Not all organizations will have such a breakdown of personnel, and depending on the size of the organization, one person may be responsible for multiple roles. All individuals do not need to be included in every aspect of the planning process, but they should be consulted in areas that affect them directly or for which they will be responsible.

GETTING THE TEAM TOGETHER

Getting all of the stakeholders in the planning process together and to take an active interest in the planning process will be an arduous task. To schedule meetings with so many participants may be even more difficult. It is critical, however, to have everyone's participation in the planning process, at least in the early stages and at critical points along the way, to draw from their expertise and ensure their buy-in to the plan.

The expertise and knowledge that participants bring of their departments' needs and resources are crucial to developing an accurate plan that considers the entire organization's needs and the resources that could be made available in an emergency. It is definitely to the organization's benefit to have the active participation of all key players.

There are several steps that can be taken to gain cooperation from participants and to gain their involvement in the planning process.

1. *Plan ahead*—Give the planning team plenty of notice of where and when the planning meeting will be held. If time permits, survey the team members to find the time and place that will work for them.

2. *Provide information about team expectations*—Explain why participating on the planning team is important to the participants' departments and to the organization as a whole. Show the team members how they will contribute to a more effective emergency response.
3. *Involve the chief executive officer*—Ask the CEO to sign the meeting notice. This will send a clear message that emergency planning is important to the organization and that the participants are expected to attend.
4. *Allow flexibility in scheduling after the first meeting*—Not all team members will need to attend all meetings. Task forces and subcommittees can complete some of the work. Where this is the case, gain concurrence on timeframes and milestones, but let the subcommittee members determine when it is most convenient to meet.

It may also be beneficial to talk to emergency managers from similar organizations in the community to gather input on how to gain and maintain interest in the planning process, as well as to see what practices they already have in place that work or have already been tested.

TEAM OPERATION

Unlike working alone, working with personnel from other departments to plan for emergencies requires some give and take—in other words, collaboration. Collaboration is the process in which people work together as a team on a common mission—in this case, the development of a company Emergency Operations Plan (EOP). Successful collaboration requires:

- A commitment to participate in shared decision making
- A willingness to share information, resources, and tasks
- A professional sense of respect for individual team members

Although collaboration among EOP planning team members may be difficult, it benefits the organization by strengthening the overall response to disaster. Collaboration can

- Expand resource availability through sharing
- Enhance problem solving through the cross-pollination of ideas

STAGES OF TEAM FORMATION

Collaboration does not come automatically. Building a team that works well together takes time and effort and typically evolves through five stages:

1. *Forming*—Individuals come together as a team. During this stage, the team members may be unfamiliar with one another and uncertain of their roles on the team.
2. *Storming*—Team members may become impatient, disillusioned, and may disagree.
3. *Norming*—Team members accept their roles and make progress toward the goal.

4. *Performing*—Team members work well together and make progress toward the goal.
5. *Adjourning*—Their task accomplished, team members may feel pride in their achievement and some sadness that the experience is ending. It is important to remember that the team should remain intact for the purpose of exercising and revising the plan.

TEAM ROLES

To keep the team focused throughout the planning stages, it is important for team members to assume roles. Perhaps the most important role is that of team leader. The team leader initiates appropriate team-building activities that move the team through each stage and toward its goal. Other team roles may include:

- The taskmaster, who identifies the work to be done and motivates the team
- The innovator, who generates original ways to get the group's work done
- The organizer, who helps the group develop plans for getting the work done
- The evaluator, who analyzes ideas, suggestions, and plans made by the group
- The finisher, who follows through on plans developed by the team

Together, all members contribute toward making the team productive.

CHARACTERISTICS OF AN EFFECTIVE TEAM

You will know that the planning team is on track when it displays the following characteristics:

- Works toward a common goal
- Accepts the leader who provides direction and guidance
- Communicates openly
- Resolves conflicts constructively
- Displays mutual trust
- Shows respect for each individual and his or her contributions

SUMMARY

Emergency planning requires collaboration from a variety of individuals and organizations. The benefits of collaboration far outweigh the difficulties that you will face during the planning process. Some of the benefits include elimination of duplication of services, expanded resource availability, and enhanced problem solving.

The planning process can be made easier by planning ahead, providing information about team expectations, and allowing flexibility in scheduling. It may be to your benefit to talk to emergency managers from other communities to gain their input on the planning process.

Successful team operation requires a commitment to participate in shared decision making, a willingness to share information, and a professional sense of respect for individual team members.

Team formation usually takes place through five stages: forming, storming, norming, performing, and adjourning.

You will know that you have an effective planning team when members agree on and work toward a common goal; provide open communication; and display constructive conflict resolution, mutual trust, and respect for other team members.

7

Developing an Emergency Operations Plan (EOP)

Michael J. Fagel

Contents

Why a Jurisdiction Needs an EOP 104
The Emergency Planning Process 105
    Performing a Hazard Analysis 106
        Developing a Hazard Profile 107
        Creating a Jurisdiction Profile 107
        Completing the Risk Analysis 107
        Creating Scenarios 111
Components of an Emergency Operations Plan 111
    The Basic Plan 111
        The Introduction 112
        The Purpose Statement 112
        The Situations and Assumptions Section 112
        The Concept of Operations Section 113
        The Organization and Assignment of Responsibilities Section 113
        Administration and Logistics Section 113
        Plan Development and Maintenance Section 113
        Authorities and References Section 114
    Annexes 114
    Hazard-Specific Appendices 115
        Implementing Instructions 115
        Standard Operating Procedures 117
        Job Aids 117
        Checklists 118
        Information Cards 118
        Forms 118

  Maps 118
  Creating Effective Implementing Instructions 119
The Emergency Planning Team 119
  Who Should Be on the Planning Team? 119
  Getting the Team Together 121
    Team Operation 122
    Stages of Team Formation 123
    Team Roles 123
    Characteristics of an Effective Team 123
Disaster Preparedness and the Law 124
  What an Attorney Can Provide as a Team Member 124
  The Potentially Applicable Laws 125
  Negligence 126
  Emergency Management Phases 127
    Planning Stage 127
    Response Stage 127
    Recovery Stage 127
    Other Areas of Potential Civil Liability Risks 127
Conclusion 129
Reference 130

A jurisdiction's Emergency Operations Plan (EOP) is a public document that does the following:

- Assigns responsibility to organizations and individuals for carrying out specific actions, at projected times and places, in emergencies that exceed the capability or routine responsibility of any one agency
- Sets forth lines of authority and organizational relationships and shows how all actions will be coordinated
- Describes how people and property will be protected in emergencies and disasters
- Identifies personnel, equipment, facilities, supplies, and other resources available within the jurisdiction, or by agreement with other jurisdictions, for use during response and recovery operations
- Identifies steps to address mitigation concerns during response and recovery activities

WHY A JURISDICTION NEEDS AN EOP

Planning to respond to emergencies and disasters is typically the responsibility of state and local governments. The elected leadership in each jurisdiction is legally responsible for ensuring that the necessary and appropriate actions are taken to protect people and property from the consequences of emergencies and disasters.

DEVELOPING AN EMERGENCY OPERATIONS PLAN (EOP)

When a disaster threatens or strikes a jurisdiction, citizens expect their elected leaders to take immediate action to deal with the situation. The government is expected to marshal its resources, channel the efforts of voluntary agencies and private enterprises in the community, and solicit assistance from outside of the jurisdiction, if necessary. The development of a comprehensive, all-hazard EOP will help ensure that all government response activities are undertaken efficiently and effectively.

THE EMERGENCY PLANNING PROCESS

In today's system of emergency management, local government must act to attend to the public's emergency needs. Depending on the size and nature of the emergency, state and federal assistance may be provided to the jurisdiction; however, local governments should not assume that this type of assistance will be available. Therefore, the local EOP should focus on the functions that are essential for protecting the public before and after a disaster. Minimally, these functions include providing warning, emergency public information (EPI), evacuation, and shelter. Emergency planning is not a one-time event. It is a continual cycle consisting of planning, training, exercising, and revision that takes place throughout the four phases of the emergency management cycle: mitigation, preparedness, response, and recovery.

The planning process does have a single purpose: the development and maintenance strategy for addressing critical needs in an emergency—to protect life and property.

Although the emergency planning process is cyclical, it does have a definite starting point, as seen in Figure 7.1. Emergency planning begins by analyzing the hazards facing the jurisdiction. Hazard analysis is the process by which hazards that threaten the community

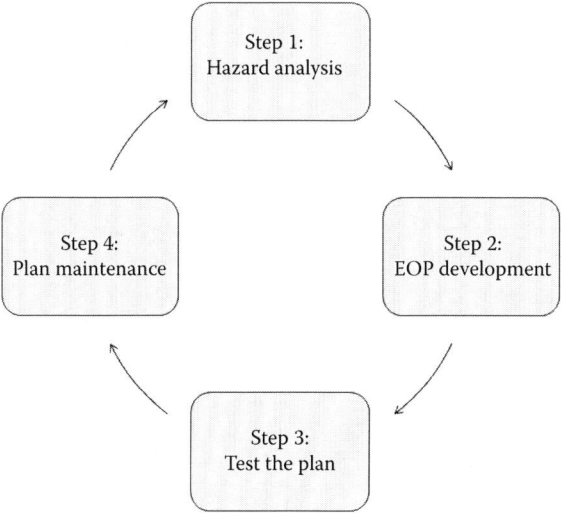

Figure 7.1 Emergency planning process. (From Michael Fagel.)

are identified, researched, and ranked according to the risks they pose, and areas and infrastructure that are vulnerable to damage from an event involving the hazards. The outcome of this step is a written hazard analysis that quantifies the overall risk to the community from each hazard.

The next step in the emergency planning process is EOP development. The outcome of this step is a completed plan, which is ready to be trained, exercised, and revised based on lessons learned from the exercises. The third step in the planning process is testing the plan through training and exercising. Exercises of varying types and complexity allow for evaluation to see what in the plan is unclear and what does not work. The outcome of this step is a series of lessons learned about weaknesses in the plan. These weaknesses can then be addressed in the final step: plan maintenance and revision. Plan maintenance and revision can be completed based on needs and resources, which may have changed since the development of the original EOP. After the EOP is developed, steps 3 and 4 repeat in a continuous cycle to keep the plan up to date. If the community becomes subject to a new hazard, however, the planning team will need to revisit steps 1 and 2.

Performing a Hazard Analysis

A hazard analysis determines:

- What can occur in the community
- How often it is likely to occur
- The damage it is likely to cause
- How it is likely to affect the community
- How vulnerable the community is to the hazard

There are five steps in the hazard analysis process:

1. Identify hazards
2. Profile each hazard
3. Develop a community profile
4. Determine vulnerability
5. Create and apply scenarios

There are many potential sources of hazard information. A starting point may be local newspapers, which can provide a more comprehensive picture of the types of hazards that a community has historically faced. However, it may be necessary to check sources such as the State Department of Agriculture, Bureau of Labor Statistics, or similar agencies; the National Weather Service (NWS); local historical societies; or long-time residents. If your community has an existing hazard analysis, start by reviewing it with an eye toward what has changed in the jurisdiction since the hazard analysis was completed.

During the hazard analysis process, it is important to keep in mind that the hazards a community faces may change over time because of new mitigation measures, the opening or closing of facilities, local development activities, or terrorist threats that were not considered before the attacks of September 11, 2001.

DEVELOPING AN EMERGENCY OPERATIONS PLAN (EOP)

There may be other long-term changes to investigate, as well. Changes in average temperature or rainfall and snowfall amounts may be harder to track but will certainly play an important role on the way to having a complete hazard analysis.

Developing a Hazard Profile
A hazard profile should consider four factors:

1. Magnitude
2. Frequency
3. Duration
4. Speed of Onset

A hazard profile should address each hazard's magnitude—or size. How strong a hazard is and the areas that it could affect could dramatically change response plans. For example, a storm that drops 2 inches of rain very quickly over a small area requires a much different response than a nor'easter that drops 20 inches of rain over a four-state area.

It is also important to consider a hazard's frequency, including whether a seasonal pattern exists. In some parts of the country, thunderstorms are a near daily occurrence. On the other hand, hurricanes are a seasonal occurrence, which may or may not present a high risk to your area. Consider each hazard's duration or how long the hazard is expected to last. For example, the duration of even the most severe thunderstorm is much shorter than that of a hurricane.

Finally, consider the speed of onset of the hazard. This is important for determining the available time for issuing a warning and is also critical to the response. The amount of damage and loss of life that an extreme hazard could cause can be mitigated if emergency personnel and the public have time to take protective action.

A profile should be completed for each hazard to which the community is vulnerable, but it is important to keep in mind that some hazards pose such a limited threat that additional analysis may not be necessary. You should not, however, ignore low-risk hazards that have a high potential for damage should they occur. These low-risk hazards may not be a planning priority but should be planned for nonetheless (see Figure 7.2).

Creating a Jurisdiction Profile
After completing the hazard analysis process, it is necessary to combine hazard-specific information with a profile of the community to determine the community's vulnerability to or risk of damage from the hazard. Because different communities have different profiles, vulnerabilities to the same hazard will vary. Table 7.1 summarizes key factors that are included in the community profile.

After gathering this information about the community, develop the community's jurisdiction profile by plotting vulnerable areas on a jurisdiction map. Table 7.2 shows the use of community factors in the jurisdiction profile.

Completing the Risk Analysis
After compiling the jurisdiction profile, the next step is to quantify the community's risk by merging the information. Risk is the predicted impact that a hazard would have on the people, services, and specific facilities in the community. Quantifying risk enables

Hazard:
Potential magnitude (Percentage of the community that can be affected):
Catastrophic: More than 50%
Critical: 25% to 50%
Limited: 10% to 25%
Negligible: Less than 10%

Frequency of Occurrence: *Seasonal Pattern:*
Highly likely: Near 100% probability in next year.
Likely: Between 10% and 100% probability in next year, or at least one chance in next 10 years.
Possible: Between 1% and 10% probability in next year, or at least one chance in next 100 years.
Unlikely: Less than 1% probability in next 100 years.

Areas Likely to Be Affected Most:

Probable Duration:

Potential Speed of Onset (Probable amount of warning time):
☐ Minimal (or no) warning   ☐ 12 to 24 hours warning
☐ 6 to 12 hours warning   ☐ More than 24 hours warning

Existing Warning Systems:

Does a Vulnerability Analysis Exist?*
Yes ☐
No ☐

* Threat & Hazard Identification & Risk Assessment (THIRA) is also a useful tool in this process. http://www.fema.gov/media-library/assets/documents/27775

Figure 7.2 Hazard profile worksheet. (From Michael Fagel.)

jurisdictions to focus on those hazards that pose the highest threat to life, property, and the environment. Quantifying risk involves:

- Identifying the elements of the community (populations, facilities, and equipment) that are potentially at risk from a specific hazard
- Developing response priorities (risk to life is always the highest priority)
- Assigning severity ratings based on the potential impact to life, essential facilities, and critical infrastructure
- Compiling risk data into the community risk profiles that show the areas of the community that are at highest risk from the hazard

In analyzing risk, it is helpful to develop response priorities, using the hierarchy in Figure 7.3 to set priorities.

Next, assign each hazard a severity rating, or risk index, that will predict, to the highest degree possible, the damage that can be expected in the community as a result of that

DEVELOPING AN EMERGENCY OPERATIONS PLAN (EOP)

Table 7.1 Key Factors in Creating a Community Profile

Geography	Property	Infrastructure	Demographics	Response Organizations
Major geographic features	Numbers	Utilities, construction, layout, access	Population size, distribution, concentrations	Locations
Typical weather patterns	Types	Communication system layout, features, backup	Numbers of people in vulnerable zones	Points of contact
	Ages	Road systems	Special populations	Facilities
	Building codes	Air and water support	Animal populations	Services
	Critical facilities			Resources
	Potential secondary hazards			

Source: Used with permission from S. Shane Stovall and Michael Fagel.

Table 7.2 Use of Community Factors in the Jurisdiction Profile

Type of Information	Used In
Geographic	• Predicting risk factors and the impact of potential hazards and secondary hazards
Property	• Projecting consequences of potential hazards to the local area • Identifying available resources
Infrastructure	• Identifying points of vulnerability • Preparing evacuation routes, emergency communication, and projecting response and recovery requirements
Demographic	• Projecting consequences of disasters on the population • Disseminating warnings and public information • Planning evacuation and mass care
Response organizations	• Identifying response capabilities

Source: Used with permission from S. Shane Stovall and Michael Fagel.

hazard. This rating quantifies the expected impact of a specific hazard on people, essential facilities, property, and response assets. Table 7.3 is an example of severity ratings that may be used.

Develop a risk index for each hazard and assign a value to each characteristic. Use the following values:

1 = Catastrophic
2 = Critical
3 = Limited
4 = Negligible

SOFT TARGETS AND CRISIS MANAGEMENT

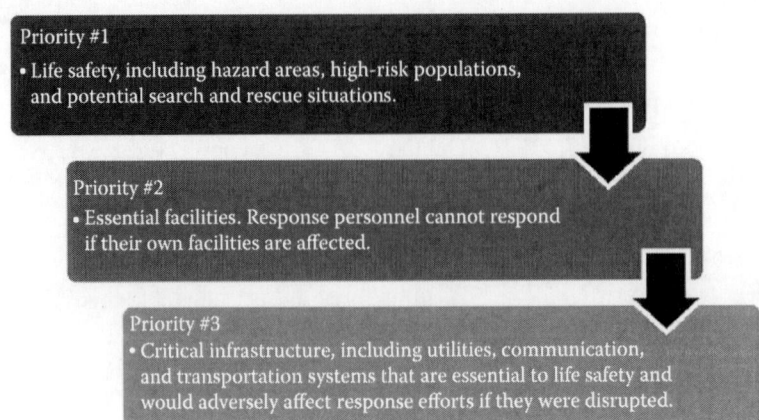

Figure 7.3 Hierarchy for response prioritization profile. (Used with permission from S. Shane Stovall.)

The ratings should be assigned for each of the following types of hazard data:

- Magnitude
- Frequency of occurrence
- Speed of onset
- Community impact
- Special characteristics

Average the value of all factors to determine the overall risk level for each hazard.

The result of this process will be a prioritized list of hazards that pose the greatest threat to the community. The planning team in your community should plan for each hazard for which the risk index exceeds a predetermined threshold.

Table 7.3 Hazard Severity Rating

Severity	Expected Impact
Catastrophic	• Multiple deaths • Complete shutdown of critical facilities for 30 days or more • More than 50% of property is severely damaged
Critical	• Injuries and/or illnesses result in permanent disability • Complete shutdown of critical facilities for at least 2 weeks • More than 25% of property is severely damaged
Limited	• Injuries and/or illnesses do not result in permanent disability • Complete shutdown of critical facilities for more than 1 week • More than 10% of property is severely damaged
Negligible	• Injuries and/or illnesses treatable with first aid • Minor quality of life lost • Shutdown of critical facilities for 2 hours or less • Less than 10% of property is severely damaged

Source: Used with permission from S. Shane Stovall and Michael Fagel.

Creating Scenarios

The final step in the hazard analysis process is to create and apply scenarios for the highest-risk hazards. Scenarios should be realistic and based on the community's hazard and risk data.

To create a scenario, brainstorm to track the development of a specific type of emergency. Each scenario should describe the following:

1. The initial notification that the event is occurring or is about to occur
2. The potential overall impact on the community
3. The potential overall impact of the event on specific community sectors
4. The potential consequences, such as casualties, damages, and loss of services
5. The actions and resources that would be needed to deal with the situation

Creating scenarios helps to identify situations that may exist in a disaster. These situations should be used to help ensure that a community is prepared if the hazard event occurs. Use Table 7.1 as a guide to make notes about key factors in your community.

COMPONENTS OF AN EMERGENCY OPERATIONS PLAN

An EOP describes actions to be taken in response to natural, man-made, or technological hazards. It also details tasks to be performed by specific organizational elements at projected times and places based on established objectives, assumptions, and assessment of capabilities.

An EOP should be comprehensive. In other words, it should cover all aspects of emergency preparedness and response, and address mitigation concerns as well. It also should address all hazards and therefore be flexible enough to use in all emergencies—even unforeseen events. Finally, the EOP should be risk based. It should include hazard-specific information based on the risks that were established in the hazard analysis.

The EOP is written to provide an overview of the jurisdiction's response organization and policies. It also should provide a general understanding of the jurisdiction's approach to emergency response for all involved agencies and organizations.

An EOP consists of three parts: the basic plan, functional annexes that address the performance of a particular broad task, and hazard-specific appendices that provide additional response. So, although the basic plan provides the general approach to emergency response, it does not stand alone. Rather, it forms the basis for the remainder of the plan.

In addition, each part of the plan may have addenda in the form of standard operating procedures (SOPs), maps, charts, checklists, tables, forms, and so forth. These addenda may be included as attachments or incorporated by reference.

The Basic Plan

Although there is no standard format, for the sake of compatibility with other jurisdictions and levels of government, it is recommended that the basic plan include the following components:

- Introduction
- Purpose statement

- Situation and assumptions
- Concept of operations
- Organization and assignment of responsibility
- Administration and logistics
- Plan development and maintenance
- Authorities and references

The Introduction
The introduction consists of five elements:

1. The *promulgation document*, which is signed by the jurisdiction's chief elected official to affirm his or her support for the emergency management agency and the planning process. It gives organizations the authority and responsibility to perform their tasks. It also mentions the tasked organizations' responsibility to prepare and maintain implementing instructions, gives notice of necessary EOP revisions, and commits to the training necessary to support the EOP.
2. The *signature page*, which is signed by all partner organizations to demonstrate their commitment to EOP implementation.
3. Dated *title page and record of changes*, which includes the date, description, and parts affected by changes to the EOP.
4. *Record of distribution*, which lists EOP recipients, and facilitates and provides evidence of EOP distribution.
5. *Table of contents*

The Purpose Statement
The purpose statement should include a broad statement about what the EOP is meant to do. It should also include a synopsis of the EOP, annexes, and appendices. The purpose statement need not be complex, but should include enough information to establish the direction for the remainder of the plan.

The Situations and Assumptions Section
The third component of the basic plan is the situation and assumptions section. The situation characterizes the planning environment, making it clear to the community why emergency planning is necessary. It draws from the hazard analysis to narrow the scope of the EOP and includes the following:

- Hazards addressed by the plan
- Relative probability and impact
- Areas likely to be affected
- Vulnerable critical facilities
- Population distribution
- Special populations
- Interjurisdictional relationships
- Maps

The assumptions statement delineates what was assumed to be true when the EOP was developed. Additionally, the assumptions statement shows the limits of the EOP, limiting liability for the jurisdiction. It may be helpful for a jurisdiction to list even obvious assumptions, such as the following:

- Identified hazards will occur
- Individuals and organizations are familiar with the EOP
- Individuals and organizations will execute their assigned responsibilities
- Assistance may be needed and, if so, will be available
- The implementation of the EOP will save lives and reduce damage

The Concept of Operations Section

The fourth component of the basic plan is the concept of operations, which explains the community's overall approach to emergency response (i.e., what, when, by whom). The concept of operations includes the following:

- The division of local, state, and federal responsibilities
- When the EOP will be activated and when it will be deactivated and—more important—by whom
- Alert levels and the basic actions that accompany each level
- The general sequence of actions before, during, and after the event
- Forms necessary to request assistance of various types

The Organization and Assignment of Responsibilities Section

The organization and assignment of responsibilities section lists, by organizations and positions, the general areas of responsibility assigned. It also identifies shared responsibilities and specifies which organization has the primary responsibility for a given function and which have supportive roles. In other words, the organization and assignment of responsibilities section specifies reporting relationships and lines of authority for an emergency response.

Administration and Logistics Section

The sixth component of the basic plan is the administration and logistics section. This section provides resource management policies and policies for augmenting response staff with public employees and volunteers, as well as a statement that addresses liability issues. It also includes the assumed resource needs for high-risk hazards, resources that are available within the community, and resources that may be available through mutual aid agreements. It is important to note, however, that the community should not rely on mutual aid agreements because neighboring jurisdictions may be faced with the same emergency or disaster that your community is facing.

Plan Development and Maintenance Section

The next component is plan development and maintenance. The responsibility for the coordination of the development and revision of the basic plan, annexes, appendices, and implementing instructions must be assigned to the appropriate persons. Therefore, this section:

- Describes the planning process
- Identifies the planning participants

- Assigns planning responsibilities
- Describes the revision cycle (i.e., training, exercising, review of lessons learned, etc.)

Authorities and References Section

The final section of the basic plan is the authorities and references section. This section cites the legal basis for emergency operations and activities, including the following:

- Laws, statutes, and ordinances
- Executive orders
- Regulations
- Formal agreements
- Predelegation of emergency authorities
- Pertinent reference materials, including related plans for other levels of government

Annexes

Annexes delineate how the community will carry out broad functions, such as issuing warnings or managing resources, in any emergency. It is important to determine early in the planning process the functions that will be included in the basic plan as annexes. When making this decision, it is important to consider the organization of the state government and that of your jurisdiction, the capabilities of your jurisdiction's emergency services agencies, and the established concept of operations. During this process, it is important to keep in mind the hazard analysis information developed for your community. What the community's planning team knows about the community's vulnerability is paramount to developing meaningful functional annexes.

Because communities vary so widely, there is no single list of functional annexes that is right for everyone. There are, however, eight core functions that FEMA recommends be addressed as annexes in every EOP:

1. The direction and control annex allows a jurisdiction to analyze the situation and decide on the best response, direct the response teams, coordinate efforts with other jurisdictions, and make the best use of available resources.
2. The communications annex provides a detailed focus on the total communications system and how it will be used.
3. The warning annex describes the warning systems in place and the responsibilities and procedures for using them.
4. The emergency public information (EPI) annex provides the procedures for giving the public accurate, timely, and useful information and instructions throughout the emergency period. It is important to note the difference between the warning annex and the EPI. Whereas the warning annex focuses on the procedures that the government uses to alert those at risk, the EPI annex deals with developing messages and accurate information, disseminating the information, and monitoring how the information is received. Because the warning system is one means for an EPI organization to get information out, the EPI annex must address coordination with those responsible for the warning system.

5. The evacuation annex describes the provisions that have been made to ensure the safe and orderly evacuation of people threatened by hazards that the jurisdiction faces.
6. The mass care annex deals with the actions that are taken to protect evacuees and other disaster victims from the effects of the disaster, including providing temporary shelter, food, medical care, clothing, and other essential needs.
7. The health and medical annex describes policies and procedures for mobilizing and managing health and medical services under emergency or disaster conditions.
8. The resource management annex describes the means, organization, and process by which a jurisdiction will find, obtain, and allocate resources to satisfy needs that are generated by an emergency or disaster.

In addition to these annexes, the planning team may want to consider annexes that make sense for the community. For example, if the community has a nuclear power plant, the planning team may want to add an annex on radiological protection. Additional functional annexes may be based on state law or jurisdictional requirements. Examples of annexes that may be added include the following:

- Damage assessment
- Search and rescue
- Emergency services
- Aviation operations

Hazard-Specific Appendices

An appendix is a supplement to an annex that adds information about how to carry out the function in the face of a specific hazard. Therefore, every annex may have several appendices, each addressing a particular hazard. The hazard-specific appendices that the planning team deems appropriate depend on the community's hazard analysis.

The decision about whether to develop an appendix rests solely with the planning team. Unlike annexes, hazard-specific appendices are not attached to the basic plan but are linked to each functional annex. Topics addressed in hazard-specific appendices include (Table 7.4):

- Special planning requirements
- Priorities identified through the hazard analysis
- Unique characteristics of the hazard that require special attention
- Special regulatory considerations

Implementing Instructions

Like the basic plan, each annex or appendix may use implementing instructions in the form of the following

- Standard operating procedures (SOPs)
- Job aids

Table 7.4 Appendix Topics for Each Functional Annex

Annex	Appendix Topics
Direction and control	• Response actions keyed to specific time periods or phases • Urban search and rescue (US&R) inspection • Protective gear for responders • Detection equipment and techniques • Laboratory analysis services • Containment and cleanup teams
Communications	• Provisions made to ensure that the effects of a specific hazard do not prevent or impede the ability of response personnel to communicate with each other during response operations
Warning	• Hazard-specific public warning protocols • Required or recommended notifications of state and federal officials
Emergency public information	• Information the public will need to know about the specific hazard (e.g., special evacuation routes and shelters, in-place protective actions, etc.) • The means by which information to the public will be conveyed
Evacuation	• Evacuation options and timing • Evacuation routes • Transportation resources to support mass evacuation
Mass care	• Shelter locations out of the hazard's vulnerable areas • Protection of shelter occupants • Food and water stocks to support extended shelter stays • Capability to decontaminate people exposed to hazardous materials
Health and medical services	• Unique health consequences and treatment options for people exposed to the hazard • Environmental monitoring and/or decontamination requirements
Resource management	• Provisions for purchasing, stockpiling, or otherwise obtaining special protective gear, supplies, and equipment needed by response personnel and disaster victims

Source: Adapted from FEMA. http://www.fema.gov/pdf/emergency/nrf/nrf-annexes-all.pdf

- Checklists
- Information cards
- Record-keeping and combination forms
- Maps
- Charts
- Tables
- Forms

Implementing instructions may be included as attachments or by reference, and the planning team may use them as needed to clarify the contents of the plan, annex, or appendix. For example, the evacuation annex may be made clearer by attaching maps with evacuation routes marked. Because these routes may change depending on the location of the hazard, maps may also be included in the hazard-specific appendices to the evacuation annex. Similarly, the locations of shelters may be marked on maps supporting the mass care annex.

Standard Operating Procedures

Standard operating procedures (SOPs) are a common method of implementing instructions. SOPs provide response protocols for carrying out specific responsibilities. They describe who, what, when, where, and how. SOPs are appropriate for complex tasks requiring step-by-step instructions, tasks for which standards must be specified, and tasks for which documentation of performance protocols is required as a protection against liability.

When developing SOPs,

1. Develop a task list.
2. Determine who, what, when, where, and how. (Remember that "who" includes who performs the activity, to whom he or she reports, and with whom he or she coordinates.)
3. Identify the steps for each task.
4. Identify the standards for task completion.
5. Test the procedures.

Job Aids

It is important to keep the SOPs up to date through review and revision. A job aid is a written procedure that is intended to be used on the job while the task is being done. SOPs may be presented as job aids.

Job aids are also appropriate for any of the following:

- Complex tasks
- Critical tasks that could result in serious consequences
- Tasks that are done infrequently
- Procedures or personnel that change often

Job aids are also useful when conformity is needed among workers or across locations. Job aids should specify information such as the following:

- The task title
- The purpose of the task
- When to do the task
- Materials needed to perform the task
- How to perform each step of the task
- The desired results
- Standards to which the task must be performed
- How to ensure that the work is done properly

A job aid may include any of the following:

- Graphics
- Flowcharts
- If–then decision tables
- Do's and don'ts

Because job aids are used during the completion of a task, they must be clear to be effective. They should use action verbs and everyday language, highlight important information, and place warnings before steps to which they apply.

Formatting is also important when creating job aids. Numbering steps and using empty space, boxes, or lines to separate steps allows users to find their place easily after looking away. Job aids may not be useful for all tasks, especially simple tasks that are performed regularly or must be accomplished quickly from memory. If a task cannot be completed while referring to a job aid at the same time, a job aid is not appropriate.

Checklists
Checklists are also useful for implementing instructions. They provide a list of tasks, steps, features, contents, or other items to be checked off as completed. They often take the form of boxes to be checked off but can be developed in any form, including as rating scales. Checklists are particularly useful for tasks that are made up of multiple steps that must be completed in sequence or for when it might be necessary to document the completion of steps. Checklists may be less useful when observations must be recorded, when calculations or evaluations must be made, or when detailed instructions are required to complete the task.

Information Cards
Information cards provide information that is needed on the job in a convenient, often graphic, form. Examples include:

- Reference forms
- Diagrams, labeled illustrations, charts, or tables
- Information summarized in matrix form

Things that might be presented in the form of information cards include:

- Call-down rosters
- Contact lists
- Resource lists
- Organizational charts
- Task matrices
- Equipment diagrams

Forms
Common forms used as implementing instructions include:

- Record-keeping forms on which calculations, observations, or other information (e.g., damage assessment) can be recorded
- Combination forms that serve multiple functions, such as checklists with record-keeping sections

Maps
Maps may be used as implementing instructions to highlight:

- Geographic boundaries and features
- Jurisdictional boundaries
- Locations of key facilities
- Transportation or evacuation routes

It is important to note that when using a map as an implementing instruction to show a particular feature, extraneous details are often eliminated.

Creating Effective Implementing Instructions
To be effective, implementing instructions must be appropriate for both the audience and the intended use. They must also be:

- Complete, in that they cover all of the components or steps
- Clear, concise, and easy to use; they should avoid jargon and ambiguity, be organized in a logical manner, and include instructions that identify the purpose and applicability of the particular implementing instruction
- Sufficiently detailed, in that they give all of the necessary information
- Up to date; the latest revision should be included
- Sufficient in scope; they must cover each function fully
- Identified in the EOP so that their existence is recorded

Implementing instructions should be incorporated by reference in the basic plan, annex, or appendix to which they refer.

Implementing instructions are used by all agency personnel who respond to disasters, whatever their function. They are developed at the agency level because agency personnel will be using them and, therefore, will know if they are effective. Implementing instructions used by the agency should support the agency's roles and responsibilities as described in the basic plan. For this reason, only some of the types of implementing instructions described will be useful to a particular agency, depending on its function in a response.

When developing any type of implementing instruction, the first step is to consider the job title or position and the tasks that go along with that position. You can then decide what type of implementing instructions would be most useful for those tasks.

Emergency planning is a continual cycle of planning, training, exercising, and revision that takes place throughout the four phases of the emergency cycle: mitigation, preparedness, response, and recovery.

THE EMERGENCY PLANNING TEAM

Emergency planning should be a team effort because disaster response nearly always requires coordination among many agencies and organizations, and at different levels of government. Planning as a team helps to ensure that everyone having the expertise to contribute in a response is a part of the process. Team planning also helps to ensure the buy-in of all key players or stakeholders.

Who Should Be on the Planning Team?

Different types of emergencies require different expertise and response capabilities. The specific individuals and organizations involved in a response will vary by the type of emergency or disaster. For example, law enforcement, fire services, and emergency medical services (EMS) will probably have a role to play in almost every emergency. On the other hand, hazardous materials (hazmat) personnel may or may not be involved in every incident but should be included in the planning process nonetheless, because they have specialized expertise that may be called upon in certain situations.

Table 7.5 presents an example of the types of agencies and personnel that may be involved in the planning process. All individuals do not need to be included in every aspect of the planning process, but they should be consulted in areas that affect them directly or for which they will be responsible.

Table 7.5 Agencies for Planning Teams

Individuals and Organizations	Contribution to the Planning Team
Chief elected official (CEO) or designee	Support for the emergency planning process, policy guidance and decision-making capability, and the authority to commit the community's resources
Fire chief or designee	Knowledge of the Incident Command System (ICS), knowledge of fire department procedures, on-scene safety requirements, hazardous materials response requirements, and search and rescue requirements; knowledge of the community's fire-related risks, and specialized personnel, and equipment resources
Police chief or designee	Knowledge of police department procedures, on-scene safety requirements, and local laws and ordinances; specialized response requirements, such as perimeter control and evacuation procedures; and specialized personnel and equipment resources
Public works director or designee	Knowledge of the community's road and utility, infrastructure, and specialized personnel and equipment resources
EMS director or designee	Knowledge of medical treatment requirements for a variety of situations, knowledge of medical treatment facility capabilities, and specialized personnel and equipment resources
Hazardous materials coordinator	Knowledge of hazardous materials that are produced, stored, or transported in or through the community; knowledge of EPA, OSHA, and DOT requirements for producing, storing, and transporting hazardous materials; knowledge of how to respond to hazardous materials and incidents
Mutual aid partners	Specialized personnel and equipment resources, and additional personnel and equipment resources
Department of health director or designee	Knowledge of community public health capabilities and limitations, familiarity with the key local health care providers, and specialized personnel resources
Department of Transportation director or designee	Knowledge of the community's road infrastructure, knowledge of the area's transportation resources, familiarity with the key local transportation providers, and specialized personnel resources

(*Continued*)

Table 7.5 (*Continued*)   Agencies for Planning Teams

Individuals and Organizations	Contribution to the Planning Team
U.S. Department of Agriculture director or designee	Knowledge of the area's agricultural sector and the associated risks
Tax assessor	Records of property in the community and their respective values
Building inspector	Knowledge of types of construction used in the community, knowledge of land use and land use restrictions, and records of planned development
School superintendent or designee	Knowledge of school facilities, and knowledge of hazards that directly affect schools
Voluntary agency directors	Knowledge of additional resources that can be brought to bear in an emergency, lists of shelters, feeding centers, and distribution centers, and volunteer resources
Air/seaport managers	Knowledge of risks associated with airport/seaport operations, specialized personnel and equipment resources that can be used in an emergency
Representatives from local industry	Knowledge of hazardous materials that are produced, stored, or transported in or through the community; facility response plans, specialized personnel, and equipment resources that can be brought to bear in an emergency
Radio Amateur Civil Emergency Services (RACES)	List of RACES resources that can be used in an emergency
Social services agency representative	Knowledge of special-needs populations in the community
Veterinarian/animal shelter representative	Knowledge of the special response needs required for animals, including livestock
City or county attorney	Knowledge of legal and liability issues; see the section titled "Disaster Preparedness and the Law" in this chapter.

Source: Used with permission from S. Shane Stovall and Michael Fagel.

Getting the Team Together

Getting all of the stakeholders in the planning process together and to take an active interest in the planning process will be an arduous task. To schedule meetings with so many participants may be even more difficult. It is critical, however, to have everyone's participation in the planning process, at least in the early stages and at critical points along the way, to draw from their expertise and ensure their buy-in to the plan.

The expertise and knowledge that participants bring of their organizations' resources is crucial to developing an accurate plan that considers the entire community's needs and the resources that could be made available in an emergency. It is definitely to the community's benefit to have the active participation of all key players.

There are several steps that you can take to gain cooperation from participants and gain their involvement in the planning process:

- *Plan ahead*—Give the planning team plenty of notice of where and when the planning meeting will be held. If time permits, survey the team members to find the time and place that will work for them.
- *Provide information about team expectations*—Explain why participating on the planning team is important to the participants' agencies and to the community itself. Show the team members how they will contribute to a more effective emergency response.
- *Involve the chief elected official*—Ask the chief elected official to sign the meeting notice. This will send a clear message that emergency planning is important to the community and that the participants are expected to attend.
- *Allow flexibility in scheduling after the first meeting*—Not all team members will need to attend all meetings. Task forces and subcommittees can complete some of the work. Where this is the case, gain concurrence on timeframes and milestones, but let the subcommittee members determine when it is most convenient to meet.

It may also be beneficial to talk to emergency managers from other communities to gather input on how to gain and maintain interest in the planning process.

Team Operation
Unlike working alone, working with personnel from other organizations to plan for emergencies requires some give and take—in other words, collaboration. Collaboration is the process in which people work together as a team on a common mission—in this case, the development of a community EOP. Successful collaboration requires:

- A commitment to participate in shared decision making
- A willingness to share information, resources, and tasks
- A professional sense of respect for individual team members

Collaboration can be made difficult by differences among agencies and organizations in:

- *Terminology*—Often, people from different organizations use different terms to mean the same thing—or use the same terms to mean different things. For example, the American Red Cross considers a house fire to be a disaster, while FEMA considers disasters to be incidents large enough in scope and impact to warrant a Presidential declaration under the Stafford Act.
- *Experience*—People's experiences lead them to respond differently in different situations. Those whose past experiences have proven them correct (or incorrect) in a given situation may have very strong opinions about what should be done, when, and how.
- *Mission*—Each agency that participates in the planning process is operating toward achieving its specific mission, which may or may not be entirely consistent with that of the emergency management agency. "Personal" missions may interfere with collaboration.
- *Culture*—Culture entails everything about a person and the agency that he or she works for, including the potential meaning of nonverbal cues. Culture may be one of the most difficult factors to overcome during the collaboration process.

Collaboration under these conditions requires flexibility to agree on common terms and priorities, and humility to learn from others' ways of doing things.

Although collaboration among EOP planning team members may be difficult, it benefits the community by strengthening the overall response to disaster. Collaboration can

- Eliminate duplication of services, resulting in a more efficient response
- Expand resource availability through sharing
- Enhance problem solving through the cross-pollination of ideas

Stages of Team Formation
Collaboration does not come automatically. Building a team that works well together takes time and effort, and typically evolves through five stages:

1. *Forming*—Individuals come together as a team. During this stage, the team members may be unfamiliar with one another and uncertain of their roles on the team.
2. *Storming*—Team members may become impatient, disillusioned, and may disagree.
3. *Norming*—Team members accept their roles and make progress toward the goal.
4. *Performing*—Team members work well together and make progress toward the goal.
5. *Adjourning*—Their task accomplished, team members may feel pride in their achievement and some sadness that the experience is ending. It is important to remember that the team should remain intact for the purpose of exercising and revising the plan.

Team Roles
To keep the team focused throughout the planning stages, it is important for team members to assume roles. Perhaps the most important role is that of team leader. The team leader initiates appropriate team-building activities that move the team through each stage and toward its goal.

Other team roles may include:

- The taskmaster, who identifies the work to be done and motivates the team
- The innovator, who generates original ways to get the group's work done
- The organizer, who helps the group develop plans for getting the work done
- The evaluator, who analyzes ideas, suggestions, and plans made by the group
- The finisher, who follows through on plans developed by the team

Together, all members contribute toward making the team productive.

Characteristics of an Effective Team
You will know that the planning team is on track when it displays the following characteristics:

- Works toward a common goal
- Accepts the leader who provides direction and guidance
- Communicates openly

- Resolves conflicts constructively
- Displays mutual trust
- Shows respect for each individual and his or her contributions

DISASTER PREPAREDNESS AND THE LAW

Why should your city or county attorney or any attorney be included on your emergency and disaster preparedness team? What does a "lawyer" know about emergency planning? What does an attorney bring to the table in the emergency planning process? Will the attorney add value to the team or be a detriment to the overall planning process?

An emergency and disaster preparedness team is assembled based upon individuals' knowledge, skills, and abilities, their expertise, and their experience with the primary goal of the team being to "do good deeds" for others in an emergency or disaster situation. Isn't it logical and prudent to have at least one member of the emergency and disaster preparedness team with a skill set that is particularly focused on the legal aspects of the team's actions and possible inactions that could inadvertently cause harm to others? From this perspective, there is a definite need for an attorney's unique legal perspective as part of any emergency and disaster preparedness team especially from the beginning of the planning process.

What an Attorney Can Provide as a Team Member

We live in a very litigious society. The good deeds that each and every member of the team wants to do to assist their community during an emergency or disaster situation may result in some form of legal action, claim, or litigation directly from their good deed, action, or even possibly from the lack of action, depending on the circumstances. An attorney on the emergency and disaster preparedness team can possibly identify potential pitfalls or potential legal risks wherein the emergency and disaster preparedness team can take proactive steps to protect against the risk, educate their members as to the risk, or even ensure against the risk to eliminate or minimize the risk from becoming a "disaster after the disaster."

An attorney on the emergency and disaster preparedness team can bring a different perspective due to the specialized skill set that was drilled into them during their legal studies. Every emergency and disaster preparedness team, no matter how well prepared, is second-guessed following any event. A skilled attorney can assist the team in analyzing the potential risks during the planning process to provide proactive alternatives to eliminate or minimize the risks. Just as there are two sides to every argument, there are two sides to every situation, depending on your perspective. A skilled attorney should be able to provide the opposing viewpoint for the emergency and disaster preparedness team, especially during the planning stage, to permit the team to carefully analyze the situation and determine if the risk is acceptable or offer alternatives to counteract the risk.

Teams are often assembled in accordance with their experience, expertise, and personality and each member is expert in his or her areas of expertise. How many team members are great at paperwork? Who on the team will double check the team's paperwork and check all of the boxes to ensure the accuracy and correctness of documents and documentation? An attorney can serve as a fail safe through which all documents can be evaluated to lessen potential liability as a result of paperwork errors.

Many of the grant proposals and other documents that may come before the emergency and disaster preparedness team may be written in legalese. The skilled attorney can interpret the legalese into a comprehensible verbiage for the emergency and disaster preparedness team members. This should permit the other team members to focus on their areas of expertise when developing the grant or other document rather than spending their time with a legal dictionary.

In addition, the media is omnipresent in any emergency and disaster situation. Although many emergency and disaster preparedness teams possess an information officer, the skilled attorney can assist in honing any statements to the press and keep the proverbial foot from entering the mouth of team members. As we are all aware, video footage from any emergency or disaster situation locks the scene in time and is often utilized to Monday-morning quarterback the actions or inactions of the team in public forums as well as in litigation in the future.

Long story short, there is very little downside to adding a city attorney or local attorney to the planning process, and the substantial "upside" is that an attorney will be able to proactively identify potential legal pitfalls during the planning process.

The Potentially Applicable Laws

Let's take a look at a few of the potentially applicable civil laws as well as the areas of potential liability or pitfalls in an emergency and disaster situation.

- Federal laws (including but not limited to):
  - Americans with Disabilities Act (ADA)
  - Occupational Safety and Health Act (OSHA)
  - Labor Management Relations Act
  - Federal Service Labor–Management Relations Act
  - Civil Rights Act of 1964 and 1991
  - Age Discrimination in Employment Act
  - Veterans' Reemployment Rights Act
  - Rehabilitation Act of 1973
  - Federal Anti-Discrimination Executive Orders
  - Fair Labor Standards Act (FLSA)
  - Minimum Wage Requirements under the FLSA
  - Overtime Pay under the FLSA
  - Employee Retirement Income Security Act (ERISA)
  - Federal unemployment laws
  - Federal privacy laws
  - Family and Medical Leave Act (FMLA)
- State laws (including but not limited to):
  - State labor relations laws
  - Anti-discrimination laws

- Protection for military personnel
- State wage and hour laws
- State safety and health laws.
- State workers' compensation laws
- Governmental immunity statutes
- Local laws (including but not limited to):
    - Building codes
    - Jurisdictional requirements
    - Land use laws
    - Traffic laws

Although these are just a few of the myriad federal, state, and local laws and requirements possessing potential civil liability, there is a completely different set of laws on the criminal side. Generally, the criminal laws are applicable to individual team members' actions or inactions, rather than the overall organization; however, many of the laws mentioned above do possess a criminal component for egregious situations.

Negligence

One specific area in which an attorney can educate your emergency and disaster preparedness team is in the frequent civil actions sounding in the area of negligence. There are numerous causes of action in many areas sounding in negligence ranging from negligence hiring to negligence supervision to negligent injury.

Generally, to prove negligence, the injured party must provide:

1. *Duty*—The emergency and disaster team possessed a duty to act.
2. *Breach*—The emergency and disaster preparedness team breached this duty through their actions or inactions.
3. *Causation*—The actions or inactions of the emergency and disaster preparedness team caused the injury to the other party.
4. *Damages*—The actions or inactions of the emergency and disaster preparedness team caused damages to the other party.

Although this breakdown is provided in a simplistic manner, negligence actions are common in federal and state civil courts against public and private-sector entities. For public-sector emergency and disaster preparedness organizations, there is the additional issue of immunity that would need to be addressed. An attorney, as a valued member of the emergency and disaster preparedness team, can educate the team as to the requirement of these laws in order to avoid risks and areas of potential liability as well as being the well-prepared advocate in case a legal action should arise against the organization or team.[*]

---

[*] This section on legal aspects taken from a lecture from Professor Thomas D. Schneid, JD, Eastern Kentucky University, 2015.

Emergency Management Phases

Let's take a look at the various stages of your emergency and disaster preparedness plan to ascertain some of the general areas where potential legal issues could arise, in which legal counsel could assist the emergency and disaster preparedness team.

Planning Stage

- There are jurisdictional issues between the state and local governmental entities.
  - Who is responsible for what activities?
- A mutual aid agreement is agreed upon between departments. One entity fails to perform the required responsibilities.
  - Who is responsible for any consequences that arise from this?

Response Stage

- The 9-1-1 system is not working.
  - Who is responsible?
  - Is the failure of the 9-1-1 system the cause of a family member dying?
- A member of the emergency and disaster team renders first aid at the scene. The victim dies and the family blames the first responder.
  - Who may have liability?

Recovery Stage

- The team is attempting to repair a gas line and the line explodes.
  - What if a team member is injured or killed?
  - What if the neighboring house is severely damaged?
- The team member is utilizing a chain saw and the chain saw malfunctions causing severe injury.
  - Is the manufacturer responsible?
  - Your organization?
  - Is third-party recovery possible?

Other Areas of Potential Civil Liability Risks

Vehicles

- A member of your team is injured while driving a vehicle during an emergency situation.
  - Who pays the medical bills?
  - Disability for those involved both on your team and otherwise?
- A family brings a legal action for a team member killed in a vehicular accidents going to the disaster scene.
  - Can the family sue?

- A team member fails to stop at an intersection on the way to the scene and kills the other driver.
  - Can the family of the other driver sue?
  - Who would they name in the action?
- The brakes on the team member's official vehicle fail and the vehicle injures a number of children when it crashes into a pizza parlor.
  - Can the children sue or will their parents sue on their behalf?
  - Who would they sue?

Employment

- A qualified 42-year-old employment candidate applies for a job with your team and is not selected.
  - Was the candidate discriminated against due to her age?
  - Is the candidate protected under the Age Discrimination in Employment Act?
- You team adopts a "no beard" policy for safety reasons.
  - Did your team discriminate against your bearded team members?
  - Can the wearing of a beard be for religious reasons?
  - Does the wearing of a beard create a safety hazard?
- A team member is not promoted due to statements he or she has made in public with regard to a controversial animal rights issue.
  - Was the team member discriminated against due to his or her public statements or stance?
  - Are the team member's public statements considered free speech under the U.S. Constitution?

Discrimination

- A male team member continually asks a female member of the team out on a date, which she refuses.
  - Is the male team member harassing the female team member?
  - Is this actionable under Title VII of the Civil Rights Act?
  - Would the action be brought against the team member or the team/department?
- The team leader promotes a white male team member with 5 years experience over a Hispanic team member with 10 years experience.
  - Did the team leader discriminate against the Hispanic team member?
  - Is this actionable under Title VII?
- A candidate for your team is in a wheelchair due to a permanent injury. The team decides that the candidate was not the most qualified for the position.
  - Did the team discriminate against the candidate under the Americans with Disabilities Act (ADA)?

Training

- One of your team members has a heart attack and dies while training.
  - Are death benefits paid under workers' compensation?
  - Can the family bring a legal action?

- A member of the team has a cold and refuses to participate in the training exercise. The team leader dismisses the team member from the team.
  - Can the team member bring a legal action against the team?
- A team member drops a pipe on the foot of a fellow team member during a training exercise.
  - Can the injured team member sue the team member who dropped the pipe?

The previous examples are but a few situations where an attorney can be of assistance in providing guidance to the team. But these are only a few of the situations that could arise as a result of the myriad federal, state, and local laws. And these are on the civil side of the aisle. There is an entirely different set of federal, state, and local criminal laws that can impact the team or individual team members.[*]

CONCLUSION

Emergency planning requires collaboration from a variety of individuals and organizations. The benefits of collaboration far outweigh the difficulties that will be faced during the planning process. Some of the benefits include elimination of duplication of services, expanded resource availability, and enhanced problem solving.

The planning process can be made easier by planning ahead, providing information about team expectations, and allowing flexibility in scheduling. It may be to your benefit to talk to emergency managers from other communities to gain their input on the planning process.

Successful team operation requires a commitment to participate in shared decision making, a willingness to share information, and a professional sense of respect for individual team members.

Team formation usually takes place through five stages: forming, storming, norming, performing, and adjourning. You will know that you have an effective planning team when members agree on and work toward a common goal, provide open communication, and display constructive conflict resolution, mutual trust, and respect for one another.

The end product of emergency planning is the Emergency Operations Plan (EOP), a document describing how citizens and property will be protected in a disaster or emergency. There are four steps in the emergency planning process:

1. The hazard analysis, or the process by which hazards in the community are identified and ranked according to the risks that they pose to the community
2. EOP development, including the basic plan, functional annexes, hazard-specific appendices, and implementing instructions
3. Testing the plan through training and exercises to determine weaknesses and strengths in the plan
4. Plan maintenance and revision, based on needs and resources that may have changed since the development of the original EOP

---

[*] This section on legal aspects was taken from a lecture from Professor Thomas D. Schneid, JD, Eastern Kentucky University, 2015.

Thoroughly completing each of these steps will help you develop an EOP that requires fewer changes and less significant changes following training and exercising (see Chapters 8 and 9 for more about training and exercising).

Last, when considering who should be on the emergency and disaster preparedness team, remember that a professional with specialized legal training can be a beneficial member. Lawyers do bring a lot of experience and expertise to the table. Every emergency and disaster preparedness team will encounter legal questions and situations where the appropriate legal advice can save time, money, and even lives.

REFERENCE

Federal Emergency Management Agency. Plan components. *Emergency Management Guide for Business & Industry*, http://www.fema.gov/media-library/assets/documents/3412

8

Exercises
Testing Your Plans and Capabilities in a Controlled Environment

James A. McGee

Contents

Introduction	132
Importance of Testing Plans and Capabilities	132
Establishing a Foundation to Exercise Plans	133
Design and Development of Exercises	134
Identify Key Personnel to Be Involved in the Exercise Process	136
Exercise Conduct	136
Design and Develop an Exercise to Include an After Action Report	141
Evaluation and Improvement Planning	142
Active Shooter Threat	143
Preface	144
Handling Instructions	144
Introduction	145
General Instructions	145
Exercise Structure	146
Exercise Objectives	147
Purpose	147
Scope	147
Participants	148
Exercise Guidelines	148
Module 1: Warning (Credible Threat)	148

Module 2: Notification and Initial Response 151
Module 3: Continued Response/Evacuation and Recovery 153
Acronyms 155
References 156

INTRODUCTION

Emergency managers should conduct exercises to test plans and capabilities while promoting awareness of various roles and responsibilities during an incident. This chapter introduces students to the value of exercises for crisis planning and emergency response. The benefits and types of exercises available for emergency responders are discussed. The knowledge of material presented in this chapter is a critical prerequisite for subsequent discussions regarding personal protection and safety, strategies, and tactics for response to a critical incident.

The financial impact and potential loss of life from disasters are significant. Many industry governing bodies have mandated preparedness training and exercising requirements. In the United States, nuclear power plants must exercise their plans annually and conduct a full-scale exercise (FSE) every 2 years, which is subsequently evaluated by the Nuclear Regulatory Commission (NRC). Airports, hospitals, and health care facilities must conduct an FSE every 2 years to maintain a license to operate. Additionally, the Occupational Safety and Health Administration (OSHA) requires many employers to develop an emergency action plan (EAP) and exercise it at least annually (Federal Emergency Management Agency, 2008a).

IMPORTANCE OF TESTING PLANS AND CAPABILITIES

Successfully conducting an exercise involves considerable coordination among participating agencies and officials. The Homeland Security Exercise and Evaluation Program (HSEEP) methodology divides individual exercises into five overarching phases (U.S. Department of Homeland Security, 2007a):

- *Foundation*—The following activities must be accomplished to provide the foundation for an effective exercise: create a base of support from the appropriate entities or senior officials, develop a project management timeline and establish milestones, identify an exercise planning team, and schedule planning conferences.
- *Design and development*—Building on the exercise foundation, the design and development process focuses on identifying objectives, designing the scenario, creating documentation, coordinating logistics, planning exercise conduct, and selecting an evaluation and improvement methodology.
- *Conduct*—After the design and development steps are complete, the exercise takes place. Exercise conduct steps include setup, briefings, facilitation, control, evaluation, and wrap-up activities.
- *Evaluation*—The evaluation phase for exercises includes a formal exercise evaluation, an integrated analysis, and an after action report (AAR) and improvement plan (IP) that identify strengths and areas for improvement in entities

preparedness, as observed during the exercise. Recommendations related to areas for improvement are identified to help develop corrective actions to be tracked throughout the improvement planning phase.
- *Improvement planning*—During improvement planning, the corrective actions identified in the evaluation phase are assigned with due dates to responsible parties, tracked to implementation, and then validated during subsequent exercises.

Establishing a Foundation to Exercise Plans

Before plans and capabilities can be tested through exercises, first responders must have the necessary foundation of training to perform appropriately during a crisis. There are three different levels of training that must be addressed in emergency management:

- *Awareness*—All emergency responders should have awareness training because the more people that are aware of potential threats and response procedures, the more efficient and effective the response will be during an incident.
- *Performance*—This training is for those who will need to execute response plans as well as those who will evaluate the conduct of others during these types of incidents.
- *Planning and management*—Management training is for officials who will have direct management responsibilities in an emergency situation.

Senior management personnel and first responders are responsible for making sure they receive the appropriate levels of training. Designated senior management personnel are responsible for developing management plans and ensuring coordinated response. Senior management must make rapid decisions when implementing an emergency response plan (ERP). They should know the basics of all hazard planning, the Incident Command System (ICS), the National Incident Management System (NIMS), the Unified Command System (UCS), and the relevant ERP. Senior management must understand how they will perform under the IC and UC systems with other jurisdictions and response agencies when reacting to and managing a terrorist event or natural disaster.

During an emergency, first responders will need to perform specific duties. Properly trained emergency response personnel will achieve the most effective response. In regard to training, exercises, and equipment, it is necessary for executive leaders to answer the following questions:

- What do we need to prepare?
- What do we need to prepare for?
- How prepared are we?
- How do we pay for these needs?

First responders who are working under a UCS structure will normally take control of the incident site and play a lead role in safeguarding lives, isolating the site, and notifying the appropriate response agencies. Training in all hazard planning and response is essential for the emergency response team and responding personnel. This information helps to identify what type of event has occurred and the procedures used when responding to specific types of human-made events or natural disasters. These emergency response team members and responding personnel should know what types of personal protective equipment (PPE) are necessary for different situations and understand how to use time, distance, and shielding to minimize exposure. Responding personnel require training in critical thinking, self-protection, proper notification procedures, chain of command, and event documentation.

Exercises are not a one-time event. They must be conducted on a continuous basis to address evolving threats, new plans and procedures, new equipment, training new personnel, and other contingencies (see Figure 8.1).

Design and Development of Exercises

The HSEEP is a capabilities- and performance-based exercise program. The program provides a standardized methodology for exercise design, development, conduct, evaluation, and assessment. Adherence to the policy and guidance presented in the HSEEP ensures that exercise programs conform to the USDHS established best practices. Furthermore, it helps provide consistent and complementary effort for exercises at all levels of government. There are seven types of exercises defined within HSEEP, each of which is either discussion based or operation based (see Table 8.1).

Discussion-based exercises familiarize participants with current plans, policies, agreements, and procedures, or may be used to develop new plans, policies, agreements, and procedures. The various types of discussion-based exercises include:

- *Seminars* use a number of instructional strategies such as lecture, panel discussions, case studies, and multimedia presentations. Seminars are informal *and* productive for small and large groups and are used for orientation to organizational policies and procedures, protocols, response resources, or concepts and ideas. Seminars normally last a maximum of 1–2 hours.

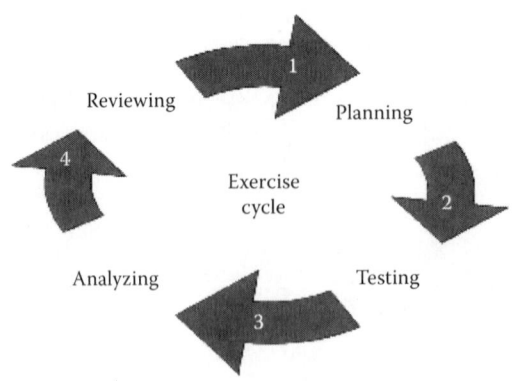

Figure 8.1 The exercise planning process. The model is cyclical, showing the importance of conducting exercises on a continuous basis and evaluating the results of each exercise. (Adapted from Hall, S.A. et al. *Security Management for Sport and Special Events: An Interagency Approach for Creating Safe Facilities*. Champaign, IL: Human Kinetics, 2012.)

Table 8.1 Types of Exercises

Exercise	Example
Discussion-Based Exercise	
Seminar	New personnel attend a lecture and PowerPoint presentation on general security procedures
Workshop	The command group attends a multiagency management and leadership workshop to develop an all-hazards emergency response plan
Tabletop exercise (TTX)	The command group tests the emergency response plan by working through a critical incident scenario
Game simulation	The command group attends a virtual learning laboratory to participate in a decision-making exercise
Operation-Based Exercise	
Drill	Facility managers test the alert and notification system before the day of an event
Functional exercise (FE)	The command group and local first responders work through a critical incident scenario to test available resources and communication capabilities (no deployment of assets)
Full-scale exercise (FSE)	The command group and local first responders work through a critical incident by deploying real-world assets and testing available resources

Source: From James McGee.

- *Workshops* increase participant interaction and are effective for solving complex problems, team building, information sharing, and brainstorming. Workshops differ from seminars in that they emphasize producing a product or goal such as a new policy or plan, mutual aid agreement, and standard operating procedures (SOP). Workshops also involve greater participant discussions and often use breakout sessions to explore parts of an issue with smaller groups.
- *Tabletop exercises* (TTXs) consist of informal facilitated discussions of simulated emergencies among key personnel. Basic TTXs involve a constant, unchanging simulation, whereas advanced TTXs present the group with injects (message updates) that progress the initial scenario. TTXs are a useful tool for first responders who want to assess current plans and identify gaps in security operations. The purpose of a TTX is to test existing plans without incurring costs associated with actually deploying resources. The TTX can involve many people and many organizations who can contribute to the planned discussion topics, typically those entities with a planning, policy, or response role. A TTX usually lasts 1–4 hours. A sample TTX situation manual is provided at the end of this chapter.
- *Games* are provided in a computer simulation of operations that often involve two or more teams, usually in a competitive environment, using rules, data, and procedures designed to depict an actual or assumed real-life situation. Game simulations conduct "what if" analysis of existing plans and potential strategies without actually deploying resources to explore the processes and consequences of decision making.

Operation-based exercises validate plans, policies, agreements, and procedures; clarify roles and responsibilities; and identify resource gaps in an operational environment. The various types of operational-based exercises include:

- Drills are coordinated, supervised activities usually employed to test a specific operation or function within the organization. Participants may gain training on new equipment, practice, and maintain skills. The time required to conduct a drill is usually 30 minutes to 2 hours.
- Functional exercises (FEs) were previously referred to as a command post exercise (CPX). An FE examines and validates the coordination, command, and control between various agencies responding to an incident. An FE involves a notional deployment of resources and personnel in a highly stressful environment requiring rapid problem solving. FEs can be used to evaluate management of emergency operation centers (EOCs), multiagency coordination centers, and command posts. An FE normally requires 3–8 hours to complete. This type of exercise does not involve any "boots on the ground" or real-world deployment of assets.
- FSEs were previously referred to as field training exercises (FTXs). An FSE is a multiagency, multijurisdictional exercise involving a functional response of assets that replicates a real-world response. Real-world deployment of assets occur in support of the exercise scenario. Participants are able to assess plans and evaluate coordinated responses under crisis conditions. An FSE may be designed to last several hours or several days.

The characteristics of discussion-based and operational-based exercises are further defined in Table 8.2.

IDENTIFY KEY PERSONNEL TO BE INVOLVED IN THE EXERCISE PROCESS

Exercise planning team members should be determined based upon the scope and type of exercise as well as the scenario or subject to be tested. Table 8.3 provides recommendations in terms of what agencies and areas of expertise might be involved in exercise planning. The lists should be modified to meet the needs of the jurisdiction or organization.

Exercise Conduct

The type of exercise selected by the entity should be consistent with the entity's multiyear training and exercise plan, which includes

- The entities' training and exercise priorities.
- The capabilities from the target capabilities list (TCL) that the entity will train for and exercise against.
- A multiyear training and exercise schedule that reflects the training activities that will take place prior to an exercise, allowing exercises to serve as a true validation of previous training.

Table 8.2 Characteristics of Exercise Types

Type of Exercise	Utility or Purpose	Type of Player Action	Duration	Real-Time Play	Scope
Discussion based	To familiarize players with current plans, policies, agreements, and procedures; to develop new plans, policies, agreements, and procedures	Notional player actions are imaginary and hypothetical	Rarely exceeds 8 h	No	Varies
Seminar	To provide an overview of new or current plans, resources, strategies, concepts, or ideas	Not applicable	2–5 h	No	Multi- or single agency
Workshop	To achieve a specific goal or build a product (e.g., exercise objectives, SOPs, policies, or plans)	Not applicable	3–8 h	No	Multiagency or multiple functions
Tabletop exercise	To assist senior officials in the ability to understand and assess plans, policies, procedures, and concepts	Notional	4–8 h	No	Multiagency or multiple functions
Game	To explore decision-making processes and examine the consequences of those decisions	Notional	2–5 h	No	Multiagency or multiple functions
Operation based	Test and validate plans, policies, agreements, and procedures; clarify roles and responsibilities; identify resource gaps	Actual player action mimics reaction, response, mobilization, and commitment of personnel and resources	May be hours, days, or weeks depending on purpose, type, and scope	Yes	Varies
Drill	Test a single operation or function	Actual	2–4 h	Yes	Single agency or function

(*Continued*)

Table 8.2 (*Continued*) Characteristics of Exercise Types

Type of Exercise	Utility or Purpose	Type of Player Action	Duration	Real-Time Play	Scope
Functional exercise	Test and evaluate capabilities, functions, plans, and staffs of incident command, unified command and intel centers, or other command or operation centers	Command staff actions are actual; movement of other personnel, equipment, or adversaries is simulated	4–8 h or several days or weeks	Yes	Multiple functional areas or multiple functions
Full-scale exercise	Implement and analyze plans, policies, procedures, and cooperative agreements developed in previous exercises	Actual	One full day or longer	Yes	Multiple agencies or multiple functions

Source: Reprinted from the U.S. Department of Homeland Security, 2007a, Homeland Security Exercise and Evaluation Program (HSEEP), Volume I.

- Reflects all exercises in which the entity participates.
- Employs a "building block approach" in which training and exercise activities gradually escalate in complexity.
- The multiyear training and exercise plan must be updated on an annual basis (or as necessary) to reflect schedule changes.

Exercise objectives should be based on capabilities and their associated critical tasks, which are contained within the exercise evaluation guides (EEGs). For example, if an entity, based on its risk/vulnerability assessment, determines that it is prone to hurricanes, it may want to validate its evacuation capabilities. The scenarios used in exercises must be tailored toward validating the capabilities and should be based on the entity's risk/vulnerability assessment. Exercise planners should develop the following documents, in accordance with HSEEP, to support exercise planning, conduct, evaluation, and improvement planning:

- *For discussion-based exercises:*
  - Situation manual (SitMan) for facilitators
  - Situation manual for participants
- *For operation-based exercises:*
  - Exercise plan (ExPlan)
  - Player handout
  - Master scenario event list (MSEL)
  - Controller/evaluator handbook (C/E handbook)

Table 8.3 HSEEP Recommended Planning Team Members for Exercises

Discussion-Based Exercises	Operation-Based Exercises
Emergency Management	**Emergency Management**
Emergency manager	Emergency manager*
Homeland security	Homeland security*
	Public health*
Public Safety	
Fire	Public works
Hazardous materials (HAZMAT)	Transportation or transit authority
Law enforcement	Public affairs
Emergency medical services (EMS)	Exercise venue management
Special operations/bomb squad	**Fire**
Federal Bureau of Investigation (FBI)	Fire department*
Public Health	Communications or dispatch*
Public health department	Special operations (e.g., HAZMAT, Metropolitan Medical Response System [MMRS])*
Communicable/infectious disease	Mutual aid fire*
Epidemiologists	**Law Enforcement**
Pathology	Police*
Poison control	Special operations (e.g., bomb squad, SWAT)*
	Sheriff's department*
Medical	
Hospital administrators	Federal Bureau of Investigation (FBI)*
Coroner or medical examiner	Mutual aid law enforcement*
Hospital infection control	**Medical**
Hospital lab managers	Hospital representatives (primary trauma center or hospital association)*
Hospital emergency room	Emergency medical services (EMS)*
Private practitioners	Mutual aid
Veterinary	Medical examiner or coroner
Other	**Other**
Public works	Volunteer organizations
Public information officer (PIO)	Subject-matter experts
Volunteer organizations	Private security
Communications or dispatch	Government officials
Government officials	Meteorologist

Source: Reprinted from the U.S. Department of Homeland Security, 2007a, Homeland Security Exercise and Evaluation Program (HSEEP), Volume I.
Note: The agencies marked with asterisks are most critical to be present during all planning conferences.

The exercise planning team determines the timelines for completion of the exercise plan, acquisition of necessary resources, and development of sufficient support. The exercise scope and statement of purpose must be clearly defined. A statement of purpose identifies the issue(s) to be addressed in detail. These issues may have been identified based on a past crisis, observation, or determined during a risk assessment.

Table 8.4 describes the important document types associated with most exercises.

A detailed *exercise scenario* must be developed based on the statement of purpose. Either in narrative format or depicted as an event timeline, the scenario provides a storyline that drives the exercise. For a discussion-based exercise, the scenario provides the backstop that is the basis for participant discussion. For an operation-based exercise, the scenario provides background information regarding the incident and is fueled by periodic injects as the exercise unfolds.

A number of factors should be taken into consideration when developing a scenario, including level of realism, type of threat or hazard, site selection, optimal date and time

Table 8.4 Exercise Document Types

Discussion-Based Exercises	Operation-Based Exercises
Situation manuals (SitMan) are individual facilitator and participant handbooks for discussion-based exercises, particularly TTXs. It provides background information on exercise scope, schedule, and objectives. It also presents the scenario narrative that will drive participant discussions during the exercise.	The exercise plan (ExPlan), typically used for operation-based exercises, provides a synopsis of the exercise and is published and distributed to players and observers prior to the start of the exercise. The ExPlan includes the exercise objectives and scope, safety procedures, and logistical considerations such as exercise schedule. The ExPlan does not contain detailed scenario information.
Multimedia presentation supports the SitMan, concisely summarizing written information. Enhances exercise realism with audio or visual depiction of the scenario. Focuses and drives the exercise.	The controller and evaluator (C/E) handbook supplements the ExPlan for operation-based exercises, containing more detailed information about the exercise scenario and describing exercise controllers' and evaluators' roles and responsibilities. Because the C/E handbook contains information on the scenario and exercise administration, it is distributed only to those individuals specifically designated as controllers or evaluators.
Exercise evaluation guides (EEGs) are necessary for all evaluated exercises. An EEG helps evaluators assess performance of capabilities, tasks, and objectives during an exercise.	The master scenario event list (MSEL) is a chronological timeline of expected actions and scripted events (i.e., injects) to be inserted into the operation-based exercise play by controllers to generate or prompt player activity. It ensures necessary events happen so that all exercise objectives are met.
Player handout	A player handout is a 1- to 2-page document, usually handed out the morning of an exercise, which provides a quick reference for exercise players on safety procedures, logistical considerations, exercise schedule, and other key factors and information.

Source: From James McGee.

for conducting the exercise, and cost. The scenario selected for the exercise should be a realistic representation of potential threats faced by the exercising entities.

Exercises should adhere to the planning timelines laid forth in HSEEP and reflect the principles of NIMS and ICS. A consistent terminology and methodology for exercises is critical to avoid confusion and to ensure that entities can exercise together seamlessly.

The list below describes the important key personnel involved with conducting the exercise:

- *Planners*—Design, develop, conduct, and evaluate exercise; determine exercise objectives, create scenarios, and develops documentation; develop and distribute pre-exercise materials; and conduct exercise briefings and training sessions.
- *Players*—Participating agency personnel who discuss their roles and responses to the scenario during the exercise; drawn from participating agencies to accomplish exercise objectives; and familiar with the agency SOP and emergency operation plans (EOPs) being tested.
- *Facilitators*—Facilitate discussion and coordinate issues between groups; focuses the group's discussions on specific areas and questions; elicit resolutions to issues; monitor the recorder and prepare notes on the group's discussions; comfortable talking in front of large groups of people; and knowledgeable on plans and policies.
- *Evaluators*—Observe and record player discussions; do not interfere with exercise play; chosen for their knowledge of a particular functional area; familiar with the jurisdictions SOPs and EOPs; and provide input to the AAR.
- *Recorders*—Record discussions during breakout sessions; should have working knowledge of commonly used terms and acronyms in the area they are recording; have no input into the exercise process; and require no formal training.
- *Subject-matter experts (SMEs)*—Add functional knowledge and expertise to the exercise planning team; help make the scenario realistic and plausible; and ensure jurisdictions have appropriate capabilities to respond.
- *Observers/VIPs*—Do not have an active role in exercise play; should be allowed to see or hear appropriate aspects of exercise play; may be senior or elected officials, neighboring jurisdictions or media representatives; observe the exercise from a designated area; require no formal training; and may attend an observer briefing.

DESIGN AND DEVELOP AN EXERCISE TO INCLUDE AN AFTER ACTION REPORT

Before a discussion-based exercise begins, the planning team must deliver the necessary exercise materials and equipment to the exercise location (U.S. Department of Homeland Security, 2007b):

- Exercise manuals (SitMans) for facilitators and participants
- Multimedia presentation (PowerPoint)
- AV equipment (including televisions, projectors, projection screens, microphones, and speakers)

- Table tents (for agency names)
- Name tents (for participants)
- Badges identifying the role of each exercise participant (e.g., player, observer, VIP, facilitator, and evaluator)
- Sign-in sheets and registration information
- Participant feedback forms

The following checklist contains all of the tasks that need to be completed for each discussion-based exercise (seminar, workshop, TTX, and game simulation). This checklist should be customized to each exercise, identifying any missing activities, and removing any redundant activities.

The setup for an operation-based exercise begins as many days before the event as necessary, depending on the scope of the scenario. The setup entails arranging briefing rooms, and testing AV and communications equipment, placing props and effects to add realism to the incident, marking the appropriate areas and their perimeters, and checking for potential safety issues. Safety is the most important consideration in planning an operation-based exercise. The following actions must take place to ensure a safe environment (U.S. Department of Homeland Security, 2007b, p. 24):

- Identify safety controllers (not to be confused with a safety officer designated by the incident commander as part of the response to the exercise scenario).
- Dedicate advanced life support or basic life support ambulance units for real-world emergencies only.
- Identify real-world emergency procedures with a code word or phrase.
- Identify safety requirements and policies.
- Consider other safety issues outside the scope of exercise control (e.g., weather, heat stress, hypothermia, fire or pyrotechnics, weapons, and traffic accidents).

The following checklist contains all of the tasks that need to be completed for each operation-based exercise (drill, FE, FSE). This checklist should be customized to each exercise, identifying any missing activities and removing any redundant activities.

Evaluation and Improvement Planning

EEGs help evaluators collect and interpret relevant exercise observations. EEGs provide evaluators with information on what tasks they should expect to see accomplished during an exercise, space to record observations, and questions to address after the exercise as a first step in the analysis process. To assist entities in exercise evaluation, standardized EEGs have been created that reflect capabilities-based planning tools, such as the TCL and the universal task list (UTL). The EEGs are not meant as report cards. Rather, they are intended to guide an evaluator's observations so that the evaluator focuses on capabilities and tasks relevant to exercise objectives to support the development of the AAR/IP.

AAR/IP is the final product of an exercise. The AAR/IP has two components: an AAR, which captures observations and recommendations based on the exercise objectives as associated with the capabilities and tasks, and an IP, which identifies specific corrective actions, assigns them to responsible parties, and establishes targets for their completion. The lead evaluator and the exercise planning team draft the AAR and submit it to conference participants prior to an after action conference (AAC). The draft AAR is distributed to conference participants for review no more than 30 days after the exercise is conducted. The final AAR/IP is an outcome of the AAC and should be disseminated to participants no more than 60 days after the end of the exercise (ENDEX).

The HSEEP methodology defines a variety of planning and AACs. The need for each of these conferences varies depending on the type and scope of the exercise. They include:

- Concepts and objectives meeting
- IPC
- Midterm planning conference (MPC)
- MSEL conference
- Final planning conference (FPC)
- AAC

Following every exercise, an AAC must be conducted in which key personnel and the exercise planning teams are presented with findings and recommendations from the draft AAR/IP. Following each exercise, a draft AAR/IP must be developed based on information gathered through the use of EEGs. AAR/IPs created from the exercise must conform to the templates provided by HSEEP. Corrective actions addressing a draft AAR/IP's recommendations are developed and assigned to responsible parties with due dates for completion. A final AAR/IP with recommendations and corrective actions derived from the discussion at the AAC must be completed within 60 days after ENDEX.

IP include broad recommendations from the AAR/IP organized by target capability as defined in the TCL. Corrective actions derived from an AAC are associated with the recommendations and must be linked to a capability element as defined in the TCL. Corrective actions included in the IP must be measurable and designate a projected start date and completion date. Corrective actions included in the IP must be assigned to an organization and a point of contact (POC) within that organization. Corrective actions must be continually monitored and reviewed as part of an organizational corrective action program. An individual should be assigned the responsibility of managing a corrective action program to ensure corrective actions resulting from exercises, policy discussions, and real-world events are resolved and support the scheduling and development of subsequent training and exercises.

Lessons learned:

1. There are three different levels of training that must be addressed in emergency management.
2. Types of discussion-based exercises.
3. Types of operational-based exercises.
4. The important key personnel involved with an exercise process.
5. A consistent terminology and methodology for exercises are critical to avoid confusion and to ensure that entities can exercise together seamlessly.

ACTIVE SHOOTER THREAT

An active shooter is defined by the U.S. Department of Homeland Security as an individual actively engaged in killing or attempting to kill people in a confined and populated area. Historically, active shooters use firearms and there is no pattern or method to their selection of victims. The shootings at Sandy Hook Elementary School in Newton, Connecticut, can be added to the increasing list of locations attacked by an active shooter. Within the last 5 years, there have been at least 15 prominent, high-casualty-producing active shooter incidents. Most of these cases have occurred in locations where the shooter has been undeterred and unobstructed from carrying out their attack. The crime scenes have been described as soft targets with limited active security measures or armed personnel available to provide protection for members of the public. In most instances, shooters have taken their own lives, been shot by police, or surrendered when forced into a confrontation with law enforcement.

A TTX facilitator's guide that includes a generic active shooter scenario is provided below. The scenario is a college campus but can be adjusted to address any location. Each TTX module concludes with questions directed to the campus and emergency responder community.

Situation Manual (SitMan)
Tabletop Exercise Facilitator's Guide
SCENARIO (Active Shooter Threat)
Place Date Here

Preface

What follows is a capabilities- and performance-based exercise experience. The TTX is based upon a standardized methodology for exercise design, development, conduct, evaluation, and assessment. The format is in adherence to the policy and guidance presented in the U.S. Department of Homeland Security Exercise and Evaluation Program (DHS-HSEEP). HSEEP ensures that exercise programs provide consistent and complementary effort for exercises at all levels of government. This TTX is designed in accordance with the DHS-HSEEP exercise methodology.

To mitigate gaps and deficiencies in preparedness for incidents, this TTX is developed to assist in training and exercising emergency preparedness for critical incidents. The TTX experience is intended to continuously develop and enhance preparedness for incident avoidance, management, response, and recovery.

This is an unclassified exercise. The control of the information is based more on public sensitivity regarding the nature of the exercise than on the actual exercise content. Some exercise materials are intended for the exclusive use of exercise planners, facilitators, and evaluators. Players may view other materials deemed necessary to their performance. All exercise participants may view the SitMan.

All exercise participants should use appropriate guidelines to ensure the proper control of information within their areas of expertise and to protect this material in accordance with current jurisdictional directives.

Handling Instructions

1. The title of this document is Situation Manual (SitMan) Table Top Exercise SCENARIO (Active Shooter Threat).

2. The information gathered in the SitMan is *for official use only (FOUO)* and should be handled as sensitive information not to be disclosed. This document should be safeguarded, handled, transmitted, and stored in accordance with appropriate security directives. *Reproduction of this document, in whole or in part, without prior approval is prohibited.*
3. At a minimum, the attached materials will be disseminated only on a need-to-know basis and when unattended, will be stored in a secure container or area offering sufficient protection against theft, compromise, inadvertent access, and unauthorized disclosure.
4. For more information, please contact the designated exercise planner.

Introduction

This facilitator manual is designed as a guidance for designated exercise controllers to facilitate this TTX. It contains general instructions to the facilitator on the overall exercise process, necessary materials, and discussion questions. Detailed notes for the facilitator's consideration are shown in bold and italicized font. A sample exercise schedule is shown in Figure 8.2.

General Instructions

This TTX begins with a PowerPoint presentation as it outlines the content of the participant manual. The presentation will detail, in the following sequence, the rules, objectives, and scenario included in this TTX. Please note that although the scenario presented is fictitious, it realistically represents a probable event affecting a campus environment.

Sample Exercise Schedule	
Facilitator Instruction: Show "Exercise Schedule" slide. Go over the exercise schedule with participants.	
8:30 am	Participant sign-in
9:00 am	Introductions
	Discuss general instructions and ground rules of the exercise
9:15 am	Exercise overview
	Discuss exercise objectives, and schedule of exercise
9:30 am	Read Module 1
9:45 am	Module 1 Discussion
10:00 am	Read Module 2
10:15 am	Module 2 Discussion
10:30 am	Read Module 3
10:45 am	Module 3 Discussion
11:00 am	After action hot wash/final comments
11:30 am	End exercise

Figure 8.2 Sample exercise schedule. (From James McGee.)

Players are strongly encouraged to participate in in-depth discussions as the primary purpose of the exercise is to evaluate and improve skills, knowledge, and ERPs. It is important for players to keep the exercise objectives in mind as all issues raised by the scenario will be thoroughly discussed.

Exercise Structure

Players will participate in the following three distinct modules:

Module 1: Warning (Credible Threat)
Module 2: Notification and Initial Response
Module 3: Continued Response/Evacuation and Recovery

Each module begins with an update summarizing the key events occurring within the time period. Following the updates, participants review the situation and engage in functional group discussions of appropriate response issues. Participants then enter into a facilitated caucus discussion in which they present their group's actions based on the scenario.

Although they are encouraged to move among tables to ensure thorough and thought-provoking discussion, participants will be divided into functional groups to discuss aspects of the situation as presented. Exchanges among functional groups will be necessary to coordinate actions and decisions. The functional groups will be determined based upon the participating agencies and their areas of expertise.

Following the functional group discussions, participants then hold a functional caucus discussion, in which a spokesperson from each group presents a synopsis of the group's actions based on the scenario.

Each exercise participant will receive a copy of the SitMan, which provides a written scenario and situation updates. Following each module is a series of questions highlighting pertinent issues for consideration. These questions are supplied as a catalyst for the group discussions. Participants are not required to answer every question, nor are they limited to those topics. Participants are encouraged to use the SitMan as a reference throughout the exercise.

Following each module, players will have a set time period to review the module and discuss the suggested issues. During this exercise, the following rules apply:

1. This TTX is conducted in an artificial environment where time compression is necessary to examine and resolve issues. Some aspects in terms of resources and response will be notionalized.
2. The scenario represents a plausible critical incident.
3. There are no trick questions or "hidden agendas" associated with this TTX.
4. Players have no previous knowledge of the scenario and will receive all information at the same time.
5. Players will respond using existing plans, procedures, and other response resources.
6. Decisions are not precedent setting and may not reflect your organization's final position on a given issue.

Note to facilitator: Before showing the slide on "Exercise Rules," brief the group on emergency exits, bathroom locations, and other relevant housekeeping items.

Facilitator instructions: Show "Exercise Objectives" slide. Read the narrative as written below.

Exercise Objectives

Exercise design objectives are focused on improving the understanding of a response concept, identifying opportunities or problems, and/or achieving a change in attitude. This exercise will focus on using the four phases of emergency management (prevention–mitigation, preparedness, response, and recovery) as a foundation to the following design objectives selected by the exercise planning team.

- *Intelligence/information gathering and dissemination*: Discuss plans, policies, and procedures for ensuring the proper gathering, analyzing, sharing, and dissemination of incident-related information during all stages of a critical incident.
- *ICS/unified command*: Responders will demonstrate the ability to implement a functional ICS, transition to unified command and effectively direct, coordinate, and manage a response to a critical incident. Responders will activate their respective ERP and all relevant annexes (evacuation plan, HAZMAT/WMD response plan, etc.). Responders will establish an incident command post (ICP) and EOC in a timely matter after the initial call for services. Responders will designate/recognize a lead agency on scene commander (OSC) for crisis response/mitigation.
- *Communications*: Understand communication channels and procedures to conduct incident management activities. Determine strengths and weaknesses in the communication of response activities. Identify critical issues and potential solutions. Identify and activate a primary and alternate communication system.
- *Threat assessment*: Assess existing hazard prevention measures by addressing threats.
- *Preparedness*: Assess preparedness, such as maintaining sufficient supplies and providing training to staff in prevention–mitigation, preparedness, response, and recovery, in anticipation of a critical incident.
- *Recovery*: Assess the ability to recover from a critical incident and restore a safe environment.

Note to facilitator:

1. These objectives should be displayed on the screen throughout the duration of the exercise if multimedia presentation capability allows.
2. Provide participants a few minutes and/or review with them the appendices and inform them of the tools they may want to reference during the scenario.

Purpose

The purpose of this exercise is to provide participants with an opportunity to evaluate current response concepts, plans, and capabilities for response to a critical incident. The exercise will focus on key emergency responder coordination, critical decision making, and the integration of assets necessary to save lives and investigate the incident.

Scope

The exercise is a 3-hour interactive exercise consisting of three modules, each portraying a milestone responding to a critical incident involving an active shooter.

This TTX will focus on the role of various agencies in response to the consequences of the critical incident. Emphasis is on decision-making emergency response processes, coordination, integration of capabilities, problem identification, and resolution.

Participants

- *Players*—Players respond to the situation presented based on expert knowledge of response procedures, current plans, and procedures in place in their agency, and insights derived from training.
- *Observers*—Observers support the group in developing responses to the situation during the discussion. However, they are not participants during the moderated discussion period.
- *Facilitators*—Facilitators provide situation updates and moderate discussions. They also provide additional information or resolve questions as required. Key planning committee members may also assist with facilitation as SMEs during the TTX.
- *Evaluators*—Evaluators observe and record the discussions during the exercise, participate in the data analysis, and assist with drafting the AAR.
- *SMEs*—The SMEs are similar to observers but may be asked specific questions about their agencies, certain policies, or areas of expertise.

Exercise Guidelines

This is an open, low-stress, no-fault environment. Varying viewpoints, even disagreements, are expected.

- Respond based on your knowledge of current plans and capabilities (i.e., you may use only existing assets) and insights derived from training.
- Decisions are not precedent setting and may not reflect your organization's final position on a given issue. This is an opportunity to discuss and present multiple options and possible solutions.
- Assume cooperation and support from other responders and agencies.
- Issue identification is not as valuable as suggestions and recommended actions that could improve response and preparedness efforts. Problem solving should be the focus.
- The situation updates, written material, and resources are the basis for discussion. There are no situational injects.
- Do not read ahead.
- List options if no plan exists.

Module 1: Warning (Credible Threat)
The scenario will present an overview of actions that may be necessary to handle an all-hazards campus emergency. It is critical for participants to understand why it is essential to plan, train, coordinate, exercise, and integrate personnel and operations for an effective all-hazards response. This multifaceted, integrated, all-hazards response will enhance

a community's ability to effectively prevent, prepare for, respond to, and recover from natural and man-made events.

One important component of a campus all-hazards response is the ability of the jurisdiction and the campus to work together. Planning, training, exercising, building relationships, and integrating operations will promote a successful management of an all-hazards campus event. These five factors, established in advance, will assist the campus community in overcoming difficult challenges. The process is continuous and constantly evolving. It will need refinement as the campus and community grow and change. Campus jurisdictions must understand that when a crisis occurs it is the established process that will enable them to successfully manage the incident. University officials and community leaders may change; however, an established process ingrained and exercised will continually assist campus communities in preventing, preparing for, responding to, and recovering from all-hazards incidents.

Friday: Game Day preparations are underway to host a Friday evening home football game. Kick-off is scheduled for 7:00 pm.

07:00 am

Neighbors hear what appears to be another loud argument coming from off-campus apartment 102 in the university apartment complex. The residents of the apartment are Jane and Tom Smith. Both are PhD candidates and employees of the local university. Neighbors call the local police to report the argument. The dispatcher advises it will be approximately 25 min before a unit can respond due to a high number of concurrent calls requiring police assistance/response.

07:29 am

A local police patrol unit arrives at the Smith apartment. Jane Smith answers the door. She is visibly upset and when questioned states that her husband and she had been arguing. She states that her husband left the premises and went to see their marriage counselor. When asked if there are any weapons in the residence, Jane Smith tells the police that her husband is a hunter and keeps a shotgun in the bedroom. Jane is provided with contact information regarding domestic violence and the police officers depart. After the police depart, Jane notices that the shotgun, normally stored in the corner of the closet, is not there.

08:10 am

Tom Smith is successful in getting an emergency appointment with his counselor. Tom vents to the counselor about his frustration with his marriage. He becomes increasingly angry as he recounts the morning argument. He talks about divorce and ultimately states, "I should end this whole mess." The counselor, who is aware that divorce has been discussed during past sessions, assumes Tom intends to file for a divorce. Tom eventually settles down and the session ends.

09:30 am

Jane Smith departs her apartment and goes to the university to attend an employee meeting at 10:00 am. Activities on the campus are preoccupied with preparations for the home football game scheduled for that evening. The weather is clear and hot. Numerous tailgaters are congregated in various designated parking areas. Other Game Day preparations are in the works to include those associated with the pregame festivities that occur adjacent to the stadium.

10:30 am
Tom Smith departs his counselor's office and travels to the university. He intends on meeting his wife to have some dialogue. When he arrives at her office, she is not there. One of her coworkers advises she left to attend a meeting with HR regarding possible changes in her employee benefits package. The coworker states that after the meeting she was going to the student union for lunch. Tom becomes visibly upset and leaves the office. He returns to his vehicle and retrieves a jacket and his 12-gauge shotgun. He places a box of ammunition in his jacket pocket. The weapon is already loaded with five rounds of buckshot. Tom then makes his way across campus to the student union on foot. As he walks, his anger intensifies. He had no idea she was planning on changing her benefits.

11:35 am
Jane has begun her lunch break with two friends from her department. They are all sitting at a centrally located table in the student union. As they begin their lunch, one of the friends notices Tom approaching their table from behind where Jane is sitting. Jane turns to greet her husband. Tom calmly continues to walk toward the table. As Tom nears the table, he reaches beneath his jacket and draws the shotgun. He then fires one round into his wife's chest. The two friends immediately dive under the table. Pandemonium erupts in the student union and Tom turns and runs out of the area.

11:37 am
Across the commons area, campus police officer Adams hears what sounds like gunfire and then notices a man running from the student union carrying a long weapon. Officer Adams calls for the man to stop. Tom stops and turns toward the officer and fires a round in the officer's direction.

Key Issues
- Game Day preparations are underway on campus for a football game that evening.
- Police respond to a domestic disturbance from people who work/attend the university.
- The disgruntled husband (Tom Smith) gets an emergency appointment with his counselor.
- Smith arrives on campus at his wife's office. He has a firearm in his vehicle.
- After retrieving the firearm from his vehicle, Smith locates his wife at the student union. He shoots her and then flees the scene.
- Smith is confronted by campus police and fires a round in their direction.

Questions
Based on the information provided, participate in the discussion concerning the issues raised in Module 1. Identify any additional requirements, critical issues, decisions, and/or questions, which should be addressed at this time.

The following questions are provided as suggested general subjects you may wish to address as the discussion progresses. These questions are not meant to constitute a definitive list of concerns to be addressed, nor is there a requirement to address every question.

University Critical Incident Response Team
1. Would coordination with surrounding agencies and jurisdictions be done at this time?
2. Would the campus police have been made aware of the earlier domestic dispute?
3. How would the university president be notified of the threat? How would the university critical incident response team be notified? Would other universities in the state and surrounding area be notified?
4. Would an evacuation of the campus be ordered at this time?
5. Would this impact Game Day preparations? In what way?
6. What interagency coordination is necessary at this point?
7. Does your university ERPs address an active shooter threat? What actions would you take at your level? What other plans do you currently have that would need to be reviewed for potential implementation?
8. What factors would support a decision to alert and/or preposition selected emergency assets based on available information? How many and what assets would be included? Would you request external assistance? What type of support would you request and from whom?
9. Is your current level of emergency response training adequate? What public affairs guidance will be provided to your personnel?
10. Would an activation of the campus EOC be directed?
11. What potential response issues should be addressed at this point (e.g., communications interoperability)?
12. Are university officials aware if the local community, city, and county have resources to respond appropriately?
13. Would the medical community be alerted to take any preliminary action? Are medical facilities adequate to handle mass casualties? Is there a local trauma center? What provisions must be made to accomplish the task?
14. Will you need to ensure that mutual-aid support can be obtained if and when necessary? What must be done to ensure this response?
15. Would you request assistance? What would you ask for and from whom?
16. What security issues will arise?

Module 2: Notification and Initial Response
Friday, 11:37 am–11:39 am
Officer Adams immediately responds in the direction of the gunfire. While responding, Officer Adams calls, via radio, that shots have been fired at the student union.

Tom Smith, having shot his wife, begins to panic. He sees the campus police officer and fires another round in the officer's direction. He begins to run and a student attempts to obstruct him. Tom fires a round at the student and the student immediately falls to the ground.

Tom makes his way toward a multistory on-campus residence hall. He enters the residence hall lobby.

Friday, 11:41 am–11:50 am
Tom enters the residence hall and confronts several students. He immediately orders them to get out of the area and states that he is going to "blow himself up."

Officer Adams sees the subject enter the residence hall. The students who were confronted by Tom inform Officer Adams of the threat made by Tom to "blow himself up." Officer Adams passes this information over the radio as well as to other officers who have arrived on scene to assist. Other Officers stop to render assistance to the down student.

Friday, 11:50 am–12:15 pm
One of the individuals leaving the residence hall informs a campus police officer that a man, armed with a rifle, was seen running into the dormitory administration area, adjacent to the lobby. The student was unsure if other people were still in that area.

Tom finds the residence hall administration office empty. He turns the shotgun on himself and fires. He dies of a self-inflicted gunshot wound to the head.

Key Issues
Officer Adams notifies dispatch, via the radio, that shots have been fired on campus.

- Officer Adams attempts to confront the armed subject/shooter.
- The subject, Tom, fires his weapon toward Officer Adams and then fires another round and shoots a student.
- Tom enters an on campus residence hall and states that he is going to "blow himself up."
- Campus police respond to the residence hall and are advised of a possible location of the subject.
- Tom turns the shotgun on himself and fires. He dies of a self-inflicted gunshot wound to the head.

Questions
Based on the information provided, participate in the discussion concerning the issues raised in Module 2. Identify any additional requirements, critical issues, decisions, and/or questions, which should be addressed at this time.

The following questions are provided as suggested general subjects you may wish to address as the discussion progresses. These questions are not meant to constitute a definitive list of concerns to be addressed, nor is there a requirement to address every question.

University Critical Incident Response Team
1. Who is in charge based on the current circumstances?
2. What resources do you anticipate will be needed from various local agencies, including Public Works, Parks and Recreation, Transportation Authority, Public Water, community service groups, and other mutual-aid or private resources? How will these resources be integrated into the response? Will there be difficulties getting resources from other municipalities?

3. Would local elected officials be notified? Who would notify these officials and how much information are they provided?
4. Would an evacuation be ordered at this time?
5. What are potential limitations in your emergency response capabilities? Will any limitations be placed on the actions of campus law enforcement personnel? What are your alternatives?
6. What capabilities do you have to conduct bomb threat and rescue operations?
7. How will responding organizations coordinate?
8. What level of response would you activate? Where would additional personnel be staged? How would transportation issues be handled?
9. What security concerns would you have as a result of this incident regarding Game Day preparations?
10. What dispatch protocols are in place for this type of incident? What mutual-aid resources could be activated?
11. Do you have a callback plan to meet needs such as this? Do you have communications interoperability capabilities with other agencies possibly involved in this incident?
12. What actions would be taken for traffic and access control in the affected area? Who would perform the functions? How would you notify the public/students?
13. How will information be exchanged to support decision making?
14. What specialized federal, state, mutual-aid, or military support do you anticipate possibly needing?
15. Are communication systems adequate if commercial and cellular systems experience overload? Does a backup communications plan exist?
16. Would coordination with surrounding communities be done at this time? Would any consideration be made for EOC activation?
17. How are you communicating with medical treatment centers or hospitals?
18. What actions would be taken to protect the public at this point? Who would perform the functions? How would you notify the public?
19. What is the closest medical facility to the incident site? Is the facility adequate?
20. What is the protocol for a shooting on campus? Would officers immediately enter the residence hall in pursuit of the shooter?

Module 3: Continued Response/Evacuation and Recovery
Recovery/Remediation
The following priorities for the post-incident phase of the crisis are identified:

- Mitigating further incidents
- Treatment and care of casualties
- Public information
- Individual and family assistance
- Site restoration
- Volunteer management
- Critical incident stress management (CISM)
- Business resumption and recovery

Key Issues
- *Prevention/deterrence/protection*: The planning and execution of this event would require significant interagency coordination.
- *Emergency assessment/diagnosis*: Actions require the delivery and distribution of vital supplies and resources.
- *Emergency management response*: Actions required include search and rescue, alerts, activation and notification, traffic and access control, protection of special populations, resource support, requests for assistance, and public information/media. Establishment of an IPC and EOC is necessary.
- *Incident/hazard mitigation*: Primary hazards include possible additional subjects, secondary threats/devices, crowd control, and traffic control.
- *Public protection*: Evacuation is required as well as additional threat assessments. The areas must be secured and cordoned.
- *Victim care*: Injuries range from gunshot victims to mental/physical trauma.
- *Recovery/remediation*: Must be coordinated with search and recovery efforts. Restoration of public confidence could take months.

Questions
Based on the information provided, participate in the discussion concerning the issues raised in Module 3. Identify any additional requirements, critical issues, decisions, and/or questions, which should be addressed at this time.

The following questions are provided as suggested general subjects you may wish to address as the discussion progresses. These questions are not meant to constitute a definitive list of concerns to be addressed, nor is there a requirement to address every question.

University Critical Incident Response Team
1. Who is in charge of the incident at this point?
2. Would the university president be kept informed about the situation? Where would this occur? How will personal property of victims be treated?
3. What protocols will be used for contact with victims' families?
4. What local resources are available for recovery?
5. What actions are necessary to restore preincident capabilities? Who will fund these measures? What process will be followed?
6. What critical incident stress support would you consider? How long would you need to use these assets? Are sufficient assets locally available?
7. Are current record-keeping requirements adequate for an event of this magnitude during the short term and the long term?
8. Is there a mechanism for updating plans, policies, and procedures as a result of this incident? Who is responsible for coordinating these changes?
9. What contingencies are in place for adapting to a response plagued by communication overload? What is your media strategy at this time? What plans are in place to establish a joint information center (JIC)? How would they be implemented?
10. What advisories would be issued to preclude widespread panic? Who would coordinate and issue these advisories?

11. What actions are necessary to restore preincident capabilities? How would you expect supplies to be replenished? Who will fund these measures?
12. Is there a plan for demobilization and release of assets? How would public inquiries be managed during the long term?
13. Is there a plan or policy in place to reconstitute supplies used in the incident?
14. Is post-incident counseling available for victims and emergency responders?
15. What are ten immediate recovery issues that need to be addressed?
    - Housing
    - Transportation
    - Parking
    - Classroom space
    - Law enforcement
    - Administrative space
    - Media relations
    - Financial
    - Registration
    - Educational

ACRONYMS

AAC	After action conference
AAR	After action report
AAR/IP	After action report/improvement plan
CPX	Command post exercise
C/E handbook	Controller/evaluator handbook
EEG	Exercise evaluation guide
ENDEX	End exercise
EOP	Emergency operation plan
ExPlan	Exercise plan
FE	Functional exercise
FPC	Final planning conference
FSE	Full-scale exercise
FTX	Field training exercise
HSEEP	Homeland Security Exercise and Evaluation Program
ICS	Incident Command System
IPC	Initial planning conference
MPC	Midterm planning conference
MSEL	Master scenario event list
NIMS	National Incident Management System
POC	Point of contact
PPE	Personal protective equipment
SitMan	Situation manual
SME	Subject-matter expert
SOP	Standard operating procedure

TCL	Target capabilities list
TTX	Tabletop exercise
UCS	Unified Command System
UTL	Universal task list
VIP	Very important person

REFERENCES

Active Shooter Booklet, U.S. Department of Homeland Security Active Shooter Response (October 2008).

Federal Emergency Management Agency, Emergency Management Institute. 2008a. Introduction to exercise design. In *Exercise Design: IS-139*.

Hall, S. A., Cooper, W. E., Marciani, L., and McGee, J. A. 2012. *Security Management for Sport and Special Events—An Interagency Approach to Creating Safe Facilities*. Champaign, IL: Human Kinetics.

U.S. Department of Homeland Security. 2007a. *Homeland Security Exercise and Evaluation Program (HSEEP)*, Volume I.

U.S. Department of Homeland Security. 2007b. *Homeland Security Exercise and Evaluation Program (HSEEP)*, Volume II.

U.S. Department of Homeland Security. *Homeland Security Planning for Campus Executives Participant Guide*. http://www.hsp.wvu.edu/r/download/27354. (Accessed August 7, 2012.)

U.S. Department of Homeland Security. *National Incident Management System* (March 1, 2004).

9
ICS/EOC Interface

Michael J. Fagel

Contents

EOC Management and Operations: Responsibilities	158
Foundations for Establishing the Emergency Operations Center	158
The Emergency Operations Plan	159
The Hazard and Vulnerability Analysis	159
Using the EOP and Hazard Analysis to Design an EOC	159
Developing Policies and Procedures	160
Communications Policies and Procedures	160
Developing Communications Standard Operating Procedures	160
Life Support Policies and Procedures	161
Identify Emergency Operations Center Life Support Requirements	161
Tracking Goods and Services Required for the EOC	161
Operating Equipment and Supplies Policies and Procedures	162
Maintenance Contracts	163
Memorandums of Understanding and Mutual Aid Agreements	164
Records and Document Retention	164
Summary	165

After action reports and studies of catastrophic disasters have identified the need to dispense additional information in emergency operations center (EOC) management. Although terrorist incidents are much less common, the same need holds true for these events as well. Major emergencies resulting from terrorist incidents can spread across multiple jurisdictions and may require a large-scale response. Emergency services must be ready and communication systems must be in place.

    Every community, large or small, urban or rural, will be able to improve its ability to centralize its flow of information during an emergency by establishing an EOC. The key to a community's disaster planning, response, and recovery lies in the EOC. In fact, the EOC is crucial to saving lives and reducing property damage. To ensure that effective coordination takes place during all phases of emergency management, the emergency managers will work closely in a team environment with other EOC personnel, elected officials, and members of private-sector groups.

EOC MANAGEMENT AND OPERATIONS: RESPONSIBILITIES

Many of the emergency-related duties or tasks that need to be accomplished at the EOC are coordinated by the emergency manager, but may be carried out by other individuals designated by the emergency manager. This person is often given primary responsibility for duties related to EOC management and operations. Other designated EOC personnel may perform non-emergency-related tasks.

Duties and responsibilities for EOC management and operations may vary according to jurisdiction; however, there are core tasks that any designated person will be required to perform. Some of these tasks may arise before the EOC is activated, whereas others are ongoing. When the EOC is activated, additional responsibilities involve the direction, control, and coordination of numerous activities that develop in an emergency situation. Figure 9.1 lists the typical duties for EOC management.

FOUNDATIONS FOR ESTABLISHING THE EMERGENCY OPERATIONS CENTER

A jurisdiction's Emergency Operations Plan (EOP) and hazard vulnerability should be in place prior to the establishment of the EOC, as much of the information contained in these two documents can be used to ensure that a survivable EOC is designed and developed. These documents will also be helpful in developing policies and procedures (see Figure 9.2).

---

1. Highlight or circle each task you perform or would be expected to perform.
    a. Assist in the location and design of an EOC.
    b. Form/convene planning team/committee.
    c. Use a hazard/vulnerability analysis to assist in locating/designing an EOC.
    d. Define functions performed in the EOC.
    e. Determine the number of personnel required to operate the EOC.
    f. Determine space requirements for the EOC.
    g. Determine funding requirements for the EOC.
    h. Assess and evaluate functional layout (i.e., operational efficiency) of the EOC.
    i. Develop contingency plan for interim operations.
2. Assist in the preparation of the EOC for operations (i.e., fully functioning capability).
    a. Assist in determining telecommunications requirements.
    b. Assist in defining life support requirements.
    c. Assist in determining operating equipment/supplies needed.
    d. Ensure that procedures are in place to maintain support systems and equipment.

Figure 9.1 Emergency management EOC duties. (From Michael Fagel.)

Figure 9.2 EOC setup. (Courtesy of FEMA, Savannah Brehmer. https://www.fema.gov/media-library#)

The Emergency Operations Plan

As described in the previous chapter, an EOP should describe a jurisdiction's approach to planning—how citizens and property will be protected during a terrorist incident—and the resource identification and management system. An EOP must be effective in turning the concept of operations into an effective emergency response.

The Hazard and Vulnerability Analysis

All geographical areas of the United States are vulnerable to a terrorist attack, due to the wide variety of means by which an attack could take place. Identification of these hazards must be accompanied by a determination of the risk each hazard poses.

USING THE EOP AND HAZARD ANALYSIS TO DESIGN AN EOC

Various terrorism-related hazards will have different effects on the ability to survive the event and to continue to direct, control, manage, and coordinate emergency operations both within a jurisdiction, with other state and local governments, and with the federal government.

The design criteria for an EOC depend on the types of terrorism-related disasters that are likely to affect a jurisdiction. Although some terrorist threats are possible in all settings, areas that are either heavily populated or heavily rural, along with those close to military and utility sites, will have different considerations in terms of designing an EOC. The hazard and vulnerability analysis provides a good basis for determining the worst-case scenario in designing the EOC. The most critical element of the EOC is its ability to survive an emergency without any interruption in the continuity of operation.

Protection needs to be an integral part of planning, building, modifying, or equipping your jurisdiction's EOC. Personnel will need to be protected against a wide range of conditions, requiring a close examination of EOC location, structural integrity, and security procedures.

DEVELOPING POLICIES AND PROCEDURES

No matter how established your EOC, if you fail to develop policies and procedures for EOC management, your EOC will not be able to sustain itself during a terrorist incident. Effective EOC management and operation is dependent upon written policies and procedures that are in place prior to a terrorism-related emergency. Without these written guidelines, coordinated and responsive efforts cannot be achieved during an incident.

Each EOC will have its own unique requirements, depending on the jurisdiction and its demographics and area; however, certain standard policies and procedures will apply to any EOC. These standard policies relate to

- Communications
- Life support
- Equipment and supplies
- Documents and records retention

Communications Policies and Procedures

Adequate communications are essential to a jurisdiction's ability to direct its emergency response personnel effectively in the aftermath of a terrorist attack, regardless of the nature of that attack. Therefore, there must be plans in place for the effective emergency use of the extensive communications systems. You should make a point to consult with communications experts to ensure that your EOC's communication system is fully functional, compatible with other systems, and able to handle the demand of increased calls during an emergency. One critical lesson learned from the events of September 11, 2001, is that a confusion in communication led to further issues at the site, instead of those communications being used to calmly and effectively monitor and lead the response.

Developing Communications Standard Operating Procedures

EOC communications and standard operating procedures (SOPs) cannot resolve all major problems, especially as the possible impact of certain types of terrorist attacks are still being studied so as to be fully understood from an emergency management perspective, but properly designed procedures can lessen their severity. Some typical SOPs include the following:

- Available communications systems
- Frequencies listing
- Call-up and alerting procedures for communications personnel
- Location and communications supplies and equipment (see Figure 9.3)
- Setup procedures for communications equipment

ICS/EOC INTERFACE

Figure 9.3 Emergency phones ready to be installed. (Courtesy of FEMA, Patsy Lynch. https://www.fema.gov/media-library/assets/images/51634)

- Tasks and responsibilities for communications personnel
- Procedures for internal face-to-face communication (e.g., relaying messages)

Life Support Policies and Procedures

In some emergency situations, EOC staff may be isolated for an extended period of time. Therefore, it is important to have life support systems to ensure that EOC personnel can sustain for a period of up to 2 weeks.

Identify Emergency Operations Center Life Support Requirements

No matter where your jurisdiction is located, there are numerous agencies, corporations, and volunteer groups that will be willing to assist you in planning for terrorist attack response operations. Some may even provide goods and services free of charge or at a reduced cost. If you work with vendors that require payment, be sure to establish a memorandum of agreement or some other type of legal agreement or procurement contract to ensure that goods and services will be available when you need them. A lack of organization or expectation of what is to be provided is particularly damaging when trying to deal with the emergency management response required from a terrorist attack.

Tracking Goods and Services Required for the EOC

Once you have identified sources of the goods and services you will require for your EOC, you should maintain a list of applicable names and phone numbers. The following is a list of minimum life support requirements for the EOC.

- *Sleeping accommodations*—Some type of sleeping arrangements for EOC personnel will need to be made. Two- or three-tier bunks can be used to conserve space. Sleeping bags and portable bedding should be stored in a location near the EOC.

- *Food service*—Plan on having enough food to sustain EOC personnel for a minimum of 14 days.
- *Water*—Arrangements for water should be in place. Faucet water may be undrinkable, especially considering the possible ways in which a terrorist attack could occur or unfold, so have bottled water or water sanitation tablets. Plan for a minimum of 10 gallons of water per day, per person. Water must also be available for showers, waste disposal systems, and so forth.
- *Sanitary facilities*—Facilities such as toilets, showers, laundry, and garbage disposal should be provided but located away from the operations area. Supplies, such as toilet paper, should be readily available to meet demand caused by having more people using the facility.
- *Medical supplies*—Medical supplies should be limited to those required for dispensary-type operation. First-aid kits and an extra supply of bandages and antiseptics should be priority items, as well as medicines to treat diarrhea, headaches, constipation, and minor respiratory problems.
- *Heating, ventilation, and air-conditioning (HVAC)*—Facilities should be comfortable at all times. Ensure that procedures are in place to compensate for failed HVAC conditioning systems. Battery-operated backup equipment should be available.

There may be some federal, state, and local standards for life support systems. You should check with your records and documents office for up-to-date information regarding these standards. If you fail to do so, you may not be in compliance. These standards help to ensure that you maintain a minimally acceptable level of performance and operations for life support systems.

Life support systems are an essential component of successful emergency operations. Therefore, it is important that a regular schedule of training and exercising be implemented. Post-incident critiques should be developed and fed back into your planning efforts.

Operating Equipment and Supplies Policies and Procedures

There are no set standards for optimum supply requirement to operate the EOC. Requirements will depend on the type of incident, the number of staff that will be working at the EOC, the size of the EOC, and other factors too numerous to list here. If historical data for your community exist, it may be a good idea to review this information as a means to estimate needs. Keep in mind, however, that estimates for fully functioning equipment and sufficiency of supplies should be based on a 2-week operational period. Although other terrorist incidents may not be as severe (and that can't safely be assumed), consider the amount of time needed for the response to the September 11, 2001, attacks and be mindful that supplies could be needed for a long period of time.

The more attention given to acquiring appropriate supplies prior to an emergency, the less you will need to rely on a backup plan for obtaining supplies. In selecting equipment, whether for initial purchase or backup plans, consider the following:

- *Mobility*—Since conditions change during a crisis, the EOC's configuration must be flexible. Bulky, heavy, or cumbersome supplies and equipment should not be used.

- *Reliability*—Equipment should be durable and reliable. If the equipment has a tendency to break down, then spare parts, instructions for repair, and training for maintenance should be included in the planning effort and—where possible—incorporated into the SOPs. Batteries should be rotated.
- *Electrical compatibility*—All electrical equipment should be tested on the EOC power supply. All equipment operating standards should be carefully compared to emergency power standards.
- *Sustainability*—Primary power sources often fail, so you should plan for backup methods to operate equipment. Similarly, you must have backup supplies to replace alternate power sources, such as batteries or generators. The following should be in place:
  - Spare parts inventory for backup lighting, communications, ventilation, and other necessary maintenance.
  - Auxiliary lighting, such as flashlights, batteries, and bulbs.
  - Office materials, including an adequate supply of forms, pencils, paper clips, tape, note pads, and so forth. Computer equipment should not be dependent on outside data banks because they could fail. Keep manual typewriters available in case of an electrical power failure.
  - Recording equipment, such as instant cameras and battery-operated recorders.
  - Specialized equipment that may be required for certain hazards, such as a hazardous materials incident that occurs as the result of a terrorist attack.

Once that initial equipment has been acquired, you should have SOPs in place that address maintenance and acquisition of backup operating equipment and supplies.

Maintenance Contracts

Inspection and maintenance schedules are used to ensure that all equipment is in operating condition, when needed. You don't need an elaborate system to keep track of inspection and maintenance schedules, but it is important to have some type of system in place to monitor compliance with these schedules.

If you have access to a computer, you should use a database tracking system for developing and maintaining the schedules. Any scheduling system you use should contain the following elements:

- Description of equipment, including model number, serial number, and manufacturer
- Names, addresses, and phone numbers for vendors
- Contract number and account information
- Date equipment was purchased or leased
- Date of last inspection or maintenance
- Date of next inspection or maintenance
- Expiration dates of contracts

If maintenance contracts are not in place for critical pieces of equipment, immediate arrangements should be made to secure them. You can purchase the equipment or other types of resources, or you can establish memorandums of understanding to ensure that you can acquire the equipment during an emergency.

Memorandums of Understanding and Mutual Aid Agreements

A memorandum of understanding (MOU) is an agreement between agencies—both internal and external—located within the jurisdiction on cooperative efforts and services that would be provided during an emergency response. The agencies involved usually maintain command of their personnel while providing specific services to the community at large and in conjunction with the normal resources available in the community.

Mutual aid agreements ensure that you have resources and logistical support available to assist you in managing the response to a terrorist incident. They must be written in accordance with existing state and local ordinances. They should include a discussion of free access across boundaries, command of resources and staff, compensation for workers, staff, support provision, and insurance. Some states even have master mutual aid agreements to which your jurisdiction may subscribe. All MOUs and mutual aid agreements should be reviewed on a regular basis to ensure currency and also that the terms are still in effect.

Records and Document Retention

It is virtually impossible to accurately and properly complete the necessary record keeping after disaster or emergency work has been completed and a period of time has elapsed. Therefore, the importance of prompt, efficient record keeping cannot be overemphasized. You must know what records to keep, how to keep them, and have someone familiar enough to start keeping these records at the onset of an emergency situation.

If the situation following a terrorist attack develops into a major disaster declaration, proper documentation will be needed to justify local expenditures for which reimbursement will be requested. Without proper record keeping, the EOC could lose considerable sums of money because claims for reimbursement will also be needed to justify expenditures for which reimbursements will not be requested.

Most EOC records retention and archiving fall into one of the following categories:

- Survivable records and databases needed to conduct emergency operations
- Survivable records needed to reconstitute the government and for recovery
- A continuity-of-government plan, including an approved succession plan

It is essential that the information requirements for disaster response in your state and local community be identified and cataloged. Primary and alternate EOCs should contain the information databases and records necessary to sustain emergency operations, with provisions made for backup of this data.

Although information needs will vary from jurisdiction to jurisdiction, there are general categories of information that each EOC should maintain. For each type of record listed, the relevant information on contact persons, procedures for contact, location, purpose, and other appropriate information should be organized so that it is easily accessible in an emergency.

- Alerting
- Notification
- EOC activation

- Sheltering
- Transportation
- Food and water supply
- Medical assistance
- Debris removal
- Damage assessment
- Disaster assistance
- Public information

The state, in collaboration with local jurisdictions, must have a system for secure storage of vital records necessary to reconstitute the government and to conduct recovery efforts.

SUMMARY

Preparing your jurisdiction's EOC may seem like a daunting task, but it is absolutely necessary to ensure an effective response to any terrorism-related incident for your jurisdiction. It is important to remember that there are resources available to you and your jurisdiction. The Emergency Management Institute in Emmitsburg (http://www.training.fema.gov/emi.aspx), Maryland, provides training to enhance U.S. emergency management practices through a nationwide program of instruction.

10
EOC Management during Terrorist Incidents

Michael J. Fagel

Contents

Preparing the EOC for a Terrorist Incident 168
Reviewing the Hazard and Vulnerability Analyses 168
Incorporating Terror Analysis into Emergency Operations Plan 168
Revisit Each EOP Annex 169
Policies and Procedures 169
Test, Train, and Exercise 170
Summary 170

After action reports and studies of catastrophic disasters have identified the need to provide additional information in Emergency Operations Center (EOC) management. Although terrorist incidents have been less common than catastrophic natural disasters, effective EOC management is perhaps more critical for the management of terrorist incidents than for even the most catastrophic natural disasters because the stakes are, potentially, much greater.

By definition, terrorist incidents are Incidents of National Significance. Incidents of National Significance include those incidents that, under Homeland Security Presidential Directive 5, require a

> coordinated and effective response by and appropriate combination of federal, state, local, tribal, nongovernmental, and/or private-sector entities ... to save lives and minimize damage, and to provide the basis for long-term community recovery and mitigation activities.

One can assume, then, that terrorist incidents will cause coordination issues within the affected jurisdiction(s) and beyond.

The local EOC plays a critical coordination role during Incidents of National Significance. The ability of EOC personnel to work effectively with command personnel at the scene as well as state and federal personnel, public- and private-sector organizations, and others throughout the response and recovery phases is critical to saving lives, protecting property and the environment, and maintaining public confidence.

This chapter will describe the role that the Emergency Manager and other EOC personnel will play following a terrorist incident. To help ensure effective EOC operations following an incident, however, it is necessary to consider the steps of pre-incident preparedness.

PREPARING THE EOC FOR A TERRORIST INCIDENT

Preparing the EOC for a terrorist incident is not much different from preparing for other high-impact disasters. To be certain, however, the Emergency Manager should convene the planning team to revisit terrorist threats and targets identified during the hazard analysis process to determine additional needs and areas or tasks that might need to be performed differently.

REVIEWING THE HAZARD AND VULNERABILITY ANALYSES

All areas of the United States are vulnerable to a terrorist attack, because of the wide variety of means by which an attack could occur. When reviewing the hazard and vulnerability analyses, the planning team should determine the type(s) of incidents that could occur, potential targets of each type of incident, and the risk each type of incident poses to the jurisdiction and its citizens.

Note that jurisdictions can be affected by a terrorist attack, even when not attacked directly. For example, an attack on the power grid could affect a large area of the country. Rural areas that are less likely to experience a direct attack may be affected by a large influx of evacuees seeking shelter in a safe area. However, rural areas that are in close proximity to a military facility may be at risk of a direct terrorist attack.

The planning team, which may need to be expanded to analyze terrorist threats, should think expansively and consider all possibilities when reviewing the hazard and vulnerability analyses. Consider "worst-case" scenarios, but also consider other scenarios that are potentially high impact, even if they might not be high damage.

INCORPORATING TERROR ANALYSIS INTO EMERGENCY OPERATIONS PLAN

Use possible scenarios to determine the possible impact on the jurisdiction's Emergency Operations Plan (EOP). Then extrapolate from the EOP, especially the Concept of Operations and the Situation and Assumptions sections, to determine if alterations may be required in EOC operations. Then, play "what if" to change the parameters of a possible terrorist attack to determine how those changes affect necessary planning. (For example, what if Timothy McVeigh would have used 2,000 pounds of ammonium nitrate–fuel oil for his attack on the Murrah Federal Building, rather than 1000 pounds? What additional damage would have occurred? How many additional casualties could have occurred? What additional resources would be required for the response? What other issues would be affected, for example, communication needs? Finally, how does this information affect the coordination function and operations at the EOC?) Keep a careful record of these discussions for later use in revising the EOP.

REVISIT EACH EOP ANNEX

Next, revisit each EOP annex to identify gaps based on the terrorist threats, potential targets, and risks. The planning team should determine what portions of each annex should work in a terrorist situation and identify areas that need revision. For example, does the current Emergency Public Information Annex describe the huge number of uncertainties that will be felt by the public? Has a procedure been developed to ensure that citizens—and the media—receive critical information accurately and in a timely way? Continue this process with each annex, in turn, until all known planning gaps and needed revisions have been identified.

When completing revisions, keep in mind the key concepts and principles required in the National Incident Management System (NIMS) for all domestic responses, which focuses on command and control using the Incident Command System, resource management, including the typing of equipment according to capability, the training and credentialing of all response personnel, interoperable and redundant communications capability, and use of multiagency coordination systems to coordinate response activity from the incident scene to the federal level, including the Federal Bureau of Investigation. If your jurisdiction has not built NIMS requirements into its planning process, the terrorism review and revision process is a good time to do so.

POLICIES AND PROCEDURES

Policy and procedure review and revision is best done at the agency level, but after revisions have been completed, the planning team should review them to ensure that they are aware of other agencies' procedures and to identify areas that may not work in concert with other agencies' procedures.

Finally, extrapolate to EOC operations. What do changes in the EOP and response procedures mean for EOC operations? Have additional layers of communications systems been added? Does the jurisdiction need to identify and equip an alternate EOC? Examine each change made to the EOP for its possible effect on the EOC. Bear in mind that the EOC plays a coordination role during a response—not a command role—and that revisions to policies and procedures should focus on support needs at the incident scene.

Do *not* change the entire EOC organization unless it has been shown through exercises or actual incident responses that it does not work. If the EOC organization must be changed, change it for all responses, rather than incorporating a unique organization for terrorism incidents alone. A less effective organization can be overcome by the comfort level and enhanced performance of personnel who are familiar with it, which is certainly preferable to requiring staff to learn two different organizations based on the type of incident.

Several specific areas to look at are included in the following job aid. A "no" response to any of the questions on the job aid means that additional planning is necessary to prepare for post-terrorism EOC operations. Use the checklist as a guide for identifying additional/different EOC operations during a terrorist response.

TEST, TRAIN, AND EXERCISE

Finally, develop a progressive test, training, and exercise program to evaluate every aspect of a potential terrorism response. Begin with orientations that describe the changes required for a terrorism response. Then, proceed to tabletop exercises, functional exercises, and full-scale exercises.

Evaluate tests and exercises honestly, carefully, and thoroughly. Lessons learned from testing and exercises may point to additional EOP revisions or additional training needs. When evaluating tests and exercises, stay focused on the jurisdiction's goal, which is to improve incident response to a terrorism incident, including improved support offered through the EOC and the hierarchy of multiagency coordination entities.

SUMMARY

Preparing your jurisdiction's EOC for a terrorism response may seem like a daunting task, but by building on the systems and procedures that are already in place and testing, training, and exercising the jurisdiction's plan, you can ensure an effective response.

Remember that resources are available to help you with your planning. Contact your State Training Officer for a complete schedule of course offerings. Other courses are offered, which have numerous resources available through the Department of Homeland Security training partners, including the Emergency Management Institute (EMI) to enhance U.S. emergency management practices through a nationwide program of instruction.

11

Emergency Management and the Media

Randall C. Duncan

Contents

Newspapers 172
Radio 172
Television 174
Social Network Sites and the World Wide Web 174
Dealing with the Media in a Crisis 176
The Public Information Officer 178
The Joint Information System/Joint Information Center 180
References 186

Understanding and working with the media is an important part of an overall emergency management system. This relationship—between the emergency manager and the media—is one that has more opportunity to excel, or fail, than almost any other.

Let's begin our examination of the relationship between emergency management and the media by defining what the media are.

Traditionally, we think of the media as consisting of newspapers, radio, and television. Newspapers have been the media staple since the modern printing press was invented in 1450 by Johannes Gutenberg. Radio and television entered the world of media much more recently, but changed the way media operated and functioned within our society by reporting news on a timelier basis (live reports) and adding the elements of voices (radio) and moving pictures (television).

More recently, the development of the World Wide Web, social network sites, and blogs have led yet another revolution in the way media impacts our lives.

In order to more fully understand the elements of the relationship between emergency management and the media, it is necessary to understand the characteristics of the various types of media. Let's begin our examination with the traditional media forms of newspapers, radio, and television.

NEWSPAPERS

Arguably, the first newspaper in the United States was called *Publick Occurrences Both Forreign and Domestick* (National Humanities Center 2006). It was published on September 25, 1690, and edited by Benjamin Harris. It only printed one issue, and was banned 4 days after publication by the Governor and Council of Massachusetts (Massachusetts Historical Society 2010). The only surviving copy of the newspaper is in the Public Record Office in London (Library of Congress 2009).

Traditionally, modern newspapers have been published on a daily or weekly basis, depending on the size of the reading audience. Circulation of newspapers varies greatly. The newspapers with the three largest average circulation in the United States in March 2013 were the *Wall Street Journal* (circulation 2,378,827), the *New York Times* (circulation 1,865,318), and *USA Today* (circulation 1,674,306) (Alliance for Audited Media 2016).

Newspapers have traditionally been viewed as providing more in-depth coverage than either radio or television because of the amount of space available in which to write the story. Newspapers also provided some of the first coverage of events far removed from the place where they were published by the mechanism of the telegraph (see section "Radio," for more details). This allowed remote correspondents to send a story from far away back to the newspaper home office, and created a style of journalistic writing known as the "inverted pyramid." The inverted pyramid style of writing called for the correspondent to relay the most important facts first, followed by those of lesser importance in the body of the story (Scanlon 2008).

Based on this information, then, we can anticipate what print organizations want in the way of news, as shown in Table 11.1.

RADIO

It is not possible to talk about the history of radio without mentioning the wired telegraph system. The telegraph was made practical within the United States by Samuel Morse, who did his first public demonstration of the device in 1838 (Smithsonian Institution 2010). In 1843, Congress provided funding to install a telegraph between Baltimore, Maryland, and Washington, DC. The Whig Party held its nominating convention in Baltimore on May 1, 1844, and selected Henry Clay as their nominee. This was the first news item relayed by telegraph (Smithsonian Institution 2010). In 1901, Guglielmo Marconi began developing what would become broadcast radio—he sent the Morse code signal for the letter S from a wireless transmitter in Poldhu, Cornwall, England, to a wireless receiver in Newfoundland, Canada (Public Broadcasting System 1998b). A few years later—on Christmas Eve, 1906—some wireless telegraph operators onboard ships heard the Christmas carol "Silent Night"

EMERGENCY MANAGEMENT AND THE MEDIA

Table 11.1 What Traditional Print Organizations Want by Way of News

Item	Explanation
Details	Print media wants to paint a picture in the reader's mind with words.
Questions	Will be more oriented toward details.
Background information	How many times did the truck roll over?
	How far away from the edge of the road did it come to rest?
	Were there flames? If so, how high?
	History related to an event.
	Has this ever happened before?
	History of individuals involved in the event.
Deadline	Traditional: usually daily. Current policy may be impacted by newspaper website.

Source: Adapted from Randall Duncan.

and a voice reading bible verses interrupted the Morse code they normally heard (Public Broadcasting System 1998a). This marked the first radio broadcast.

From these humble beginnings, radio had an impact on the way we listened to news and found out about other events. We could sit in our living rooms and hear the voices of presidents, dictators from overseas, and Hollywood stars endorsing commercial products.

Unlike newspapers, radios could bring us the sounds and words of a news event as they happened.

Modern radio stations are separated into various interest groups called "formats." Some of the formats in today's radio broadcasting include news/talk stations, music stations, public radio, and non-English radio. Table 11.2 explains some of the features and news items as pertain to radio media.

Table 11.2 Features and News Items That Pertain to Radio Media

Item	Explanation
Details	Radio news utilizes short, concise information in the voice of the newsmaker.
	Typically, they are 10- to 15-second "actualities" or "sound bites."
News/talk	More stories; a little more depth than other radio formats.
Public radio	Uses "natural sound." Records background of event happening with open mike.
Non-english stations	Help to reach those who speak a language other than English within the community.
Music stations	May or may not carry news. If they do, it typically consists of only short news items.
Deadline	Hourly—depending on schedule of newscasts.

Source: Adapted from Randall Duncan.

Table 11.3 Features and News Items That Pertain to Television Media

Item	Explanation
Details	Video of the event may determine whether there is a story.
	A story that otherwise would not make the news may become a story if there is video.
	Similarly, a story of real importance may not make the news if there is no video.
Types of news	Local news includes spot news; regular local news; investigative reports. Also, feature news programs (either local or network); national news; and international news.
Deadline	Varies depending on the type of news that will be broadcast. Major deadlines are typically for the evening news and late night news broadcasts.

Source: Adapted from Randall Duncan.

TELEVISION

The first authorized broadcast of a television in the United States started on July 2, 1928, in Wheaton, Maryland, a suburb of Washington, DC, by C.F. Jenkins (Popular Mechanics 1928). The heyday of television may have come on the evening of March 7, 1955, when one in two Americans watched Mary Martin's portrayal of "Peter Pan" on live television (Bogart 1958, p. 1). Other significant events that marked the impact of television on the way Americans received news include the coverage of such live events as the Kennedy–Nixon debates and mankind's first step on the moon. The addition of live images to go with sound literally brought the world into our homes every night. There are various types of news broadcasts on local television stations (Table 11.3). They may range from spot news/breaking news of particular activities currently in progress to the regularly scheduled news programs. In addition, some local television stations may air special investigative reports or programs. Typically, local television stations will have an affiliation with a network, and will present a network-originated program of national and international news.

SOCIAL NETWORK SITES AND THE WORLD WIDE WEB

No examination of media would be complete without exploring the impact of the World Wide Web and social media on individuals, as well as the traditional media of newspaper, radio, and television.

The World Wide Web, as we know it today, first became reality in 1990 with the release of a point-and-click hypertext editor called "World Wide Web" (Berners-Lee 1998). In the ensuing years, the number of websites, their functionality, and the pure amount of information has literally exploded. Naturally, with such rapid expansion, there is a need for a "buyer-beware" approach by users. It is as easy to find an academically reputable and accurate source on the World Wide Web as it is the lunatic ravings of fringe elements. It is up to the user to find and place the appropriate value on sources available through this medium.

Aside from the ease of accessing information on the World Wide Web, people soon found that it was becoming a tool for personal communication and networking between friends, leading to the development of social network sites. Social network sites are defined as

> ... web-based services that allow individuals to (1) construct a public or semipublic profile within a bounded system, (2) articulate a list of other users with whom they share a connection, and (3) view and traverse their list of connections and those made by others within the system. The nature and nomenclature of these connections may vary from site to site. (Boyd and Ellison 2007)

We also think of social network sites as social media—a way to convey information to friends and learn about information from friends and others with opinions. Sites such as Twitter, Facebook, MySpace, and blogs have become a way to share information, opinions, and even reflect on news events.

These same sites have had an impact on the way traditional media—newspapers, radio, and television—interact with their viewers and listeners. As a result, newspapers now also shoot video of news stories and provide it to their readers through the mechanism of their website. Radio stations do the same thing. Television stations now write news stories and publish them, similar to newspapers, on their websites. This has had a major impact on the traditional deadlines for the various forms of media.

Yet another World Wide Web–based phenomena has emerged recently—the Weblog, or Blog. This phenomenon has blurred the distinction between traditional press and bloggers.

> In the media world, it used to be clear who was in the news business and who was not. News businesses provided news, nonnews businesses did not. Reporters worked for companies who were in the business to provide news. News businesses got paid, usually by advertisers, to collect, package and distribute information of interest to news audiences. Nonnews businesses or organizations exist for other purposes—perhaps to deliver public service such as environmental protection, or to produce commercial goods such as fertilizer. Providing news for these businesses is simply not their reason for existence. In an instant news world, that distinction is becoming increasingly fuzzy.
>
> One of the most significant trends to come out of the collection of technologies we call the Internet, is the emergence of citizen journalists. "Blogging," from the term "Weblog," which used to describe people who would record and publish what they discovered on the Internet, reflects the ease with which almost anyone who writes today can also publish. As mentioned earlier, some of the bloggers have accumulated audiences in the millions and have influence as great as any of the celebrity journalists that used to be staples of our early evening hours at home. (Baron 2006, p. 47)

Because of these factors, the Emergency Manager or spokesperson for the jurisdiction has to keep in mind that there are now more audiences for the information they prepare than the traditional media. There are the families of those directly impacted by the event or emergency, those in the immediate area of the emergency or disaster, and the "traditional media" along with the citizen journalist.

DEALING WITH THE MEDIA IN A CRISIS

To begin our discussion about dealing with the media in a crisis, we need to understand some of the basics about communication. We first learn to communicate as babies—before we can even begin to say words. We communicate with gestures and sounds, through a thing called nonverbal behavioral clusters. We will discuss this in more detail shortly.

Communications is extremely complicated for such a seemingly simple thing. The act of communications starts as an idea in our brain. That idea wishes to be expressed or communicated to someone. It must then make its way through the filters of our belief system and perceptions. Then it must be encoded (either in speech or writing) and then broadcast to a receiver (a reader or listener). The receiver has to get the message, decode it, and run the decoded material through their own filters of belief systems and perceptions in order to understand the idea we originally wished to communicate to them. An understanding of the complications associated with the process of how we communicate allows for a new appreciation of a statement as seemingly simple as, "Pass the salt, please."

Nonverbal behavioral clusters associated with how we say and express things are more important in conveying meaning than the words we actually say (Blatner 2009). Because these nonverbal behavioral clusters typically convey a larger percentage of our communication than the specific choice of words do, we tend to place more faith in the way the message is expressed. When the message being conveyed to an audience by a speaker's words is in conflict with their nonverbal behavioral clusters, the audience will not believe the speaker. As a simple thought experiment, recall the last time you observed a person on television and your reaction to that person was the thought that you did not believe a word they said. The odds are, you felt that way because there was a conflict between the words of the speaker and their nonverbal behavioral clusters.

The normal process of communications takes a slight detour under a crisis situation. When a crisis is in progress, we need to provide assistance to the elected official or spokesperson to make sure we do not allow circumstances to take away from the messages we need to communicate to the public typically through the media. In other words, we need to avoid media pitfalls. Table 11.4 is adapted from unpublished material from Dr. Vincent Covello, founder and director of the Center for Risk Communication.

In a crisis situation, the media follows certain patterns, which include:

- Searching for background information on the incident
- Dispatching reporters to the scene
- Obtaining access to the scene or the official spokesperson
- Dramatizing the situation
- Expecting a briefing complete with written information
- Expects *you* to panic
- Becomes confused by technical information
- Exhausting resources
- Sharing information among themselves
- Acting professional and expecting the same
- Providing filler for stories if credible information is not available

Table 11.4 Media Do's and Don'ts

Do	Don't
Define all technical terms and acronyms (jargon).	Use language that may not be understood by even a portion of your audience.
If you use humor, direct it at yourself.	Use humor in relation to safety, health, or environmental issues.
Refute negative allegations without repeating them.	Refer to national problems—"This isn't Love Canal."
Use visuals to emphasize key points.	Rely entirely on words.
Remain calm. Use the question or allegation as a springboard to say something positive.	Let your feelings interfere with your ability to communicate positively.
Ask whether you made yourself clear.	Assume you have been understood.
Use examples, stories, and analogies to establish a common understanding.	Talk only in abstractions.
Be sensitive to nonverbal messages you are communicating.	Allow your body language or your position in the room to be inconsistent with your message.
Make them consistent with what you are saying.	Dress inconsistently with your message.
Attack the *issue*.	Attack the person or the organization.
Promise only what you can deliver.	Make promises you cannot keep.
Set, then follow strict deadlines.	Fail to follow up on those items that you promise to follow.
Emphasize achievements made and ongoing efforts.	Say there are no guarantees.
Refer to the importance you attach to health, safety, and environmental issues—your moral obligation to public health outweigh financial considerations.	Refer to the amount of money spent as a representation of your concern.
Use personal pronouns (e.g., I, we).	Take on the identity of a large organization.
Take responsibility for your share of the problem.	Try to shift blame or responsibilities to others.
Assume everything you say and do is part of the public record.	Make side comments or "confidential" remarks.
Discuss risks and benefits in separate communications.	Discuss your costs along with risk levels.
Use risk comparisons to help put risks in perspective.	Compare unrelated risks.
Stress that the true risk is between zero and the worst-case estimate.	State absolutes or expect laypersons to understand.
Emphasize performance, trends, and achievements.	Mention or repeat large, negative numbers.
Focus your remarks on empathy, competence, honesty, and dedication.	Provide too much detail or get drawn into protracted technical debates.
Keep presentation to 15 minutes total.	Ramble or fail to plan the time well.
Keep answers to 2 minutes maximum.	Tell people more than they want.

Source: Adapted from Vincent Covello, Center for Risk Communication.

THE PUBLIC INFORMATION OFFICER

The Public Information Officer (PIO) typically has the responsibility for coordinating the collection, verification, and dissemination of information to the public. These duties may occur as a part of day-to-day organizational operations, or on an emergency basis. Since the focus of deliberations in this discussion is "Emergency Management and the Media," let's concentrate on the roles and responsibilities of a PIO in an emergency.

The PIO has responsibilities to a number of different constituencies. These include

- *The Public*—This segment is the largest user group for emergency messages. This implies that the PIO should be aware of any special demographic characteristics of the community being served, and have familiarity with the best media channels to distribute information to those who need it.
- *The Media*—This is one of the most important relationships to establish to make sure information is distributed to those who need it. The PIO will need to understand the traditional and social media outlets within the jurisdiction.
- *The Agency*—This relationship is the basis of trust within the jurisdiction. The PIO has a duty to positively portray the efforts and successes of the agency they represent. This relationship will be especially important when navigating an agency through the dangerous shoals and reefs of a negative news story.
- *The Other Responding Agencies*—The PIO needs to have a good working relationship with other agencies responding to the emergency or disaster. These relationships are especially important in helping to avoid conflicting stories or statements. If the incident becomes large enough to engage the Joint Information System (JIS) or the Joint Information Center (JIC), the PIO needs to be able to function in that environment. In addition, the PIO needs to be aware of the possibility that there will be differing priorities among the agencies responding to the emergency or disaster and how to deal with those differing priorities so that "mixed messages" are not given to the public.

We generally make the statement that the PIO provides public information. It would be helpful to provide a definition of what public information is, with respect to emergencies and disasters. Generally, we can conclude that public information is used by people to save lives, reduce injury and harm, and protect property (FEMA 2009). Given that public information covers such a wide territory, it is understood that almost every piece of information coming from your agency or emergency operations center (EOC) could result in the public taking some type of action to protect themselves or others from the effects of a disaster or emergency. This also emphasizes the criticality of the accuracy and timeliness of your information (FEMA 2009).

We generally expect that information communicated to the public through our PIO will result in action by people, provide information, change behaviors or attitudes, or create a positive view of our agency or EOC within the community. Some examples of public information could include:

- The current status of the emergency or disaster
- Agency response actions to the disaster
- Information or warnings as conditions change

- Important locations (i.e., where food, water, and shelters are located)
- Specific evacuation information or directions
- Other pertinent information:
    - What is open or closed
        - Government facilities
        - Stores
        - Roads
        - Schools
    - Status of lifeline systems—electricity, gas, water, and sewer
    - Volunteer recruitment
    - Where people can find aid or assistance
    - Public inquiry telephone numbers

In order to perform the job of PIO well, the person in that position needs to have a number of qualities, some of which include:

- Knowledge of the organization they represent. This allows them to speak with credibility about the operations of the organization. It demonstrates to the media and the public that the PIO has access to agency leadership. It also provides the media with opportunities for interviews and briefings about the agency.
- A good working relationship with the organization or EOC. This is a necessary quality for the PIO to have access to the information and resources everywhere within the agency or EOC they are representing.
- A certain amount of aggressiveness. The PIO may need to be able to go directly into the organization and get to decision makers and leaders with minimal delays. In addition, the PIO will undoubtedly be called upon to provide advice to leadership—making it necessary for the PIO to be in the inner circle of the organization or EOC.
- A high level of trust and ability to strategize. The PIO needs to be able to establish trust within the organization or EOC. Essentially, the PIO becomes an advisor that understands how things will be viewed outside of the agency or EOC, and understand the implications of information to which the public will have access. It will be important to understand what the potential negative consequences of these issues are and how to present them, truthfully, in as positive a light as possible.
- Community relations skills are necessary for the PIO. The PIO needs to understand the demographics of the jurisdiction—who lives and works there and what the prevailing local values, concerns, and interests are. The PIO must also know about organizations within the community and how they work and interact.
- The PIO needs to have good media relations skills. This includes a level of credibility that is usually only developed over time and through hard work.

It is also important for the PIO to have several sets of skills, including:

- Writing abilities—organizes clear thoughts in a written format (whether electronic or printed). This includes the ability to develop talking points, guidance,

strategy papers, speeches, and general information for management. It is especially important that proper grammar and spelling be utilized in these pieces. This should probably include familiarity with Associated Press Stylebook (http://www.apstylebook.com). Generally, the PIO needs to be able to produce quality documents, whether electronic or paper.
- Other abilities—the PIO should be able to understand the basics of using video as a means of communication. This would include knowledge of the basic elements of photography. In addition, the PIO should be able to clearly communicate and outline ideas in a manner allowing the PIO or other spokesperson to communicate effectively with an audience; the ability to speak effectively and persuasively in front of an audience; and, an awareness of the impact of nonverbal behavioral clusters on the delivery of a message.

THE JOINT INFORMATION SYSTEM/JOINT INFORMATION CENTER

As you have read previously, the National Incident Management System (NIMS) was developed at the direction of Homeland Security Presidential Decision Directive (HSPD)-5. The original NIMS document was developed in 2004, updated in 2006, and updated again in 2008 (FEMA 2008). Public information, and the process of establishing the system to collect, integrate, and coordinate it, is defined as a part of NIMS Component IV—Command and Management (FEMA 2008, pp. 70–74). The overall system is called the JIS, and the specific place where this process happens is called the JIC.

Typically, this process involved a PIO, who supports the incident command structure, and who is also a member of the Command staff. The responsibilities of this position typically include:

- Responding to inquiries from the media, public, and elected officials
- Supervising the process of collecting, integrating, and coordinating information for emergency public information
- Supervising the process of collecting, integrating, and coordinating information for warning information
- Monitoring for rumors and responding to them
- Relations with the media
- Creating coordinated and consistent messages through
  - Identifying key information to be communicated to the public
  - Creating the message that provides the key information in a clear and easily understood method
  - Prioritizing messages so the most important gets out first and that the public is not overwhelmed with the amount of information
  - Verifying the accuracy of information
  - Making sure the message gets out through the most effective means available

The overall JIS provides the means to coordinate the messages being released to the public by all elements of government involved in the disaster response—whether multiple

Table 11.5 Noteworthy Details on JIS and JIC

JIS Provides a Structure and System For	JIC Provides a Place To
Developing and delivering coordinated interagency messages	Centrally facilitate operation of the JIS during and after an incident
Creating, recommending, and executing public information plans and strategies	Increase information coordination
Advising Incident Commander about incident relevant public affairs issues	Reduce misinformation
Monitoring and correcting erroneous information circulating among the media or the public	Maximize resources for dealing with the public and the media
Be adaptable to the size and scale of the incident—from three PIOs at the scene to 150 PIOs at a major disaster from multiple locations	Provides "one-stop shopping" for the media

Source: Adapted from Randall Duncan.

local jurisdictions, or local, state, and federal governments; multiple disciplines involved in the response; nongovernment organizations involved with the response; and the private sector. This coordination is particularly important because all the voices involved in the disaster should be providing substantially the same message.

One of the important elements to keep in mind when dealing with a large emergency response situation is that different disciplines may have their own spokesperson or PIO present and different jurisdictions may have their own spokesperson or PIO present as well. It is possible that these PIOs may serve as the basis for the JIC staff. Remember that each of these officials will have the primary responsibility for making sure the story—as it relates to their agency, discipline, or jurisdiction—gets out to the media. But there is no reason they cannot work together and collaborate in order to establish the JIS and provide the personnel for the JIC.

The JIC is basically an instrument to help facilitate the processes that take place within the JIS—much like the EOC is an instrument to help facilitate the processes that take place within the Local Emergency Operations Plan (LEOP). As a result, there are other parallels between these two elements of Public Information. The elements of the JIS must be worked out well in advance of the occurrence of a disaster. The system plans and processes, much like the roles and responsibilities within the LEOP, must be worked out and understood in advance of their application.

There should also be consideration given as to what kind of triggers might initiate the activation of the JIC. Some suggestions might include:

- The creation of a standard operating procedure or guideline that defines the opening of the facility. This SOP/SOG could be modeled after the existing document for the activation of the Emergency Operations Center.

Table 11.6 Different Types of JICs

JIC Type	Description
Incident	Typically, an incident-specific JIC is established at a single, on-scene location in coordination with federal, state, tribal, and local agencies or at the national level, if the situation warrants. It provides easy media access, which is paramount to success. This is a typical JIC.
Virtual	A virtual JIC is established when a physical colocation is not feasible. It connects PIOs through e-mail, cell/landline phones, faxes, video teleconferencing, Web-based information systems, etc. For a pandemic incident where PIOs at different locations communicate and coordinate public information electronically, it may be appropriate to establish a virtual JIC.
Satellite	A satellite JIC is smaller in scale than other JICs. It is established primarily to support the incident JIC and to operate under its direction. These are subordinate JICs, which are typically located closer to the scene.
Area	An area JIC supports multiple-incident ICS structures that are spread over a wide geographic area. It is typically located near the largest media market and can be established on a local, state, or multistate basis. Multiple states experiencing storm damage may participate in an area JIC.
Support	A support JIC is established to supplement the efforts of several Incident JICs in multiple states. It offers additional staff and resources outside of the disaster area.
National	A national JIC is established when an incident requires federal coordination and is expected to be of long duration (weeks or months) or when the incident affects a large area of the country. A national JIC is staffed by numerous federal department and/or agencies, as well as state agencies and nongovernment organizations.

Source: Adapted from FEMA. 2009. *G290 Basic Public Information Officer Training*. Washington, DC: Federal Emergency Management Agency.

- An analysis of the potential impact of the incident.
- An analysis of the potential media interest in the incident.
- The potential duration or the response and recovery phases of the emergency or disaster.

Other noteworthy items about the JIS and JIC are outlined in Table 11.5.

Table 11.6 presents a number of different types of JICs. It is adapted from information found in FEMA G290 Basic Public Information Officer Course (FEMA 2009).

We close this chapter with a useful JIC Readiness Assessment form in Tables 11.7 through 11.9.

Table 11.7 JIC Readiness Assessment Form—Plans

Plans		
Do you have systems and procedures for:	Yes	No
✓ Developing an emergency response or crisis communications plan for public information and media relations?	☐	☐
Does your emergency response or crisis communications plan have systems and procedures for:	Yes	No
✓ Designating and assigning line and staff responsibilities for the public information team?	☐	☐
✓ Identifying and updating current contact numbers for PIO staff and other public information partners in your plan?	☐	☐
✓ Identifying and updating current contact numbers for regional and local news media (including after-hours news desks)?	☐	☐
✓ Establishing the JIC at the Emergency Operations Center (if activated)?	☐	☐
✓ Securing needed resources (space, equipment, people) to conduct the public information operation during an incident 24 hours a day, using such mechanisms as Memorandums of Understanding, contracts, etc.?	☐	☐
✓ Creating messages for the news media and the public under severe time constraints, including methods to clear these messages within the emergency response operations of your organization (including multijurisdiction and/or agency cross-clearance)?	☐	☐
✓ Disseminating information to news media, the public, and partners (e.g., website capability 24/7, listservs, broadcast fax, printed news releases, door-to-door leaflets)?	☐	☐
✓ Verifying and clearing/approving information prior to its release to the news media and the public?	☐	☐
✓ Operating a public inquiry hotline with trained staff available to answer questions from the public and control rumors?	☐	☐
✓ Activating the Emergency Alert System, including the use of prescribed messages?	☐	☐
✓ Coordinating your public information systems planning activities with other response organizations?	☐	☐
✓ Testing the plan through drills and exercises with other response team partners?	☐	☐
✓ Updating the plan as a result of lessons learned through drills, exercises, and incidents?	☐	☐

Source: Adapted from FEMA. 2009. *G290 Basic Public Information Officer Training.* Washington, DC: Federal Emergency Management Agency.

Table 11.8 JIC Readiness Assessment Form—People

People		
Do you have systems and procedures for:	Yes	No
✓ Identifying staffing capabilities needed to maintain public information operations for 24 hours per day for at least several days? (Note: Staff may include regular full- and part-time staff as well as PIOs from other agencies or departments, disaster employees, volunteers, etc.)	☐	☐
✓ Establishing and maintaining agreements for acquiring or borrowing temporary staff? (Note: Such agreements may be mutual aid arrangements or Memorandums of Understanding.)	☐	☐
✓ Granting emergency authority to hire or call up temporary staff or those on loan from other organizations?	☐	☐
✓ Establishing and maintaining job descriptions and qualifications for individuals serving as your organization's PIO and other roles during an incident?	☐	☐
✓ Assigning a staff member and at least one to alternate the role and responsibilities of PIO?	☐	☐
✓ Determining if the assigned PIO(s) is qualified? Sample qualifications include: • Experience and skills in providing general and emergency public information • Ability to represent your organization professionally (can articulate public information messages well when dealing with the media and the public, and can handle on-camera interviews) • Written and technical communication skills (writing/editing, photography, graphics, and Internet/Web design proficiency) • Management and supervision experience and skills needed to run a JIC	☐	☐
✓ Establishing and maintaining a list of language translators available to assist with public information? (Note: Such a network should include sign language interpreters and individuals capable of writing and speaking the non-English language(s) used by individuals in your jurisdiction.)	☐	☐
✓ Establishing and maintaining working relationships with PIO partners from other organizations that you might need to work with during an incident (e.g., PIOs from other jurisdictions, other government agencies or departments, nongovernmental organizations, and private entities)?	☐	☐
✓ Developing and maintaining working relationships with your local and regional media, and established procedures for providing information to those media entities effectively and efficiently during incidents?	☐	☐

Source: Adapted from FEMA. 2009. *G290 Basic Public Information Officer Training.* Washington, DC: Federal Emergency Management Agency.

Table 11.9 JIC Readiness Assessment Form—Logistics

Logistics		
Do you have a Go Kit for PIO use during an incident, including:	Yes	No
✓ Laptop computer capable of linking to the Internet/e-mail?	☐	☐
✓ Cell or satellite phone, pager, and/or PDA/palm computer with wireless e-mail capability?	☐	☐
✓ Digital camera, photo storage media, and charger/backup batteries?	☐	☐
✓ Flash drives, CDs and/or disks containing the elements of the crisis communication plan (including news media contact lists, PIO contact lists, and information materials such as topic-specific fact sheets, backgrounders, talking points, and news release templates)? REMEMBER: Redundancy is important in case the computer you are using doesn't have a USB port, CD, or floppy drive.	☐	☐
✓ Office supplies such as paper, pens, self-stick notes, etc.?	☐	☐
✓ Manuals and background information necessary to provide information to the media and the public (e.g., your Smart Book)? (Note: A Smart Book is a compilation of factual information assembled about your jurisdiction, such as population, number of schools and hospitals, size and description of geographic or infrastructure features, etc.)	☐	☐
✓ Hard copies of all critical information?	☐	☐
Do you have systems for:	Yes	No
✓ Acquiring and maintaining Go Kits with a funding mechanism (e.g., credit card) that can be used to purchase operational resources? (Note: A Go Kit is a mobile response kit that allows PIOs to maintain communications in the event that they are working outside of their normal place of operation.)	☐	☐
✓ Ensuring PIOs can access the Go Kit when serving at an incident?	☐	☐
✓ Acquiring and maintaining portable communications equipment, critical up-to-date information, and supplies?	☐	☐
✓ Acquiring and maintaining essential media production equipment (cameras, digital storage, laptops, etc.)?	☐	☐
✓ Acquiring and maintaining a Smart Book (or equivalent technologies) to assist PIOs in accurately informing the media and the public during an incident?	☐	☐
✓ Identifying a dedicated location to house the JIC? (Note: The location selected must be wired for telephone, internet access, cable, etc.)	☐	☐
✓ Securing and maintaining the necessary JIC equipment and supplies to allow information to be disseminated to the media and the public?	☐	☐
✓ Inventorying and restocking the PIO Go Kit after an incident?	☐	☐
✓ Inventorying and restocking JIC equipment and supplies after an incident?	☐	☐
✓ Periodically updating your Smart Book with current information?	☐	☐
Do you have equipment and supplies needed for a JIC, including:	Yes	No
✓ Computers on a LAN with Internet access and e-mail listserves designated for news media and partner entities?	☐	☐
✓ Laptop computers?	☐	☐

(*Continued*)

Table 11.9 (*Continued*)   JIC Readiness Assessment Form—Logistics

Logistics		
✓ Electric and manual typewriter(s) in case of power outage or other problems that interfere with computer/printer usage?	☐	☐
✓ Fax machine preprogrammed for broadcasting fax releases to news media and partner entities?	☐	☐
✓ Printers and copy machines, with supplies such as toner and paper?	☐	☐
✓ Paper shredder and trash bags?	☐	☐
✓ Televisions with access to cable hookups and VHS VCRs or other recording media?	☐	☐
✓ Cell or satellite phones, pagers, and/or PDAs/palm computers with wireless e-mail capability?	☐	☐
✓ Digital camera, photo storage media, and charger/backup batteries?	☐	☐
✓ Audio recorder and batteries?	☐	☐
✓ Flash drives, CDs, and/or disks containing the elements of the crisis communication plan (including media contact lists, PIO contact lists, and information materials such as topic-specific fact sheets, backgrounders, talking points, and news release templates)?	☐	☐
✓ Office furniture/accessories such as desks, chairs, file cabinets, bulletin boards, whiteboards, trash cans, lights, in/out baskets, landline phones, clocks, large calendars, etc.?	☐	☐
✓ Audio equipment and furniture necessary for conducting news conferences (e.g., wireless microphones, lectern, multibox, etc.)?	☐	☐
✓ Office supplies (e.g., white and colored paper, pens, self-stick notes, folders, blank tapes, binders, overnight mail supplies, tape, poster board, erasable and permanent markers, chart paper, easels, staplers and staples, press kit folders, binders, computer disks/CDs, hole punch, organization logo on stickers, letterhead, postage stamps, etc.)?	☐	☐
✓ Manuals, directories, and background information necessary to provide information to the media and the public (e.g., your Smart Book)?	☐	☐
✓ Hard copies of all critical information?	☐	☐

Source: Adapted from FEMA. 2009. *G290 Basic Public Information Officer Training*. Washington, DC: Federal Emergency Management Agency.

REFERENCES

Alliance for Audited Media. 2016. Average Circulation at the Top 25 U.S. Daily Newspapers. Retrieved June 7, 2016 from Alliance for Audited Media: http://asuditedmedia.com/news/reseasrch-and-data/top-25-us-newspapers-for-march-2103.aspx.

Baron, G. R. 2006. *Now is Too Late 2: Survival in an Era of Instant News*. Bellingham, WA: Edens Veil Media.

Berners-Lee, T. 1998. The World Wide Web: A Very Short Personal History. Retrieved June 7, 2016, from World Wide Web Consortium (W3C): http://www.w3.org/People/Berners-Lee/ShortHistory.html.

Blatner, A. 2009. *About Nonverbal Communications.* Retrieved April 21, 2010, from Adam Blatner's website: http://www.blatner.com/adam/level2/nverb1.htm.

Bogart L. 1958. *The Age of Television.* New York: F. Unger Publishing Company.

Boyd, D. M., and N. D. Ellison. 2007. Social Network Sites: Definition, History, and Scholarship. Retrieved April 21, 2010, from Journal of Computer-Mediated Communication: http://jcmc.indiana.edu/vol13/issue1/boyd.ellison.html.

FEMA. 2009. *G290 Basic Public Information Officer Training.* Washington DC: Federal Emergency Management Agency.

FEMA. 2008. National Incident Management System. Retrieved April 23, 2010, from Federal Emergency Management Agency: http://www.fema.gov/pdf/emergency/nims/NIMS_core.pdf.

Library of Congress. 2009. *Eighteenth-Century American Newspapers in the Library of Congress.* Retrieved April 15, 2010, from Library of Congress: http://www.loc.gov/rr/news/18th/200.html.

Massachusetts Historical Society. 2004. Premier issue of the Boston News-Letter. Retrieved April 15, 2010, from Massachusetts Historical Society, http://masshist.org/object-of-the-month/objects/the-boston-newsletter-2004-04-01.

National Humanities Center. 2006. *Publick Occurrences Both Forreign and Domestick.* Retrieved April 15, 2010, from National Humanities Center: http://nationalhumanitiescenter.org/pds/amerbegin/power/text5/Publickoccurrences.pdf.

Popular Mechanics. 1928. What television offers you. *Popular Mechanics* 50(5): 820–824.

Public Broadcasting System. 1998a. KDKA begins to broadcast 1920. Retrieved April 16, 2010, from A Science Odyssey: People and Discoveries: http://www.pbs.org/wgbh/aso/databank/entries/dt20ra.html.

Public Broadcasting System. 1998b. Marconi receives radio signal over Atlantic 1901. Retrieved April 16, 2010, from A Science Odyssey: People and Discoveries: http://www.pbs.org/wgbh/aso/databank/entries/dt01ma.html.

Scanlon, C. 2008. The Inverted Pyramid Structure. Retrieved April 16, 2010, from Purdue Online Writing Lab: http://owl.english.purdue.edu/owl/resource/735/04/.

Smithsonian Institution. 2010. History Wired: A few of our Favorite Things. Retrieved April 16, 2010, from National Museum of American History, Smithsonian Institution, http://historywired.si.edu/detail.cfm?ID=324.

12

Deterring and Mitigating Attack

Jennifer Hesterman

Contents

Introduction	190
Effects-Based Hardening (EBH)	191
Physical Security	193
Parking Lots	194
Pre-Positioned Vehicles	194
Traffic Duty	196
Security	196
Locks	197
Alarms	198
Visitor Access and Badges	200
Closed Circuit Television (CCTV)	200
Public Address Systems	201
Ventilation Systems	202
Dining Halls	202
Emergency Preparations	203
Hold Room	203
Emergency Response Team and the Command Center	203
Bomb Threat	205
Medical Program	206
Shelter in Place	207
Elementary Schools in Focus	208
Hardening the College Campus	209
Building Relationships	209
Students and Staff as Force Multipliers	210
Exercising Due Diligence	211

Churches 211
Hospitals 212
Malls 213
Sports and Recreation Venues 214
Red Teaming Soft Targets 216
Conclusion 217
References 217

The supreme art of war is to subdue the enemy without fighting.

Sun Tzu

INTRODUCTION

The goal of soft target hardening is simple: Deter any would-be attackers through the presence of a secure facility and if they breach your access points or strike from inside, engage with the ability to mitigate the attack and save the lives of your staff and occupants. As discussed in Chapter 2, hardening begins with you: the acceptance the threat exists and your operation and facilities are vulnerable. You likely have already taken on some amount of risk by not being able to expend the resources to protect the operation fully, whether due to insufficient resources, lack of support from your leadership, or for business-related reasons. However, there is a spectrum of hardening actions you can take from nothing to everything, from inexpensive to exorbitantly expensive. The key is to understand the desired effect and plan to use your resources in the best possible manner to lower your risk.

DHS's "Soft Targets Awareness Course" (DHS 2016) is a great place to start grasping the overview of the threat and your vulnerabilities. The 4-hour curriculum provides facility managers, supervisors, and security and safety personnel with baseline terrorism awareness, prevention, and protection information. The course enhances individual and organizational security awareness and participants gain insight as to why it is important to engage in proactive security measures and define their roles in deterring and detecting terrorist activity, and defending their facilities from it. The class is held all over the country and administered by qualified contractors.

Although your security plan should be an all-hazards approach, meaning that it is suitable no matter the emergency situation, you should focus on preparedness, rather than the specific kinds of weapons or tactics the bad actor may bring to your facility. With this in mind, you should first assess your vulnerabilities. Prior to creating or improving upon your security plan, complete the FBI's Vulnerability Assessment in Appendix B. Note this document must be safeguarded because it spells out all of your vulnerabilities. If the total score for the organization exceeds 256, and if local law enforcement has not been involved in the assessment, notify them at once. Hand-carry the document, never e-mail or send through a postal service. After you identify and think through your vulnerabilities, you can create a plan to reduce risk in your operation.

EFFECTS-BASED HARDENING (EBH)

We simply cannot apply all resources toward all threats; a methodology is necessary to ensure actions and resources are directed in the most effective manner to cover your vulnerabilities against the threat and lower your risk. Prior to Operation Desert Storm, the Air Force's aerial campaign strategy was that of attrition—to bomb targets repeatedly and shoot down as many aircraft as possible until the enemy lost either the will or the firepower to fight. However, in order to prosecute the war against Iraq successfully, with restraint to spare civilians and not destroy the infrastructure of Baghdad, Air Force strategic thinkers devised a new approach: effects-based operations (EBO). An EBO approach is one where "operations against enemy systems are planned, executed, and assessed in order to achieve specific effects that contribute directly to desired military and political outcomes" (Carpenter 2004). EBO provides a strategy for the application of resources, phased in a particular way to achieve the desired cumulative effect.

EBO provides a good theoretical foundation for our efforts to harden soft targets, and for these purposes we call it EBH—effects-based hardening. This proposes a new (or improved) way of thinking and a specific process on both physical and psychological planes. Several axioms must be accepted prior to implementing this approach:

- Actions cause results.
- Inaction also causes results.
- Not seen does not mean not there.
- The goal is to remove the enemy from the fight before it starts.
- Actions are not universally applicable and must be tailored to your situation.
- The plan is fluid; you must constantly assess and adjust based on changes in the environment.
- The "fog of war" means you do not know everything about the threat; there are inescapable unknowables.
- You have no experience with the situation that might occur in your organization; nothing that happened in the past can prepare you.

EBH provides a system for visualizing violent scenarios which may happen to your organization in an unemotional, data-driven way. In order to identify your "Achilles heel" and other vulnerabilities increasing your risk and susceptibility to attack, you have to "go there" and visualize and map out the worst possible scenario in your facility. At the least, consider an enraged outsider, a plotting insider, an active shooter, and a kidnapping and hostage situation.

The good news is that you are fighting this "battle" on your own territory. No one knows the vulnerabilities and strengths of your operation better than you; this puts you in the position of power over bad actors when it comes to deterring or mitigating their attacks. No matter how much preplanning or surveillance takes place, or even if you face an insider threat, you have the upper hand. Accordingly, keep certain details about your plan close, think of your employees' "need to know," and draw a diagram of concentric circles with the critical operations and people at the center and continue outward to the periphery, where you may have building custodial staff and volunteers. They need security and response training, but do not need to understand the specific security apparatus in place or your plan for protecting the facility and its occupants.

There is an art and science to security. The science part is physical: barricades at certain locations, walk-through metal detectors to keep out weapons, and security personnel stationed at entrance points for presence. The art is using your resources efficiently and effectively to achieve strategic security objectives. EBH can complement (or replace) your current efforts. Any new security processes should be incrementally phased in to shape the behavior of your people (and the enemy), perhaps in a time-phased manner or from most to least critical. However, security is not just a program; it must infiltrate everyday operations and decision making. By baking security into the organization and not just leaving it to the security guard at the front door, you tap the full spectrum of your assets: people, equipment, and building location. You may never know what types of terrorists or violent criminal acts you have thwarted using this methodology, but at the very least you will have a data-driven plan effectively using your resources to cover vulnerabilities.

Move forward *unapologetically* with your security plan. Leaders at soft target locations often covey their sense of regret about how efforts to tighten security inconvenience their staff and visitors. Those operating a for-profit operation are often concerned with customer satisfaction and whether measures will drive patrons away. They should imagine, for a second, a horrendous attack at their facility. It could be an angry ex-spouse exacting revenge, a fired employee, a disturbed teenager, or a terrorist or group seeking to make the news and further their religious or political cause. As the leader or head of security, you must face the devastated family members and explain how you failed to protect their loved ones. Being a leader means taking responsibility, not only for the good things happening at your place of work, but also for the bad, and the very, very bad.

EBH, by its very nature, encourages the harmonizing and synchronizing of actions. For example, during an active-shooter event, the front office has a plan: one predesignated person calls 9/11, one makes an announcement on the loudspeaker, and one locks and barricades the door. These types of actions take training and practice. Similar to pilots who are thrown impossible situations to handle in the flight simulator, if you practice for the worst possible scenario, small security issues will be handled effortlessly by the staff, working together as a team. They will also be confident of their ability to handle a large-scale emergency, and this confidence makes them force multipliers to you and your security team.

Matrixing your vulnerabilities, desired effects, the means to lessen your vulnerability, and the capabilities you both have and need help with the decision-making process. Figure 12.1 is an example of an EBH decision matrix for a church or school.

How do you know if your EBH efforts are working? You can test your system by having an outside security company do a red teaming exercise on your property, a tactic addressed later in the chapter. Also, you should ask the people who work in and use your facilities if they feel safe and if not, how you could do better. Not only will you glean valuable information, but the mere process of asking for and then acting on their ideas will strengthen your relationship and open the lines of communication about vulnerabilities.

The rest of the chapter is devoted to best practices and ideas harvested from industry experts—information to guide your EBH efforts.

DETERRING AND MITIGATING ATTACK

Prioritized Scenario	Desired Effect	Means	Capabilities and Cost	Implement/Partially Implement/Table
1. Highly visible location on busy highway draws opportunists	Lower "heat"	Remove external signage facing road.	In house, volunteers, free	Implement
2. Too many people with keys to the main door	Restrict building access.	Install electronic key lock on main door and obtain keying equipment and cards.	Contracted, $3,000	Partially implement, re-key current lock, reissue keys, budget electronic key system for summer 2015
3. Holding meetings after hours for outside groups, attendees wandering in building	Restrict access to the rest of the building.	Install locking door between basement and upstairs office.	Contracted, $1,500 with labor	Implement

Figure 12.1 Effects-based hardening matrix. (From Jennifer Hesterman.)

PHYSICAL SECURITY

(Thank you to Brian Gallagher, former physical security specialist at the U.S. Secret Service and owner of Security at Church [http://www.securityatchurch.com] for his assistance with the following section.)

We cannot easily interview terrorists or those who perpetrate violent crimes about deterrents to their activity; however, convicted thieves are accessible and give valuable insight. The chart in Figure 12.2 shows responses from 360 burglars regarding physical security measures that serve as the biggest deterrents to their activities.

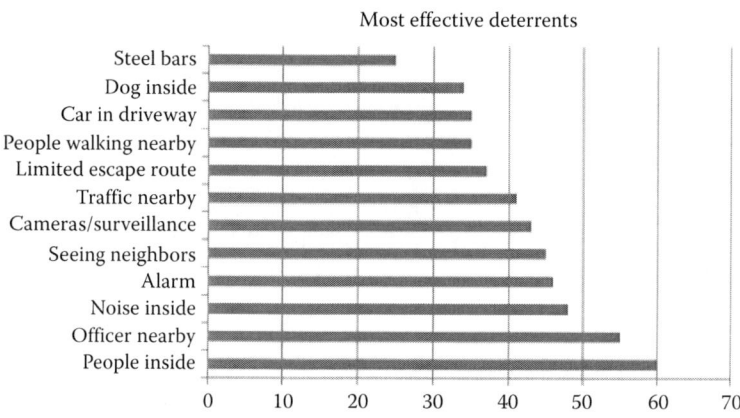

Figure 12.2 Perception of effectiveness of burglary deterrents according to burglars. (From Brian Gallagher, Security at Church. With permission.)

Note the presence of people is the number one deterrent, followed by an officer nearby (which may be substituted at night by a parked patrol car). Alarms are effective, as well as surveillance cameras, dogs, and steel bars on windows. The limited escape route is something to consider as well.

The exterior of your building, the grounds, and the parking lot are all critical to the security of your occupants. Lowering the profile of your building with less signage will deter opportunists from attacking your church or school. Physical hardening of your property could be as simple as installing a security fence or raising the height of current fencing to conceal your building and its occupants (such as children in a schoolyard) and to keep out intruders. Always check the perimeter fence for breaches or the stacking of wood or objects that could allow someone to climb over the fence. Industry experts recommend the following fence standard: 7 feet high, with three strands of barbed wire, 6 inches apart. Shrubbery should be no higher than 3 feet, and set back 1 yard from buildings, and tree branches trimmed 8 feet above the ground. Outdoor lighting not only illuminating buildings, but also the surrounding property, is important to deter trespassers. Motion lights will keep your electricity bill lower and startle any would-be intruder. There are industry standards for external security lighting; see IESNA, ASNI, and OSHA guidelines online.

Parking Lots

As illustrated in previous case studies, several international and domestic terrorists used parking lots to plant the primary bomb, stage a shooting, or place secondary devices aimed at injuring response personnel and evacuees. Therefore, you must secure your parking lot as an extension of your building. The preferable situation is to have the parking lot located inside the fence line, with a greeting area/entry point to control visitors. Churches are especially vulnerable to burglary, with service times posted outside on signage and a lot full of unattended vehicles for an hour or more. A roving parking lot security team provides an extra layer of protection.

Operations with large parking lots such as megachurches, large schools, malls, and sports and recreational venues might consider training for parking attendants called "First Observer" (https://www.tsa.gov/for-industry/firstobserver). The program is jointly operated by the Department of Homeland Security (DHS) and the Transportation Security Administration (TSA) to train parking attendants to identify a potential threat. The program also educates the parking attendants with background information on terrorist groups, their tactics, and trends as well as an understanding of weapons. Training of this type turns a parking attendant into a force multiplier for your organization.

Pre-Positioned Vehicles

Perhaps you could ask your local police department for a marked police cruiser to be placed at the entrance of your building. You could offer parking to an officer who could leave his or her personal vehicle at your facility and swap out with the cruiser when off duty. The marked vehicle serves as a visual deterrent for those who might be performing surveillance on the facility. If the marked cruiser is not possible, perhaps a member of

DETERRING AND MITIGATING ATTACK

Photo 12.1 An SUV repurposed as a mock security vehicle at a compound in the Middle East.

your church, school, hospital, or organization will give up an old SUV or truck, which you can mark to look like a security vehicle from afar using fluorescent tape and other tools (Photo 12.1).

There is no data to prove the pre-positioning of law enforcement vehicles deters attack; however, anecdotal stories support the theory. For instance, there is reason to suggest that Sandy Hook Elementary School shooter Adam Lanza's initial target was actually Newtown High School. According to a source familiar with the investigation, Lanza's car was identified on the school surveillance footage circling the school parking lot. The official believed Lanza saw two police cars parked in the lot and decided to move on (Lysiak 2013). The absence of such procedures has been used in several "negligent security" lawsuits; for instance, in 2007, a man was killed by a shooter with an AK-47 in a case of mistaken identity in a Waffle House parking lot in Pensacola, Florida, at 3:30 am. In the successful lawsuit filed by the family: "The plaintiff alleged that the restaurant was located in a high crime area. The plaintiff's security experts opined that the defendant was negligent in failing to have an armed, uniformed sheriff's officer on the premises with a marked police car which would have deterred the crime" (Rose, Vangura, and Levin 2009).

If your facility is hosting a special guest—advertised to the public and may be controversial or draw protest, or even a popular celebrity—know that a bull's-eye has just been painted on your facility. You should pre-position an escape vehicle near an emergency exit closest to the place where the person is addressing the crowd. A member of your staff or a trusted volunteer should be positioned at the vehicle as the driver in case of emergency.

Traffic Duty

Soft target facilities have predictable schedules: schools have drop off/pick up, church services start and end at the same time every week, sporting and recreational events have traffic issues at the beginning and end of events, and malls are typically their busiest on Friday and Saturday nights. Choke points outside your facility present a security hazard and you need to keep traffic moving. Mir Aimal Kansi, a Pakistani citizen residing in the United States with phony immigration papers and a forged green card, became enraged while watching CNN news coverage of U.S. operations in Iraq and CIA involvement in Muslim countries. A courier, he often drove by the entrance to CIA headquarters, noting the two lanes of traffic waiting at the light to turn left onto the agency's ground. At 8:00 am on January 25, 1993, Kansi drove his courier vehicle to that very spot, emerged from his vehicle with an AK-47 semiautomatic rifle, and walked up and down the lines of vehicles, firing a total of 10 rounds, killing two CIA employees, and injuring three others. Kansi escaped the country and was arrested by FBI agents in Pakistan in 1997; he was convicted and sentenced to death by lethal injection, which was accomplished in 2002. Sayed Farook of the San Bernardino attack also had plans to attack stationary cars in rush hour traffic. The military actively tries to avoid chokepoints, such as long traffic lines at gates, where people are vulnerable to attack and are blocking emergency response vehicles, if required.

Many states and counties require some sort of intersection control for larger churches, schools, and sports and recreation events. Although it is tempting simply to use volunteers with reflective vests, it is important to have a uniformed police officer directing traffic along with a marked vehicle with flashing lights, not only to slow vehicles but also to show presence to any opportunist who may decide to strike during the congested, chaotic time. A possible solution is to use a member of your church congregation or the spouse of one of your school, hospital, or sports/recreation venue employees who happens to be a law enforcement officer. However, if you have to hire an officer, go directly through the county or local police departments instead of using a contractor, which is costlier and who will likely outsource to the same organization.

SECURITY

We accomplish three goals through robust physical security features: portray to would-be bad actors that the facility is hardened and deter their actions, protect our property and its occupants in the event of penetration or attack, and make them feel safe, improving staff productivity and providing a better experience for users whether learning, worshipping, or healing. In for-profit organizations, a strong security infrastructure and program will positively affect your bottom line, especially for customers who make decisions about what sporting and recreational venues they will entrust with their lives and those of family members.

DETERRING AND MITIGATING ATTACK

Locks

Some facilities have only a few doors and others have many; no matter your situation, know that locks are critical to securing your property. Locks can be very easy to defeat if they are not made or installed properly. For instance, most locks installed on home doors can be defeated in a matter of minutes by someone with basic knowledge, which can be easily gleaned from the Internet. A simple padlock can be cut with a bolt cutter. A file cabinet can be breached by a crowbar. So how do you keep out those with bad intentions?

Although we want the visitor's first impression of the facility to be favorable, always choose security over aesthetics. Exterior doors must have, at the least, a dead bolt and never contain glass or be surrounded by glass window panes. In the Newtown, Connecticut, shooting at Sandy Hook Elementary School, the school's security procedures had just changed, requiring visitors to be admitted individually through a set of security doors after visual and identification review by video monitor. Doors to the school were locked at 9:30 am after morning arrivals. Adam Lanza arrived at the school at 9:35 am and the door was locked. He simply shot through the glass panel next to the door and stepped inside the school (Photo 12.2). According to the official report: "The doors to the school were locked, as they customarily were at this time, the school day having already begun. The shooter proceeded to shoot his way into the school building through the plate glass window to the right of the front lobby doors" (Office of the State's Attorney Judicial District of Danbury 2013).

Once inside the school, Lanza had a distinct advantage because he had attended Sandy Hook and had intimate knowledge of the inside of the building. However, there were infrastructure weaknesses detailed in the Newtown report, disadvantages allowing Lanza to kill twenty innocent children and six teachers and staff members in just 5 minutes before turning the gun on himself when the police arrived. For instance, the report discusses the office and classroom doors: "The doors in the hallway all locked from the outside with a

Photo 12.2 Front entrance of Sandy Hook Elementary School.

key. The interior door handles had no locking mechanism. All of the doors opened outwardly toward the hallway. All doors were solid wood with a circular window in the upper half of the door." Although investigators did not discuss how these doors failed to protect the building occupants, the report is replete with examples of how Lanza was walking up and down the hallway trying doors and looking through windows. Perhaps, if the doors also locked from the inside, with a double key lock, lives may have been spared. Also, having windows on the doors provides visual advantage to the shooter and another way to get through the door if it is locked from the inside. Photographs taken at the scene show the classroom windows were the type that opened up and inward just a few inches, with not enough space for a child to escape. The classrooms were on the first floor and perhaps people could have fled the scene through the windows if they slid on a track. The Sandyhook tragedy gives much to consider when assessing external and internal building security.

Who has access to your building? How many people have keys to your building? Do they all need a key to the front door? Typically, too many people have keys to a building. Keys should be numbered, issued by signature, and stamped "do not duplicate." The types of locks are also important; for instance, a standard lock has a cylinder inside with a series of pins that move up and down. When you insert a key into the cylinder, it moves the pins up and down vertically; when the pins line up correctly, the cylinder rotates and the lock opens. Locks with more pins are harder to defeat; for instance, a filing cabinet lock may have three pins whereas the lock on your front door may have six or more. The depth of the pins also helps make the lock more secure, to prevent opening by mere jiggling of the cylinder using a pin. Professional locks have pins that are not merely vertical, making them extremely difficult to defeat.

Locks can also have "smart cards" or electronic chips in the key, allowing for remote access and the ability to activate or deactivate a key and to control and track access to a room. For example, using keys and locks with electronic chips allows the facility manager to give access only on certain days (church services or school days). Naturally, we should not assume people with keys to our facility have bad intentions; however, keys are routinely stolen or lost. In government facilities, lost keys often mean locks must be changed and new keys issued. Electronic keying allows for instantaneous changes that immediately prevent access in the case of nefarious intent by a thief or a person finding a lost key. The basic principle is to limit access.

Interior doors should have locks, including supply closets, which are notorious hiding places for a bad actor and/or the stockpiling of supplies for an operation. Business offices must be further hardened to protect not only from the theft of private information, but also the occupants who might be the target of a criminal act or an attack. Keys to these doors must be strictly kept to those who need daily access. Although some large hospitals now have sophisticated systems automatically locking all doors and stopping elevators in the case of a lockdown, consult with law enforcement and fire officials prior to installing such a system, as it could trap people trying to flee.

Alarms

What is the backup plan if your key system is defeated? Some soft target facilities have sophisticated alarm systems and others have none. If you do not have an alarm system, why not? Security companies can install alarm systems at your facility for minimal cost and charge modest monthly monitoring fees. If you do have an alarm system, is it the right one and are all of your bases covered? It may be time for a security assessment and upgrade.

An alarm system obviously protects your building and its occupants, but acts as a powerful deterrent to many would-be intruders. When 422 convicted burglars were surveyed, approximately 83 percent of the offenders said they would attempt to determine if an alarm were present before the burglary and about 73 percent said they would seek an alternative target if so. Among those who discovered the presence of an alarm while attempting a burglary, half reported they would discontinue the attempt, and another 37 percent said they would leave the property. About one-third of the respondents planned their crime; the rest were opportunists, mostly looking for money or drugs (Kuhns and Blevins 2013).

If you have a central alarm system, it can be wired or wireless. It can be monitored by a central location or "bells only" in which you hope the loud alarm is a deterrent to the penetrator, who then departs the property. Remember the alarm code is just as valuable as the master key and you should try to limit distribution of this code to staff who come to work early to open the building and stay late to close, or those who need to enter at odd hours. However, all staff should have the silent duress code number they can punch into a panel, alerting the alarm company to call law enforcement, a feature only possible with monitored systems.

If you choose a monitored alarm system, sensors will be placed on doors and windows. You may also add motion sensors to catch movement in the building and glass breakage detectors on sliding doors. When triggered, the alarm signal travels via a telephone or Internet line to a central monitoring location where an employee receives information such as your address and possibly a schematic of your building with the exact location of the breach. The alarm company will have several contacts on file and will first call you before calling law enforcement, to prevent false alarms. Typically, you will have a prearranged, easy to remember telephone password confirming your identity. You must think redundancy when deciding on an alarm system for your facility; if the person is hiding inside the building when you lock up for the day, the motion sensor will detect him or her. If the intruder can breach the external door lock and alarm (pressure pads are easy to defeat, magnetic pads, not so easy)—again, the motion sensor will work. Glass breakage detectors are important on sliding doors, as the actor may try to avoid the sensor on the track by entering through the broken glass.

As previously mentioned, if your system is monitored, an employee can enter a duress code into the panel to alert authorities; in this case, no calls will be made to protect those in the situation and law enforcement will immediately respond to the location. Also, panels typically have hot buttons to push for medical help, fire, and police; these are timesavers during an emergency. An alarm system is a great investment and most insurance companies offer discounts for their use.

You should also have a panic alarm built into the reception area; this is a simple button located under the desk that can be discreetly pushed in the event of an emergency to signal silently for help. The signal can be set up in different ways; it can be wired so the duress call will go straight to the local alarm company, which will immediately send the police. Churches, schools, and hospitals should consider having a duress button in areas where counseling takes place and possibly also in conference rooms. The panic signal button does not have to be programmed to notify the local alarm company or police department; it is possible to have it signal for backup: for a colleague to come to the area and provide backup. The button should be pushed only in a situation where there is the danger of physical violence; the staff could be trained not to enact the duress signal if someone is merely acting demanding or overly emotional. Banks are served very well by these silent alarms, with police typically arriving while the robbery is being committed or shortly thereafter. Naturally, bank robbers know banks have these alarms and yet are undeterred. Similarly, we need to realize security features will not keep determined bad actors from engaging.

Visitor Access and Badges

Having a solid visitor management program to control access to your facility is incredibly important. If there are several access points to the building, have clearly marked signage funneling visitors through one set of doors. Have the visitor sign a log with name, time, and cell phone number and produce photo identification for entry, preferably a government-issued card such as a driver's license. You may hold this identification in exchange for a visitor's badge. Many secure government buildings such as the Pentagon require two forms of government-issued photo identification. The best practice is to have the visitor wait in a designated area while you contact the individual he or she came to see; do not let him or her wander aimlessly around the building. Having positive control of the visitor is important. Secure schools do not let parents walk items to classrooms (forgotten lunches, books, musical instruments); they may leave the items in a designated area in the front hallway and the child comes later to collect. Sign-in logs must be kept for a designated period of time, as they will provide evidence if the visitor commits a crime on the property, is in a parental custody battle, and so forth. The rosters also allow other due-diligence activities, such as running the name through the sex offender database. Although these activities may feel uncomfortable or intrusive, remember you are responsible for the protection of hundreds, if not thousands, of innocent people. Inconveniencing a visitor is a small price to pay.

Closed Circuit Television (CCTV)

We have already discussed cameras in terms of violent criminals, who barely see them as a deterrent, hoping they aren't taping or their image is obscured enough to cause doubt of their identity. On the other hand, over 40 percent of convicted burglars indicated the presence of surveillance cameras would deter them from a target. We have discussed how some terrorists often conduct surveillance of their targets while planning, while others simply hit a target of opportunity. A key tenet of this book is facility owners should not base hardening activities on guesses as to who might attack or how. Therefore, our opinion is all soft target locations should have CCTV that is preferably monitored, but at the least, is taping for later reference or evidentiary purposes. Many systems refresh on a cycle; 48 hours is probably a good point to start taping over old data. CCTV acts as a set of eyes when people are not available, and it helps gain convictions and even win court cases for you and your insurance company in the event of a lawsuit or claim.

Admittedly, CCTV does not deter a suicide terrorist who plans to lose his or her life or an enraged violent criminal who is not thinking rationally. However, if the actor believes the cameras are being monitored, they know there will be an immediate law enforcement response, compressing the timetable for the criminal act. Camera systems should be used to watch not only the front door, but also other exterior doors, hallways, classrooms and other multipurpose rooms, and the parking lot. There are state laws against filming in certain areas such as bathrooms, locker rooms, changing areas, or nursing mothers' rooms. New camera systems transmit images over the Internet, so you can watch the feed from your computer or even wireless devices such as a Blackberry or an iPhone. Another great feature of CCTV systems is pan, tilt, and zoom (PTZ) cameras that can be remotely controlled if you need to get a closer look at an individual. For outside cameras, you could also employ infrared (IR) technology to see better in the dark; in fact, many companies are now installing IR cameras indoors to capture criminal activity that may be occurring in dark corners or with flashlights. Many systems may be purchased over the counter and installed by the user; however, security system companies would be happy to bring their experts to your property to discuss CCTV coverage.

Public Address Systems

We strongly recommend facilities have a public address system for broadcasting emergency information to the entire building. During the Sandy Hook Elementary School shooting event, one of the injured staff members unknowingly tripped the loudspeaker system when dialing 911 for help. Her phone conversations were heard throughout the building, as were the gunshots in the front office area. This alerted the teachers and staff to the crisis and they had precious seconds to lock doors and hide students. The librarian called the front office and the staff to see what was happening and the staff member told her there was a shooter; again, this was broadcast to the entire school.

The Winnenden School, located in southwestern Germany, was the scene of a mass shooting when former student Tim Kretschmer entered the school with semiautomatic weapons on the morning of March 11, 2009. Immediately following the start of the attack, the school's headmaster broadcast a coded announcement saying "Mrs. Koma is coming," which is the word "amok" spelled backward. The message was a safety measure installed to alert the teachers of a school shooting and give them a chance to help students escape or shelter in place; it was established in Germany after a previous school massacre at Erfurt.

Intercom systems are also excellent ways to transmit specific threat and emergency information immediately. Hospitals have elaborate code systems that are easy to adapt to other situations; for instance:

1. *Code blue/code 99:* CPR team
2. *Code red/red alert/Dr. Firestone:* Fire alarm, activate department fire protocol (close fire doors, move people past fire zones, evacuate if ordered)
3. *Code orange/code purple/code silver:* Internal incident (psychiatric patient missing, active shooter in building, active bomb threat, etc.; activate case-specific disaster plans)

4. *Code black/code yellow/code 10:* External incident (natural or man-made disaster, mass casualties; activate department disaster plan)
5. *Code pink/code Adam/Amber alert:* Missing infant/child (lock down all exits, be on lookout for suspicious persons)
6. *Code green/code 00/all clear:* All clear, resume normal duties

In one school district, a "code 303 meeting" was known universally, even among the students, as the code for a bomb threat. "Mr. Falkes" and his parents being in the office meant a bomb threat, "Professor Norris" needing to meet his wife in the teacher's lobby meant weapon/stranger on site, and an "ROTC Club meeting" being canceled meant something very bad was happening that required immediate staff-wide attention (via e-mail or intercom). At an exhibition center in London, loudspeaker calls include, "Will Mr. Goodfellow report to the security suite"—the code for a fire. A report that, "Mr. Goodfellow has left the building" is all-clear code for the fire situation. "Staff call 100" is the code for a bomb threat, and, "Staff call 100 has been canceled," is the code for the bomb threat passing. At one college, instructors use a phone code to alert security that a student they are meeting in private may turn violent. If they want a security guard to come to the office as a precaution, they call the outer office and say "I'll be a little late for our meeting with 'Dr. Barry'"; if they want a guard to come in immediately, they say the appointment needs to be postponed.

A combination of an intercom and a code system can save lives.

Ventilation Systems

We discussed the threat of weapons of mass destruction (WMDs) and, for closed buildings and venues, it is important to understand your central air conditioning and ventilation systems. For example, find out if your building has fresh air intakes or recycled air intakes; newer systems are likely a combination of both. Are the units located on the roof, where they may be harder to reach (ideal) or are they on the ground and easily accessible? If an attacker is using a chemical or biological weapon against the property, he or she will likely introduce it through the intake system for maximum dispersal in the building. Even over-the-counter pepper spray will affect building occupants if introduced this way. If the unit is on the roof, make sure hatches are padlocked. If the unit is on the ground, consider building a fence around the unit with a lock on the gate. If possible, try to monitor these locations with security cameras and, if your organization cannot afford security cameras, then include these locations in a walk-through for your security team or administrative staff.

DINING HALLS

For facilities with kitchens, consider whether you have a "clean" environment; for instance, when they are not in use, do you secure items which could be used as weapons such as large knives? Are food items secured in tamper-proof containers or locked refrigerators so a bacterial agent cannot be introduced? In churches, kitchens may only be used once a week, so they may be a good place for an insider or potential attacker to store supplies.

DETERRING AND MITIGATING ATTACK

Dining halls are particularly vulnerable to a mass shooting or kidnapping/hostage situation because there can be many people in a confined space; several attacks already profiled in this book have occurred in a cafeteria.

EMERGENCY PREPARATIONS

Hardening your facility includes being prepared for any contingency, whether there is an emergency in your town or city and you are receiving scared or hurt people, or whether you have a security incident on your property and need to provide medical care and possibly shelter in place until help can get to you. The first action is to take the applicable free FEMA courses listed in Appendix A, such as IS-360: "Preparing for Mass Casualty Incidents: A Guide for Schools, Higher Education, and Houses of Worship" or IS-362.a, "Multi-Hazard Emergency Planning for Schools."

Specifically, you should have a four-pronged approach to cover all bases: a "hold room" to secure your leadership, a command center for you and your key staff, a strong medical program, and the ability to shelter in place.

Hold Room

A hold room is a place where you and other key staff or important visitors can retreat in the event of an emergency and hold for an undetermined amount of time. The room should be located in an evacuation area or point of the building where law enforcement can most easily get to you and you can get out as needed. The hold room can be as simple as a room with a door and lock or as elaborate as an underground facility. It may be worth the money to purchase a security door for the hold room—whether an "intruder" door that will resist being breached or a ballistic door that protects from forced entry but also repels 30.06 caliber rounds. At the very least, you want to make sure the door can lock from the inside and it has no windows. Although it seems counterintuitive, it is best to select a room with no exterior windows for additional security. The hold room must have a landline phone in case cellular service is not working, and you may also want to have a charged cell phone in the room in the event the phone lines are not working. Also pre-position a case of water and energy bars in case you need to sustain those in the hold room for more than a few hours. Flashlights are also a good idea in case the electricity goes out, as well as a battery-powered radio so you can get news updates on the unfolding situation. The goal in an emergency is to get out of the facility; however, if escape is not possible, the hold room will buy you time until law enforcement arrives. Be sure to add checking the hold room phones, supplies, and door lock mechanism to your facility walk-through checklist.

Emergency Response Team and the Command Center

In the event of a major incident in the surrounding community or at your facility, you may want to stand up an emergency response team and have an area to serve as your command center. The command center might be the same as the hold room. You will need landlines, flashlights, and a radio with extra batteries, water, and a supply of nonperishable food, such as energy bars. Handheld radios that are charged and ready are a plus if someone needs to leave and communicate with the command center. Pads of paper, pens, and even a whiteboard with dry erase markers would be helpful. This is the central repository for information and checklists.

Next, assess your potential emergency team. You may want to poll your employees to evaluate the additional skills and resources they bring to your operation. For instance, do they have a conceal carry permit and do they regularly carry a weapon? Do they have medical training, private security or a military background, or experience as a volunteer firefighter or paramedic? When formulating the emergency team, think about the skillset and also the mindset: can he or she handle an extremely upsetting, emotional event and be an asset to the team, instead of a liability? Once your emergency team is identified, create an "alert roster" with their names and cell, home, and e-mail contact information. Make sure to check the information regularly to ensure currency.

Who should be around your table? If your staff has directors or heads of specialized sections, they should each be there—security, facilities, communication, human resources, legal, etc. Also, if you have any counselors on staff, one should be designated to sit on the emergency response team. Additionally, have someone on your team in charge of handling the press and social media. If your building has been attacked or there is a mass hostage situation, others will blog, tweet, and Facebook; take control of the "message" and transmit what you want people to know about what is happening inside. Also, information you transmit could save the lives of others; for instance, during the strong earthquake on the Eastern seaboard on August 23, 2011, citizens in New York City saw the Twitter alerts about an earthquake originating in Virginia 15–20 seconds before seismic waves struck the city. According to Facebook, the word "earthquake" appeared in the status updates of three million users within four minutes of the quake. Twitter said users were sending as many as 5,500 messages ("tweets") per second.

For the purpose of keeping control of the message and transmitting data that could warn others of danger, you may want to have Twitter and Facebook shell accounts established to use as needed. Naturally, do not transmit any information about law enforcement activity in the building—which responders are on site, numbers, and their plan. In many cases, such as the Westgate Mall siege, the terrorists are also using the media to transmit and television and radio to assess the activities of law enforcement outside. Also, never transmit the number of injured or fatalities, or any pictures that could be used as propaganda by the bad actors or disturb family members.

We recommend you have a binder with checklists for each situation that could arise. One thing you *must* have on hand are maps of the building and/or blueprints as these are critical for first responders and law enforcement, especially if there is a hostage or active shooter situation. These floor plans must have cardinal directions and specific distance by feet (and even steps); walk your buildings and grounds and then map it out. Law enforcement will want to know everything about the ventilation system ducts, whether doors open in or out, and location of light switches. This is why you must know every single inch of your property; you will be the expert they turn to for answers.

Perhaps your binder will include checklists for different contingencies your facility might face. The best way to ensure your team is prepared and ready to handle an emergency is to practice through exercises. Take the FEMA courses listed in Appendix

A, specifically IS-120.a, "An Introduction to Exercises"; IS-130, "Exercise Evaluation and Improvement Planning"; and IS-139, "Exercise Design." In addition to full-scale practices, crisis "tabletop" exercises where you and the team report to the command center and simply talk through the event are also a must, perhaps quarterly. There are experts who can create scenarios tailored to your organization, or you can simply draft the scenario yourself based on past events such as the Beslan school and Moscow theater sieges, the Boston Marathon attack, Columbine, and so on. Run through it with the team and keep throwing in unexpected problems such as the phones dying, a family member breaching the cordon and entering the building looking for a loved one, or the electricity going out. Remember the "pilots-in-the-simulator" approach; make it very difficult or even impossible for your team to handle and they will rise to any future challenge and succeed. During contingency operations drills in the military, a card may be handed to the leader within the first few minutes of the exercise indicating he or she is out of the fight—either killed in the attack or otherwise incapacitated. This forces the team to consider who would step up and lead and gives the person the opportunity to sit in the "hot seat." Of course, you are not on your property 24/7, and there is a possibility a crisis will strike while you are away. Do not be the single point of failure for your team.

Some things to ponder are the trigger points for action. Let's say there is an emergency in the community and it is infringing on your property, threatening your facility and the people within. When should you abandon the facility and how? During the 2012 uprisings in Libya, several U.S. hotels were caught in the middle of the crisis in which a U.S. ambassador and his team had already been killed in Benghazi and the embassy burned to the ground. One hotel unexpectedly was the recipient of an influx of people after the embassy made an announcement (without coordinating) that all Americans should go to this particular hotel. The hotel manager looked out the window and saw a line of cars and Americans outside his property's gate, which he had locked down. Naturally, he took all of the Americans in and bedded them down with the rest of the hotel customers and staff. But at some point, he had to make the difficult decision to leave the property, possibly forever because it would likely be overrun and destroyed by militants. They needed more buses than originally planned, and people could only take one small bag, leaving the rest behind. Travel arrangements had to be made with airlines and the now-large convoy of Americans to the airport required security. A bad situation quickly degraded to the worst scenario. Fortunately, they all made it out of the country; however, the property still sits vacant.

The manager of the hotel in Libya participates in security conferences to tell his story, and states that he wishes he had three plans on the shelf that day, instead of just one. He called them the "alpha, bravo, and zulu plans," Zulu was the worst-case scenario of no rescue possibility and ditching the facility, something he never planned for and had to execute. The account of the hotel manager and others profiled in this book teaches an invaluable lesson: expect the unexpected and plan accordingly.

Bomb Threat

Organizations all over the country receive bomb threats every year by phone, mail, e-mail, or note at the facility. Most of these threats are geared toward one main purpose: to disrupt everyday activities, whether final exams, a church service, or a major event. Although many threats turn out to be benign, attention-getting mechanisms, you should not guess—call law enforcement immediately and they will conduct a thorough sweep of the building, possibly bringing dogs that can detect explosives and other high-tech equipment to ensure the building is clear. The bomb threat sheet provided in Appendix D should be close to all telephones so the call recipient can immediately write down the details from the call. Train those who answer your phones to stay calm and keep the caller on the phone as long as possible to gather information about gender, background noises, and so on. The conversation might even allow your staff member to ask when the bomb will go off, to include the time and date, and to query about whether the bomb has already been placed in the building and, if so, where. Practice is key to ensuring your employees respond properly when receiving a phoned-in bomb threat. Appendix D also contains the evacuation distance chart you can use to move people, vehicles, and the like away from the location immediately. Have an evacuation plan, post it, and practice often.

Medical Program

A good medical program is an important part of your security plan. During a medical emergency, unless at a hospital, your staff, building occupants, and visitors may not be able to think clearly and respond while being part of an intensely stressful situation, such as violent shooting. Therefore, it is important to have medical protocols in place that everyone is familiar with and to train to the worst possible scenario.

There have been cases of people freezing and being unable to dial 911. It was not because they did not know the number "911" but rather because an additional digit was needed to get an outside line, like dialing 9 first. To prevent these kinds of situations, simply place a sign or sticker on every phone to remind the caller to dial 9 for an outside line and then 911 in the event of an emergency. Recall that part of your hardening efforts will be to ask the staff if they have special skills, including first aid, CPR, or advanced emergency medical care. During an emergency, have a plan to find these people and get them to the scene. Your organization may want to sponsor training days with on-site classes delivered by the American Red Cross, American Heart Association, or the National Safety Council. Check with your staff, church membership, and parents of your students; you may have certified trainers who can teach one of these classes. CPR training for all is a must, and your organization should also consider purchasing automated external defibrillators, known as AEDs, now located in public areas such as airports and shopping malls. These devices are used to shock a person's heart back into a healthy rhythm and are designed for the lay person who has no training, as the unit gives verbal commands to the user on how to use the apparatus. Based on the size of your facility, you may want more than one AED. Make sure you perform routine checks to ensure the battery is charged; add this to your security checklist.

There are many helpful smartphone applications on the market with CPR and tourniquet instructions you and your staff could preload into your phone or tablet computer. We have compiled a list of emergency-related apps in Appendix E for your consideration.

A first aid station or a central repository for supplies is a good idea. Based on the size of the facility, you may want one on each floor. These range in all types of sizes, shapes, and prices. However, it is important to have at least a rudimentary first aid supply available in the event of an emergency, with many gauze pads and dressings, roller bandages, adhesive bandages including butterflies for deep cuts, tape, scissors, antibiotic and antiseptic salve, "space" blankets, and latex gloves. You may also want to have breathing barriers with one-way valves in case you need to administer CPR to multiple victims. Some kits have the equipment needed to run a peripherally inserted central catheter (PICC) line IV with fluids, as necessary. We recommend keeping these supplies packed in backpacks, as "Go Kits" for easy transport to the scene of the emergency. Finally, your organization should consider joining the Red Cross Ready Rating Program, which is outstanding and free. The program offers an online 123 point readiness evaluation for businesses, churches, schools, and other organizations to assess preparedness for an emergency and help to address vulnerabilities (American Red Cross 2014).

Finally, consider purchasing hoods for your employees. A product used extensively in the military is now available for purchase by individuals and companies. The victim rescue unit (VRU) escape hood is a head and respiratory protective hood that can be donned quickly and is compact, lightweight, and totally enclosed with its own oxygen supply. The VRU will protect the user from smoke, chemicals, biological agents, and radiation. It is available for order online (http://myescapehood.com/). You may want to place hoods in your vehicles, and pre-position enough for your ERT in the command center.

Shelter in Place

During emergencies happening either outside your property or on it, you may choose to direct the staff and building occupants to "shelter in place." This term is widely used by the military and government agencies; the concept originates from procedures taken during a nuclear, chemical, or biological attack. You should prepare for this worst scenario where you need to create a barrier between yourself and potentially contaminated air outside, a process known as "sealing the room" as a matter of survival. This type of sheltering requires prior preparation, planning, and practice.

First, designate a room or rooms with the least numbers of windows, a landline phone, and are stocked with water and nonperishable food supplies such as energy bars. Have a pre-positioned emergency kit with flashlights and a battery-powered radio, and do not forget extra packs of batteries. Gather all building occupants into the room(s), bringing in your first aid backpack "Go Kits," and then lock the doors and close the windows and all air vents. Turn off fans, air conditioners, and forced-air heating systems. Seal all windows, doors, and air vents with thick plastic sheeting and duct tape. Consider measuring and cutting the sheeting in advance to save time; it must be wider than the opening you are trying to cover. Secure the four corners first with duct tape, pulling the sheeting tightly across the opening. Then tape down all edges to form a seal.

"Shelter in place" is now a term commonly used during active shooter or other violent situations, as a way to keep people inside locked rooms. If sheltering during a violent attack, first lock the door from the inside and cover any windows on the door. Turn off the lights. Move as much heavy furniture as you can in front of the door to prevent it from being kicked in. Move to the back of the room. Tip over long tables and hide behind them in the far corner of the room, lying flat on the ground to lower your profile. Silence all cell phones and pagers and be absolutely quiet. If you are hit by debris or a stray bullet, do not cry out; remain as quiet as possible and play dead. For more on how to prepare your staff and building occupants for an active shooter situation, we recommend reading DHS's active shooter booklet (DHS 2014) and viewing the Houston Police Department's video "Run, Hide, Fight" (City of Houston 2013).

ELEMENTARY SCHOOLS IN FOCUS

School children are particularly vulnerable due to their size and inability to protect themselves, other than running fast and away from the scene, which several children in the Sandy Hook shooting were able to do. Children have more difficulty than adults discerning between real life and fiction; in the middle of a hostage situation or an active shooter event, they are likely to freeze in a state of suspended reality while trying to decide if it is a staged event or real life. A teenager or adult in today's society is more likely to realize the gravity of the situation and take some type of action to run, hide, or fight. Ralph Fisk, who works in the emergency management field and has experience with school preparedness, compiled the checklist in Appendix F to offer unique tactics for hardening elementary schools.

Marisa Randazzo is a former chief research psychologist for the U.S. Secret Service who applied a "threat assessment" model to examine the behavior of forty-one school attackers over the previous 26 years. She found there was no good "profile" of the type of person who becomes a school shooter. However, there were similar patterns of behavior. School shooters did not just "snap" and begin shooting impulsively; they planned. The attacker was vocal about his or her angst or intentions, trying to procure weapons, writing about the situation in journals and schoolwork. Randazzo states: "paying attention to changes in kids' behaviors and regularly conferring with one another about smaller threats is key to heading off bigger ones." Of the shooters she profiled in the study, Randazzo found: "These are not kids who were invisible—they actually were on multiple radar screens" (Toppo 2014). In 2014, there were several cases of concerned parents and friends reporting odd and worrisome behavior to school officials and law enforcement, a practice we must encourage, even though the informant likely does not want to get involved. Social media is not only a place for disgruntled staff to rant, but an outstanding recruiting tool. An unhappy employee who is vocal about his or her dissatisfaction with your organization makes them a target for bad actors looking for an insider to help plan and leverage an attack. Stay vigilant and monitor social media and the Internet through keyword searches to see if anyone is discussing your facility and operation.

As a result of the spate of school attacks in the recent year, law enforcement officials have stepped up efforts to hold active shooter drills at schools—sometimes, in a controversial move, using students as role players. On May 19, 2014, at Jefferson Middle School in Tennessee, emergency responders converged on the school after a "report" of shots fired and injuries. The school system fully participated and practiced the evacuation of hundreds of students by bus to a nearby church. Officers and paramedics had to find and subdue the shooters, as well as tend to twenty-two casualties, and school officials had to "lock down" the school and evacuate the students (Marion 2014). This type of realistic exercising is the best possible preparation for the school staff, teachers, students, and local law enforcement in the event of an active shooter situation.

HARDENING THE COLLEGE CAMPUS

Although all of the physical security procedures covered before can be applied, target hardening is not just about barricades and cameras, but also relationships, psychological preparation, and resiliency. Soft target hardening must also harness and apply the soft sciences.

Building Relationships

Several opportunities already exist for schools to partner with law enforcement to address vulnerabilities on campus and harden against threats; however, few engage. This is also a good model for other countries wishing to increase partnering activities between academia and government security organizations. The local FBI office can assist schools with points of contacts for these programs:

1. *The National Security Higher Education Advisory Board (NSEAB).* In response to increased concerns about security on college and university campuses and to open the lines of communication between academe and law enforcement, the FBI created the NSHEAB in 2005. The NSHEAB consists of nineteen university presidents and chancellors who meet on a regular basis to discuss national security matters that intersect with higher education. Previous panels included discussion on protection of weapons of mass destruction research and laws regarding domestic terrorism investigations on campuses. A new cyber subcommittee is addressing computer vulnerabilities on campuses.
2. *The College and University Security Effort (CAUSE).* FBI special agents in charge meet with the heads of local colleges and universities to discuss national security issues and share information and ideas. CAUSE is a conduit for schools to understand how to harden against the counterintelligence threat.
3. *National Counterintelligence Working Group (NCIWG).* NCIWG was designed to establish strategic interagency partnerships at the senior executive level among the U.S. intelligence community, academia, industry, and defense contractors.
4. *Regional Counterintelligence Working Group (RCIWG).* The RCIWG is a subset of the NCIWG and focuses on special vulnerabilities of local institutions and the threat.

Productive dialogue between education and law enforcement leadership will enhance security efforts. Colleges and universities would be better informed on the threat and mitigation opportunities and can in turn educate government security officials on the rights and protections afforded by the First Amendment in academia and the unique challenges facing our schools. In terms of procedure and policy, college administrators responsible for creating and executing human resources, training, and other programs designed to reduce vulnerability to infiltration and recruitment on campus will benefit. The dialogue will also educate legal personnel in higher education and technology transfer offices at higher education institutions responsible for the execution of sensitive government contracts. The following are suggested questions for colleges and their off-campus law enforcement counterparts:

1. What is the health of this relationship, perceived and real?
2. What is working between the two "tribes?"
3. What are the perspectives regarding which entity is ultimately responsible for protecting the higher education enterprise?
4. How do internal and external factors contribute to the relationship?
5. What policies, procedures, and training would enable a healthy partnership?
6. How can we ensure faculty and students are part of the solution through increased awareness and decreased vulnerability?
7. How can we open the lines of communication between higher education and the intelligence community?

Students and Staff as Force Multipliers

Often, we fail to share vulnerability information with those who can help us the most: the population we serve. As previously stated in the book, civilians are now the target and therefore have the right to engage in protective activities. Rationale for withholding threat information ranges from not wanting to scare people to not making the organization's weaknesses or vulnerabilities public for business or accreditation reasons. The culture must shift to one in which having a vulnerability and threat dialogue with customers and staff is seen as a sign of strength, not weakness.

As a result of our unwillingness to convey the threat, the "see something/say something" campaigns are largely ineffective and can flood the system with useless data. If citizens do not know the specifics, we will not fully leverage this incredible tool. A better approach would be: If you see (*fill in the "what"*), you say (*fill in what data we want them to collect*) to (*agency they should contact*).

For instance, with respect to meth labs on campus, we may want town pharmacies to be cognizant of repeat pseudoephedrine buyers and give the information to a local drug task force. Agriculture schools have a special resource coveted by bomb makers: fertilizer. Therefore, they must know how to protect the material and report any theft or suspicious activities. Small local airports should be trained to recognize suspicious drug or human trafficking activities. Schools hosting sensitive government research and development (R&D) contracts should ask law enforcement experts to train professors to understand their value to countries of interest and to recognize elicitation attempts by students and report them to the local FBI office. Students participating in sensitive R&D activities, in military ROTC units, and degree programs such as criminal justice and national and homeland security might also receive specialized training to make them force multipliers on campus. The administrative staff that handles the J-1 visa process should know who to contact if a student is a no-show for classes or leaves the university. Clearly, just starting the conversation is a hardening method.

Exercising Due Diligence

If you ask a college president who is responsible for protection from foreign theft, terrorist threats, or a criminal element on campus, he or she may point toward the local police department and FBI office. If you ask law enforcement, they may point to the campus leadership. In the past, schools have very much acted like victims when a national security or major criminal incident occurs, instead of accepting any type of responsibility. When a school reports a violation of government rules regarding R&D programs, it is typically only "slapped on the hand" and funding is not pulled. In April 2011, the FBI Counterintelligence Unit issued a white paper indicating the escalation of targeting and collection activities, asking universities to please engage to protect their programs (FBI 2011). Who is responsible for the existence of a major spy ring on a research campus, a meth lab in a dorm, or a student whose J-1 visa has expired, yet is still on campus? There is an urgent need for an honest dialogue about responsibility and establishment of punitive action against those who fail to engage properly. This could be the withdrawal of government R&D funds from a campus or a higher level government investigation into failure of local law enforcement to protect the school. As it stands, the lines of ownership are blurred—another vulnerability.

Colleges and universities are soft targets and extremely vulnerable to nefarious activities ranging from misdemeanor crimes to drug trafficking to infiltration by agents of foreign governments. As routine targets become less accessible, domestic and international terrorist groups might also prey on the open campus environment to recruit, spread propaganda, or even stage an attack. In addition to physical hardening, soft targets can be further protected by activities to educate the populace on the threat and build relationships to open lines of communication and ensure unity of effort during a crisis. The successful partnering of academia and law enforcement is essential for both to meet their critical missions. The overarching goal is a balanced and rational approach preserving our tenets of academic freedom and accessibility yet protecting colleges and universities from exploitation.

CHURCHES

All of the preceding hardening tactics certainly apply to churches. For the best example of securing worship services, we might turn to synagogues to gather their ideas and perspectives. Those of the Jewish faith have a history of persecution and Israel is actively targeted by Islamist extremists who have the stated goal of annihilating the country. Naturally, in the face of so many enemies, the Jewish people have unique concerns about securing their facilities. Many synagogues have always employed armed plainclothes security officers, who not only work the perimeter and entrances, but are also seated among the service attendees. Security was tightened with the shooting at a Jewish community center in Los Angeles in 1999, the al-Qaeda attacks of 9/11, and the shootings on July 28, 2006, by Naveed Haq, a self-proclaimed "Muslim American, angry at Israel," at the Jewish Federation office in Seattle. The attack at the Holocaust Museum in Washington, DC, on June 10, 2009, and the ambush of people in the parking lot of Jewish Community Center in Kansas City, Missouri, on April 15, 2014, further reinforced the need to protect those of the Jewish faith not only from radical Islamists but also American white supremacists.

The Sikh Temple shooting in Wisconsin in 2012 and the Mother Emmanuel AME church massacre in Charleston, South Carolina, in 2015 were both carried out by white supremacists. Mosques in the United States have been defaced and torched. Basically, every religion has an enemy that may lash out; therefore, all places of worship should take steps to harden their facility.

The SAFE Washington program (http://www.safewashington.com/) provides an outstanding model of cooperation between religious facilities and federal, state, and local law enforcement. For its eighty Jewish entities, SAFE also "develops best practices for disaster response, community security, community preparedness, and provides low cost or no cost training for community partners through annual training." The website has a password-protected area with secure files only for SAFE members. Another source is an outstanding tutorial entitled "Synagogue Security: The Basics" (Moses 2014) gives five pages of security pointed at those leading or securing synagogues. Guidelines for handling visitors, suspicious packages, and security incidents are thorough and tailored to the religious environs.

Following the attack on the Sikh temple, the Sikh community also took actions to better secure their facilities. Using security "sevadars" or volunteers, the temples are protected by trained individuals and the guidelines are to "act without fear, act without anger, act to defend the weak, act to protect the innocent" (Sant Sipahi Advisory Team 2014). As with the Jewish community, much can be gleaned from the activities of those previously targeted to secure their facilities and congregations. There must be a balance between security and the open, welcoming environment part of religious doctrine; however, ignoring risks and vulnerabilities is not prudent in today's world.

HOSPITALS

Most hospital planning efforts are centered on response to a natural disaster or mass casualty incident in the local community. Also, there is in-depth planning and training for staff on how to react to common hospital crimes such as violent outbursts in the emergency room, attempts to steal drugs, and domestic situations with spouses and parents. OSHA 3148 requires hospitals and health care organizations to do annual workplace violence assessments, and more than thirty-three states also require enhanced protection of hospital and health care staff, typically from enraged patients.

However, depending on their operations, hospitals themselves are targets for domestic terrorists such as antiabortion and animal rights activists. Examples presented in Chapter 3 show al-Qaeda and splinter groups are actively targeting first responders at the scene of the incident and then later at the hospital when victims and family members arrive for care. After a few near-misses at its own facilities, the United Kingdom seems to be leading research on the threat of terrorist attacks against hospitals. Recommended studies include "The Vulnerability of Public Spaces: Challenges for UK Hospitals under the 'New' Terrorist Threat" (Fischbacher-Smith and Fischbacher-Smith 2013), which started the conversation in England regarding the vulnerability of health care facilities.

Similar to churches, hospital culture dictates the doors are always open to the masses and restricting entry usually is not possible or desirable. Hospitals not only carry great liability for patient care but also many state patients' bill of rights/state licensing regulations direct patients "receive care in a safe environment." This naturally extends to protection from criminal and terrorist elements and attack. DHS recognizes nonprofit soft targets such as hospitals need financial assistance to bolster their security and include money in their budget for security enhancements; however, this money is usually apportioned to the states for further dissemination. FEMA has a direct funding program through the Nonprofit Security Grant Program (NSGP). In FY 2014, NSGP was funded to $13 million and plays an important role in the implementation of the national preparedness system by supporting the development and sustainment of core capabilities.

Core capabilities are essential for the execution of each of the five mission areas outlined in the national preparedness goal. The FY 2014 NSGP's allowable costs support efforts to build and sustain core capabilities across the goal's prevention, protection, mitigation, response, and recovery mission areas (FEMA 2014). Several hospitals received NSGP grants in 2014, including John T. Mather Memorial Hospital in New York, which used $75,000 to upgrade its security system, including a new camera system and cards and a card reader for one section of the hospital. Researching how hospitals are using the NSGP grants is a good idea before applying, in order to better tailor the request and posture the hospital for success.

Emergency room entrances are not the only concern; hospital loading docks also present vulnerability. International Association for Healthcare Security and Safety President Lisa Pryse describes loading docks as "volatile" and "often overlooked" and notes there have been two instances where active shooters entered hospitals through unsecured loading dock doors (Canfield 2013). Security cameras should be installed on loading docks as an absolute minimum and a vehicular access control system is also desirable.

MALLS

Naturally, as businesses trying to attract customers and make a profit, malls prefer to avoid heavy-handed security measures like metal detectors, armed guards, and bag screenings. They gravitate toward more passive measures, such as mass crowd surveillance and using human behavior theory to identify would-be troublemakers.

However, the Westgate Mall massacre served as a wake-up call to malls worldwide, which are now upgrading cameras and adding layers of security to protect their businesses and shoppers. Several lessons learned from the strategic response must be addressed in communities hosting malls. First, Kenyan officials did not act to protect their lucrative soft targets despite intelligence reporting on increased capability and threat of active, known, and capable al-Qaeda groups and chatter about their targeting of malls and other civilian venues. Therefore, the mall staff and security officers had no idea the threat was high and likely did not raise the security posture. Second, the police and military had no ability to coordinate and had never practiced communicating or working through thorny first-response issues such as who is the incident commander at a mass shooting event. This lack of coordination resulted in police being fired upon by the military while trying to rescue shoppers. Furthermore, military forces had only exercised a rescue scenario one time and this was their first real-life experience with a mass hostage situation; they were late to the fight and wholly unprepared, lacking even basic equipment such as night-vision goggles. As noted in the case study in Chapter 14, the behavior of security forces at the Westgate Mall lacked professionalism and discipline. Their rampant looting that trumped finding the terrorists prolonged the siege and exposed a degree of corruption that shocked the public and tarnished confidence in the forces' ability (and desire) to keep the population safe. There are many lessons learned from the Nairobi disaster to incorporate into our training and exercises. In light of the Nairobi attack and specific threats by ISIS and al-Qaeda, malls are now conducting large counterterrorism and mass casualty exercises.

In Portland, Oregon, joint training between local law enforcement and mall personnel paid off during an active-shooter event at the Clackamas Town Center in December 2012. Responding officers knew the mall layout from the training session and were able quickly to corner and stop the shooter, who had already killed two shoppers and seriously wounded a third in a random act of violence. The training initiative is one of many positive developments driven by the International Council of Shopping Centers (ICSC), which, in conjunction with the Department of Homeland Security, Federal Bureau of Investigation, and police, dramatically improved readiness throughout the mall and shopping center industry. Mall security, formerly ridiculed and scoffed at in pop culture, is now a highly trained, professional force (Bradley 2013). They are faced with a rising number of violent incidences ranging from assault to gang violence and mass shootings and are learning and sharing best practices globally.

A new study examined the infrastructure of the Westgate Mall and how it allowed for a successful asymmetric attack by a small group of men against a large group of first responders and military personnel and apparatus. For instance, the open atrium allowed the shooters to get a high position on top floors and shoot down at fleeing customers and arriving police and military. However, the atrium also provided an advantage for store owners on top floors, some of whom could see the carnage below and were able to lower their security doors and barricade themselves in the store. Enclosed areas, such as the casino and the cinema, were used as holding areas for hostages (Butime 2014). Studying the Westgate Mall attack from the perspective of the element of surprise achieved by the attackers, the vulnerability of shoppers, the physical layout of the mall, and the poorly coordinated response is critical for all who operate, secure, and visit malls.

SPORTS AND RECREATION VENUES

Sporting and recreational venues are actively targeted by terrorist groups who appreciate the large, dense crowds and televised coverage to ensure a ripple effect of fear across the populace. The main stadium in Paris was a target during the 2015 ISIS attack; however, the suicide bombers were unable to gain entry and forced to detonate their vests in the parking lot. Current security procedures in the United States include limiting entrance points, limiting the size of bags and thoroughly searching, and using CCTV and facial recognition technology.

Unfortunately, most venues rely on part-time, low-paid security guards who are both the first line of defense and the weakest link in the sports venue security infrastructure. In 2013, California revoked 154 security guard licenses, often due to criminal convictions discovered after the license was issued, and Florida revokes an average of more than 350 security licenses annually for criminal records. Compounding the problem, there is a "county option" approach to licensing and training of part-time security guards. States vary wildly in their procedures and seven require no security-guard licensing at all. Among those states requiring licensing, several, including Massachusetts, do not require training. In Florida and California, perhaps the strictest states, 40 hours of training are required, including a course on terrorism awareness and weapons of mass destruction. Some companies have classified employees as "event staff" in security roles at stadiums to avoid training requirements and to increase profits (Schrotenboer 2013). Also, in the quest to increase profits, sports venues often award security guard contracts to companies that are the lowest bidders. We need to remember part-time staff and volunteers provide vulnerability in terms of facility access and the ability to stockpile supplies. They can also glean an intuitive understanding of the infrastructure and its vulnerabilities.

According to an Israeli security consultant, there is another inherent problem with security in the United States: Security personnel "don't watch the race, they watch the crowd. That's what they didn't do [at the finish line of the Boston Marathon]" (Schrotenboer 2013). The same consultant explains how, in Israel, unattended packages and backpacks are given about 10 seconds before a security official engages. The next time you attend a sporting event, look at the security guards. Are they looking up into the stadium and scrutinizing people walking by? Or are they watching the game, concert, or event? A quick Internet search yields many pictures from major sporting events such as football and baseball games where professional and volunteer security team members are facing in the wrong direction—the most distressing at the Boston Marathon finish line, seen as the first bomb exploded (Photo 12.3).

Photo 12.3 CCTV capture: The first bomb explodes at the finish line.

In January 2005, the Department of Homeland Security launched the first online vulnerability self-assessment tool (ViSAT) for public venues such as large stadiums. The online tool incorporates industry safety and security best practices for critical infrastructure to assist in establishing a security baseline for each facility. Modules focus on key areas such as information security, physical assets, communication security, and personnel security. As part of the National Infrastructure Protection Plan, DHS also offers site visits and other helpful assistance to partner with owners, operators, and security at commercial venues (DHS 2014). There is a delicate balance between providing security and an enjoyable experience for participants in and spectators of sports and recreation events; with technology, training, and practice, this goal is attainable.

RED TEAMING SOFT TARGETS

As mentioned in Chapter 2, a healthy dose of imagination is needed to protect your facility from threats and to expose your vulnerabilities. According to the 9/11 Commission Report: "It is therefore crucial to find a way of routinizing, even bureaucratizing, the exercise of imagination. Doing so requires more than finding an expert who can imagine that aircraft could be used as weapons" (National Commission on Terrorist Attacks upon the United States 2004).

Every security measure has the opportunity to work, but if it fails, it works for the offender. Unfortunately, you, as the facility manager, operator, or security professional, may not see the flaws in your security plan, or be too close to imagine how your techniques could be defeated. In order to have the best plan and make sure your methods work, you truly need to think like the bad guy. Red teaming may be an answer. The term "red team" comes from American military war gaming, where the blue team was traditionally the United States and, during the Cold War, the red team was the Soviet Union. Defined loosely, red teaming is the practice of viewing a problem from an adversary's or competitor's perspective (Mateski 2014). The goal of most red teams is to enhance decision making, either by specifying the adversary's preferences and strategies or by simply acting as a devil's advocate. Red teaming may be more or less structured, and a wide range of approaches exists. In the past several years, red teaming has been applied increasingly to issues of security, although the practice is potentially much broader.

Superior red teams tend to (Mateski 2014):

- View the problem of interest from a systems perspective.
- Shed the cultural biases of the decision maker and, as appropriate, adopt the cultural perspective of the adversary or competitor.
- Employ a multidisciplinary range of skills, talents, and methods.
- Understand how things work in the real world.
- Avoid absolute and objective explanations of behaviors, preferences, and events.
- Question everything (to include both their clients and themselves).
- Break the "rules."

A red team can undermine a decision maker's preferred strategies or call into question his or her choices, policies, and intentions. As this might be uncomfortable, it is important to put the security of the innocent people who occupy your facility ahead of any ego, sunk cost, or group think about the security of your facilities and organization.

In particular, the Homeland Security Act requires DHS to apply red team analysis to terrorist use of nuclear weapons and biological agents. As terrorists seek to exploit new vulnerabilities, it is imperative the appropriate tools be applied to meet those threats. Therefore, most red teaming effort currently lies in the WMD spectrum, with professional teams trying to penetrate nuclear facilities, chemical and biological weapons labs, and even military installations owning or operating these sensitive activities. A red team is a group of subject-matter experts (SMEs) with various appropriate backgrounds that provides an independent peer review of your processes, acts as a devil's advocate, and knowledgeably role plays the potential enemy. Red teaming can be passive and serve to help you understand the threat and expose your biases and assumptions. Or, activity can be active as the red team attempts to probe and test your security to expose your strengths and flaws. Training is another aspect of red teaming.

There is cross-application of methodology, currently employed by the U.S. government, to soft targets including schools, churches, and hospitals. Perhaps begin with the cross-inspection of security procedures by trusted colleagues from other facilities and the cross-pollination of ideas. Or, you may ask them to test your security by sending someone in to try and penetrate your system. One red teaming exercise recently shared by a colleague included the "perpetrators" wearing a shirt bearing the symbol and name of a famous soft drink company. Holding a clipboard and stating the purpose was to check the soda machines, the red teamer had unlimited access to a school.

For more about the red teaming concept, please see the homepage of *The Red Team Journal*, http://redteamjournal.com/, started in 1997 by Dr. Mark Mateski, the industry expert on the topic.

CONCLUSION

Unfortunately, the world has changed drastically since 9/11 and the places we should feel the safest and go for relaxation and recreation are in the terrorist's target book. Fortunately, other sectors such as the military provide a model for soft target hardening activities. Case studies of soft target attacks provide rich examples of successes and failures, enabling you to accurately assess your own preparation activities.

You are no longer helpless against this rising threat: you can now confidently move forward and prepare your facility, staff, and users for the unthinkable.

REFERENCES

American Red Cross. Ready Rating Program. http://www.readyrating.org/. 2014.
Bradley, Bud. State of U.S. Mall Security Post 9-11. December 17, 2013.
Butime, Herman R. The lay-out of Westgate Mall and its significance in the Westgate Mall attack in Kenya. *Small Wars Journal*, May 10, 2014.
Canfield, Amy. Hospital loading docks rival ERs for security concerns. *Security Director News*, November 25, 2013.

Carpenter, Mike. Evolving to Effects Based Operations. http://www.dodccrp.org/events/9th_ICCRTS/CD/presentations/8/092.pdf. March 2004.
City of Houston, Texas. Run, Hide, Fight. http://www.youtube.com/watch?v=5VcSwejU2D0. 2013.
DHS (Department of Homeland Security). Active Shooter: How to Respond. http://www.dhs.gov/xlibrary/assets/active_shooter_booklet.pdf. 2008.
DHS (Department of Homeland Security). National Infrastructure Protection Plan: Commercial Facilities Sector. http://www.dhs.gov/xlibrary/assets/nipp_commerc.pdf. n.d.
DHS (Department of Homeland Security). Soft Target Awareness Training for Facility Managers, Supervisors, and Security and Safety Personnel: Course Syllabus. https://www.dhs.gov/xlibrary/assets/nipp_commerc.pdf. 2016.
FBI (Federal Bureau of Investigation). Higher Education and National Security: The Targeting of Sensitive, Proprietary, and Classified Information on Campuses of Higher Education. 2011.
FEMA (Federal Emergency Management Agency). Fy 2014 Urban Areas Security Initiative and Nonprofit Security Grant Program. http://www.fema.gov/fy-2014-urban-areas-security-initiative-uasi-nonprofit-security-grant-program-nsgp. 2014.
Fischbacher-Smith, Denis, and Moira Fischbacher-Smith. The Vulnerability of Public Spaces: Challenges for UK Hospitals under the "New" Terrorist Threat. *Public Management Review* 15(3): 330–343, 2013.
Kuhns, Kristie R., and Joseph B. Blevins. *Understanding Decisions to Burglarize from the Offender's Perspective*. The University of North Carolina at Charlotte Department of Criminal Justice & Criminology, Charlotte, North Carolina, 2013.
Lysiak, Matthew. *Newtown: An American Tragedy*. New York: Gallery Books, Simon and Schuster, 2013.
Marion, Steve. Mock school attack tests readiness. *The Standard Banner*, May 20, 2014.
Mateski, Mark. Red Team Journal: Understand, Anticipate, Adapt. http://redteamjournal.com/. 2014.
Moses, Manfred. Synagogue Security. 2014.
National Commission on Terrorist Attacks upon the United States. 2004: 344.
Office of the State's Attorney Judicial District of Danbury. Report of the State's Attorney for the Judicial District of Danbury on the Shootings at Sandy Hook Elementary School and 36 Yogananda Street, Newtown, Connecticut. 2013.
Rose, Roger, M., A. Vangura, Jr., and M. Levin. Wrongful Death, Failure to Deter Crime on the Premises. *Zanin's Jury Verdict Review & Analysis*. http://www.jvra.com/verdict_trak/article.aspx?id=187168. 2009.
Sant Sipahi Advisory Team. Security and Risk Assessment. http://www.harisingh.com/SantSipahiAdvisoryTeam.htm. 2014.
Schrotenboer, Brent. Holes in Stadium Security, May 2, 2013.
Toppo, Greg. Nerves Fray as Anniversaries of April Attacks Arrive, April 19, 2014.

13
Soft Target Threat Assessment
Schools, Churches, and Hospitals

Jennifer Hesterman

Contents

Introduction 220
Vulnerability in the United States 220
Small Cities and Rural Areas in the Crosshairs 221
How Bad Actors Choose Targets 222
The Soft Targets 223
    Schools 223
        Stabbings: A Growing Concern 224
        K–12 in the Crosshairs 226
        Colleges: Campus Culture and Vulnerability 234
    Churches 242
        The Proliferation of Church Attacks 243
        U.S. Churches as Targets 245
    Hospitals 246
        Nefarious Use of Ambulances 248
References 250

> Kazbek Misikov stared at the bomb hanging above his family. It was a simple device, a plastic bucket packed with explosive paste, nails, and small metal balls. It weighed perhaps eight pounds. The existence of this bomb had become a central focus of his life. If it exploded, Kazbek knew, it would blast shrapnel into the heads of his wife and two sons, and into him as well, killing them all.
>
> C. J. Chivers
> *"The School" (New York Times 2007)*

INTRODUCTION

Kazbek Misikov and his family started September 1, 2004, on a wonderful note; it was the first day of school and he, his wife, and two sons dressed up and went to the Beslan School for the special "Day of Knowledge" welcome-back ceremony that began at 9:00. At 9:20, with 1,200 children, teachers, and staff in the gym, Chechen separatists attacked. The next 4 days were hell on earth for those who were unable to escape; Mr. Misikov and his family were injured, but fortunately, came out alive. Sadly, 330 innocent people lost their lives in the siege, including 108 children.

Every day, somewhere in the world, a school, church, or hospital is attacked by a terrorist or insurgent group, usually one that has also threatened the United States or its citizens. The psychological vulnerabilities covered earlier are only part of the challenge related to protecting people from these types of heinous soft target attacks. There are also physical vulnerabilities that make soft targets more attractive to would-be attackers—some that can be mitigated, and others that cannot. Prior to discussing individual venues, we start with a strategic view of the United States to assess vulnerability of certain regions and cities and then we explore the targets themselves.

VULNERABILITY IN THE UNITED STATES

If you were asked to name the top four U.S. cities most vulnerable to a terrorist attack, the natural response is those with large populations and symbolic targets such as Washington, DC, and New York City. Would it surprise you to know that a government-funded study, conducted by mathematicians and statisticians, produced a wholly different list?

In 2007, a groundbreaking report rated 132 U.S. cities on their vulnerability to terrorist attack using a newly developed statistical method. "Benchmark Analysis for Quantifying Urban Vulnerability to Terrorist Incidents" was written by Dr. Susan Cutter, a hazards and vulnerability expert and Dr. Walter W. Piegorsch, a leading statistician and environmental risk expert (Piegorsch, Cutter, and Hardisty 2007). The study was funded by the Department of Homeland Security and yielded some unanticipated results.

The overarching goal was to analyze the relationship between location vulnerability and terrorist outcome. Taking a new approach, the study calculated the susceptibility of areas to attack by assessing socioeconomic factors and demographics as a way to predict the impact of an attack on the populace, and to assess the likely response of residents. Natural geographic features and environmental hazards were also considered, as well as critical industries, ports, railroads, bridges, tunnels, water/sewage systems, and the age and fragility of the existing infrastructure. Finally, the team analyzed and factored in historical data from the terrorism knowledge base and the global terrorism database. Although most Americans can only recount a handful of terrorist attacks if asked, the research considered over one thousand unique terrorist-related incidents in U.S. cities, spanning a 30-year period. A surprising result is that, contrary to typical threat assessment criteria, areas with nuclear power plants and military facilities did not come up as "high risk" in the analysis.

The final computation of factors resulted in a place vulnerability index, or PVI, rating. Overall results presented on the chart in Figure 13.1 indicate the eastern and southern seaboards of the United States are at greatest risk and also show a large swath of vulnerability

The cities that scored the highest:

Cities Scoring the Highest Place Vulnerability Index (PVI)

City	PVI
New Orleans, LA	3.110
Baton Rouge, LA	3.061
Charleston, SC	2.543
Norfolk, VA	2.326
New York/Newark area	2.154
Washington, DC, area	1.978
Houston, TX	1.844
Philadelphia, PA	1.737
Boise, ID	1.696
Atlanta, GA	1.683
Chicago, IL	1.404

Figure 13.1 A place vulnerability index.

from Texas to Ohio. Cities scoring the highest place vulnerability index (PVI) can be seen in Figure 13.1.

Other major areas typically thought of as vulnerable were much lower in ranking, even falling below Columbia, South Carolina, which had a PVI of 1.117. For example, Los Angeles, California, had a PVI of 0.421; Dallas, Texas, a PVI of 0.447; and Denver, Colorado, a PVI of 0.171. Surprisingly, Seattle, Washington's PVI was very low, at –0.315. To view the map with color code results and more detailed computation information, please access the study (Piegorsch et al. 2007).

SMALL CITIES AND RURAL AREAS IN THE CROSSHAIRS

Cities previously hit or targeted by terrorists have been hardened with hundreds of millions of dollars of taxpayer money. But those few urban areas represent less than 5 percent of the U.S. population; what about everyone else? Not only are the top four cities identified in the study as vulnerable to attack surprising, but Boise, Idaho, with a population of 220,000, emerged as a new area of concern, the lone red "blip" in the northwest. Although funded by the government, the study's results were not widely disseminated to the public or covered by the news media. Boise city officials acknowledged surprise at the results, engaging with both Piegorsch and the state's Homeland Security officials for clarification. In 2008, I blogged about the study's results on the Internet and was contacted by several concerned citizens from Baton Rouge and chided by one for "painting a bull's-eye" on his city. Instead of responding emotionally to studies such as this one, it is more helpful to appreciate how the study illuminates a pathway to greater awareness, inspires businesses and community institutions to review security procedures and review terrorism insurance coverage, and may even be used as a vehicle to lobby the government for target-hardening funding. Therein lies the conundrum: Do citizens truly want to know their vulnerability to terrorist attack? Bad actors hope we become passive bystanders or, worse, remain in blissful denial so they will not have to fight us on the way to the target.

The standard perception in our country is that smaller cities and rural areas are not vulnerable to attack by international or domestic terrorists. For example, the Terrorism Risk Insurance Act (TRIA) was enacted in 2002 as a response to the 9/11 terrorist attacks as a public–private partnership requiring private insurance companies to provide terrorism risk coverage, with federal funding as a backstop if costs to insurance companies exceed $100 million. The law was renewed in 2005 and 2007 and is set to expire in December 2014. Critics argue that not only does the law place an unnecessary burden on taxpayers, it puts an undue cost on those living in rural areas with "less of a terrorism threat" (Novak 2014). The critics are not looking at vulnerability studies or data from our law enforcement agencies that show rural areas are now being preyed upon and exploited by violent gangs, drug cartels, and organized crime.

For instance, in 2009, the DEA discovered during Operation Xcellerator that the Sinaloa's drug cartel hubs were not in major metropolitan areas as previously thought, but rather the cartel was using unsuspecting rural areas to move product. One distribution center was located in Stow, Ohio, a quiet community of 35,000. The Sinaloas were using the small airport to move drugs between Stow and distribution hubs in California (U.S. Department of Justice 2009). In rural Rhode Island, the La Cosa Nostra (Mafia) allowed the Hells Angels outlaw motorcycle group to continue illegal trafficking activities on their turf without interfering, in return for the bikers paying a percentage to the mob. According to the FBI, the unlikely, dangerous relationship between the Hells Angels and the Mafia continues today (WPRI Newport, Rhode Island, 2008). Who would predict quiet Stow and rural Rhode Island would be host to such violent transnational actors? The presence of organized crime or cartel activities in a rural area is a red flag for an even more vexing threat: terrorists often prey on established routes, and groups with different goals are increasingly working together.

The place vulnerability study teaches us much about risk assessment, and its methodology can be applied in any country to similarly illuminate the gap between perceived and actual vulnerability. The study's surprising results indicate that probability, not vulnerability, is driving the government's security focus and funding streams.

HOW BAD ACTORS CHOOSE TARGETS

Before discussing the soft targets themselves, it is essential to get inside the minds of criminals, especially killers. Dr. Martin Gill is a renowned British professor of criminology and author of several outstanding publications that view crime through the eyes of the criminal (Gill 2014). His field work includes interviews with convicted killers about their crimes, to assess motives and the real versus perceived usefulness of closed circuit television (CCTV), store security guards, alarm systems, and other physical security apparatus. When posed the question regarding how criminals choose their targets, Dr. Gill's answer is simple: because they are easy. Gill found that CCTV does not affect the way criminals commit their offenses, as they merely wear a disguise or pull a hat down low and look at the ground. They bet the cameras are not working or if they are caught and facing prosecution, the burden will be on the prosecutors to prove it is them in the video. In fact, some criminals have shared that if they believe during the course of the crime that their image has been captured on CCTV, the severity of the crime may escalate. One convicted murderer commented that if he was going to spend a long time in prison anyhow, he might as well kill, saying "They've taken away my incentive *not* to kill." Another stated, "If I'm going to jail for armed robbery and I need to kill someone in order to get away, I may take that next step." In fact, one man in prison for murder said that if it is a risky crime, the decision to shoot and kill a lone security guard is actually easy. It eliminates the one piece of evidence that will always stand in court: visual recognition of the perpetrator. Criminals report being more concerned about being stopped by people than any type of technology. They also note favoring large, bulky security guards because they can be outrun. As shown in this example regarding the potential ineffectiveness of CCTV, seeing the potential crime scene through the eyes of a criminal is invaluable to the discussion of target hardening.

THE SOFT TARGETS

Although successfully hitting a hard target such as a government building, military base, or a symbolic target brings credibility to a terror group, a soft target attack would certainly damage the national psyche and discredit the government's ability to protect its people. Factor in the use of a chemical or biological agent, or radiation, and the impact could be immense on sectors such as tourism, shopping, and recreation.

All citizens have the right to learn, worship, and receive medical care in a safe environment. However, we all can easily slip into a false sense of security in these facilities and become complacent about safety. Security is also not the primary goal of these institutions, which are typically resource constrained and do not have money to spend on extra security measures or guards, adding to their vulnerability. Also, these are typically "gun-free" zones, so the only resistance a bad actor will meet is a security guard or two, typically not armed. This combination of factors makes schools, churches, and hospitals targets of choice for terrorists or homicidal killers.

Schools

On April 7, 2011, a 24-year-old man named Wellington Oliveira traveled to the Tasso da Silveira Municipal School in Rio de Janeiro where, as a former student, he had been subject to bullying. He methodically killed twelve students. A firefighter who responded to the scene told newspapers "There is blood on the walls, blood on the chairs. I've never seen anything like this. It's like something in the United States" (Johnson 2012). His statement illustrates the prevailing worldview toward the escalation of violence in our country, especially with the recent epidemic of school shootings and stabbings that garner wide press coverage.

Perhaps nothing more deeply affects the American public than an attack at a school. It is never expected that innocent children would be targeted by anyone, be it a fellow student, a member of the community with a mental illness, a criminal, or a terrorist. Therefore, we are wholly unprepared, shocked, and deeply saddened. The ripple effect of school attacks is also immense, traumatizing students, teachers, and first responders who view the scene, inducing posttraumatic stress and panic disorder in many.

At any given time, there are at least 75 million Americans attending some type of school from kindergarten (K) through doctorate level courses, and there are five million teachers, administrative, and support staff on campuses (DHS 2012). Many schools also serve community needs and can be used as places for meeting or polling, or as shelters in times of emergency. Even if schools are not the terrorists' intended target and their act of violence is a city building or mass transit, children must be protected from physical and emotional side effects of being in proximity to such horrific violence. For example, there were four elementary schools and three high schools located within six blocks of the World Trade Center on 9/11, and thousands of children were exposed to the toxic dust clouds from the collapsing buildings. Children in at least three states had parents working in or around the World Trade Center that day; in the Washington, DC, area, school children faced similar stress when the Pentagon was attacked (CDC 2003).

The top two deadliest mass shootings by a single person in U.S. history both occurred on school campuses. On April 16, 2007, 23-year-old Seung-Hui Cho killed two students in his dormitory and then went to a classroom building, barricaded himself inside, and shot fifty-three students and teachers, killing thirty in just 9 minutes. On December 14, 2012, Adam Lanza killed his mother in their home and then went to the Sandy Hook Elementary School in Newtown, Connecticut. There, he bypassed the security door and shot through a plate glass window to gain entrance to the building. He killed twenty first-graders and six staff members in only 6 minutes. Since the 1999 Columbine attack by students Eric Harris and Dylan Klebold, which killed fifteen students and injured twenty-four, there have been thirty other major school shootings in our country. The terrorists are watching and see the relative ease of attacking a school and attaining high casualty counts.

Stabbings: A Growing Concern
Gun attacks are obviously the most feared assault, since mass casualties are inflicted in a short period of time. There were at least 165 school shootings in the United States between 2013 and 2015 (EveryTown Research 2016). However, as our focus naturally centers on keeping guns out of schools with recent mass shooting incidents, we can't forget that knives are easier to conceal and transport and can also inflict significant bodily harm in a short period of time.

We confiscate more knives in schools across the country than guns, and knifing incidents are on the rise across the nation. Between April 2012 and November 2015, there were 24 serious or fatal stabbings at schools in the United States and many more worldwide in surprising locations such as Ireland and Sweden. In the 2010–11 school year, according to the Department of Education, public schools reported 5,000 cases of student possession of a firearm or explosive device, and an alarming 72,300 cases of possession of a knife or other sharp object.

Since school behavior is reflective of societal trends, it is also important to note that knifing deaths in the United States are prevalent and on the rise; there are over 1,600 stabbing murders annually, constituting 16 percent of all homicides. One factor driving a perpetrator to use a knife instead of a firearm is that knives are much easier to obtain. In many recent attacks, perpetrators simply used kitchen or hunting knives. Also, for those hoping to escape anonymously following an attack, knives are usually untraceable. After pregnant Brenda Paz, 17, was stabbed 20 times by an MS-13 gang member and left to die in a creek near my home in Virginia, I researched the gang's penchant for using knives and machetes in attacks versus firearms. Often, the stabbing or slicing weapons used by MS-13 were purchased at hardware stores or mall kiosks, and paid for with cash. Police were often frustrated that even if they had the weapon in an MS-13 attack, it was virtually untraceable. Unless a knife has a decorative handle, is an antique or unique in another way, it is very difficult to trace to the sale point and buyer. Also, MS-13 prefers knifings since they are more up close and personal, appealing to a certain type of killer who wants to inflict prolonged suffering or a leave a lingering reminder of the attack. This type of information is helpful as we psychologically profile potential attackers.

Knife attacks are also fast, unexpected and can be devastating in terms of injuries. On December 14, 2012, just hours before the Sandy Hook shooting, a 36-year-old villager in the village of Chenpeng, Henan Province, stabbed 23 children and an elderly woman at the village's primary school during morning arrival. The attack took only a few minutes and the perpetrator was eventually restrained by teachers and arrested. All of the victims survived, however many lost fingers or ears, and some sustained serious internal damage to organs, requiring long-term care. Attacks on elementary schools are of great concern since young children are basically defenseless and unable to protect themselves.

High school attacks can also be devastating; on April 9, 2014, 16-year-old Alex Hribal used two kitchen knives to stab 22 victims in their stomachs and lower backs at Franklin Regional High School outside of Pittsburgh. The attack took less than 5 minutes and Hribal was eventually subdued by brave students and a gym teacher. Stabbings at colleges are also on the rise; in November, 2015, ISIS sympathizer Faisal Merced used a hunting knife with a 10-inch blade to quickly and gleefully stab four students at the University of California. He was shot and killed by responding police when refusing to drop the weapon.

Therefore, all types of schools have proven vulnerable to stabbing attacks, which can be very fast and lethal. Another challenge is the presence of a knife in school may not be alarming. A good example comes from a recent stabbing attack at a primary school in Sweden in October 2015. One witness, a student, told broadcaster SVT the attack was initially believed to be a joke and attracted students, pulling them in close to the killer: "He wore a mask and black clothes. There were students asking to take pictures with him and touch the sword." Obviously a student wielding a gun would be treated quite differently.

The physical response to stabbings presents yet another challenge. In stabbing events, victims often do not realize what is happening until they start bleeding, delaying their "fight-or-flight" response and allowing the attacker more time to engage. During a knife attack that involves puncture wounds, with the attacker plunging in and out of the body, victims often don't feel pain. There will be a cold, icy feeling at the stabbing site as the body goes into shock, and a person may only realize they are injured with the actual presence of blood. Panic, disbelief, and confusion will set in at the large volume of blood lost from the penetration site, rendering the victim even more unable to save themselves from an ongoing violent situation. Interestingly, there aren't many pain sensors inside our bodies; in fact, a large blade penetrating a critical organ results in death so quickly that most people pass out and die without much pain. Slashing attacks are quite different and painful, since the skin has many nerve endings. However, most premeditated stabbing attacks have the perpetrator plunging the knife or knives (one in each hand is a popular tactic) deeply into victims. The superficial slashing only begins when the attacker is confronted and shifts to the defensive, vice offensive mode.

In reviewing stabbing case studies, it appears people are more willing to engage and try to subdue an attacker with a knife, while choosing to run or hide from a gunman. However, physically confronting an enraged assailant wielding large knives is very difficult, and those who approach are likely to be severely injured or killed. In stabbing attacks, the use of committed, overpowering force by more than one person is necessary, or neutralizing the attacker with a disabling (or more) potent weapon like a firearm. This is where "bringing a gun to a knife fight" might actually be a good tactic.

In terms of mitigation, knives are easily detected at school entrances with metal detectors or, in their absence, by random wanding in hallways. Unannounced locker inspections are also a good way to stay on top of non-firearm weaponry such as knives, nunchucks, brass knuckles, and batons. Regularly look in closets and restrooms for hidden knives, perhaps in ceiling tiles, taped in toilet tanks, and behind moveable objects. Also discuss the rising possibility of knife attacks when discussing active shooter scenarios or conducting desktop or live exercises in your schools. Think about response and ways to engage and disable the attacker, as well as render first aid to stabbing victims when precious minutes can mean the difference between life and death. Finally, remember teachers, students, and staff are the new first responders, as most attacks are over and damage inflicted before the arrival of law enforcement and medical personnel.

The kindergarten through twelfth grade vulnerability differs from that on a college campus. First, a younger populace cannot defend themselves as readily and are more likely to slip into suspended disbelief as the situation unfolds instead of experience a fight-or-flight response. College campuses have their own vulnerabilities related to the culture of openness, physical location in the community, poor partnering with or distrust of external law enforcement, and nefarious elements to which they are already exposed such as drugs and even espionage. Examining the unique vulnerabilities associated with K–12 and colleges and universities helps to understand trends, the risk of attack, and mitigation challenges better.

K–12 in the Crosshairs
The first K–12 school attack in the United States was the Enoch Brown School massacre, which occurred July 26, 1764. On this date, four American Indian warriors entered a white settler's log cabin school in Greencastle, Pennsylvania, and used a tomahawk to kill and scalp the teacher and ten students. Throughout the years, primary and secondary schools have been the sites of revenge murders, racial attacks, gang violence, suicides, workplace violence, and lover's quarrels. They have also been used by domestic terrorists as a way to express rage and garner attention to their cause. For example, on May 18, 1927, the Bath Consolidated School was the scene of the deadliest act of mass murder in a school in U.S. history, a lone-wolf, antigovernment attack. Andrew Kehoe, upset with policies and tax law he believed led to his farm foreclosure, murdered his wife at home and then detonated three dynamite bombs at the Maine school, where he worked as the accountant. Kehoe spent months planting explosive material throughout the building in a premeditated act that stunned the country. When confronted at the scene by law enforcement, he detonated

a vehicle bomb, killing himself and the school superintendent. In all, the attack killed thirty-eight school children and five adults.

Modern-day schools have been used as political targets by international terrorist groups and embattled governments. Students have been the victims of bombings, shooting, and kidnapping and hostage situations. In the last 40 years, there have been massacres at the Ma'a lot school in Israel, the Bahr el-Baqar school in Egypt, the Beslan school in Russia, and the Nagerkovil school in Sri Lanka. Schools in the Gaza Strip, Iraq, and Afghanistan are routinely targeted by insurgents, and, in Syria, school children are bombed by their own government, which is engaged in a worsening civil war. Mass student kidnappings became a new fear when terrorists from the al-Qaeda–linked group Boko Haram posed as soldiers to gain trust and then kidnapped more than 270 girls from their boarding school in Nigeria on April 16, 2014. Boko Haram leaders threatened to sell them into marriage and the sex trade for $12 apiece, to raise money for the cause. Boko Haram next attacked the village where the girls were from, killing 150 family members and search-and-rescue team personnel. The Nigerian army is unable to repel Boko Haram attacks and protect citizens.

The 2004 Beslan school attack in Russia is difficult to read about, and the pictures are extraordinarily painful to view. However, to glean valuable information for hardening efforts, we must confront the harsh reality of these attacks. This case, in particular, provides insight into the mentality of terrorist groups and actors that will target defenseless children, as well as how poor response by the government to a mass hostage situation can lead to even more bloodshed.

Case Study: Siege at Beslan School Number 1—Innocence Lost

Since the advent of modern terrorism in 1968, hostage taking is a tactic often used in terrorist operations, whether as a bargaining chip or to generate additional public fear. September 1, 2004, again proved another employment of the tactic, as over thirty Chechen Islamic militants took control of Beslan School number 1, in Beslan, North Ossetin-Alania, Russia. Terrorists stormed the school and quickly took 1,200 hostages, including 705 children. Unfortunately, the siege had a horrific ending, but only by revisiting the case can we learn how to prevent a similar massacre in our own country.

Beslan is a picturesque town of 30,000 near the beautiful Caucus Mountains. Unfortunately, the North Caucasus region also serves as a base of operation for al-Qaeda operatives, among other jihadists. The "Day of Knowledge" is the first day of the new school year in Russia, when students attending for the first time are greeted in a festival-like atmosphere, and those entering their last year are given flowers by the younger children as a sign of congratulations and good luck. On this fateful day, parents were also in attendance, pushing the population inside the school to maximum capacity. School started in Beslan at 9:00 am that morning, the celebration having been pushed up an hour because of the forecasted high temperatures. At 9:20 am, terrorists seized the school, gathering those who could not escape and herding them inside the gymnasium. The terrorists were wearing camouflage uniforms and appeared to be Russian security forces conducting a counterterrorism drill, so the adults were initially unsuspecting of their activity. Overwhelmed by the size of the terrorist group and their firepower, the adults could do nothing but obey the orders of the hostage takers and sit through three days of unimaginable terror with the children.

SHAMIL BASAYEV

To understand the Beslan School siege, one needs to understand the group's leader, Shamil Basayev, who approved, planned, and ordered the Beslan attack. Basayev was no stranger to Russian security forces, leading guerrilla operations against the Russian military in both the first and second Chechen Wars, and as the leader of the radical wing of the Chechen insurgency. He is in a unique position to plan attacks, intuitively understanding how forces will respond and their tactics and vulnerabilities. His goal is recognition of the independence of Chechnya and UN and Russian withdrawal from Chechnya. No stranger to large civilian hostage operations, Basayev also planned the soft target attack against 916 civilians at the Dubrovka Theater on October 23, 2002, a case study covered in Chapter 6 of this book. In addition to attacks on government buildings, Basayev also ordered the suicide bombing of two Russian Civil Aviation aircraft on August 24, 2004. He directed an attack at the Rizhskaya metro station in northeast Moscow on August 31, the day before his Beslan operation. In all, over one hundred persons were killed in these three attacks.

TACTIC: BLITZ ATTACK

On the morning of the attack, children and adults dressed in their finest clothes and arrived at Beslan School number 1. Waiting in the distance were thirty-five Chechen militants in military vehicles, ready to conduct a surprise attack on the school and, with overwhelming force, seize as many hostages as possible. With violence of action and confusion as to whether they were members of the Russian military, their plan was successful and they corralled most of the students and adults into the school's gym. Fifty people escaped and alerted police, and several were able to hide in the boiler room. Upon responding, police, who were unaware of the scope of the incident, wildly exchanged gunfire with the terrorists; five police officers and one terrorist were killed. Realizing the situation exceeded their capabilities, a call was put in to the Russian army for assistance.

The terrorists moved a majority of the hostages into the gymnasium. Then, with tactical proficiency, they began to construct defensive fortifications. They established sniper positions and had male hostages move desks and chairs around the entrances of the building to block potential counterattacks. The terrorists also placed improvised explosive devices (IEDs) in a "daisy-chain" configuration in the gymnasium, placing some in basketball hoops and above the heads of the hostages for the maximum casualty-producing effect (Figure 13.2). They tactically positioned terrorists in strategic locations throughout the school to repel an outside assault. Later in the morning, hostages, mostly children, were staged at the windows as human shields.

BLACK WIDOW INVOLVEMENT

As discussed in Chapter 3, the Chechen Black Widows are an asymmetric and formidable force in the terrorist group's arsenal. Another tactic used at the Moscow Theater

SOFT TARGET THREAT ASSESSMENT

Figure 13.2 IED placement in the Beslan School Gym.

siege and during the Beslan operation was the use of female terrorists rigged as suicide bombers. By the conclusion of day 1, it is believed three of the female suicide bombers were dead. One was killed by a fellow terrorist after protesting the treatment of the hostages, specifically the children, and another died during the malfunction of her bomb. A third may have been used to detonate a bomb to kill the strongest 20 male hostages, selected by the terrorists from the group in the gymnasium. Their bodies were dumped out of a second floor classroom into the schoolyard (Chivers 2007).

PSYCHOLOGICAL TERROR

At the beginning of the siege, two male hostages were killed in front of the others in the gymnasium, their bodies dragged through the crowd. Additional threats were made by the terrorists to the government response team assembled outside the school: If any terrorist was killed by Russian security forces, fifty hostages would be executed and, if one was wounded, twenty hostages would be killed. Cell phones were confiscated to reduce the risk of hostages transmitting information to the outside, and the threat was made that anyone caught with a cell phone would be executed, as well as anyone around that person. The hostages were told that if disobedience or resistance was suspected, the hostage being disobedient would not be killed, but everyone around him or her would. Only Russian was allowed to be spoken by the hostages. To show their intent to follow through on these threats, the terrorists allowed a male to address the group. In an attempt to calm the hostages, he did not speak in Russian, and instead spoke in the local Ossetian language; he was shot and killed in front of the other hostages (Terror Operations: Case Studies in Terrorism 2007).

LOGISTICS

From the terrorists' perspective, food, water, and medical supplies were robust because they had total control of the school and its assets. They also had a large stock of ammunition and weapons; every terrorist was armed with a Kalashnikov rifle, and there was a cache of sniper rifles, RPG-7 grenade launchers, and machine guns, hand grenades, pistols, dynamite, and at least one hundred IEDs (Terror Operations: Case Studies in

Terrorism 2007). The sheer amount of weaponry and the terrorists' apparent familiarity with the building led some to speculate there was on-site reconnaissance and storing of supplies during the summer while the school was being renovated (McDaniel and Ellis 2009).

The hostages were deprived of food, water, and medical supplies by the terrorists during their captivity; children and adults even had to resort to drinking their own urine. By day 3, a significant number of children and adults began showing signs of exhaustion, dehydration, and food deprivation. Many of the hostages began fainting as a result of not having any food or water and being forced to stand for lengthy periods of time. This deprivation and mental confusion was evident on day 3 when the attack and counterattack started, as some of the escaping hostages were seen running back into the school during the fighting. The terrorists were also growing weary of the standoff and became more brutal to the hostages.

DEMANDS

One of the survivors recalled her conversation with a terrorist; when she asked him why they had seized the school: "One of the terrorists said that a Russian plane flown from our airfield had killed his entire family. Now, he wanted to kill, and didn't care that it was women and children" (Leung 2005). Initial attempts to engage the terrorists in negotiations were futile; the terrorists only wanted contact with senior-level Russian officials. A request was sent for three politicians and a pediatrician as part of the negotiation team. In addition to this demand, the terrorists had made three other demands: for Russian forces to end operations and withdraw from Chechnya, for the release of militants arrested in the raids on Ingushetia, and for the United Nations to recognize their independence. On the second day of the standoff the former president of Ingushetia, Ruslan Aushev (the only one of the four people requested by the terrorists to come to the scene who did so), arrived in Beslan. Aushev had also worked on the negotiations during the Moscow Theater hostage crisis in October 2002 and was well respected in the Caucasus region.

During his meeting with the terrorist leader he secured the release of 26 hostages: 11 women (nursing mothers) and their infants. One nursing mother who had other children in the school was allowed to leave with her baby, but refused because her other children had to remain behind. She handed her baby to a terrorist, who in turn gave the baby to Aushev (Leung 2005). Aushev was also handed a written note from Shamil Basayev with his demand. This was the only face-to-face negotiation conducted during the siege.

CNN and other news networks carried the Beslan siege live, multiplying the fear felt not only by children and parents in Russia, but also around the world. This kind of publicity is a terrorist group's dream, as they and their cause are catapulted onto the world stage. Horrific images of the aftermath of the bloody final assault are widely available on the Internet, keeping the event alive.

THE BATTLE

During the late morning on September 3, the terrorists agreed to allow medical workers into the schoolyard to retrieve the bodies of the dead thrown out of the second floor window on day 1. At approximately 12:50 pm, four people moved toward the school yard to begin the process of retrieving the bodies; reports indicate that the terrorists fired shots at the workers, an event that had a cascading effect that would end the Beslan siege.

At 1:05 pm, two explosions occurred in the gym, killing hundreds. It was never made clear if the explosions were caused by the terrorists or the Russian army. Fortunately, many children escaped through holes in the walls and broken windows. A full-out firefight ensued. Russian security forces detonated explosives as breaching charges to enter the gym and attempt to rescue hostages. The terrorists detonated their IEDs, setting the roof of the gym on fire; the roof collapsed, killing more of the hostages. Another breach would be conducted using a BTR-80 on the gym's western wall; this armored infantry tank fired its 14.5 millimeter machine gun as it rolled toward the school and breached the wall and the windows, likely killing more hostages.

THE CAFETERIA

While the situation in the gym unfolded, terrorists were moving the remaining children and adults, first to the hallways of the school and then to the cafeteria, which gave slight tactical advantage for the terrorists because the windows were covered with security bars. Terrorists made the women and children stand in front of the windows as human shields while Russian security forces and police fired at the building. Children and adults who were not standing in front of the windows hid in and under anything they could find; some children even hid in large pots and pans. Although they could clearly see these human shields, and women were waving white napkins and screaming not to shoot, Russian security forces, police, and local armed civilians continued to fire on the school (Chivers 2007).

The fight for the school continued, and Russian military tanks took up positions near the school and, on orders from security forces, fired 125 millimeter main gun rounds into the building locations still occupied by the terrorists. Thirteen of the terrorists, including two women, escaped the school and hid in an outbuilding on the school property. Using tanks and Shmel rocket infantry flame-throwers, the building was destroyed and all thirteen terrorists killed. Some of the terrorists escaped the fight and tried to blend into the local population. At least one was beaten to death and another lynched by outraged family members. Some reports indicate that some of the terrorists escaped, but Russian authorities say these are rumors. Only one terrorist was captured alive; he was convicted and sentenced to life in prison.

In the end, over 330 people died at the Beslan School, including 108 children. At least 700 were wounded and required medical care. All survivors received deep psychological wounds that will stay with them a lifetime. Second only to the attacks on 9/11, the Beslan attack had the highest death toll of any terrorist attack in history.

EPILOGUE

Some of the casualties at Beslan came at the hands of the terrorists, but hundreds more occurred at the hands of responding forces. Aggravating factors included

- The apparent total disregard of hostage safety
- The use of high-caliber weapons

SOFT TARGETS AND CRISIS MANAGEMENT

- Firing at human shields
- Failure to establish a perimeter cordon and allowing parents and the local population to stand near the school and even arm themselves and fire at the building
- Not having the appropriate first-response assets such as medical and firefighting support

These actions and inactions played a major role in the final, disastrous outcome of this attack.
(*Source:* Fisk, Ralph R. Siege at Beslan School Number One—Innocence Lost, unpublished. 2014.)

Could the Beslan siege happen in the United States? Although the likelihood of an incident on this scale, with attackers approaching a school in tanks with heavy firepower, is remote, it is quite plausible that a large-scale attack and hostage situation could occur with the help of insiders who pre-position supplies and accomplish preoperational planning. Considering that the massacre acts at Columbine, Virginia Tech, and Sandy Hook were perpetrated by amateurs, it is irresponsible to think that a preplanned, coordinated attack by a group of professionals would not result in more destruction and casualties. All types of schools, in both urban and rural areas are vulnerable. Local, county, and state law enforcement and first-responder support organizations should exercise the possibility of a mass hostage incident in a school, reviewing the Beslan case study and fine-tuning their own response. The National Incident Management System (NIMS) is a valuable tool for preparedness, and the incident command system must be understood to prevent confusion at the scene. It is likely that if a terrorist organization is involved in a school siege, federal law enforcement will respond; therefore, prior coordination and relationship building would be extremely helpful. A list of valuable FEMA courses regarding the coordination of response is located in Appendix B.

Passive security measures, such as those discussed in Chapter 8, would be helpful at schools, including a comprehensive identification check system, and extensive background checks on all workers employed by the institution, as well as those that are contracted and subcontracted to support educational operations such as bus drivers and construction workers. Entry control points and securing points of egress are critical to repelling a school attack and catching the perpetrators. The arming of teachers might be a step toward hardening this soft target, as they could act as force multipliers during active shooter or kidnapping and hostage situations.

K–12 Vulnerability

Why are K–12 schools more vulnerable to attack? First, the student populace is younger and easier to overpower. Security measures are typically in place, but not consistent. For instance, as violence in our country began to rise in the 1980s, many schools began installing metal detectors at entryways. Although they work extremely well to catch weapons, school administrators found this type of screening to be time consuming, especially when trying to move hundreds of students to their classrooms every morning. Also, operating detectors or individual wands is extremely manpower intensive, so most schools abandoned the idea. The concept of school resource officers (SROs) took hold in the 1970s when protests and unrest related to the Vietnam War spilled over into school systems. SROs are sworn law enforcement officers who are detailed to the school system and work to enhance security at their institution. They may be armed and can make arrests.

However, SROs can also be of limited help when facing a determined gunman or gunmen with a practiced, solid plan and heavy firepower. For example, in the Columbine High School attack, Eric Harris and Dylan Klebold managed to kill fifteen people and injure twenty-four despite the presence of an armed guard. Jefferson County Sheriff's Deputy Neil Gardner, a 15-year veteran of the Sheriff's Office, usually ate his lunch with the students in the cafeteria, his car parked in front of the cafeteria doors between the junior and senior parking lots. On the day of the attacks, Deputy Gardner was eating elsewhere on campus, watching an area frequented by smokers. When shots were fired inside the school, he pulled up to the indoor/outdoor cafeteria area where Harris and Klebold had tried to set off two bombs and had already started killing students. Gardner engaged them in a gun battle; however, he was unable to hit the perpetrators. One injured teacher and a student were able to escape during the chaos and Gardner was responsible for later saving other students as he protected them when they were fleeing. He also exchanged gunfire with the shooters when they were killing students in the library, before they committed suicide. He likely saved lives in the end, but Gardner's daily presence on the school grounds obviously did not deter the shooters from their operation. In fact, they may have purposely chosen the area where Gardner typically had lunch to start the operation, intending to kill him first and remove their only obstacle to success.

The Red Lake school massacre occurred on March 21, 2005. On that morning, 16-year-old Jeffrey Weise killed his grandfather, a tribal police officer, and his girlfriend at their home. Weise then took his grandfather's police weapons, vest, and vehicle, driving to Red Lake Senior High School, where he had been a student some months before. Weise first shot and killed the unarmed security guard at the entrance of the school and then targeted a teacher and five students. After the police arrived, Weise was undaunted and exchanged gunfire with them; he was wounded and then committed suicide in a vacant classroom.

In May 2014, police in Waseca, Minnesota, arrested 17-year-old John David LaDue on charges related to an elaborate plan to carry out a massacre at a nearby school. According to his 180-page diary, which police found in his bedroom, LaDue plotted to kill his family members, start a diversionary fire to distract first responders and then go to a nearby school. He was first going to kill the school resource officer and then set off bombs and shoot students and staff. A resident living next to a storage facility worker tipped off police to the suspicious teen; contents of his locker revealed a pressure cooker, pyrotechnic chemicals, steel ball bearings, and gunpowder. He had also stockpiled three completed bombs, an SKS assault rifle, a Beretta 9 millimeter handgun, hundreds of rounds of ammo, and a safe with several other guns at his home. LaDue tested his devices at a local elementary school playground and intended to attack the school on the anniversary of Columbine; however, it fell on Easter Sunday and school was not in session. Locals described LaDue as a polite boy who did well in school and had plenty of friends (Ford and Brumfeld 2014).

Finally, the Marysville Pilchuck High School shooting occurred in Marysville, Washington, on October 24, 2014. Jaylen Fryberg, a 15-year-old freshman who upset over a broken relationship, shot five other students in the lunchroom, fatally wounding four, before fatally shooting himself. A review of social media shows that Fryberg was depressed and spiraling downhill, making suicidal statements. His father was convicted of illegally obtaining the firearm that Fryberg used in the shooting.

Often, the perpetrators of K–12 violence are known—current or former students, staff, or teachers. They know the school layout and class schedule, the school resource officer's habits, and when and where to strike for the least resistance and most effect. Deterring school violence under these circumstances is very difficult.

Religious elementary schools in the United States have also been the target of terrorists. In August 2011, federal law enforcement officers arrested Emerson Winfield Begolly in New Bethlehem, Pennsylvania. Begolly was a moderator and supporter for the internationally known Islamic extremist web forum Ansar al-Mujahideen English forum (AMEF). He produced and distributed a 101-page document with instructions for constructing chemically based explosives and a target list including Jewish schools (Investigative Project on Terrorism 2011). Religious schools must be especially vigilant since religious terrorism is the most dangerous, with actors believing their violent actions are sanctioned by their God and just.

Colleges: Campus Culture and Vulnerability
Every type of bad actor now sees the university campus—facilities, programs, students, and staff—as a conduit to their illicit activities. Many cultural factors are at play that resulted in this exploitation. For instance, a basic higher education tenet is academic freedom, which allows professors and students to explore topics in an environment free of repression or censorship so they may fully participate in the learning process. Except for some instances of religious universities, this freedom also applies to research activities, where open inquiry means no query or methodology is out of bounds. Freedoms on campus are typically extended to student activities, school newspapers, and protest events, and universities are unable or unwilling to contain inciting or seditious speech. Also, formerly able to operate in a disconnected manner from society in their "ivory tower," the higher education system is more transparent than ever. Skyrocketing tuition rates and campus scandals have made parents, taxpayers, and political representatives angry and demanding disclosure. The combination of these internal and external dynamics has contributed to a very open campus environment that is rich for learning, as well as exploitation.

For example, due to the government's problems with cost overruns by major defense contractors, universities are increasingly relied upon for research and development (R&D) in sensitive areas such as cyber warfare and weapons of mass destruction, including the testing and development of chemical, biological, and nuclear weapons and counterweapons (American Association for the Advancement of Science 2014). Often, the projects involve future cutting-edge technology; therefore, schools compete for the work and associated funding and prestige. However, the mere presence of R&D activities makes the campus a target for antigovernment anarchists or single-issue domestic terrorists such as animal or environmental rights activists. Annual government reports listing R&D funding to specific schools may paint a target for foreign actors who may also attempt to penetrate the campus (American Association for the Advancement of Science 2014). Nation-state and non-state actors are stepping up collection activities against the U.S. industrial base, which includes campuses conducting government research (Office of the Director of National Intelligence 2011). According to the FBI, "The open environment of a university is an ideal place to find recruits, propose and nurture ideas, learn, and even steal research data" (FBI 2011).

Prevailing negative attitudes regarding security measures and agencies among our citizens may also contribute to the increase in nefarious activity in and around campuses. As time passes from the 9/11 tragedy and subsequent terrorist plots have failed or been interrupted, intelligence and law enforcement agencies fear a sense of complacency. Complicating the situation, it is important to note that the American citizen's trust in the government is at a historic low point (19 percent), and one in three people believe the government is threatening their personal freedoms with post-9/11 security activity and laws (Pew Research Center 2013). Certainly these attitudes are conveyed on campus with the younger, more activist-minded generation and faculty trying to enforce principles of academic freedom. Therefore, any security measures put into effect on campus may be met with resistance, not the compliance we expect. The school may even back down to a lower security posture to pacify students and faculty, resulting in more vulnerability.

Also complicating matters is the fact that security fatigue has set into the population as a whole, with citizens tired of looking for threats and being vigilant. Wake-up calls such as the tragic Virginia Tech murders, an event which could be characterized as "higher education's 9/11," and subsequent campus shootings have shifted security focus to weapons and student lockdown/notification procedures. Although the lone-shooter threat is compelling, campus security must be careful not to lose sight of other vulnerabilities and threats. In other words, while we are specifically looking for guns at the entrance of a classroom building, a crude but effective bomb may be under construction in the basement.

Campuses Already under Attack
Make no mistake: our schools are already under persistent passive attack. For instance, foreign intelligence services (FISs) and non-state actors continuously prey on campuses to steal intellectual property and bypass expensive research and development. Also, FISs are actively recruiting students and professors for espionage and to use later as sources, with multiple cases on college campuses in the last 5 years and actors exhibiting increasingly bold tactics (FBI 2011).

The case of University of Tennessee professor Dr. John Reece Roth illustrates lack of policy and procedure, and how the open academic environment is penetrable by "countries of interest." Despite statutory requirements, Roth was able to hire foreign national students from China and Iran to work on his classified, Department of Defense–funded plasma-related aerospace project on campus. The students recruited him, and he traveled at least eight times with sensitive material to China (FBI 2009). Roth is not the type of professor we might expect would be susceptible to this type of FIS activity; he was in his late sixties when approached by the students, an emeritus professor of electrical engineering who taught at Tennessee for nearly 30 years after his time at NASA. He also holds eleven patents and testified before Congress on nuclear fusion. Roth was coerced into thinking that intellectual property should be open and available to all countries. Roth's academic colleagues feel that his ego was stroked by attention from the students and host government and that he believed he was making a significant intellectual contribution to the world (Golden 2012).

Intellectual property theft is not the only goal of foreign governments operating on our campuses. The 2009 case of Russian spy Lidiya Gurveva (using the name Cynthia Murphy) is particularly illustrative of how the campus environment can be exploited merely to collect information on professors and students. Gurveva was pursuing an MBA degree at Columbia Business School, Columbia University, in 2008 when her handlers gave her the following assignment:

> [S]trengthen ... ties w. classmates on daily basis incl. professors who can help in job search and who will have (or already have) access to secret info ... [r]eport to C[enter] on their detailed personal data and character traits w. preliminary conclusions about their potential (vulnerability) to be recruited by Service.

They also directed Gurveva to "dig up personal data of those students who apply (or are hired already) for a job at CIA" (U.S. Department of Justice 2010). In yet another campus spying case, Andrey Bezrukov (also known as Donald Hatfield) was arrested in June 2010 for being an agent of Russia. He graduated from Harvard's Kennedy School of Government and, while a student, developed associations with professors at George Washington University and Oxford University. Bezrukov/Hatfield targeted a professor who was once former Vice President Albert Gore's national security advisor, as well as other policymakers at Kennedy School reunions and think-tank events (Srivatsa and Zu 2010). These cases clearly illustrate the vulnerability of professors and students to recruitment and the ease with which foreign nations can penetrate our universities.

Cybercriminals are also present on campus; data theft is an emergent challenge as servers are breached to steal student information protected by law, putting the university at great risk and liability. For example, the University of Maryland regretfully announced in February 2014 that information had been stolen from its database since 1998 on 300,000 students and faculty who were issued identification cards (University of Maryland 2014). Although Maryland's breach may have been due to failure to protect onsite servers and recognize ongoing penetration activities, many universities are now moving their electronic student, faculty, and staff records to the virtual servers offered in the "cloud"—which presents new and significant vulnerabilities.

As if spying and cybertheft are not enough of a challenge for university security officials, campuses are increasingly seen as a "safe haven" for drug dealers. The activity is not what we might expect, such as students attempting to grow marijuana in their dorm rooms or selling dime bags at parties. Police are now discovering methamphetamine labs on college campuses, which not only pose health dangers to students, but also place explosive materials in high-occupancy buildings and bring an unwanted, dangerous element to campus property. In 2013, the cases were not just confined to urban settings, and it seems no university is now immune from this emergent threat. For example, meth labs were discovered in dormitory rooms at Georgetown University, in a music practice room at Southern Methodist University, and on the roof of an on-campus student apartment building at the College of Charleston.

Finally, the J-1 Student Visa Program has been repeatedly misused and exploited, notably by nineteen convicted terrorists since 1990, including Mohammed Atta, leader of the 9/11 cell (Kephart 2005). Over 100,000 foreign students enter the United States annually on student visas, but not all of them actually show up or stay at school. Despite new, rigorous post-9/11 laws to scrutinize requests and follow up after visa expiration, many still slip through the cracks. In 2006, eleven Egyptians with J-1 visas did not appear for class at the University of Montana. The school waited 48 hours after classes began to report the absence to the FBI, which then started a resource-intensive manhunt to find the eleven (workpermit.com 2006). Not all were captured.

The J-1 visa has also been used as a conduit to attain asylum. The case of Ibragim Todashev, friend of Boston Marathon bomber Tamerlan Tsarnaev and admitted murderer of three people in Massachusetts, is compelling. A Chechen, Todashev came to the United States on a J-1 visa and, instead of attending college, immediately applied for asylum and, later, citizenship. He was on a downward spiral with a lengthy criminal record, yet his newly obtained citizen status protected him from deportation. Todashev was killed by law enforcement when he attacked investigators during an interview to collect evidence regarding the Boston bombings and the triple murder in Massachusetts. The University of Massachusetts-Dartmouth also became part of the bombing investigation because coconspirator Dzhokhar Tsarnaev was living in his dorm room and storing evidence in his closet at the time of the event. Although Tsarnaev had failed seven classes in the previous three semesters, was carrying $20,000 in student debt, and was a known marijuana smoker/distributor, he was allowed to remain on campus, a decision the university likely regrets today (Goode and Kovaleski 2013). These cases show how university policy decisions can result in vulnerabilities.

In another case, Khalid Ali-M Aldawsari, a Saudi student studying chemical engineering at Texas Tech University, was arrested in February 2011 on a charge of attempted use of a weapon of mass destruction. A notebook was found at Aldawsari's residence indicating he had been planning to commit a terrorist attack in the United States for years. One entry describes how he sought and obtained a particular scholarship because it allowed him to come directly to the United States and helped him financially, which he said "will help tremendously in providing me with the support I need for Jihad" (U.S. Department of Justice 2011). The journal goes on to say: "And now, after mastering the English language, learning how to build explosives and continuous planning to target the infidel Americans, it is time for Jihad." Aldawsari had the blueprints of chemical IEDs and had already started purchasing the components for his weapons. Targets included former U.S. military personnel, a nightclub, dams, and nuclear power plants.

The Escalating Number of Attacks on College Campuses

Perhaps inspired by successful school attacks abroad, terrorists are now turning their sights toward our universities and exploiting lax and unenforced J-1 visa rules. For instance, El Mehdi Semlali Fahti, 27, came to the United States from Morocco in 2008 on a J-1 student visa. However, he flunked out of Virginia International University and the visa expired. Fahti traveled across the county and was arrested for a trespassing charge that was dismissed, turned over to immigration officials in Virginia, and placed in custody. However, he was able to stay in the United States after manufacturing a story about imprisonment and beatings by Moroccan police. A judge was persuaded to approve his request for political asylum in 2011. Fahti traveled to the West Coast, and was arrested for theft in California; while serving his sentence, his case was again turned over to the immigration agency, but a judge released him from custody in August 2013 (Investigative Project on Terrorism 2014d).

Fahti moved to Connecticut and by January was plotting to attack a university and a federal building using remote-controlled hobby planes packed with explosives, components of which he had started to gather. Fahti was living in an apartment in Bridgeport, Connecticut, with a man he met while incarcerated in Virginia. Fahti told undercover FBI agents, who thankfully became aware of his plot during a sting operation, that he had studied the plan for months, had learned how to pack the explosives while he was a student in his home country, and would be able to obtain the materials at the Mexican border. He also said the plot would be funded through "secret accounts" that used laundered cash and drug-dealing profits. In a conversation with the FBI, Fahti says "the more he thinks about the case, he laughs because he cannot believe the judge believed him" and allowed him to stay in the United States for political reasons (Mayko 2014). He also stated that three things cause fear in the American people: causing harm to schools, the economy, and their sense of security (Investigative Project on Terrorism 2014c).

The J-1 "summer study program" visa is also exploited by human traffickers and so-called labor recruiters to bring young aspiring students to the United States to participate in illegal or immoral activities. The problem came to light in 2010, when law enforcement officials found would-be students in Myrtle Beach, South Carolina, working in strip clubs, crammed into dirty apartments, eating at food kitchens, and even begging on the street. Summer study students are also recruited by the adult entertainment industry and have been used to smuggle cash and other goods back to their home country by criminal groups (Mohr et al. 2010). The hopeful students' passports are seized upon entry to the United States and they are threatened into complying with the trafficking groups. Some students do attend programs on campuses, possibly introducing the criminal element.

Terrorist attacks on college campuses are increasing worldwide. Recent terrorist attacks in the United States, such as the San Bernardino shooting, included plans to target a college campus. In a successful terrorist attack in 2006, University of North Carolina-Chapel Hill honors graduate Mohammed Reza Taheri-azar sought to "avenge the deaths of Muslims worldwide" and to "punish" the U.S. government. He drove a rented sports utility vehicle into a crowd of students with the intent to kill. Although no one was killed in the attack, nine people were injured (none seriously). Taheri-azar was born in Iran, but moved to the United States at the age of two, growing up in North Carolina. He pled guilty to nine counts of attempted first-degree murder, and in 2008 was sentenced to 33 years in prison, on two counts of attempted murder. In one letter, Taheri-azar wrote, "I was aiming to follow in the footsteps of one of my role models, Mohamed Atta, one of the 9/11 hijackers, who obtained a doctorate degree" (Schuster 2006).

Deadly al-Qaeda splinter Boko Haram (name loosely means "Western education is sinful") actively targets colleges in Nigeria in a campaign of fear. The group strikes at night, when students and staff are asleep. In 2014, fifty-nine boys were killed at the Federal Government College of Buni Yadi in Yobe State, Nigeria and all twenty-four buildings of the school burned down. In 2013, gunmen entered the male dormitory in the College of Agriculture in Gujba, killing forty-four students and teachers in their sleep. Boko Haram burned down 14 schools in 2012, and was also responsible for a night massacre at a technical college in 2010.

On April 2, 2015, Somali Al-Qaeda offshoot al Shabaab, which has inspired attacks in the United States and recruits from our cities using social networking sites, killed 150 students at the Garissa University in Kenya, located 90 miles from its border with Somalia. According to terrorists, the university was targeted because it was educating Christian students in Muslim land.

Bottom line: groups that have actively targeted the United States are actively engaged in attacks on universities overseas, and we should harvest lessons learned from these operations and protect our schools accordingly.

Case Study: Umpqua Community College Shooting

On the morning of October 1, 2015, at approximately 10:38 am Pacific Daylight Time, a student entered his English classroom in Snyder Hall on the campus of Umpqua Community College (UCC), in Roseburg, Oregon, and proceeded to shoot his teacher and fellow students. At the end of the shooter's 10-minute spree, nine people were dead, with an additional nine wounded. The gunman, wounded by responding Law Enforcement Officers, took his own life.

Official After Action Reviews are not yet available, there are lessons learned based on the available Umpqua Community College, Emergency Action Plan, media reports, interviews, and recorded radio traffic.

- At 10:38 am—The Roseburg Police, start response to 911 calls reporting an active shooter at Umpqua Community College. Roseburg Police en-route with an approximate response time of 5 minutes.
- 10:39 am—Dispatchers update possible location of shooter is in the Science building.
- 10:40 am—Dispatcher updates that UCC has activated their Emergency Response Plan and has placed all UCC buildings in lock down.
- 10:41 am—Dispatchers update that shots are now being reported from "Snyder Building." (Update issued approximately 3 minutes after dispatch.)
- 10:42 am—Oregon State Police receive reports of active shooter.
- 10:44 am—First two Roseburg Police officers and Oregon State Police Trooper arrive on scene. (Approximately 6 minutes after first dispatch received.)
- 10:46 am—Officer reporting shots being exchanged with suspect.
- 10:48 am—Officer relays to dispatch "Suspect down." (Time from first dispatch to suspect down report; approximately 10 minutes.)
- 10:48 am—First ambulance sent into scene.
- 10:50 am—Dispatch requests all local ambulances to report to UCC.
- 11:02 am—Designated Operations Radio Channel established as "fire2."
- 11:02 am—Incident Command Established at Administration Building.
- 11:14 am—Dispatch notifies Mercy Medical of ten deceased, six critical to transport.
- 11:26 am—Request for bomb squad response.
- (Umpqua Community College Shooting Public Safety Response recording, https://www.youtube.com/watch?v=BDSjVJ1cigg).

From the time of the initial dispatch call to "subject down" report, approximately 10 minutes transpired and there were eighteen people down. EMS was not sent to the site until the officer relayed "Suspect down" and no further units were sent until that EMS crew sent a transmission at 11:14 requesting transport for sixteen people. Incident Command wasn't established until 15 minutes after initial dispatch was sent out. There was no initial primary channel for communications, with all communications being sent on the primary channel until a fire channel was designated at 11:02 am.

A review of the Umpqua Emergency Action Plan reveals that it addressed very basic actions for a "Violent Situations" but no plans for response to an Active Shooter Incident. This was surprising considering the sheer number of shooting incidents on campuses, particularly the Virginia Tech massacre.

Interviews conducted after the shooting were also very telling. Former (recent) President of Umpqua Community College Dr. Joe Olson, a former deputy sheriff, cited a previous debate as to whether there was a need for an armed security presents at the school. "We talked about that over the last year because we were concerned about safety on campus," he said. "The campus was split 50–50. We thought we were a very safe campus, and having armed security officers on campus might change the culture. If you want to come on the campus and you want to shoot five people, you are going to do that before our security would arrive" (Williams 2015). He said the college had three training exercises with local law agencies in the past 2 years, "but you can never be prepared for something like this" (CBS 2015). Dr. Rita Cavin, who was the interim president of UCC during the shooting, was asked that afternoon if there would be changes in security at the school in light of the attack and she responded. "No. This was an anomaly" (Press interview, October 1, 2015).

The campus has participated in some type of emergency drills with local law enforcement and those agencies have standardized plans, responses, and tactics to neutralize threats upon responding. But the burden is on leaders of soft target locations like college campuses to have an all hazards, worst case plan to help prevent and mitigate attacks, and best partner with first responders in the event those actions fail. We need to change the way we think, the way we prepare, the way we respond to mass casualty incidents. The approach cannot be lackadaisical; it must be aggressive and meaningful. A sense of denial or thinking "it can't happen here" or "there is nothing we can do" is not the leadership perspective needed these violent, unpredictable times. (*Source:* Fisk, Ralph R. Case Study: Umpqua Community College Shooting, unpublished. 2016.)

Bad Blood: Intelligence, Law Enforcement, and Higher Education

There is a possibility action (or nonaction) by security agencies may be contributing to campus vulnerability to domestic and foreign bad actors. Unfortunately, the historically poor relationship between academia and intelligence and law enforcement agencies has led to difficulty with their partnering to address vulnerabilities and threats on campuses. This "bad blood" dates back to 1956 with the FBI's Counterintelligence Program (COINTELPRO), a series of covert projects in which the Bureau conducted surveillance and disruption activities against domestic political organizations on campuses (Churchill and

Vander Wall 2002). COINTELPRO infiltrated college campuses to monitor antiwar activity during the Vietnam War and, when this activity came to light in 1971, massive protests erupted on campuses, impacting the relationship between academia and the government.

Activities on campus by the Central Intelligence Agency (CIA) from 1967 to 1973 further strained the relationship. Operation CHAOS was the code name of a CIA domestic espionage program to unmask possible foreign influences on the student antiwar movement on campuses (Rafalko 2011). Project RESISTANCE was a parallel operation, in which the CIA worked with college administrators, campus security, and local police to identify antiwar activists and political dissidents without any infiltration taking place (Church Committee Reports 1975).

A review of official documents regarding COINTELPRO, CHAOS, and RESISTANCE, especially the Church Committee proceedings, yields great data regarding interaction between higher education and intelligence agencies during a period of extreme domestic unrest. Present-day concerns about government and law enforcement activities on campus re-emerged in 2011, when it was revealed the New York Police Department had infiltrated colleges to monitor Muslim students and activities (Hawley and Apuzzo 2011). This case was a lightning rod for the media and led to state-by-state cases of intelligence community activity on college campuses documented by the American Civil Liberties Union (ACLU) in 2011 (ACLU 2014).

The literature regarding attitudes toward, perceptions of, motivation for, resistance to, and barriers to the relationship between higher education and the intelligence community highlights the cultural disconnects of two enterprises that essentially share the same goal: protecting First Amendment rights and national security. In terms of the intelligence perspective, the FBI 2011 white paper entitled "Higher Education and National Security: The Targeting of Sensitive, Proprietary, and Classified Information on Campuses of Higher Education" gives great insight into the subject. The higher education view on the tenuous relationship comes into view through an essay in the *Chronicle of Higher Education* entitled "Academics and National-Security Experts Must Work Together" (Gansler and Gast 2008). In this article, the authors discuss legislative and environmental changes since the 9/11 terrorist attacks and the impact on higher education. They also address the cultural divide between higher education and intelligence agencies, asserting that it is "so often at the root of poor policies and unnecessary roadblocks for both groups" (Gansler and Gast 2008). This balanced essay asserts that experts in the intelligence field don't fully understand how communication with other countries is crucial to scientific research, and universities and research institutions fail to appreciate the importance of securing technology and protecting sensitive information. Finally, an analysis of the impact of the Economic Espionage Act of 1996 gives insight into how what was supposed to be a framework for success is not working to protect the academy (Brenner and Crescenzi 2006).

The institutions of higher education and the intelligence community have both witnessed great change and turmoil in light of the 9/11 terrorist attacks on our country. The events of 9/11 and subsequent investigations into the campus-related activities of the hijackers, new statutory reporting requirements, and continued, persistent collection on campuses by "countries of interest" have potentially aggravated an already strained relationship between the two enterprises. Both have undergone institutional reform, forced by uncontrollable environmental conditions, and both are inherently resistant to change.

Due to the urgent nature of national security concerns following 9/11, there was no time to plan strategically for the change, to seek mutual agreement, or to communicate cross enterprises properly. The situation continues to evolve with current events and emergent challenges; therefore, an opportunity exists to bridge the gap and to partner strategically for future change events.

In response to increased concerns, the FBI created the National Security Higher Education Advisory Board (NSHEAB) in 2005. The NSHEAB consists of nineteen university presidents and chancellors and strives to meet on a regular basis to discuss national security matters that intersect with higher education. Previous panels included discussion on protection of weapons of mass destruction research and laws regarding domestic terrorism investigations on campuses. However, the NSHEAB has become much less engaged over time. The Department of Homeland Security instituted the Homeland Security Academic Advisory Council (HSAAC) in March 2012 to discuss matters related to homeland security and the academic community, mainly issues regarding security-related curriculum, recruitment of students to DHS, academic research, and cybersecurity on college campuses.

One of the subcommittees deals with campus resilience. The Resilience Pilot Program (CR Pilot) was launched in 2013 for seven select U.S. colleges and universities to help them take proactive steps to enhance preparedness and campus resilience. The CR Pilot is a joint initiative of DHS's Federal Emergency Management Agency (FEMA), the U.S. Immigration and Customs Enforcement Student and Exchange Visitor Program, and the Office of Academic Engagement. The CRP will harvest best practices, lessons learned, and resources to help schools identify their vulnerabilities and be more resilient and prepared. Drexel University, Texas A&M, and others are participating in the project (DHS 2013).

In Chapter 3, we discuss hardening efforts for colleges that include strengthening partnerships with law enforcement and harvesting best practices from around the country.

Churches

Churches are another soft target where violent attacks are met with disbelief, especially when perpetrated by an actor with an extreme religious ideology. How can anyone's God condone the killing of another human being? We also now have a "black swan" phenomenon of Christians killing Christians and Muslims killing Muslims in the name of their extremist religious ideology. The worship space, formerly respected and shielded from violent crime, has been infiltrated by criminals and murderers, whether as a place to exact personal revenge against a parishioner, to steal money or artifacts to pay for drug buys, or to carry out an attack in which the institution and ideology is a target.

Despite these cultural changes, clergy may not see security of their building and faith community as a pressing issue. A survey conducted by LifeWay Christian Resources posed questions to more than 1,300 evangelical leaders from around the world to gather information about what they considered their most urgent concerns. The results of the survey, which included topics related to faith and life in modern society, did not return any concerns about safety, security, or terrorism. Carl Chinn, a church security expert, has been tracking U.S. church-related crimes and statistics since 1999. Deadly force incidents (DFIs) include abductions, attacks, suspicious deaths, suicides, and deadly force intervention/protection. Between January 1, 1999, and February 1, 2014, Chinn reports 792 DFIs at faith-based organizations, with 286 episodes resulting in deaths. Almost 54 percent of DFIs were related to domestic violence, personal conflict, and robbery; guns were the weapon of choice (472 incidents) followed by knives (141 incidents) and explosive devices/Molotov cocktails (48 incidents). There were forty-two incidents related to what Chinn labels "religious bias" (Chinn 2014).

Chinn also points out there is no government organization responsible for tracking crimes specifically associated with churches. Following up on the LifeWay survey and data compiled by Chinn, Brian Gallagher, a former member of the U.S. Secret Service and church security expert, interviewed a church planting organization about its perspective on safety and security related to its family of 100 churches in the United States. This particular organization, which asked to remain anonymous, revealed less than 10 percent of their churches invested funds to provide proper security for their facility. The majority of the churches in the organization had never considered security or terrorism as a concern to their congregations (Gallagher 2014).

The advent of "megachurches" in the United States has presented a unique vulnerability: there are fewer worship sessions, but they are attended by thousands of attendees. Lakewood Church in Houston, Texas, is the largest megachurch, with over 16,800 worshippers attending popular author Joel Osteen's services at the former Compaq sports arena. Although the church has a high-tech security camera system and around-the-clock guards including off-duty law enforcement officers, it is still vulnerable. On March 9, 2014, possibly during services, a safecracker stole over $600,000 in collection money. Lakewood generates $75 million in revenue annually, making it a lucrative target for theft. Nationally, thefts of money and securities constitute a significant problem for churches. Wisconsin-based Church Mutual Insurance Co., which insures about 95,000 churches, reported 178 such thefts in 2013. That total declined from 240 cases in 2011, but the value of money stolen rose to more than 9 percent of total theft claims paid (Turner and Hlavaty 2014).

Churches also have a unique vulnerability: they invite wayward members of society to worship, for counseling, or to attend support groups. This exposure is difficult to mitigate, although Chapters 8 and 9 discuss additional steps clergy and staff can take to secure the facility and populace. However, unlike school attacks, many church crime perpetrators have no association at all with the church; they have never attended a service and do not know the clergy members or parishioners. Therefore, the church attack may be more about opportunistic crime or could just serve as a symbolic target. The outward appearance of a hardened facility is therefore extremely important so that the terrorist or violent criminal shifts his or her intentions elsewhere.

The Proliferation of Church Attacks

Church attacks are common around the world, and many are perpetrated by enemies common to the United States. For instance, in Kenya, Christian churches are routinely targeted by radical Islamist terrorists from the Somalia-based al-Qaeda splinter group al-Shabaab. Tactics include use of bombs, semiautomatic weapons, or even machetes. Kenya is a natural target for terrorists; it has East Africa's biggest economy, is Westernizing, and is a recipient of U.S. counterterrorism funding. Also, the Nairobi government has sent troops into neighboring Somalia as part of an African Union force (AMISOM) to combat al-Shabaab, drawing their wrath. Christianity is the predominant religion in Kenya also making it a target for those with a radical Islamist ideology. In Nigeria, Boko Haram militants routinely target churches, often walking in doors closest to the altar and using semi-automatic weapons on stunned worshippers for maximum death in a short period of time.

A recent case in England perfectly illustrates how a terrorist can enter a foreign country and quickly move to the execution stages of an attack against churches. Twenty-five-year-old white supremacist Pavlo Lapshyn, a Ukrainian citizen, was a PhD student who won a competition for a work placement with Delcam, a specialist software firm in Birmingham, UK. However, Lapshyn never intended to work; the competition merely facilitated his plan to carry out hate crimes against Muslims in the United Kingdom. He practiced building and detonating hexamethylene triperoxide diamine (HMTD) devices packed with nails in the Ukraine and researched how to buy similar bomb components in Birmingham. He flew to England on April 24, 2013, on his work visa and had been in the country only 5 days when he started a campaign of terror. His first victim was 82-year-old Mohammed Saleem killed by Lapshyn as he walked home from evening prayers at a mosque in Birmingham. On June 21, Lapshyn planted explosive devices in a child's lunch box at the mosque gates, and 7 days later, he placed a bomb on a roundabout near Wolverhampton Central Mosque. The most serious attack was on July 12 at the Tipton mosque, where Lapshyn packed hundreds of nails in a bomb on a railway embankment next to its car park and detonated it during what he thought was the prayer service. Thankfully, mass casualties were averted because, unbeknownst to Lapshyn, prayers were held an hour later than usual, so parishioners had not yet arrived. Using forensics from the scene and a shadowy picture of Lapshyn from CCTV, police were finally led to his workplace at Delcam, and his apartment and a trove of evidence. Lapshyn is serving a forty-year prison sentence for the killing and the bombings. Interestingly, he was on police radar in Ukraine, where his flat was damaged in 2010 by a chemical explosion that he told authorities was related to his school work in a lab project gone bad. Yet no travel restrictions were imposed and he was not being watched by authorities. The case illustrates how a perpetrator can go from planning stages to execution in a short period of time, even days after entering a foreign country. Police in the United Kingdom were on Lapshyn's trail during his attack spree, but the investigation was difficult and he was able to continue his operation (Lumb and Casciani 2013).

Shared religion or nationality is not a deterrent when selecting targets. For example, in 1993, members of the Italian Mafia (presumably Catholic) bombed two of Rome's most venerable Catholic churches—San Giovanni and San Giorgio—to further their political goals. In our country, right-wing religious extremist Christians have killed other Christians without second thought to advance their goals. Islamist extremists have bombed mosques in Libya, Nigeria, and Mali. Even historic mosques and tombs, revered by Muslims, have also been targeted. For instance, the world was stunned in July 2012 when terrorists from the al-Qaeda splinter group Ansar Dine bombed the Yahya mosque in Timbuktu, a UNESCO World Heritage site. They even pried open a fifteenth century door, which many Muslims believed was supposed to be closed until the end of the world and the return of the prophets. In 2006, Wahhabi militants who bombed the Askariyyah shrine in Samarra, Iraq, almost ignited a civil war. The shrine houses the graves of the 10th and 11th holy imams, descendants of Muhammad. In 2014, members of al-Qaeda splinter ISIS destroyed the tomb of the prophet Jonah (in the Old Testament as being swallowed by a whale) in

Mosul, Iraq. These "crimes against history" are shocking, get widespread press coverage for a group, and can shift the core of a conflict. Therefore, asymmetric attacks on a church by members with similar religious beliefs should not be discounted.

U.S. Churches as Targets
Unfortunately, terrorists have taken a page from the international playbook and are now targeting U.S. churches. Levar Haley Washington, Gregory Vernon Patterson, Hammad Riaz Samana, and Kevin James were arrested in August 2005 and charged with conspiring to attack synagogues and other targets in the Los Angeles area. Kevin James allegedly founded Jamiyyat ul-Islam Is-Saheeh (JIS), a radical Islamic prison group, and converted Levar Washington and others to the group's mission. The JIS allegedly planned to finance its operations by robbing gas stations (Zuckerman 2014). The men were convicted and their sentences ranged from 5 to 22 years in federal prison.

In the 2009 "Bronx plot," James Cromitie and his three accomplices separately converted to Islam while in prison. After meeting at the Masjid al-Ikhlas mosque in Newburgh, New York, following their release, the men devised a complex plan to bomb the Riverdale Temple and Jewish Center in New York City and, using Stinger surface-to-air guided missiles, shoot down military planes flying out of a nearby air base. Each man is serving a 25-year sentence (Investigative Project on Terrorism 2014c).

In October 2010, two packages shipped from Yemen to Chicago-area synagogues contained explosive materials of the same type used by airline attackers Richard Reid ("shoe bomber") and Umar Farouk Abdulmutallab ("underwear bomber"). The packages contained printer cartridges filled with explosive material and were intercepted while in transit on cargo planes in the United Kingdom and Dubai from intelligence tips from Saudi Arabian authorities. The Yemen-based al-Qaeda in the Arabian Peninsula (AQAP) claimed responsibility (Zuckerman 2014).

In May 2011, Ahmed Ferhani of Algeria and Moroccan-born Mohamed Mamdouh, a U.S. citizen, were arrested by the New York Police Department after attempting to purchase a hand grenade, guns, and ammunition to attack churches and synagogues in Manhattan. The men planned on disguising themselves as Orthodox Jews in order to access the facilities. Ferhani, the mastermind of the plot, also sold Percocet, cocaine, and marijuana to finance weapons for the attack. Both pled guilty and are serving 5- to 10-year terms (Investigative Project on Terrorism 2014a).

Amine El Khalifi, a Moroccan citizen illegally in the United States, was arrested in February 2012 on charges of plotting to attack the U.S. Capitol. Before choosing the Capitol building as a target, Khalifi had proposed targets including DC office buildings, restaurants, and synagogues. He believed undercover FBI agents were al-Qaeda operatives and revealed his plots, even showing them his practice detonating test bombs with cell phones. He was arrested as he left his parked car next to the Capitol building with a bomb that was, unbeknownst to him, inert (Investigative Project on Terrorism 2014b). Khalifi was sentenced to 30 years in prison.

The worst attack against a place of worship in the United States occurred on August 5, 2012, when Wade Michael Page fatally shot six people and wounded four others at a Sikh temple in Oak Creek, Wisconsin. Page was an American white supremacist and U.S. Army veteran with ties to white supremacist and neo-Nazi groups. He was part of the white power music scene and openly spoke and sang about the impending racial holy war. Page committed suicide after he was shot by a responding police officer. Sikh men are often confused with Muslims due to their beards and turbans, and they have been the targets of multiple hate crimes in the United States since the 9/11 terrorist attacks (BBC News 2012).

Throughout history, churches have been a place of sanctuary and refuge for those in need. They have been targeted before, however; during World Wars I and II, hundreds of churches and cathedrals in Europe were destroyed, whether purposely hit by the enemy or destroyed by collateral damage. During the ongoing religious and political wars in myriad foreign countries, churches have similarly been affected. However, the mere idea of a church being affected by crime and terror seems new and surprising to the U.S. populace. The unwillingness to believe a church could be a target creates a large blind spot and gives the advantage to those who wish to do us harm in shocking ways. Our churches are being targeted; therefore, security must move closer to the top of concerns for U.S. clergy and staff.

Hospitals

Violent crimes in hospitals are typically along the lines of assault, drug theft by armed perpetrators, and gang-related violence. In addition to these criminal activities, there is discussion in the counterterrorism community about the possibility of terrorists stealing radioactive material found in hospitals in order to make a "dirty bomb." But what about the hospital itself as a target?

Hospitals are considered a critical infrastructure/key resource (CI/KR) under the national response framework, making them high-priority targets. CI/KRs are defined as individual targets whose destruction "could create local disaster or profoundly damage our nation's morale or confidence." In a multistage attack, taking hospitals and medical personnel out of play would surely lead to an increased casualty count. A terrorist attack against a hospital has not happened in the United States yet; however, these heinous assaults routinely happen in other parts of the world. Either the hospital itself is targeted or it is targeted after an event as a secondary or tertiary attack when the wounded are being rushed into the building.

As covered in Chapter 2, terrorists are often underestimated in terms of intelligence, and many have formal education. In this new type of warfare, those entrusted to save and protect lives have actually been the trigger pullers. Consider the perpetrators of the June 30, 2007, attack at Scotland's Glasgow Airport and the failed bombings in London's nightclub district two days before. Four of the seven perpetrators in the terrorist group were physicians, two were in medical school, and one had a PhD in engineering. A hospital was also in their crosshairs; on the afternoon of July 1, police carried out a controlled explosion on two cars in the parking lot of the Royal Alexandra Hospital, where the suspects worked and had apparently intended to attack.

On November 29, 2008, members of the al-Qaeda affiliate, Lashkar-e-Taiba (LeT), began their 4-day assault in Mumbai, India. The twelve coordinated shooting and bombing attacks shocked the world and lasted four days, killing 164 people and injuring hundreds. Targets included a train station, a Jewish community center, the Taj Mahal Palace Hotel, and the Trident-Oberoi Hotel. Overshadowed by the horrific events at these locations, not much has been written about one of the main targets in Mumbai: the Cama Hospital for Women and Children. The LeT terrorists specifically attempted to gain access to the maternity ward, after killing two security guards with an AK-47 and grenades at the entrance. The security doors of the ward kept them from entering, and scared patients barricaded and refused to open up, despite threats from the terrorists. The shooters wandered other hospital floors, killing other security guards and workers. When directly confronted, a frightened hospital orderly offered the shooters water; they took a drink before killing the man (IBN Live 2008).

The Tehrik-i-Taliban Pakistan (TTP), or the Pakistani Taliban, is a designated terrorist group that specializes in hitting soft targets. TTP has already tried to attack soft targets inside the United States, most notably through the failed Times Square bombing. In an example of a secondary attack against a hospital, TTP militants in Pakistan's restive province of Baluchistan first bombed a bus carrying female college students, killing fourteen. The perpetrator was a female suicide bomber, who boarded the bus with the students and then carried out the attack. At the Bolan Medical Complex, a second suicide bomber sat in the reception area of the emergency room, waiting for the right moment to attack. As dead and injured bus victims were brought into the emergency room and parents and local officials gathered, he detonated his vest, killing eleven (*The Telegraph* 2013). Heavily armed militants then stormed the hospital, taking hostages and shooting from the windows, killing several police officers and local officials.

Other coordinated, violent attacks against hospitals by the TTP in the last few years include the 2008 DI Khan Hospital attack (thirty-two dead) and the 2010 attack against the Jinnah Postgraduate Medical Center in Karachi (thirteen dead). The resurgent Taliban in Afghanistan has also attacked that country's hospitals; the worst attack to date occurred in 2008, when a lone suicide bomber struck in the cafeteria at the Kabul hospital, killing six people and injuring twenty-three during a surgery training session. The hospital is located in the "green zone," heavily guarded, and very close to the U.S. embassy. Also in Kabul, three Americans were killed on April 24, 2014, when a security guard opened fire at a hospital funded by a U.S. Christian charity, killing a father and son who were visiting from the United States and a doctor.

A brazen attack by al-Qaeda in the Arabian Peninsula (AQAP) against a hospital in Yemen on December 5, 2013, stunned the world.

The primary target was a nearby defense building; however, stray militants entered a nearby hospital, killing 63 staff and patients. The attack was captured on closed-circuit television footage and broadcast by state media, causing widespread outrage among Yemenis, as AQAP has portrayed itself as fighting for the average citizen against foreign drone strikes. Whether outraged or sensing a rift between the citizens and AQAP, a member of the core al-Qaeda group issued the following apology: "We do not fight in this way, and this is not what we call on people to do, and this is not our approach. We warned fighters not to attack the hospital (Figure 13.3)."

Here in the United States, we have had a few indications that hospitals might be in the terrorist crosshairs. In November 2002, the FBI issued an alert to hospitals in San Francisco, Houston, Chicago and Washington, DC, warning of a vague, uncorroborated terrorist threat. In August 2004, the FBI and DHS issued a nationwide terrorism bulletin that al-Qaeda might attempt to attack Veteran's Affairs hospitals throughout the United States, and in April 2005 the FBI and DHS investigated unusual incidents of imposters posing

SOFT TARGETS AND CRISIS MANAGEMENT

Figure 13.3 An al-Qaeda gunman lobs a live grenade at a group of civilians in a Yemeni hospital. (McCluskey, Brent. Graphic CCTV Footage of Militant Attack in a Yemeni Hospital. Podcast audio. http://www.guns.com/2013/12/14/graphic-cctv-footage-militant-attack-yemeni-hospitalvideo/, 2013)

as hospital accreditation surveyors. The Joint Commission on Accreditation of Healthcare Organizations sent a security alert to hospitals.

Nefarious Use of Ambulances

Ambulances are being used in conflicts as vehicle-borne (VB) IEDs (Libya in 2016; Pakistan in 2013; Syria and Iraq in 2015), as shooting platforms (Gaza in 2001), to move terrorists (Pakistan in 2004), to hide suicide belts (Gaza in 2002), and as getaway vehicles (Gaza in 2004). We have much to learn from these case studies in terms of hardening our hospitals to protect ambulances from theft, enhancing our laws regarding re-selling first-response vehicles, and educating first responders on the threat.

Terrorists have used ambulances as VBIEDs in creative ways. January 12, 2013, was a deadly day in Quetta, Pakistan, with 103 people killed and over 200 wounded in four separate, coordinated, and planned soft target attacks by the TTP. First, there was an attack at a mosque that killed twenty-two worshippers, followed by a mass shooting at a recreational area used by Pakistan's military commandos, with another twelve dead. Next, a suicide bomber detonated his vest at a billiards hall packed with families. As police officers, journalists, and rescue workers rushed into the building, a second blast emanated from a bomb hidden in a stolen ambulance that had folded itself into the pack of response vehicles. The total death toll at the billiards hall was sixty-nine (Associated Press 2013).

The Taliban in Afghanistan has also used ambulances extensively in its attacks. For example, on April 7, 2011, Taliban suicide attackers used an ambulance VBIED to attack a police training center in the southern province of Kandahar. If actual ambulances are not available, groups will simply create one for the attack. In January 2010, the threat of cloned first-response vehicles came to light when, during coordinated attacks in downtown Kabul, the Taliban detonated a VBIED in a van disguised as an ambulance.

The international terrorist group Hamas is expert at using ambulances. On May 11, 2004, an Israeli television station aired footage of armed Arab terrorists in southern Gaza using an ambulance owned and operated by the United Nations to support Palestine refugees. Palestinian gunmen commandeered the emergency vehicle as getaway transportation after murdering six Israeli soldiers in Gaza City. The footage shows two ambulances with flashing lights pulling onto a street. Shots and shouts rang out during the nighttime raid, and then a gang of militants piled into one of the ambulances, clearly marked "UN" with the agency's blue flag flying from the roof, and sped away from the scene (Malkin 2004).

On November 17, 2015, just 4 days after the Paris attacks, a German soccer stadium was evacuated in Hannover after officials learned of a credible ISIS threat against the event, attended by Prime Minister Merkel. A German newspaper reported that an unattended vehicle disguised as an ambulance outside the venue contained a bomb; however German authorities deny the presence of any incendiary devices near the stadium.

Consider how often first responders leave ambulances idling at the scene or at the hospital entrance. There have been reports of ambulances stolen in the United States and some were never recovered. But if a bad actor does not want to steal an ambulance or build a clone, it is now possible simply to buy one on the Internet. Cash-starved communities often sell their old emergency response vehicles to fund a new fleet. Websites such as Ambulance Trader and Fire Truck Trader could be exploited by terrorists and criminals flush with cash who want to make a quick buy. Ambulances are now for sale on eBay and payable in full through PayPal, a nonbank that can be funded anonymously (Hesterman 2013).

Recognizing the growing threat, DHS issued an advisory in 2013, entitled *Terrorist Tradecraft: Impersonation Using Stolen, Cloned or Repurposed Vehicles*, with the following guidance:

- Mitigating the risk:
  - Secure station or facility entrance and exit points, including apparatus bay doors
  - Limit or lock unattended emergency vehicles
  - Establish a policy for decommissioning vehicles
  - Stay current on the "branding" of vehicles used by neighboring jurisdictions and mutual aid companies
  - Consider using holograms on emergency vehicles for authentication
  - Establish a stolen vehicle reporting process that includes "be on the lookout" warnings for high-interest vehicles
- Possible indicators:
  - Improperly marked emergency vehicles
  - Driver of emergency vehicle not knowledgeable about area of responsibility or service
  - Incorrect vehicle decal verbiage, colors, word font, and size
  - Visible identifiers such as phone numbers, license plates, or call numbers that are inconsistent with the vehicle's operating area or mission
  - Heavily loaded vehicle, possibly beyond capacity

In conclusion, schools, churches, and hospitals are places where citizens should feel safe and be protected from danger. Unfortunately, world events indicate an escalation in the number of attacks and scale of violence against these soft targets. We are also no longer safe in the places we go for relaxation and entertainment; the next chapter explores the rising threat against shopping, sporting, and other recreational venues.

REFERENCES

ACLU (American Civil Liberties Union). Spying on First Amendment Activity—State-by-State. 2014.
American Association for the Advancement of Science. Guide to R&D Funding Data—R&D at Colleges and Universities. 2014.
Associated Press. Bombings Kill 103 in Pakistan, news release, http://www.nydailynews.com/news/world/bombing-pakistan-billiard-hall-kills-69-article-1.1237569. 2013.
BBC News. Profile: Wisconsin Sikh Temple Shooter Wade Michael Page. http://www.bbc.co.uk/news/world-us-canada-19167324, August 7, 2012.
Brenner, Susan W. and Anthony C. Crescenzi. State-sponsored crime: The futility of the Economic Espionage Act. *Houston Journal of International Law*, 28(2), Winter 2006.
Cavin, Rita. Umpqua Community College president speaks about shooting. https://www.youtube.com/watch?v=6iQIeRNrfY0.
CBS. No Armed Security at Umpqua Community College, Says Former President. October 1, 2015.
CDC (Centers for Disease Control and Prevention). Schools and Terrorism: A Supplement to the National Advisory Committee on Children and Terrorism Recommendations to the Secretary. Atlanta, GA, 2003.
Chinn, Carl. Ministry Violence Statistics. http://www.carlchinn.com/Church_Security_Concepts.html, 2014.
Chivers, C. J. The School. *New York Times*, March 14, 2007.
Church Committee Reports. 1975.
Churchill, Ward and Jim Vander Wall. *The Cointelpro Papers: Documents from the FBI's Secret Wars against Dissent in the United States*. Cambridge, MA: South End Press, 2002.
DHS (Department of Homeland Security). Homeland Security Academic Advisory Council Minutes. 2013.
DHS. FEMA 428, Primer to Design Safe School Projects in Case of Terrorist Attacks. 2012.
DHS. Terrorist Tradecraft: Impersonation Using Stolen, Cloned or Repurposed Vehicles. https://publicintelligence.net/dhs-fbi-cloned-vehicles, April 3, 2013.
EveryTown Research. 165 School Schootings in America since 2015. http://everytownresearch.org/school-shootings/, 2016.
FBI (Federal Bureau of Investigation). Former University of Tennessee Professor John Reece Roth Sentenced to 48 Months in Prison for Illegally Exporting Military Research Technical Data. 2009.
FBI (Federal Bureau of Investigation). Higher Education and National Security: The Targeting of Sensitive, Proprietary, and Classified Information on Campuses of Higher Education. 2011.
Fisk, Ralph R. Siege at Beslan School Number One—Innocence Lost, unpublished. 2014.
Fisk, Ralph R. Case Study: Umpqua Community College Shooting, unpublished. 2016.
Ford, Dana and Ben Brumfeld. Police: Minnesota Teen Planned School Massacre, http://www.cnn.com/2014/05/01/justice/minnesota-attack-thwarted/, May 2, 2014.
Gallagher, Brian. The Terrorist and WMD Threat in the Place of Worship, unpublished. 2014.
Gansler, Jamie and Alice Gast. Academics and National-Security Experts Must Work Together. *Chronicle of Higher Education* 54, no. 44, 2008: A56.
Gill, Martin. How Offenders Say They Get around Security Measures: Why They Say It Is Easy. Dubai, UAE: ASIS Middle East 5, 2014.
Golden, Daniel. Why the Professor Went to Prison. *Business Week*, November 1, 2012.

Goode, Erica and Serge F. Kovaleski. Details of Tsarnaev Brothers, Boston Suspects Emerge. *New York Times*, April 19, 2013.

Hawley, Chris and Matt Apuzzo. NYPD Infiltration of Colleges Raises Privacy Fears. Associated Press, October 11, 2011.

Hesterman, Jennifer. *The Terrorist–Criminal Nexus: An Alliance of International Drug Cartels, Organized Crime, and Terror Groups*. Boca Raton, FL: CRC Press, 2013.

Homeland Security Academic Advisory Council Minutes. Department of Homeland Security, 2013.

IBN Live, Terrorists Kill Man Who Gave Them Water. November 27, 2008.

Investigative Project on Terrorism. *USA v. Begolly, Emerson*. 2011.

Investigative Project on Terrorism. *State of New York v. Ferhani et Ano*. 2014a.

Investigative Project on Terrorism. *USA v. El-Khalifi*. 2014b.

Investigative Project on Terrorism. *USA v. Cromitie, James et al*. 2014c.

Investigative Project on Terrorism. *USA v. Fahti, El Mehdi Semlali*. 2014d.

Johnson, Bryan. Top 10 Chilling Quotes During School Shootings. http://listverse.com/2012/05/09/top-10-chilling-quotes-during-school-shootings/, 2012.

Kephart, Janice. Immigration and Terrorism. Center for Immigration Studies, 2005.

Leung, Rebecca. New Video of the Beslan School Terror, edited by 48 Hours Documentary CBS News. 2005.

Lifeway Christian Resources Survey: Prayer Is Most Urgent Concern for Churches. 2005.

Lumb, David and Dominic Casciani. Pavlo Lapshyn's 90 Days of Terror. BBC News, 2013.

Malkin, Michelle. The Ambulances-for-Terrorists Scandal. *The Philadelphia Enquirer*, June 7, 2004.

Mayko, Michael P. FBI: Drone-Like Toy Planes in Bomb Plot. *Connecticut Post*, April 7, 2014.

McCluskey, Brent. Graphic CCTV Footage of Militant Attack in a Yemeni Hospital. Podcast audio. http://www.guns.com/2013/12/14/graphic-cctv-footage-militant-attack-yemeni-hospital-video/, 2013.

McDaniel, Michael C. and Cali Mortenson Ellis. The Beslan hostage crisis: A case study for first responders. *Journal of Applied Security Research* 4: 21–35, 2009.

Mohr, Holbrook, Mitch Weiss, and Mike Baker. J-1 Student Visa Abuse: Foreign Students Forced to Work in Strip Clubs, Eat on Floor. *Huffington Post*, December 10, 2010.

Mohr, Holbrook, Mitch Weiss, and Mike Baker. U.S. Fails To Tackle Student Visa Abuses. Associated Press, December 6, 2010.

New Video of the Beslan School Terror. Edited by Leung, Rebecca, 48 Hours Documentary CBS News, 2005.

Novak, Sophie. Should Congress Speed Up Its Push for Terrorism Protection? *National Journal*, April 14, 2014.

Office of the Director of National Intelligence. Annual Report to Congress on Foreign Economic Collection and Industrial Espionage. Washington, DC, 2011.

Pew Research Center. Trust in Government Nears Record Low. 2013.

Piegorsch, Walter W., Susan L. Cutter, and Frank Hardisty. Benchmark analysis for quantifying urban vulnerability to terrorist incidents. *Risk Analysis*, 27(6): 1411–1425, 2007.

Rafalko, Frank J. *Mh/Chaos: The Cia's Campaign against the Radical New Left and the Black Panthers*. Naval Institute Press, 2011.

Schuster, Henry. What Is Terrorism? CNN, 2006.

Srivatsa, Naveen and Xi Zu. Alleged Russian Spy Blends into Harvard. In *The Harvard Crimson*, 2010.

The Telegraph. Gunmen in Pakistan Bomb Female Students' Bus Then Attack Hospital, June 15, 2013.

Terror Operations: Case Studies in Terrorism. *TRADOC G2 Handbook No. 1*, July 25, 2007: 6–19.

Turner, Allan and Craig Hlavaty. Thieves Take $600,000 from Lakewood Church Safe. *The Houston Chronicle*, March 12, 2014.

University of Maryland. President's Task Force on Cybersecurity. http://www.umd.edu/datasecurity/, 2014.

U.S. Department of Justice. Operation Xcellerator Press Conference. http://www.dea.gov/pr/speeches-testimony/2012-2009/xcellerator.pdf, 2009.

U.S. Department of Justice. Affidavit, *US v. Christopher R. Mestos* et al. 2010.

U.S. Department of Justice. Texas Resident Arrested on Charge of Attempted Use of Weapon of Mass Destruction. February 24, 2011.

U.S. Department of Justice. workpermit.com. Foreign Students to the U.S. Skip out on J-1 Visas, Become National Criminals. http://www.workpermit.com/news/2006_08_10/us/student_visa_criminals.htm.

Williams, Timothy. Plan for Armed Campus Security Guards Was Dismissed. *The New York Times*. October 1, 2015. http://www.nytimes.com/live/shooting-at-umpqua-community-college/plan-for-armed-campus-security-guards-was-dismissed/.

WPRI Newport, Rhode Island. Unusual Mob Partnership: Inside the Mafia Reveals Relationship with Gangs. https://www.youtube.com/watch?v=BYRreNJ-b9k, 2008.

Zuckerman, Jessica. 60 Terrorist Plots since 9/11: Continued Lessons in Domestic Counterterrorism. Heritage Foundation, 2014.

14

Soft Target Threat Assessment
Malls, Sporting Events, and Recreational Venues

Jennifer Hesterman

Contents

Introduction	253
Shopping Centers	254
Other Significant Mall Attacks	259
Terrorist Threats against U.S. Malls	259
Vulnerable Open Air Marketplaces	261
Sports Venues	261
Stadiums and Arenas in the Crosshairs	261
Performing Arts and Recreation Venues	269
Tourist Sites: A Vulnerable Target	275
References	278

"I swear to God, as we struck France in its stronghold Paris, we will strike America in its stronghold, Washington."

ISIS fighter, video released 2 days after the Paris attack

INTRODUCTION

Just weeks before the Paris attack, FBI Director James Comey revealed the Bureau was investigating ISIS-related activity in all 50 states and that "ISIS is recruiting here 24 hours a day." (Sewell 2015) He later revealed there were 900 investigations currently in progress against suspected ISIS operatives, recruits, and individuals inspired by ISIS, with the number of investigations slowly growing. The George Washington University's Center for Cyber and Homeland Security formed a team, the Program on Extremism to explore homegrown radicalization. Their landmark report, *ISIS in America: From Retweets to Raqqa*, is a must-read for those responsible for investigating threats, protecting soft target venues, and responding to soft target attacks. The report, by counterterrorism and counter-radicalization experts Dr. Lorenzo Vidino and Seamus Hughes, consists of two parts. The first examines all cases of U.S. persons arrested, indicted, or convicted in the United States for ISIS-related activities with case information and empirical evidence for identifying demographic factors such as age, gender, and location. The second part of the report examines various aspects of the ISIS-related mobilization in America, analyzing individual motivations; the role of the Internet and, in particular, social media, in their radicalization and recruitment; and whether radicalization took place in isolation or with other, like-minded individuals, called clustering (http://cchs.gwu.edu/reports). In the words of Sun Tzu, "Know the enemy."

Not only are Americans targeted in the places where they study, worship, and heal, but shopping and recreational venues are also in the crosshairs. The terrorist's main goals are achieved: instant notoriety for the group and the cause, a ripple of fear in the community, immediate impact to a business sector, and making the government appear to be unable to protect communities and civilians. Naturally, profit-taking soft targets do not want to become fortresses and typically struggle with the challenge of balancing security with a pleasant customer experience. Unfortunately, the decision to lower the security construct in exchange for customer satisfaction and loyalty presents increasing risk to those who own and operate the venues. For those who are hesitant or unwilling to expend funds, or sacrifice aesthetics or convenience for security, ask yourself one question: what is the cost of NOT securing your facility?

SHOPPING CENTERS

The modern shopping mall is extremely large and glitzy, hosting more than just retailers; many also have crowd-drawing attractions. For example, most new malls in the Middle East have amusement parks with small roller coasters, Ferris wheels, laser tag, bumper cars, arcades, and bowling alleys. The Mall of America in Bloomington, Minnesota, is the largest in the country; it has a seven-acre indoor theme park, theaters, and nightclubs, as well as hosting the state's aquarium. The megamall has more than 500 stores, employing close to 12,000 workers and hosting over 35 million visitors yearly, or 95,000 per day. On Black Friday, the after-Thanksgiving visitor total can top 200,000 shoppers. This business owner's dream is also a security nightmare.

Malls have a unique vulnerability: there are many entrances versus limited or controlled checkpoints, making them extremely difficult to protect. As the opening quote in the chapter alludes, the mall's name alone may entice attack. For instance the "Mall of America" and "Pentagon City" are not only venues that are soft targets with thousands of unsuspecting, vulnerable shoppers, but also could be symbolic targets. As Maureen Bausch, executive vice president, Mall of America stated: "I think our name, Mall of America, is attractive to people that want to hurt America" (Schulz et al. 2011).

The addition of attractions in or near a mall, such as a casino, could further anger radical Islamists, as was likely the case in the Westgate Mall attack in Nairobi. A collocated aquarium, zoo, or circus could draw the ire of single-interest terrorists such as ALF, putting the mall on their radar. Indeed, intense protests against these venues happen worldwide on a regular basis by groups such as the Toronto Aquarium Resistance Alliance (TARA), Marineland Animal Defense (MAD), and Grassroots Ontario Animal Liberation (GOAL). Children's amusement park areas and daycare centers pose additional concerns as targets themselves or if the mall is attacked. These areas are typically located in a far recess of the mall and are especially vulnerable. One of the worst daycare fires in history occurred at the new, luxurious Villagio Mall in Doha, Qatar, in May 2012, killing thirteen children and four employees, as well as two firefighters who tried to rescue them.

Michael Rozin, an Israeli security expert, was formerly employed at the Mall of America as their special operations security captain and gave an informative interview about security at the facility. The mall calls its counterterrorism unit RAM, or risk assessment and mitigation. Rozin explains that, although detecting weapons is essential, assessing intent is equally if not more important. Therefore, his team was trained on behavioral recognition techniques to identify suspicious actors before they engage in a criminal or terrorist act. Rozin brought his experiences working security at the extraordinarily secure Ben-Gurion International Airport in Israel, where he learned behavioral detection and interviewing techniques (*American Jewish World* 2011). Human factors and intelligence are extremely important in soft target situations where there are many access points such as malls and where technology, such as metal detectors and bag checks, is not available. Rozin's SIRA behavioral detection techniques are covered in Chapter 3 as an option for soft target venues with no ability to control entry points.

Mall violence has dramatically increased in the last few years, with violent fights, shootings, and even bombings. During the holiday shopping season of Christmas 2015, 14 malls across America experienced violent mob attacks. On December 26th, at the Mall St. Matthews in Louisville, Kentucky, and a group of teens, estimated between 1,000 and 2,000, terrorized the mall the day after Christmas. This type "flash mob" attack, fueled by the use of social media, is on the rise in the United States and of concern in large venues. "Flash mob" attacks at shopping venues and movie theaters occurred in several U.S. cities since 2013—Los Angeles, Chicago, Cleveland, Washington, DC, Philadelphia, Baltimore, Milwaukee and at the Mall of America in Minnesota.

Terrorism experts are concerned about ISIS's ability to recruit and motivate using social media, similar to tactics used by civilian mob organizers. A jihadist "call to action" could quickly result in a mob terrorism type of attack. In September 2014, Australian authorities stopped a plot by ISIS to mobilize a large group and grab random people off the street, beheading them on videotape. The massive raid, said to be the largest counterterrorism operation in Australia's history, involved more than 800 police officers and raids of at least 12 properties (Goldberg 2014). ISIS often instigates virtual flash mobs on social media to target and "attack." For instance, in September 2014, an Air Force pilot and his teenage son were subjected to a Facebook ISIS attack after the service member posted a picture from a mission flown over Iraq. A post on one ISIS-linked Twitter account stated: "We have a raid on an American pilot account at 9:00 evening Mecca time who is participating in the crusaders' bombing. Retweet this if you are ready to be part of the raid." The family's Facebook pages were then defaced with a barrage of derogatory messages (Herridge 2014). This type of virtual attack could certainly be considered psychological terrorism, attacking state of mind and sense of security, making people feel uneasy or afraid without causing physical harm.

On the other end of the spectrum from acts of spontaneous violence we have the 2014 terrorist attack against the Westgate Mall in Nairobi, Kenya. This attack showed a new level of sophisticated planning and methodical execution in attacks against shopping centers, raising the bar for mall security officials worldwide.

Case Study: The Nairobi Mall Attack

Saturday, September 21, 2013, was Kenya's annual International Day of Peace. However, the day was anything but peaceful; at noon, armed militants affiliated with the Somali terrorist group al-Shabaab stormed the Westgate Mall in Nairobi, Kenya. Over the next four days, at least sixty-seven people died, 175 were injured, and the mall was destroyed. Reports at the time were disjointed and confusing and after months of investigation by U.S., British, and Israeli experts, there is still little clarity about exactly what transpired, who was involved, or how the Kenyan security forces' response may have exacerbated the problem. However, the successful soft target attack provides insight to security experts and first responders regarding this new page in the terrorists' playbook.

THE ATTACK

The upscale Westgate Mall, owned by an Israeli businessman, sits across a park from the U.S. and Canadian embassies and is frequented by Westerners and affluent Kenyans, making it a lucrative target. Even after a wave of bombings throughout Kenya in 2012 at the hands of al-Hijra, an al-Shabaab-affiliated group, security measures at the mall were described as cursory. Confidential documents accessed by the *Sunday Telegraph* newspaper showed the United Nations had been warned, in the previous month, that the threat of an "attempted large-scale attack" in Kenya was "elevated" (Pflanz and Alexander 2013). One week before the attack, Kenyan police claimed to have disrupted a major attack in its final stages of planning after arresting two people with suicide vests packed with ball bearings, grenades, and AK-47 assault rifles. A manhunt was also launched for eight more suspects (Pflanz and Alexander 2013). After the incident, Nairobi senator Mike Sonko claimed he warned the security services of a possible attack against the mall three months previously (*Jambo News* 2013).

The attack commenced when gunmen entered the mall and began shooting and throwing grenades. Initial reports of fifteen gunmen simultaneously entering the building from the main entrance, the rooftop parking lot, and a ramp into the basement (BBC 2013a) were disproven after reviewing CCTV footage. It appears there were likely only four to six gunmen, most of whom entered through the main entrance and then split into teams (Pflanz 2014). They carried hand grenades, AK-47 assault rifles, and ordinary munitions. Panic ensued as unsuspecting shoppers tried to flee or take cover. The initial attack was ruthless, including the killing of children who were attending a cooking competition in the parking lot. According to survivors, the attackers attempted to separate Muslims and non-Muslims and allowed most Muslims to leave unharmed. One woman declared to the attackers she was a Muslim; however, she was shot for not being dressed conservatively enough. Hostages were gathered and held in the cinema and casino on the second floor and in the basement. The situation was extremely complex, with responding law enforcement units likely unaware of the severity of the situation or the sophistication of the actors.

Despite the first calls for help, the police did not arrive on scene until 12:30 p.m. Thirty minutes later, they entered the mall to attempt rescue operations for the unknown number of people trapped inside. Kenyan Defense Forces (KDF) military personnel arrived at 3:00 pm and also entered the mall, engaging the shooters. The transition of command authority at the scene from police operations to military was poorly coordinated; the police later asserted that while helping people escape, they were mistaken for the attackers and fired upon by the military (BBC Africa 2013b). The police and military were using different radio frequencies, neither had mall blueprints, and the Kenyan military scrambled to fly their best-trained soldiers back from Somalia (Kulish, Gettleman, and Kron 2013). CCTV footage shows Kenyan security forces abandoning the search for the shooters and, instead, looting stores within hours of the siege. A jewelry store was emptied, and security forces looted watch and clothing shops, grabbed cash from tills and ATMs, and tried to shoot their way into a casino safe. At about 10:30 pm, footage shows four attackers holed up in an office, one of them injured, while security forces stroll around the mall with bags full of merchandise and cash (Dixon 2013).

By Sunday, over 1,000 people had been rescued from the mall, but government forces did not have control, and gunfire continued throughout the day. The death toll was up to fifty-nine and the attackers were still loose in the mall. Israeli commandos and specialized police officers arrived, along with American-supplied night vision goggles for the Kenyan forces. On Monday the government launched another offensive. There were loud explosions and sounds of gunfire, and smoke began to pour out of the building as the rear of the building collapsed; the death toll was up to sixty-two and it was believed there were more hostages inside. On Tuesday, the police declared the operation over, although there were still reports of sporadic gunfire as they cleared the mall. The final toll was at least sixty-seven people dead, including six security personnel and several terrorists who were in the rear of the shopping center when it collapsed. Some remains were burned beyond recognition and not capable of yielding DNA evidence; authorities believe the death count could have been as high as ninety-four (Butime 2014).

THE PLANNING

The attackers rented a small shop in the mall to gain access to otherwise prohibited areas such as storage rooms and service elevators. This also would have given them access before and after shopping hours, and they used this time to pre-position weapons and ammunition in a ventilation shaft (BBC 2013a; BBC Africa 2013b). Their presence in the mall provided opportunity to assess which areas would yield the highest casualties, and whether there were scheduled events to target, such as the children's cook-off outside the grocery store.

Western investigators from the United Kingdom, the United States, and Israel were called to help with the investigation. At one point the FBI had over eighty investigators on scene, including members of the Evidence Response Team (ERT), who helped process the scene (FBI 2014). Despite initial reports of fifteen heavily armed attackers, investigators all concluded the attack was probably carried out by a group of four to six people with only a basic plan and light weapons, grenades to gain entry to the mall, and a few AK-47 s. Kenyan officials disagree, insisting a larger force was involved in a detailed, intricate plan. A team of four to six gunmen thwarting Kenyan government control for 4 days makes the government look ineffectual and ill prepared, especially given the known threats against Kenya. Therefore, it is easy to see why officials in Nairobi failed to embrace the investigative reports.

There is also confusion over the status of the attackers. Although it is possible some of the attackers may have put down weapons and left with the fleeing customers, there is no evidence to prove this theory. To add further confusion, after reviewing thousands of hours of CCTV footage, investigators from the New York Police Department said there is no evidence the militants were in the mall after the first day and they all most likely escaped in the confusion (New York City Police Department 2013). The FBI disagrees with this assessment, with their Nairobi LEGAT stating:

> We believe, as do the Kenyan authorities, that the four gunmen inside the mall were killed. Our ERT made significant finds, and there is no evidence that any of the attackers escaped from the area where they made their last stand. Three sets of remains were found. Also, the Kenyans were on the scene that first day and set up a very secure crime scene perimeter, making an escape unlikely. Additionally, had the attackers escaped, it would have been publicly celebrated and exploited for propaganda purposes by al-Shabaab. That hasn't happened. (FBI 2014)

Months after the attack, precious little information has been added to the initial investigations and few security forces have been held accountable for their ineffective and even deplorable behavior at the scene. There are four men on trial in Nairobi for aiding the attackers; all deny the charges (Pflanz 2014). Pseudonyms were released by Kenyan authorities less than a month after the attacks; however, a *New York Times* article in January 2014 listed what are believed to be the men's given names, including Hassan Abdi Dhuhulow, a Somali-born Norwegian national (BBC 2013a). Al-Shabaab claimed three Americans were involved in the Westgate Mall attack, naming Ahmed Mohamed Isse of St. Paul; Abdifatah Osman Keenadiid of Minneapolis; and Gen Mustafe Noorudiin of Kansas City, Missouri, via their Twitter account, @HSM_Press (http://jihadology.net/2013/09/21/new-statement-from-%E1%B8%A5arakat-al-shabab-al-mujahidin-claiming-responsibility-for-the-westgate-mall-attack-in-nairobi/).

Al-Shabaab's claims have not been confirmed by American law enforcement. In addition, additional Twitter accounts claiming to be al-Shabaab–run have suggested other Americans were also involved.

The threat in Kenya has not subsided, and U.S. citizens are cautioned by the State Department regarding travel to the country and the threat of attacks against soft targets and kidnapping by al Shabaab. For example, on March 17, 2014, a SUV packed with 350 pounds of explosives was found in the port of Mombasa (Pflanz 2014). Between June 15 and June 17, 2014, more than 60 people were killed in mob-style attacks in Mpeketoni. The April 2015 early morning attack on the Garissa School by al Shabaab

killed at least 148 students. Finally, complicating the security situation in Kenya is the emergent ISIS wing of al-Shabaab; their first attack was the hijacking of a bus outside of Nairobi in December 2015.
(*Source*: Kinzer (2014). Ms. Kinzer is a Reserve Air Force intelligence officer and senior intelligence consultant at Patch Plus Consulting.)

Other Significant Mall Attacks

Hamas is also planning attacks against shopping malls. Shin Bet, Israel's intelligence agency, foiled a bomb attack plotted by Hamas timed for the High Holy Days in September 2013. Two men who were employed as maintenance workers at the upscale Mamilla Mall were planning to smuggle the bomb into the shopping center and to hide it in a closet. Their handlers ordered them to plant the bomb in a restaurant, store, or trash can and to cover it with wrapping paper, in order to make it look like a gift in preparation for the upcoming Jewish holidays, and to detonate it when the mall would be packed with shoppers. One scenario included wrapping the device in a box of chocolates and placing it in a garbage can, according to the indictment (Lappin and Bob 2013). The Irish Republican Army (IRA) has also used targeted shopping areas; on the morning of June 15, 1996, a cargo van filled with 3,300 pounds of explosives exploded in the middle of a busy shopping center in Manchester, England, injuring over 200 people. England was hosting the 1996 European Football Championship that year, and a match between German and Russian soccer teams was scheduled for the next day in Manchester's stadium.

Terrorist Threats against U.S. Malls

The United States has 800 shopping malls and Allied Barton Security Services is one of the largest suppliers of shopping mall security personnel, with some 12,000 officers. Their assessment on the possibility of a mall attack in the United States is that it can happen, and we need to prepare (Bradley 2013). In fact, several plots have already been intercepted by law enforcement since 9/11. Nuradin M. Abdi, a Somali immigrant living in Columbus, Ohio, was arrested in November 2003 and charged in a plot to bomb a local shopping mall and shoot the evacuating shoppers. Abdi traveled to Ethiopia purposely to receive training at an Islamist camp in construction of explosives for his operation (Investigative Project on Terrorism: *US v. Nuradin Abdi* 2014b). Abdi recently finished his 10-year term in federal prison and was deported to Ethiopia.

Derrick Shareef was arrested in December 2006 on charges he planned to set off hand grenades in a shopping mall outside Chicago. Shareef acted alone and was arrested after meeting with an undercover agent. FBI reports indicated the mall was one of several potential targets, including courthouses, city halls, and government facilities. Shareef, however, settled on attacking a mall in the days immediately preceding Christmas because he believed it would cause the greatest amount of chaos and damage (Investigative Project on Terrorism *U.S. v. Shareef* 2004). Shareef was sentenced to 35 years in prison without the possibility of parole.

Tarek Mehanna, previously indicted for lying to the FBI about the location of terrorist suspect Daniel Maldonado, was arrested on October 21, 2009, on allegations of conspiracy

Figure 14.1 Screen still from the al-Shabaab video "The Westgate Siege; Retributive Justice." (From https://videopress.com/v/BYQBQQAY)

to kill two U.S. politicians, American troops in Iraq, and civilians in local shopping malls (Investigative Project on Terrorism: *U.S. v. Mehanna* 2014a). He was sentenced to 18 years in jail.

In February, 2015, al-Shabaab released a video entitled "The Westgate Siege; Retributive Justice" targeting on London's Oxford Street and the Westfield shopping centers, as well as U.S. and Canadian shopping malls. Figure 14.1 shows an al-Shabaab terrorist discussing the merits of attacking the Mall of America, along with its longitude and latitude (https://videopress.com/v/BYQBQQAY).

Omar Faraj Saeed Al Hardan, 24, a Palestinian born in Iraq, was arrested in Houston in January 2016, charged with providing material support to ISIS (namely himself). According to testimony at his arraignment hearing, an FBI agent said al Hardan was planning attacks on two malls, including the upscale Galleria. His plan was to put bombs into trash cans and detonate the devices using cellphones (Walker 2016).

Malls seem irresistible to violent actors with no apparent motive, other than homicidal or suicidal intentions. Typically, the perpetrator commits suicide or is killed by law enforcement; therefore, it is difficult to gather enough forensic information to profile mall attackers. On January 26, 2014, a suicidal 19-year-old man, Darion Marcus Aguilar, went to his favorite store, popular with skateboarders, and shot two employees and then himself at a mall in Columbia, Maryland. He also carried two makeshift bombs in his backpack cobbled together with fireworks. In other random mall attacks, a 22-year-old man killed two people and then himself at a mall near Portland, Oregon, in December 2012; a 19-year-old man killed eight people and then himself at an Omaha, Nebraska, mall in December 2007; and an 18-year-old man killed five people before he was killed by police at a mall in Salt Lake City, Utah, in February 2007.

Following the Nairobi mall attack and other violent incidents, the FBI and Department of Homeland Security stepped up exercises at major malls in the United States as part of the Complex Mall Attack Preparedness Initiative established in late 2013 to promote preparedness and strengthen public/private partnerships. Exercises range from tabletop to full-scale "attack" scenarios to test response. Typically, shoppers have departed the stores and cleared the parking lots, taking away from the realism of difficulty of approach and dealing with panicked citizens; however, the exercises include actors fronting challenges to law enforcement that would be present during busy shopping hours.

Vulnerable Open Air Marketplaces

As the center of economy for poor villages and towns, open-air marketplaces are a soft target routinely singled out in war-torn countries. The open-air market is an attractive target because overwhelming firepower and planning are not necessary to cause mass casualties. For example, on February 5, 1994, during the Bosnian war, a 120 millimeter mortar shell fired by Serbian forces landed in the Markale marketplace in Sarajevo during the busy lunch hour. UN observers reported 168 people were killed and 144 more were wounded. These types of attacks against innocent civilians as they go about their daily routine can even change the course of a war. Markale was struck 18 months later, on August 28, 1995, again during lunch hour. This time, Bosnian forces fired five mortar shells, killing forty-three people and wounding seventy-five others. The event pulled an angry NATO into the conflict, and its punishing air strikes against Bosnian Serb forces brought them to the table for the Dayton Peace Accords and a negotiated peace.

ISIS is actively attacking open markets through its proxies throughout the Middle East. On November 2015 in Beirut, two ISIS suicide bombers detonated vests within 5 minutes of each other in a busy marketplace, killing forty-three and wounding two hundred thirty-nine. ISIS struck again on January 13, 2016, with an open air attack in a tourism area, Sultanahmet Square, in Istanbul, Turkey. Ten German visitors were killed and eleven injured. Indicating the ripple effect of these attacks, travel advisories were immediately issued, putting Istanbul on the "no travel" list for several countries. Turkey heavily relies on the business generated by tourists, with over 38 million each year. Other terrorist groups actively participating in open market attacks include Hamas, the Taliban and Boko Haram.

We must consider the vulnerability of open-air markets in our country; for instance, farmers' markets, flea markets, and similar venues are popular on weekend days in large U.S. cities, and they, too, are completely unprotected. As with enclosed malls, detecting bad actors and their weaponry is difficult; therefore, deterrence, mitigation, and response are critical to contain the threat.

SPORTS VENUES

Large sporting events are of increasing interest to terrorists, and this vulnerable soft target serves two of their needs. First, there is a crowd of unsuspecting, vulnerable people and, second, there is the added benefit of live television coverage with the amplifying effect of terrorized viewers.

Stadiums and Arenas in the Crosshairs

The November, 13, 2015, ISIS attack in Paris included a plan for three suicide bombers to attack the main stadium, Stade de France. The first bomber, who had a ticket to the game, was prevented from entering the stadium after a security guard patted him down and discovered the suicide vest; a few seconds after being turned away, the terrorist detonated the vest, killing himself and a bystander. Investigators later surmised the plan was for the first bomber to detonate his vest inside the stadium, causing a stampede and pushing the panicked crowd into the streets where two other bombers were lying in wait (Robinson and Landauro 2015). Ten minutes after the first bombing, the second bomber blew himself up near the stadium. Another 23 minutes later, the third bomber's vest detonated at a nearby McDonalds. Fortunately, only one bystander was killed in the melee.

Football stadiums are also high on al-Qaeda's list of targets. The *Encyclopedia of Afghan Jihad* (2004), found in the London residence of Islamic cleric Sheikh Abu Hamza al-Masri, proposed football stadiums as possible terrorist attack sites. In July 2002, the FBI issued an alert warning suspects with links to al-Qaeda affiliates were downloading stadium images from http://www.worldstadiums.com, including the Edward Jones Dome in St. Louis and the RCA Dome in Indianapolis (Grace 2002). In response to these and other threats, DHS developed an extensive national planning scenarios document to cover possible terrorist attacks specifically aimed at stadiums. The 2005 document specifically addressed the potential of a biological attack on a sports arena, stating that the spreading of pneumonic plague in the bathrooms would potentially kill 2,500 people. DHS's 2006 planning scenarios included a light aircraft spraying a chemical agent into a packed college football stadium, contaminating the stadium and generating a downwind vapor hazard (DHS 2006).

Over 1,300 professional sport stadiums and arenas in the United States are used for a variety of events from school graduations to concerts, political events, and evacuation and sheltering locations in natural or national emergencies. As such, the Department of Homeland Security (DHS) has identified sport stadiums/arenas as critical infrastructure/key resource (CI/KR) key assets. However, as with all soft target locations, the government relies on the venue owners and operators to secure their properties and users properly. College stadiums have the most vulnerability, as they rely on part-time and seasonal help to man events, and background screening is limited due to expense. Dr. Stacey Hall, an expert on spectator sports safety and security, has written several outstanding articles and books on the topic, worth review by all who own, operate, or secure these venues. According to Dr. Hall (2010), college stadium vulnerabilities include:

- Lack of emergency and evacuation plans specific to sport venue
- Inadequate searching of venue prior to event
- Inadequate searches of fans and belongings
- Improperly secured concessions
- Dangerous chemicals stored inside the sport venue
- No accountability for vendors and their vehicles
- Inadequate staff training in security awareness/response to
- WMD attacks

The good news: entrance points to most sporting and recreating venues can be limited, bags checked, and behavior assessed. However, the cost to secure a stadium of 75,000 people or an Olympic venue is exorbitant, especially if protecting against broad spectrum of threats. For example, in response to threats against its games, the NFL instituted mandatory pat-downs at entrances, upsetting the venues and teams, which had to pay extra for security and handle irate fans. The host city for the annual Super Bowl, designated a special event assignment rating (SEAR) 1 event, by DHS, typically loses $4-5 million overall due to cost of augmenting federal security measures. Super Bowl 50, held in February, 2016, had 60 agencies coordinating security in Santa Clara, California, including F-15E combat air patrols and a large presence of federal law enforcement agencies and equipment on the ground.

U.S. taxpayers were outraged at the $300 million price tag for the 2002 Winter Olympics in Utah, yet the 2014 Sochi Olympics cost $50 billion, including extreme security measures to protect from terrorist attacks. The first major terrorist attack against a sporting event was at the Olympics, the 1972 Munich Games, when the Palestinian terrorist group Black September took the Israeli national team hostage, eventually killing eleven athletes and coaches and one West German police officer. The world was transfixed on the event, as the kidnappers and their hostages could be seen on television through their hotel windows. German officials devised a sound rescue plan, Operation Sunshine; unfortunately, members of the International Olympic Committee were interviewed on television and gave away the details. Unbeknownst to them, the terrorists were watching and learned of the plan. After more failed negotiations, a new rescue strategy was developed involving escape helicopters, with snipers targeting terrorists as they walked across the airfield with the hostages. However, the number of terrorists and their firepower were underestimated; the snipers were unable to gain control of the situation, and all of the athletes were killed in the rescue attempt.

The Olympic scene was quiet until Eric Robert Rudolph, a former U.S. Army soldier and member of Christian Identity (a white nationalist group), attacked the 1996 Atlanta Games. At 1:20 am on July 27, 1996, a bomb was detonated in Centennial Olympic Park, an entertainment and vendor area. Two people died and over a hundred were injured in the blast. According to Rudolph, in a riveting *Sports Illustrated* oral history of the bombing, the attack was meant to shut the Games down, not kill people:

> The plan was to clear the park, and hopefully after clearing the park and the explosion, this would create a state of instability in Atlanta, potentially shut the Games down or at least eat into the profits that the Games were going to make. The idea was to use them as warning devices, not to target people ... In retrospect, it was a poor decision. (Zaccardi 2012)

After placing the bomb, which had a 55-minute timer, Rudolph called in two bomb threats to 911, anticipating the park would be cleared. As he watched from a distance, security officers inspected the bag and determined it was a bomb. Much to his surprise, people were still milling around the area when the device detonated: the bomb threat information was never transmitted to the scene. As Rudolph left the park, he detonated four unexploded devices in a trash can outside the park instead of planting them. Unfortunately, the security guard who originally noticed the suspicious bag containing the bomb, Richard Jewell, was named as the prime suspect while Rudolph slipped out of Atlanta. After 7 years on the run and bombings at abortion clinics and a lesbian bar, Rudolph was caught by a rookie police officer in 2003 while scavenging food from a dumpster; he is currently serving four life terms at the supermax prison in Colorado. The only attack he ever apologized for was the one against the Olympics.

The Basque separatist group ETA targeted the 2002 European Champions Soccer League final in Madrid, Spain. On May 1, 2002, a car bomb exploded near Santiago Bernabeu stadium hours before the game, and a second car bomb exploded a half hour later about one mile away. Seventeen people were injured in the attacks. However, in a show of defiance to the terrorist group, officials went ahead with the game with 75,000 fans in attendance. Also, the Boston Marathon was not the first running race targeted by terrorists. On April 6, 2008, a suicide bomber detonated a device at the start of a Sri Lankan marathon, killing fifteen people and injuring a hundred. The terrorist group Liberation Tigers of Tamil Eelam (LTTE) took responsibility for the attack (*New York Times* 2008).

There have been several sporting event threats or attacks in the United States. In June 2003, a bomb threat was made against Continental Airlines Arena during game 5 of the NBA finals, and responding police found ten cars on fire in the arena parking lot. In October 2005, an Oklahoma University student suicide bomber strapped a detonating device to his body outside the school's stadium during a football game, killing only himself. His apartment had jihadist and bomb-making material, as well as a cache of the explosive triacetone triperoxide (TATP), a favorite of al-Qaeda terrorists. In October 2006, the National Football League received an uncorroborated threat indicating the use of radiological "dirty bombs" against seven National Football League stadiums.

Finally, in the worst attack against a sporting venue in the United States, the Boston Marathon bombings on April 15, 2013, were carried out by brothers Tamerlan and Dzhokhar Tsarnaev, naturalized US citizens of Chechen descent. The attack killed three people, and wounded more than two hundred. Tamerlan was clearly the ringleader—a vicious killer who was posthumously implicated in a prior unsolved triple homicide of his best friend and two others in Boston. At some point, he was (at the very least) communicating with and inspired by radical Islamists. The brothers used an edition of al-Qaeda in the Arabian Peninsula's *Inspire* magazine to construct the pressure cooker bombs, which, interestingly, are used in many Chechen rebel attacks. However, in a show of psychological resiliency, the city hosted a marathon a few weeks later for those unable to complete the main event. In April 2014, Boston held a full-up marathon event with thousands of security officers and the event itself went off without incident. However, a week before, a man dressed in black, screaming "Boston strong," walked down the middle of Boylston Street with two backpacks containing rice cookers; the bags were detonated by law enforcement. Although not a terrorism-related event, it rattled the nerves in the still healing city.

Other sporting venues have been targeted by the al-Qaeda splinter group, al-Shabaab, in both Somalia and Uganda. The Mexican cartels are particularly cruel, targeting children's soccer matches and sports-themed birthday parties. So the concept of sporting events as targets is not new and is being leveraged by everyone from petty criminals to major cartels and terrorist organizations.

Exploring Russia's security laydown for the 2014 Sochi Olympics provides the framework for discussion regarding the delicate balance between security, venue aesthetics, and comfort for attendees. Not only lives are at stake, but also world opinion of the host's ability to secure itself, athletes, and visitors from attack. This pressure to secure the vast venue led to heavy-handed security decisions and scrutiny, as well as exorbitant cost.

Case Study: Putin's Ring of Steel—How the Russian Federation Constructed the Most Extensive and Controversial Security System in Olympic History

In a January 2014 interview, President Vladimir Putin asserted: "[i]f we allow ourselves to be weak, feel weak, let our fear to be seen, by doing that we'll assist those terrorists in achieving their goals" (Radia 2014). The 2014 Winter Olympics in Sochi, Russia, was historic because, for the first time, the global discussion was more focused on the escalating potential of terrorist activity against unprotected "soft" targets and the requisite security for the Games' venues rather than on the athletic competitions. With serious threats of terrorist attacks shrouding the Games, the Russian Olympic Committee generated the most sophisticated and extensive security regime in history. From deploying thousands of troops to positioning hundreds of antiaircraft missiles and implementing a pervasive surveillance system, the security at the Sochi Olympic Games was unprecedented in depth and extent. However, despite these defensive measures and Russian assurances, the world nervously speculated as to whether the Games in Sochi would be safe. Agitated by statements from terrorist leaders that they were explicitly targeting the Olympics, and suicide bombings occurring only months before the Games, world leaders, Russian citizenry, Olympic athletes, and international tourists feared the advisories, precautions, and implemented security could not protect against potential terrorist attacks. Yet, despite the numerous threats and all of the concerns, the 2014 Winter Olympic Games in Sochi were not attacked. President Putin accomplished his goal of protecting the Games, maintaining his reputation, and promoting Russia as a powerful and impenetrable global entity. But, what was the ultimate price for Russia's Olympic success?

THE VENUE

The warmest place in Russia, and a common vacation spot for natives during the long winter, Sochi was a seemingly strange location to host the 2014 Winter Olympics. However, Sochi is a favorite locale of President Putin, who owns a personal retreat in the resort city. Throughout the history of the former Soviet Union, Sochi offered peace and quiet to many of the politically elite, including Joseph Stalin, whose former dacha (vacation home) can still be visited by tourists today. Despite the weather and the Soviet legacy in Sochi, the most surprising aspect of the city's selection was its close proximity to the volatile region of the northern Caucasus. Situated in the southwestern-most corner of Russia, nestled near the Black Sea, Sochi is separated only by the Caucasus Mountains from a hotbed of Islamic militant groups—deeply embedded, actively engaged in terror attacks, and passionately intent on attacking Putin's Russia. Sochi is just 300 miles from Chechnya and only 185 miles from Kabardino-Balkaria, site of more than sixty-nine attacks in 2013. Unquestionably, the host city of the 2014 Winter Olympics was located on the edge of a war zone.

The long and historic conflict in the northern Caucasus dates back to 1817, when Russian forces invaded the Persian-controlled region and sparked a war with the local tribes. After nearly 50 years of fighting and more than 290,000 deaths, Russia captured the opposition leader and annexed the Caucasus. Relations were further antagonized during Joseph Stalin's reign in the Soviet Union. Relying on accusations of collaboration with Nazi Germany, Stalin deported the entire Chechen population and the majority of the Ingush population to Siberia and Central Asia. The historical and cultural clash between Russia and the northern Caucasus culminated in the First and Second Chechen War during the 1990s, which took over 150,000 lives. Today, almost 200 years after the first spark of violence in 1817, relations between the Kremlin and the northern Caucasus continue to be strained and the region remains volatile.

The violence and cultural tension associated with the northern Caucasus is predominantly embedded in the republics of Dagestan, Chechnya, and Ingushetia. The North Caucasus Federal District, with a population of approximately ten million, is the smallest of Russia's eight federal districts. However, it is the most diverse district, with approximately forty ethnic groups, and the only district in which ethnic Russians are actually in the minority. Unlike the rest of the Russian Federation, Sunni Islam is the dominant religion in the northern Caucasus, with most practicing Sufism, or mystical Islam. In 2005, a group of Sunni Islam extremists in this region ignited an uprising against the Russian Federation in an effort to spurn Russian rule and establish an independent Islamic state.

The Islamist insurgency's umbrella group, known as the Caucasus Emirate or Imarat Kavkaz, imposes its religious tenets through various acts of terrorism, including suicide bombings and army ambushes. Its mission to establish an Islamic caliphate is led by Chechen warlord Doku Umarov, whose first attack was the 2010 suicide bombing in the Moscow metro, killing thirty-nine people. He also claimed responsibility for the 2011 suicide attack at Russia's Domodedovo International Airport killing thirty-five. Numerous other terrorist acts against soft targets have recently been perpetrated throughout Russia under the Caucasus emirate banner. In October 2013, a female suicide bomber from Dagestan killed six people and injured over thirty others on a passenger bus in the Russian city of Volgograd. Two months later, another Dagestani female terrorist attacked the same city, killing sixteen people at a local train station.

The Volgograd terrorist attacks, which occurred a mere 6 weeks before the 2014 Winter Olympics, caused global concern and speculation about Russia's ability to stem the increasing threats of similar attacks to soft targets surrounding the Olympic Games. These concerns were not unfounded. In July 2013, Umarov summoned his Islamist followers to make the Sochi 2014 Olympics a target of extremist activity, claiming these "Satanic Games" were being held on the "bones of our ancestors," alluding that Sochi was the last stand for Muslims who were slaughtered and buried there during the Russian–Circassian war that ended in 1864 (*Moscow Times* 2013). Some historians agreed, stating that Russia's hosting the Olympic Games at Sochi was akin to Germany hosting games at Auschwitz (Somra and Watson 2014). So Putin's choice of Sochi was not only based on personal preference, but also incited the Chechens and brought forth memories of genocide that wiped out a generation of Russian Muslims. Media outlets around the world discussed the blatant threats of terror and ruminated over whether Russian security at the Winter Olympics would be sufficient to protect the city, the citizens, the coaches and athletes, and the tourists. The issue of security soon dominated worldwide media discussions regarding the 2014 Winter Olympics in Sochi.

THE PLAN

In response to the blatant terrorist threats and the increasingly negative media coverage of Russia's security, the Sochi Olympic Committee and President Putin implemented a public relations scheme to generate a positive perception of the 2014 Winter Olympic Games. The chief Sochi organizer, Dmitry Chernyshenko, confidently proclaimed Sochi was the "most secure venue on the planet" (Wilson 2014). President Putin adamantly affirmed the Sochi Games would showcase the greatness of the Russian Federation and guaranteed that "our job—needless to say, the job of the Olympics host—is to ensure security of the participants in the Olympics and visitors to this festival of sports and we will do whatever it takes" (Radia 2014).

To the Russian president and the Russian Federation, the Games were indicative of much more than just the ability to host a sporting event. More than 30 years after Moscow hosted the controversial 1980 Olympic Games; the 2014 Winter Olympics in Sochi were intended as an exposition of Russia's greatness. President Putin articulated the significance of the Sochi games stating:

> [F]ollowing the collapse of the USSR, following hard and, let's put it bluntly, blood-soaked developments in the Caucasus, the overall state society was depressing and pessimistic. We need to cheer up, we need to understand and feel that we are capable of pulling off major, large-scale projects and do so on schedule and with good quality. (Radia 2014)

As a result of the significance he placed on the Olympic Games, the sporting event soon earned the moniker of "Putin's Games." Achieving Putin's goal and protecting his reputation came at a price tag of $50 billion, more than what was spent on all previous winter games combined. The exorbitant cost overrun of the Sochi Games is widely attributed to Russia's continued political corruption and construction-industry fraud, and one construction worker openly complained about the $50 million bribe paid to a Kremlin official to secure work at Sochi.

Kremlin officials described the impenetrable fortress of security around the Games as Putin's "ring of steel." The architect of this "ring of steel" was the appointed chairman of security for the Sochi Games, Oleg Syromolotov. Syromolotov serves as the deputy director of Russia's Federal Security Service (FSB) with particular expertise in counterintelligence. Central to the security laydown was the most extensive, sophisticated, and controversial surveillance system in operation (Matthews 2014). The 2014 Olympics became a monumental experiment for comprehensive and invasive surveillance, making Sochi the most heavily policed environment in history (Matthews 2014).

At the core of the operation was a sophisticated technical platform, the system of operative-investigative measures (SORM), which, by all accounts, makes the U.S. National Security Agency's controversial PRISM system look elementary. First developed in the 1980s by the Soviet KGB, SORM is a nationwide electronic interception and monitoring program that authorizes numerous Russian agencies to intercept all electronic transmissions legally. The latest version, SORM-3, has been described as, "a giant vacuum cleaner which scoops all electronic communication from all users all the time" (Matthews 2014). Designed to search metadata and content, SORM-3 intercepts, collects, examines, and stores all information and communications (e.g., e-mails, social networks, phone calls) from electronic devices, storing them for future use.

The breadth of FSB's authority and SORM-3's technical ability to gather and analyze information by reading and listening to electronic communications far surpasses the capabilities of the National Security Agency's PRISM system, which Edward Snowden exposed in 2013. Sochi Olympic officials worked diligently to ensure all Internet service providers (ISPs) installed the SORM-3 devices on their networks to provide the FSB unrestricted surveillance capability of tourists, coaches, athletes, and journalists. The Olympic Committee justified this unbridled surveillance as necessary to protect against terrorism. Furthermore, ISPs that did not have the system installed were fined by the FSB.

In furtherance of the government's antiterrorism security scheme, Russia's leading telecom operator, Rostelecom, promised to provide the fastest Wi-Fi in Olympic history by launching an extensive 4G LTE network throughout Sochi, which was offered for free to all visitors to the Games. However, with this free, rapid Wi-Fi access, all Sochi Internet users were subject to deep packet inspection (DPI), which allowed ISP providers and other intermediaries to collect and analyze the Internet communications of millions of users simultaneously. Rostelecom required the installation of deep packet to facilitate the FSB's ability to monitor and filter all communication traffic.

The Olympic Committee's conventional security was also extensive. More than 40,000 police were on duty throughout the Games guarding the Olympic city (Figure 14.2).

Two sonar systems were purchased to protect against sea-launched terror attacks and, to repel a possible air attack, there were antiballistic missile batteries around the city. Finally, with more than 4,000 surveillance cameras and closed-circuit television (CCTV) technology installed, Sochi became one of the most watched cities in the world. With cameras watching at every corner and missile defense units visible to the public, many argued the spirit of the Sochi Games was shrouded in a blanket of extensive and intimidating security measures (Bender 2014).

Figure 14.2 Russian military officers on patrol in Sochi during the Olympic Games.

At first, the Sochi security officials were reserved and uncooperative, not only about sharing details regarding the scope and magnitude of the implemented security, but also about sharing the duty of security enforcement. Over time, the relationship between the FBI and the Russian FSB improved, and a month before the Games, FBI Director James Comey announced he would send about a dozen U.S. federal agents to the Sochi Games and even more agents to Moscow. This announcement came amidst the growing concern of terrorist attacks on "soft" targets and questions about the security of the U.S. team and others. Cognizant of the increasing terrorism threats against soft targets, a top U.S. counterterrorism official warned a Senate panel prior to the Games that terrorists might strike targets on the outskirts of Sochi (Madhani 2014). Much to the world's relief, the two weeks of Olympic competition were carried out successfully and without any reported breach of security.

The monetary cost to protect the games was high, but the real cost for the unparalleled security at Sochi was felt at the personal level. In and around the Sochi Olympic venues, athletes and coaches, journalists, and tourists were subjected to intrusive electronic surveillance and burdensome physical security measures. In addition to posting photos of the poor sanitary and safety conditions at Sochi via the Twitter account @SochiProblems, tourists also complained (and offered proof) that their hotel rooms and bathrooms were bugged. In angry response to questions posed by reporters regarding poor facility conditions such as the lack of hot water, Russian Deputy Prime Minister Dmitry Kozak may have slipped regarding surveillance, stating: "We have surveillance video from the hotels that shows people turn on the shower, direct the nozzle at the wall, and then leave the room for the whole day" (Sridharan 2014). The price of admission to participate in, to be a spectator of, or to report on an Olympic event was the complete surrender of one's privacy. Russia believes the cost for creating an adequate defense for "soft" targets against terrorist attacks is the complete relinquishment of personal space and private information. Perhaps for those attending the Olympic Games in Sochi, the price was not too high.

(*Source*: Jorns (2014). Ms. Jorns lived and studied in Russia, and she worked at the U.S. embassy in Moscow in the months leading up to the 2014 Olympic Games.)

PERFORMING ARTS AND RECREATION VENUES

The Kennedy Center for the Performing Arts is a beautiful facility located on the Potomac River, just two miles from the White House and the Capitol. On a typical Friday or Saturday evening, there are dignitaries and public figures in attendance, sitting in front or perhaps viewing from the president's or vice president's box, as I was fortunate to do some years back. The annual star-studded Kennedy Center Honors event is even televised live and the holiday concert is often attended by both the president and vice president. It does not take a security expert to realize this venue, which also has an underground parking garage, is a prime soft target for a bold and shocking terrorist attack. As the details are still gathered on the Bataclan concert hall attack in Paris, looking back at a similar attack in Moscow in 2002 provided lessons learned and a departure point for securing concert facilities in the United States.

SOFT TARGETS AND CRISIS MANAGEMENT

Case Study: The Dubrovka Theater Attack

The Dubrovka Theater in Moscow is a similar venue to the Kennedy Center and was the scene of a horrific Chechen terrorist attack on October 23, 2002 (Figure 14.3).

Around 9:00 pm, during Act II of a sold-out performance, forty-one heavily armed and masked men and women drove a bus directly into the theater's main hall, firing assault rifles. Within seconds, 916 people were taken hostage, including a Russian general, while ninety people managed to flee the building or hide. The terrorists had grenades, improvised explosive devices strapped to their bodies, and they wired large bombs throughout the theater. The terrorists stated they had no anger toward foreigners and promised to release anyone who showed a foreign passport. About 200 hostages were released, including some children, pregnant women, and foreigners. The rest were kept in the theater at all times, using the orchestra pit as a lavatory. As the situation moved into its second day, both an outsider who entered the theater to find his son and a hostage who stormed a terrorist were killed.

The mass death toll resulting from the siege actually came at the hands of the Russians themselves during a poorly executed "rescue" on the third day of the standoff. After Special Forces troops stormed the theater, President Putin ordered security forces to inject a substance into the ventilation system, an aerosol anesthetic that was possibly weaponized fentanyl or Kolokol-1, an artificial, powerful, opium-like substance. Instead of merely going to sleep as the government hoped, people suffocated as their tongues relapsed in their mouths, blocking the airway passage. Medical personnel at the scene did not have the proper antidote, and in all, 130 hostages died. In a stunning admission, Moscow's health committee chairman announced all but one of the hostages killed in the raid had died from the gas (BBC News 2002). Those who survived the attack face a lifetime of health-related issues. The terrorists leveraged mass media to broadcast the event live to the world and allowed those in the theater to use their cell phones to call family and friends. Prior to the theater attack, the Chechens were merely fringe actors, an irritant to the Russian government. However, the siege brought Moscow to its knees as it negotiated with the terrorists and the world critically judged

Figure 14.3 A CCTV still from inside the Dubrovka Theater, Moscow.

the government's handling of the crisis. The Chechen Rebels were now a formidable terrorist organization.

Other concert events have been targeted by terrorist groups. Recent, substantiated terrorist threats were made against overseas venues hosting U.S. singers such as Madonna (St. Petersburg and Moscow, 2012), Lady Gaga (Indonesia, 2013), and Aerosmith (Indonesia, 2013). On July 3, 2013, New York City canceled a Jay-Z concert due to security concerns including possibility of a terrorist attack. Also, intercepted al-Qaeda communications indicated the Lollapalooza music event in Chicago was targeted in 2013.

Large outdoor music festivals are also a concern. For instance, the Bonnaroo Music and Arts Festival in Tennessee is now ranked near the top of the list of potential terrorist soft targets in the state. With over 80,000 music fans packed into a rural area and thousands of vehicles, the event is a security challenge. One solution has been to increase the number of undercover officers working the crowd.

Amusement parks present yet another security concern and challenge. Since 9/11, Disney properties have been assessed by terrorist groups as potential targets or have assumed an increased security posture due to an overall increase in threat levels in the United States. Certainly, amusement parks are vulnerable to attack, and many have increased bag inspection and limitation of items that may be brought into the park. Because bringing weapons through the gate is unlikely, the threat of chemical or biological attack might be more probable. Also, similarly to the Westgate Mall attack, the insider threat to an amusement park is a distinct possibility. Park employees are able to carry out reconnaissance on possible targets and pre-position materials needed for an attack. Therefore, screening of potential workers and vigilance concerning their activities is paramount to preventing a deadly attack from within.

ISIS-Inspired Orlando Nightclub Shooting: Will America Ever Be the Same?

In the early morning hours of June 12, 2016, 29-year-old Omar Saddiqui Mateen entered the Pulse gay nightclub in Orlando, Florida, and opened fire with his recently legally purchased AR-15 assault rifle. Before being neutralized by police, he carried out the worst mass shooting in U.S. history with 50 people dead and another 53 seriously injured. In what was a well-planned attack, Mateen also carried one handgun and a suspicious device on his body. He apparently walked into the club unchallenged.

Sadly, Omar Mateen was another "known" wolf. An American citizen, the son of Afghani immigrants, Mateen was on the FBI's radar in 2013 due to suspicious coworkers. The FBI opened an investigation on him, and a second in 2014. However, the case was closed due to a lack of evidence of any terroristic activities. In 2016, he declared his allegiance to ISIS, but was not on law enforcement's radar. His father was a Taliban sympathizer with YouTube videos and online posts to that effect. Mateen had both a concealed carry license and a security officer license, and was employed as a security guard G4S, screening visitors at the Port St. Lucie (Florida) Courthouse. His ex-wife stated that he used to regularly abuse her, although it is not known if she reported this activity to law enforcement.

ISIS sounded the alarm in the weeks prior to the shooting. On May 21st, Islamic State accounts on Twitter distributed a warning from the spokesman of the Islamic State, Abu Muhammad al-Adnani. The message was a call-to-action for ISIS followers to launch attacks on the United States and Europe during the Islamic holy month of Ramadan, which began in early June. "Get prepared, be ready ... to make it a month of calamity everywhere for the non-believers ... especially for the fighters and supporters of the caliphate in Europe and America." Al-Adnani further suggested attacks on military and civilian targets. Three days before the Orlando attack, ISIS released a "kill list" to their followers with the personal information of 8,000 Americans, including at least 800 from Florida's Atlantic coast. The FBI was still working on the kill list issue when the nightclub shooting took place.

As we try to get better prepared for horrific events such as these, we must look at the entire effect an event of this magnitude has on the "Whole of the Community."

Police, fire departments, EMS, and the hospitals are all firsthand recipients of the aftermath of the tragedy (allegedly purported to have been carried out by one lone offender).

EMS and fire department responses to these horrific incidents require a vastly different approach than utilized on a daily basis. The complexity increases with the variations in EMS delivery nationwide. These range from strictly fire-based to private providers. In any case, responses need to be based on extensive planning and exercises to develop the tactical coordination required. Application of routine standard operating procedures will be inadequate to manage the number of patients, severity of wounds, and an evolving scene. Responders will need to focus on the most severely injured based on rapid triage assessments. "Shopping for reds" will require bypassing patients with less severe injuries in order to address hemorrhaging in viable victims.

Coordinating with law enforcement will require determination of the tactics to be utilized. Rescue Task Force (RTF) concepts have become common but even they have variations. In some models, ballistic-protected EMS providers accompany police officers into hot zones to remove victims. In other variations, the scene is divided into hot, warm, and cold zone designations with the threat located in the hot zone. The warm zone has been cleared for immediate threats, allowing EMS and fire to enter under protection from officers. No matter what model a jurisdiction chooses to follow, extensive planning and exercising is required. A plan without exercise is just a theory. Moving theory to operational capability can begin with a tabletop and advance to a full-scale exercise. Utilization of an HSEEP framework is critical to ensure the maximum benefit is realized of the collaborative efforts.

The entire local or regional medical system will be stressed as patients begin to be transported. Notification of events and estimates of victims is essential to allow emergency rooms to implement triage and disaster plans. Notification through on-scene commanders allows adjustments to be made at hospitals to the extent that routine care patients can be relocated to urgent care or other facilities. Stresses placed on the health care system are likely to last for days beyond the initial event as medical personnel provide critical care for the severely injured. Public safety agencies need to include health care providers in the planning and exercise cycle.

After incident care for responders and others professionals involved in the incident has to be considered. Professionally Critical Incident Stress Debriefings (CISDs) should be conducted for all responders, hospital personnel, and telecommunication operators involved. With a general rise in responder suicides and other conditions, providing care is critical. Monitoring sick leave use and other trends should be conducted post-incident in a non-punitive manner. Personnel affected may attempt to cope by increasing sick time usage or other methods to minimize potential post-incident personal survival methods.

EMS and fire will be required to search for victims as the scene is rendered safe, and they must begin the horrific task of triage and transport of those injured after given the "all clear" by appropriate law enforcement officials on the scene.

The task of looking for viable victims in the "Red Tag" category may go on for hours, looking for those who can be safely evacuated.

The hospital systems must go into "TRAIGE and DISASTER" mode immediately. Canceling all elective hospital events and moving out the daily ER visits to those who are only mortally injured. This puts pressure on the entire public health care system. Hospitals must invoke Triage and Disaster Codes.

The effects of one event such as this can bring community health care to a very fragile breaking point. Ancillary services must be moved and patient care in the urgent situation may be compromised due to the sheer number of presenting patients. ER staff, x-ray staff, and the physicians required to care for this large influx of victims is certainly an overwhelming event. Think of *all* of the hospital services that will be impacted, to name a few.

- Medical Surge
- Nursing
- Pastoral
- Pharmacy
- Radiology
- Housekeeping (as the rooms must be cleaned rapidly to move in the next victim)
- Transportation (internal and external)

The entire community is affected by this event, and the need to move to a recovery mode is extremely urgent.

This was a horrific event, but not shocking. The world we live in has become extremely fragile, and we are sitting on a powder keg that could ignite at any moment. Our first responder community will be forced to face this type of event again and again. The magnitude of what a lone perpetrator can do is astounding.

Law enforcement will be faced with a daunting task, clearing the scene of assailants so that the medical and rescue staff can enter. Law enforcement can no longer stand by and wait precious moments to engage, while assembling a SWAT-type operation. The first due officers are totally engaged in the firefight and in an attempt to neutralize the assailant(s).

The difficult work of Victim Identification, clearing the scene for booby traps, and getting the scene ready for removal of the victims is intense. The teams of fire, law enforcement, and the medical community *must* be totally coordinated and cooperative in the recovery operations. Response must be practiced and planned for, as the event may go down very quickly. If a hostage situation ensues, then it will be a timed, planned response as was shown in Orlando on June 12, 2016.

In an incident such as this, the Emergency Manager must maintain his focus on the "big picture." Although this is a law enforcement/homeland security situation, there are many "moving parts" that cannot be lost in the shuffle. These may include consistent messaging to the media and the public, rumor control, handling victims' family inquiries, managing mutual aid resources, cost and resource tracking in case there is a financial reimbursement stream (many communities cannot afford the cost of resources needed to handle such an incident), managing medical surge capacity, handling of potential volunteers and donations, and providing psychological and counseling needs for victims and first responders. These are a few of the areas that need to be kept "on the radar" of the Emergency Manager as he helps to lead community recovery efforts. The Emergency Manager must lead the jurisdiction's tactical and strategical planning of such situations that may not be directly part of the law enforcement efforts in such an incident.

Strategically, if these items are not handled effectively, the community runs the risk of affecting its "brand" and reputation. In this case, Orlando depends on tourism as an element of the city, and Orange County, Florida's, economy. Negative publicity associated with an incident, such as the June 12, 2016 massacre at the Pulse nightclub, could affect not only the economy, but also the reputation of the city as a "tourist-friendly destination."

The Emergency Manager must think of him/herself as a "symphony conductor," whereas the conductor can see the "big picture," yet may not have the expertise to handle all of the moving parts. He/she must manage all of the subject matter experts to reach the community goals of an efficient and effective "piece of music," or recovery. Accomplishing this task, whether it be this incident or any future incident, will assist the community in returning to a sense of normalcy, despite what an enemy may try to accomplish through terroristic threats or events.

5 THINGS TO REMEMBER

1. Mass public attacks require a coordinated response. There is no room for politics, egos, or turf wars across response disciplines.
2. Assessment of soft targets in the community is critical for protecting people. You can't protect what you don't know.
3. Planning should be based on risk and not fear. A mediocre plan today is better than a perfect one next week.
4. Start slow and turn up the heat on multidiscipline and jurisdictional exercises. Don't just check the box, practice like you will work in the street.
5. Work to increase public capability to provide immediate care before responders arrive.

A new response reality has emerged. Public safety agencies need to recognize and accept this reality and commit to evolve.

Planning, Preparation, and Practice are the keys to survival.

Jennifer Hesterman, Mike Fagel, Greg Benson, and Shane Stovall
June 12, 2016

TOURIST SITES: A VULNERABLE TARGET

Attacks on tourists are also a tactic in the ISIS and al-Qaeda playbook that could be used at any number of locations in the United States. One of the worst terrorist attacks on tourists occurred on November 17, 1997. At 8:45 am, during peak tourist entrance into the park, six Islamist terrorists from Jihad Talaat al-Fatah, masquerading as Egyptian security forces, killed the two armed security guards and then stormed the Deir el-Bahri, an archaeological site in Luxor, Egypt. Tourists were trapped inside the ancient buildings, and the terrorists went on a rape, mutilation, and killing spree, taking sixty-two lives. The terrorists escaped the scene in a tourist bus; at a checkpoint, they scattered into the hills. Some were later found dead, having committed suicide. The tourist industry in Egypt, especially at Luxor, was deeply affected and has never fully recovered. The massacre did serve to turn Egyptian public opinion against al-Fatah and forced the government into action against the group, which later claimed it did not carry out the attack and blamed Osama bin Laden. Several tourist sites in the United States could lend themselves to a similar attack, particularly where a few guards could be overtaken and visitors trapped and taken hostage.

A recent attack by ISIS in Libya exposed their tactics for attacking a tourist site.

Case Study: Attack at the Bardo National Museum, Tunis, Tunisia

The Bardo National Museum is one of the top tourist destinations in Tunis, the capitol city of Tunisia, with an extensive collection of Roman mosaics. The museum is comprised of an 18th century palace with a new modern wing and main entrance hall. The buildings are located in a large complex adjacent to the Tunisian Parliament buildings in central Tunis. The museum is a popular destination for Western cruise ship passengers on shore excursions.

On March 18, 2015, at least three gunmen wearing dark apparel and armed with Kalashnikovs and hand grenades passed unnoticed through an initial security check onto museum grounds where they waited for the arrival of tourist buses. Some accounts suggest the gunmen slipped past exterior security while the guards were conducting security checks on a tour bus. Other accounts suggest the security guards were on a coffee break and not at their posts at all. At around 1230 local time, as the buses from cruise ships began unloading, the gunmen brandished their weapons and began shooting into the crowd killing, several tourists instantly. The crowd scattered. It only took the police minutes to arrive, but the gunmen threw grenades at the police to impede their movement, fatally shot and officer and his dog, and then followed a group of tourists through the glass doors leading into the museum's entrance hall. (Gall 2015) There were approximately 200 tourists present on the property at the time of the attack.

The attackers spent the next 3 hours inside the museum holding the tourists hostage (Figure 14.4). They spared the lives of Tunisian workers, targeting only foreign visitors. Accounts from an Italian tourist interviewed after the attack described how the tourists tried to find cover as the terrorists sprayed bullets around rooms. CCTV footage released by the Tunisian authorities shows two gunmen roaming the hall,

Figure 14.4 A still photo of hostages inside the Bardo Museum. (From https://www.youtube.com/watch?v=RdCUnyHFq0M)

hunting the tourists, and speaking to a third man. The standoff with police ended when Tunisian Special Forces secured the building. The death toll was twenty-three, including twenty tourists from Japan, Poland, Italy, and Colombia; one Tunisian police officer; and two of the three gunmen. There were 50 people seriously injured.

THE RESPONSE

Tunisian police immediately evacuated the area, to include the Tunisian Parliament, which was in session at the time. Television footage showed law enforcement escorting civilians to safety, and emergency medical response arriving to tend to causalities. Police and Special Forces snipers took positions surrounding the building, and a helicopter circled overhead. Counterterrorism police garrisoned nearby arrived on the scene quickly.

When they entered the building, the team quickly killed two terrorists and secured the building. The Tunisian government identified the two dead gunmen at the scene as Yassine La Abidi, 20, and Hatem Khachnaoui, 26. Abidi was from Tunis, and well liked in his neighborhood. He recently became religious, and his name was known to Tunisian authorities, although friends and family could now believe he was involved in such a brutal attack. Khachnaoui was from Kasserine, near the Algerian border. In the days following the siege, the Tunisian authorities arrested four of Khachnaoui's relatives in connection to the attack. Both men had illegally left Tunisia to attend a jihadi training camp in Libya in December of 2014 (Stephen 2015). A third gunman escaped, probably with fleeing tourists. Maher Ben Moudli Kaidi, who Tunisian officials believe coordinated the attack and may have been the third gunman is still at large. On March 29th, Tunisian security forces killed the commander of the group responsible for the massacre, Khaled Chayeb, also known as Lokman Abu Sakhr, along with eight other Islamist militants.

THE PLANNING

The Bardo attack was a strategic move to hurt the newly formed secular government and disrupt tourism—the third largest sector of the Tunisian economy at 7.4 percent of the GDP in 2014. From a tactical perspective, the planning was easy to plot and execute. The simplicity of the actions belies the impact.

The gunmen, and most likely several associates with planning experience conducted reconnaissance missions to determine the museum lay out and security practices. They obviously concluded there was not sufficient security to simultaneously admit tour busses and ensure pedestrians were properly searched—or even noticed. If, as suggested in at least one media report, most of the guards were habitually away on a coffee break at a specific time each day, that would have been apparent to any surveillance team noting patterns of life. The same team could easily have noted peak arrival times for tourist busses full of cruise line passengers. Obtaining Kalashnikovs and hand grenades would have been easy across the porous borders with Libya or Algeria. One Tunisian official lamented the entire attacked could not have cost more than 4000 Dinars—approximately $2,000 (Kirkpatrick 2015).

The results were dramatic for the Tunisian economy, and coupled with the Sousse resort attack in June of 2015, devastating. There is no dollar figure associated with lost revenue yet, but as of September 2015, tourism from Britain alone was down 80–90 percent. (Stephen 2015)

THE THREAT

Tunisia was the only country to benefit from the Arab Spring uprisings of 2011, and was enjoying relative peace and expanded democratic rights. In December 2014 national elections put a new, largely secular government in power. In contrast, Tunisia is also the largest supplier of foreign fighters to Syria, with an estimated 3,000 currently in the fray. In February 2015 alone over 30 returning jihadis were arrested. Tunisia also shares long and porous borders with both Libya and Algeria, hotbeds of lawlessness and extremism. It is easy for militant recruiters to cross into Tunisia to persuade disaffected youth to join their cause, and just as easy, would be recruits can cross to attend jihadi training camps in Libya.

ISIS threatened Tunis in December with a video via the twitter handle "ghazwat Tunis" or "Raid of Tunis" and they were quick to praise—and on some social media sites take credit for—the Bardo Museum attack. However, Tunisian law enforcement ultimately determined the Okba Ibn Nafaa Brigade—a local splinter group of Al-Qaeda in the Islamic Maghreb—was to blame (*Financial Times* 2015).

LESSONS LEARNED

The Prime Minister of Tunisia dismissed six police commanders, including the Tunis police chief, for security failures brought to light after the attack (Gall 2015).

The fact the Islamic State and other jihadists groups to claim the attack in Tunis "demonstrates the interconnectedness of these loosely affiliated networks" (Kirkpatrick 2015). Lack of a clear recruitment or funding path for the gunmen is another problem for intelligence and law enforcement agencies trying to detect amateurs planning terrorist attacks. The interest of individuals and small groups to engage in attacks on soft targets is very visible on social media, where some aspiring jihadists have begun using the hashtag, in Arabic, "Lone Wolf." They are literally "hiding in plain sight." (Kirkpatrick 2015)

Small groups are much harder for law enforcement to infiltrate and stop, and as was demonstrated here, a few men with little planning and scant resources hitting a soft target had an outsized effect on the national economy and grabbed worldwide headlines for their cause.

Stopping lone wolf or small non-affiliated group attacks is extremely difficult for intelligence and law enforcement agencies, making bolstered security and increased vigilance at soft target destinations the first and only line of defense.

(*Source*: Kinzer (2016). Ms. Kinzer is a Reserve Air Force intelligence officer and senior intelligence consultant at Patch Plus Consulting.)

Recent history is replete with examples of attacks on or threats against shopping, sports, and recreational venues by groups also threatening the United States. We are wise to draw from the many lessons learned abroad to factor into our detection, prevention, mitigation, and response plans.

REFERENCES

Al-Shabaab Twitter messages during the Westgate Mall Attack. http://jihadology.net/2013/09/21/new-statement-from-%E1%B8%A5arakat-al-shabab-al-mujahidin-claiming-responsibility-for-the-westgate-mall-attack-in-nairobi/, 2013.
American Jewish World. Israeli Counters Terror at Mall of America, February 2, 2011.
BBC. Nairobi Siege: How the Attack Happened. 2013a.
BBC Africa. Q&A: Westgate Attack Aftermath. October 21, 2013b.
BBC News. Gas "Killed Moscow Hostages." October 27, 2002.
Bender, J. Security Measures That Will Put a "Ring of Steel" around the Sochi Olympics. January 31, 2014.
Bradley, Bud. State of U.S. Mall Security Post 9–11. December 17, 2013.
Butime, Herman Rujumba. The Lay-out of Westgate Mall and Its Significance in the Westgate Mall Attack in Kenya. *Small Wars Journal*, May 10, 2014.
DHS (Department of Homeland Security). National Planning Scenarios. 2006.
Dixon, Robyn. Video Shows Kenyan Soldiers Looting Besieged Mall. *Los Angeles Times*, 2013.
Encyclopedia of Afghan Jihad, 2004.
FBI. On the Ground in Kenya—Part 2: Terror at the Westgate Mall, 2014.
Financial Times Reporters. At Least 37 Killed in Attack on Tunisian Beach Resort. *Financial Times*, June 26, 2015.
Gall, Carlotta. Tunisian Museum Attack Leads to Firing of Chiefs. *New York Times*, March 23, 2015.
Goldberg, Jonah. Islamic State's Flash-mob Terrorism a Worrying Success. Tribune Media Services Inc., September 23, 2014.
Grace, Francis. FBI Alert on Stadiums. CBS News, July 3, 2002.
Hall, Stacey. "Securing Sport Stadiums in the 21st Century: Think Security, Enhance Safety." *Journal of Homeland Security*, 2010.
Herridge, Catherine. Source: Air Force father, son targeted online by ISIS followers. http://www.foxnews.com/politics/2014/10/07/source-air-force-father-son-targeted-online-isis.html.
Investigative Project on Terrorism. *US v. Shareef, Derrick*, 2004.
Investigative Project on Terrorism. *US v. Mehanna, Tarek*, 2014a.
Investigative Project on Terrorism. *US v. Nuradin Abdi*, 2014b.
Jambo News. Nairobi Senator Mike Sonko Reveals What He Knew about Westgate Mall Terrorists, October 10, 2013.
Jorns, Eileen. Putin's Ring of Steel: How the Russian Federation Constructed the Most Extensive and Controversial Security System in Olympic History. 2014.
Kinzer, Sarah. Attack at the Bardo National Museum, Tunis, Tunisia. 2016.

Kinzer, Sarah. The Nairobi Mall Attack. 2014.

Kirkpatrick, David. Militants, ISIS Included, Claim Tunisia Museum Attack. *New York Times*, March 19, 2015.

Kulish, Nicholas, Jeffrey Gettleman, and Josh Kron. During Siege at Kenyan Mall, Government Forces Seemed Slow to Respond. *New York Times*, October 1, 2013.

Lappin, Yaakov and Yohah Jeremy Bob. Shin Bet Foils Hamas Bomb Attack Planned for Jerusalem Mall. *Jerusalem Post,* September 2, 2013.

Madhani, Aamer. U.S. Worried About Attacks on Soft Targets near Sochi. *USA Today,* January 29, 2014.

Matthews, Owen. Russia Tests "Total Surveillance" at the Sochi Olympics. *Newsweek*, February 12, 2014.

Moscow Times. Islamist Umarov Vows "Maximum Force" to Stop Sochi Games, July 3, 2013.

New York City Police Department. Analysis of Al-Shabaab's Attack on the Westgate Mall in Nairobu, Kenya, 2013.

New York Times. Suicide Bomber Kills Official and 13 Others at Sri Lanka Race, April 7, 2008.

Pflanz, Mike. Nairobi's Westgate Mall Attack: Six Months Later, Troubling Questions Weigh Heavily. *The Christian Science Monitor,* March 21, 2014.

Pflanz, Mike and Harriet Alexander. Nairobi Shopping Mall Attacks: Britons among Those Caught Up in Terrorist Assault. *The Telegraph,* September 21, 2013.

Radia, Kirit. Putin: "Can't Feel Weak" in the Face of Terror Threats to Sochi Olympics. ABC News and World Report, January 17, 2014.

Robinson, Joshua and Inti Landauro. Paris Attacks: Suicide Bomber Was Blocked From Entering Stade de France. *Wall Street Journal*, November 15, 2015.

Sewell, Dan. FBI head: Homegrown Terrorist Recruitment 24-Hour Threat. Associated Press, October 14, 2015.

Somra, Gena and Ivan Watson. Circassians: Sochi Olympians "Are Skiing on the Bones of Our Ancestors." CNN, February 18, 2014.

Sridharan, Vasudevan. Sochi Winter Olympics 2014: Hotel Rooms and Bathrooms "Bugged." *International Business Times,* February 8, 2014.

Stephen, Chris. Tourists Desert Tunisia after June Terror Attack. *The Guardian*, September 25, 2015.

Stephen, Chris. Tunis Museum Attacks: Police Hunt Third Suspect in Shootings. *The Guardian*, March 22, 2015.

Tunis Museum Attack. YouTube video from Tunisia 1, 1:10, posted by ReBlop TV. https://www.youtube.com/watch?v=RdCUnyHFq0M, March 18, 2015.

Vidino, Lorenzo and Seamus Hughes. ISIS in America: From Retweets to Raqqa. December, 2015.

Walker, Lauren. Iraqi Refugee Accused of Planning to Bomb Houston Malls. *Newsweek*, January 15, 2016.

Wilson, Stephen. Sochi Chief: City Is World's "Most Secure Venue." Associated Press, January 29, 2014.

Zaccardi, Nick. An Oral History of the Bombing That Rocked the 1996 Atlanta Games, July 24, 2012.

Zwerdlng, Daniel, G. W. Schulz, Andrew Becker, and Margot Williams. Under Suspicion at the Mall of America. National Public Radio, September 7, 2011.

15
Hospital Business Continuity

Linda Reissman and Jacob Neufeld

Contents

Historical Perspective 281
   Early Hospital Preparedness 281
Why Business Continuity? 283
What Is the Business Impact Analysis? 286
   Business Impact Analysis 286
   Physical Risk Assessment Process 287
   Advantages of Using a Business Continuity Planning Tool 288
      Level 1: Self-Governed 288
      Level 2: Supported Self-Governed 288
      Level 3: Centrally Governed 289
      Level 4: Enterprise Awakening 289
      Level 5: Planned Growth 289
      Level 6: Synergistic 289
      Key Continuity Definitions 290

HISTORICAL PERSPECTIVE

Early Hospital Preparedness

Prior to the events of 9/11 and Hurricane Katrina, traditional hospital preparedness planning was focused on mass casualties and emergency department readiness. Very little planning focused on scenarios in which the hospital was directly impacted by the disaster, impeding the ability to continue care. Internal and community-level planning was limited in scope, and concepts such as the Hospital Incident Command System (HICS) were not employed.

    Since 2002, federal support for hospital emergency planning has been provided through the Department of Health and Human Services Hospital Preparedness Program

(DHHS-HPP) grants. Early funding was primarily directed at bolstering hospital capabilities to manage the threat of terrorism and pandemic. Then, in 2005, the devastating consequences of Hurricane Katrina explicitly bore out the need for improved hospital and public health preparedness. The traditional hospital disaster plan was not effective for catastrophic events that impacted the hospital, the community, and the critical infrastructure supporting both.

After Hurricane Katrina, a number of governmental reports issued assessments and recommendations for methods for ensuring the continuity of operations, the coordination of communications, and standards for strengthening mitigation measures. A report by the ANSI Homeland Security Standards Panel concluded that organizations of all sizes, in both the public and private sectors, would be well served by complying with NFPA 1600 and using it as a guideline for their disaster/emergency management and business continuity programs.

Subsequently, there were dramatic changes made to the scope of hospital preparedness. As an example, in 2008, the Joint Commission revised its standard on a hospital's ability to sustain in place. Previously, the standard had been that a hospital Emergency Operations Plan (EOP) must address an assessment of its ability to sustain in place for 72 h. After the revision, this was increased to 96 h. This standard does not require the hospital to sustain itself for 96 h, but rather have a process to assess capabilities during escalating incidents to determine its ability to do so, and then plan and respond accordingly. In addition, the hospital EOP must address the following critical areas:

- Communications
- Resources and assets
- Safety and security
- Staff responsibilities
- Utilities
- Patient clinical and support activities

Joint Commission Accreditation Requirements: Emergency Management (EM)

EM.02.02.01 As part of its Emergency Operations Plan, the hospital prepares for how it will communicate during emergencies.

EM.02.02.03 As part of its Emergency Operations Plan, the hospital prepares for how it will manage resources and assets during emergencies.

EM.02.02.05 As part of its Emergency Operations Plan, the hospital prepares for how it will manage security and safety during an emergency.

EM.02.02.07 As part of its Emergency Operations Plan, the hospital prepares for how it will manage staff during an emergency.

EM.02.02.09 As part of its Emergency Operations Plan, the hospital prepares for how it will manage utilities during an emergency.

EM.02.02.11 As part of its Emergency Operations Plan, the hospital prepares for how it will manage patients during emergencies.

Although the term "business continuity" is not specifically termed in the Joint Commission Emergency Management standards, the elements of performance do address it in terms of continuity of operations, patient care, and other critical functions. The Joint Commission's standard does address disaster recovery in IM.2.30, and requires that the hospital develop and maintain a disaster recovery plan identifying the most critical information needs for patient care, treatment, and services and impacts if information systems are disrupted. The plan should also outline alternative means for processing and providing data recovery.

Joint Commission Accreditation Requirements: Information Management (IM)

IM.01.01.03 The hospital plans for continuity of its information management processes.

INTRODUCTION TO STANDARD IM.01.01.03

The primary goal of the information continuity process is to return the hospital to normal operations as soon as possible with minimal downtime and no data loss. The hospital needs to be prepared for events that could impact the availability of data and information regardless of whether interruptions are scheduled or unscheduled (due to a local or regional disaster or an emergency). Interruptions to an organization's information system can potentially have a devastating impact on its ability to deliver quality care and continue its business operations. Planning for emergency situations helps the organization mitigate the impact that interruptions, emergencies, and disasters have on its ability to manage information. The hospital plans for interruptions by training staff on alternative procedures, testing the hospital's Emergency Operations Plan, conducting regularly scheduled data backups, and testing data restoration procedures.

The health care industry has become increasingly dependent on technology to computerize almost all aspects of patient care, ranging from automated provider order entry to sophisticated image-guided surgery systems, billing and accounts management, and center-wide communications. The unavailability of clinical and communication applications is untenable to clinicians left without the critical electronic resources they rely on to provide care. New generations of technology-oriented clinicians will have little, if any, experience with handwritten documentation and downtime processes. Therefore, a hospital's information technology (IT) infrastructure must be available to provide an uninterrupted flow of data and is a critical component in the delivery of care, the well-being of its patients, and its reputation. Therefore, it is critical for health care organizations to conduct ongoing assessments, disaster recovery, and clinical and business continuity planning.

WHY BUSINESS CONTINUITY?

In addition to the continuity of information management, hospitals must ask themselves, "What if facilities or their contents became inaccessible" or "What if 30 percent of your workforce is unavailable (e.g., pandemic)?" or "What if key suppliers and partners can't fulfill obligations?" or "What alternate process do we use if IT systems are unavailable?" These are the hard-line questions a health care institution needs to ask itself. Some of these events may not even be precipitated by a disaster per se. Today's health care environment relies on IT to support clinical and business operations. In addition to protecting human capital, institutions must consider the protection of research assets and other intellectual property that are irreplaceable. The critical infrastructure provided by external or municipal sources can fail despite all efforts of an organization to insulate itself from such events through mitigation practices.

A business continuity program can considerably improve clinical and business recovery (CBR) capabilities for health care organizations. Knowing the upstream and downstream impacts of critical systems provides leadership with key information required to make decisions quickly that will minimize disruptions and impacts to life safety and lessen financial losses.

Business continuity, emergency planning, disaster recovery, and IT security management are mutually exclusive planning processes, yet interdependent. A comprehensive business continuity program incorporates all three planning platforms and is directed at maintaining an organization's ability to sustain an acceptable level of care despite the emergency. Although there are similarities in terminologies, processes, and response steps, it is important to articulate the differences between emergency management, disaster recovery, and business continuity when attempting to implement a program and presenting a proposal to leadership. Statements such as "but we have an emergency plan or we have a disaster recovery plan" are commonplace when introducing the concept of business continuity to an organization. The organization's leadership must believe there is a value rather than another project burden that has no "end game." That being said, cultural acceptance of a business continuity program is integral to its success, so hospital leadership and the organization as a whole must understand the value of such a program. You will need to provide clear and concise definitions and descriptions to differentiate the three as well as include the value of such a program in day-to-day operations. The following definitions clarify these differences:

- *Emergency planning*: The procedures and steps taken immediately after an event. It is a component of a comprehensive business continuity program.
- *Disaster recovery*: The steps to restore some functions to resume some level of services (IT), including security management.
- *Continuity*: Restoration planning to get the organization back to where it was before an interruption.

A hospital EOP is usually designed under one of two different models: (1) all-hazards planning or (2) event-specific planning. Business continuity planning is "class specific" and applies mitigation and response strategies that can be applied for any vulnerability, regardless of the event that caused it. These can be classified as the 5 S's of clinical and business continuity planning. They include

1. Space
2. Stuff

3. Staff
4. Systems
5. Services

A clinical and business continuity program focuses on the ability of a health care institution to fully identify and sustain core operations, functions, and services through a formal recovery process. Unlike short-term crisis planning, continuity planning addresses *extended* disruptions. This will require specific and detailed planning and recovery strategies not within the purview of traditional emergency management. They include

- Maintaining/rapidly restoring critical systems/services
- Developing the capability to s elf-sustain when resources are scarce
- Providing a safe level of care, treatment, and services, and/or developing strategies to prioritize or safely discontinue care, treatment, and services
- Identifying, documenting, and exercising practical work around procedures

A business continuity program and its focus on the 5 S class vulnerabilities dovetail with the Joint Commission's six critical areas for hospital emergency management. Essentially, a business continuity plan is a road map for recovery. However, during the implementation of a continuity program and plan development, it will become apparent that equally important to establishing documented and codified processes is the planning process itself. The process provides a platform for information sharing among individuals who may have never had the opportunity to interact and discuss interdependencies and decisions that could impact their respective department and/or functions. Participants will soon see the everyday operational value, and the process flow of day-to-day business usually improves as a result. At the departmental planning level, specific informational components must be collected.

These can be done via a survey process manually or with the business continuity tool:

- Identify critical services and functions
- Gauge disruption impacts
- Identify the length of time until an impact becomes unacceptable (recovery time objective [RTO])
- Identify dependencies/interdependencies
- Codify/validate recovery strategies and tasks
- Contact information (staff, vendors, etc.)
- Identify needed recovery resources

A high-level "business continuity program design checklist" is presented below, which can be used to initially scope a clinical and business continuity program.

1. Scope business continuity/disaster recovery project.
2. Collect business continuity/disaster recovery data and information.
3. Complete business impact analysis (BIA).
4. Formulate business continuity/disaster recovery strategies.
5. Design and activate recovery and crisis management organization.
6. Document agreements and action plan to institutionalize business continuity and finalize pilot preparations.

7. Prepare a business recovery template using a business continuity planning tool or process.
8. Prepare pilot launch support material.
9. Execute a pilot and report on results.

WHAT IS THE BUSINESS IMPACT ANALYSIS?

Business Impact Analysis

Through the enterprise BIA, you will communicate with specific senior leadership representatives (stakeholders) about scalable options of how CBR planning can be accomplished across the institution:

- What granularity of planning detail?
- Which functions and services must participate in the program?
- Which resources will be doing the work?
- In what time frame are initial CBR plans completed?
- At what cost (time and money) to the organization?

Here, you set your hospital's goals and lay out what the hospital will need to reach those goals. The work carried out here allows leaders to see the effect that scaling up goals will have on resources.

Through the CBR program design workshop, you get management to commit to a course of action that gets a sustainable CBR program underway by developing a practical, achievable implementation plan composed of

- *Program scope*—Who and what are included in the CBR program?
- *Program operations*—How will the CBR program be conducted?
- *Program rollout*—When will the CBR program be deployed?

Stakeholder conversations during the enterprise BIA provide necessary insights into what is appropriate within your institution. You can now roll out a program in which it is inherent that reassessment and adjustments are made much like the cycle of continuous improvement.

Below is a detailed outline of specific tasks and processes associated with the full implementation of a business continuity program.

1. Engage leadership teams and employees with business continuity responsibility to conduct a self-assessment to establish or validate organizational business continuity objectives, develop a business continuity planning structure, and provide appropriate resource allocation.
2. Prepare for pilot deployment of an integrated business continuity program.
   a. Conduct an enterprise BIA.
      i. Engage executives to define critical impacts and "RTO" guidelines.
      ii. Review BIA results and engage leadership in developing an implementation plan utilizing enterprise standard methods and tools.

  b. Conduct a business continuity program design workshop.
    i. Assemble a business continuity program design team composed of a selected subset of participants from the business continuity self-assessment workshop.
    ii. Define who, what, when, and where of business continuity program deployment. Develop the draft charter for business continuity.
3. Develop a formal business continuity awareness program and delivery vehicle for new employees (as part of orientation) and for existing employees. Delivery vehicles can include websites (with online training modules), periodically scheduled training classes, and access to business continuity knowledge sources.
4. Use the results of the BIA and the agreed-upon business continuity program scope to confirm and authorize resource requirements for implementing and sustaining the business continuity program (budget, staff, and tools).
5. Develop, document, and distribute enterprise-level business continuity policies, standards, and procedures. Standards will cover such things as business continuity plan content, testing and maintenance requirements, and vendor and supplier expectations.
6. Establish a business continuity program governance office responsible for enforcing business continuity policies and standards, and assuring that business continuity goals will be achieved.
7. Develop a system of metrics (reviews, reports, and enforcement) to assure compliance in meeting organizational and enterprise-wide business continuity requirements, objectives, and goals. This includes business continuity-related regulatory requirements.
8. Consider including manager/employee performance relative to business continuity-related goals and objectives on their annual performance reviews.
9. Require that existing change control processes include checks for any changes that may have impacts on clinical or business, incident management, technology recovery, or security management-related recovery plans or strategies.
10. Utilize the emergency notification system to develop departmental- and enterprise-level business continuity calling trees for notifying employees, internal and external business partners, and service providers following a disruptive event.

Physical Risk Assessment Process

The scope of this assessment is to identify internal and external vulnerabilities, the availability and use of resources and controls to eliminate or mitigate risks, and plans and procedures to address emergency events. The following threat parameters provide the baseline for risk identification:

- *Acts of nature*: Earthquakes, rain, wind, ice, and so on that threaten facilities, systems, personnel, utilities, and physical operations, as well as hamper or deny access to the sites.

- *Hazardous conditions*: Fire, chemical and nuclear spills, biological events, structural instability, and so on, that threaten facilities, systems, personnel, and operations. These may be the result of natural events, environmental control failures, human errors, and/or violent acts, as well as hamper or deny access to the sites.
- *Dependency failures*: Failure of a system or service outside the direct control of your organization that harms the system and/or affects its ability to perform. Examples include public utility failures, or the failure of a service or system controlled by an external company or public agency.
- *Environmental failures*: Failure or lack of a protective control that disrupts, harms, or exposes the system to harm or loss. Examples include lack of heating/air conditioning/cooling (HVAC) for cooling servers, which can be affected by loss of power and generator.
- *Public safety actions*: Actions taken by law enforcement, fire, regulatory, administrative, and/or other parties to safe-guard the community that may inadvertently result in harm to the organization or the system. Examples include mandating the complete shut-down of IT systems and generator to safeguard occupants during external or internal hazardous incidents.
- *Prior events*: An analysis of prior events that threaten facilities, systems, personnel, and operations, as well as activities to mitigate the effects of prior events.

Advantages of Using a Business Continuity Planning Tool

One of the most challenging steps in the development of a new clinical and business continuity plan or improving upon one that is already implemented is assessing the maturity of a business continuity management (BCM) program and is achieved by assessing the organization's current level of BCM capability by identifying gaps and risks and establishing a process improvement process. There are a number of tools and processes on the market that can do this. Two examples are the Gartner IT Score and the Virtual Corporation's Business Continuity Maturity Model® (BCMM®). The evaluations usually consist of rating scales with descriptive and detailed elements of assessment. In the below example, upon the completion of the assessment, the organization will achieve a level to determine their organizational resiliency and program maturity.

Level 1: Self-Governed

- *Business continuity planning attribute*: No formal planning; the business reacts to disruptive events.
- Business continuity has not been formally implemented anywhere within the organization. Few, if any, documented business continuity plans exist for any of the three business continuity disciplines. The organization reacts to disruptive events when they occur. The state of preparedness is low across the enterprise.

Level 2: Supported Self-Governed

- *Business continuity planning attribute*: Planning is limited to a few areas that prepare alone.

- A few functions or services, on their own, have developed and maintained business continuity plans within one or more of the three business continuity disciplines. There is little or no cooperation or coordination of planning activity between these groups. The state of preparedness may be moderate for participants, but remains low across the majority of the enterprise.

Level 3: Centrally Governed

- *Business continuity planning attribute*: Participating functions share methods and tools.
- Participating functions or services have instituted common business continuity practices, tools, and support resources in one or more of the business continuity disciplines. The interest in leveraging the work of these groups is being promoted as a business driver for launching an enterprise business continuity program. Some functions or services have achieved a high state of preparedness. However, as a whole, the enterprise is still largely ad hoc.

Level 4: Enterprise Awakening

- *Business continuity planning attribute*: All functions implement initial, self-interested plans.
- An enterprise business continuity program is deployed using standard methods and tools integrating all three business continuity disciplines, typically supported by a centralized support business continuity program office. These initial business continuity plans encompass most critical functions and services across the enterprise, although with little attention on protecting critical dependencies. Most business continuity plans are tested and updated routinely, as dictated by enterprise business continuity policy.

Level 5: Planned Growth

- *Business continuity planning attribute*: Most critical dependencies are incorporated into plans. All critical enterprise functions and services have developed and tested business continuity plans, including their critical internal and external dependencies. A communications and staff training program continuously measures business continuity planning competency. Audit reports no longer highlight business continuity shortcomings. Competitive advantages achieved from business continuity planning are highlighted in internal and external communications.

Level 6: Synergistic

- *Business continuity planning attribute*: Comprehensive plans are tied to changing business needs. Continuous process improvement keeps this organization at an appropriately high state of preparedness to stay current with the dynamic business environment. Over time, innovative policy, practices, processes, and technologies are piloted and incorporated into the business continuity program. Cross-functional business continuity capabilities are measured and codified.

- How will employees maintain open lines of communication?
- Where will employees go if the disruption involves an inability to use the existing office space?
- Fundamentally, how will employees be able to keep doing their jobs?
- How will the company's employees continue to be paid on a regular basis, with no interruptions?

Key Continuity Definitions

Business continuity: A formal planning process that minimizes or eliminates the impact of disruptive events on critical business operations, functions, and services.

Clinical and business recovery (CBR): The discipline of business continuity planning that provides advanced planning and preparation to help ensure the continuity of critical clinical and business functions in the event of a disaster. It includes identifying impacts to the business environment, implementing viable risk mitigation and recovery strategies, and developing business disaster preparedness plans.

Business impact analysis (BIA): A technique that identifies both tangible and intangible impacts on a business function, service, or program usually over time based on given criticalities. Provides senior management with information to devise a recovery strategy and prioritization.

Technology recovery (TR): The discipline of business continuity planning that provides advanced planning to help ensure the ability to recover and restore critical assets, including IT delivery systems, voice and data networks, business applications, and other critical information technology components within defined RTOs.

Incident management (IM): The discipline of business continuity planning that provides advanced planning to help ensure health and safety of people, limit environmental impacts, and protect company assets. IM includes emergency response, crisis management, and emergency operations.

Recovery point objective (RPO): The term used to identify where in the data stream a company or organization needs to recover lost information. Simply put, how much data can be lost if the primary data storage device is destroyed?

Recovery time objective (RTO): The term used to identify how quickly a company or organization must recover a disrupted business function, service, or program.

So, what should the hospital consider when embarking on a clinical and business continuity program and in what order? The following is an eight-step process that begins with a basic analysis of the present capabilities and an identification of known gaps in capabilities or infrastructure. Then, after the limitations and potential weaknesses in the system are identified, plans can be put in place to improve the issues, train staff, conduct exercises, and evaluate the results. This will lead to a high-quality approach to developing contingency and continuity plans (see Figure 15.1).

1. The organization must commit to an enterprise-wide culture of continuity planning.
2. Know the capabilities and limitations of the institution's mission-critical systems. The first and most critical step is for all department heads and senior hospital

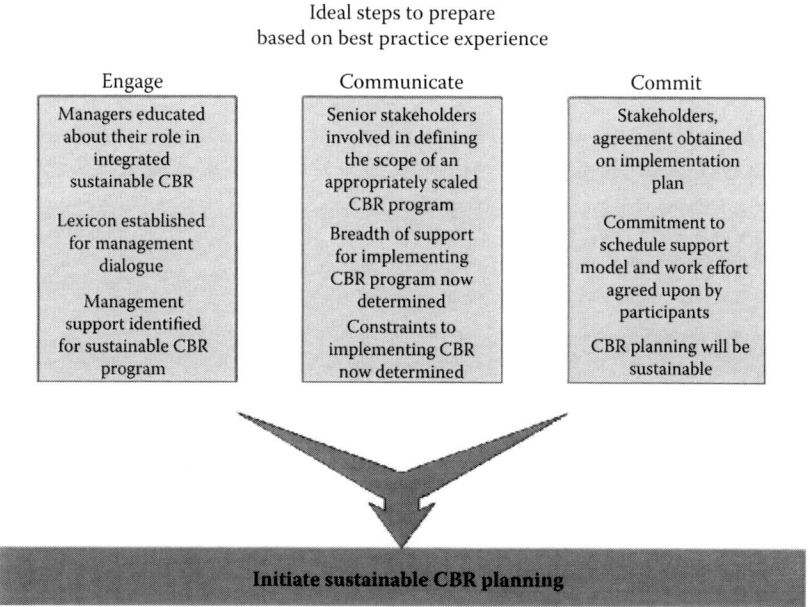

Figure 15.1 Steps used to prepare contingency and continuity plans. (Copyright Virtual Corporation, http://www.virtual-corp.com/)

management (including clinical leadership) to be aware of these capabilities and limitations. All too often, this understanding is limited to the hospital engineer or facility manager.

3. Ensure that a risk assessment is conducted for all mitigation projects and should be based on the hazard vulnerability analysis, as required by The Joint Commission (TJC). It should ensure that activities to mitigate effects from specified hazards are prioritized. It is also important to consider the impact that the surrounding community may have upon the hospital's critical infrastructure such as proximity to an industrial facility. Once the hospital administration and clinical department heads are aware of the capabilities and gaps, contingency plans can be developed to minimize risks and sustain operations.
4. Develop mitigation plans and projects to enhance resiliency to ensure continued operations in times of outage. Examples include the capability to provide 100 percent backup generator power and alternate communications capabilities.
5. Develop contingency plans and redundancies; if mission-critical systems fail, these may include conservation for such things as water, fuel, or food shortage, illumination devices in cases of power outages, and downtime procedures such as use of paper records when IT systems are impacted.
6. Assure that service providers are aware of mission-critical systems to prioritize the restoration. Do not assume that hospitals or health care agencies receive the highest priority restoration without a discussion and written agreements with providers.

7. Educate leadership, staff, and department heads. Providing emergency management and continuity training so that staff understand their roles and responsibilities so that the plan will be executed in an efficient and timely manner.
8. Exercise plans and the mission-critical systems to fail. Only then will gaps in planning and training be identified and process improvements can take place. Develop a formal process-improvement program that tracks gaps identified in exercises and real events. Update plans accordingly and conduct yearly reviews.

16

Soft Targets, Active Shooters, and Workplace Violence

Lawrence J. Fennelly and Marianna A. Perry

Contents

What Are Soft Targets? 294
    Solutions 295
San Bernardino, California, at 11:00 am PST on December 2, 2015 296
    A Possible Motive 298
    Farook's Home 298
What Is Workplace Violence? 298
Work Environment 299
Violence and Crime in the Workplace 299
    Solutions 300
Seven Things You Need to Know about Soft Targets 301
    U.S. Department of Homeland Security Guidance 302
Conclusion 303
Ten Things You Need to Know 303

Many organizations have vaguely defined security programs that attempt to protect unknown vulnerabilities from unknown threats.[*] In order to effectively protect people, assets, and property, organizations have to understand what their risks are and where they are vulnerable. Unfortunately, we have found that most organizations do not have these critical pieces of information. The logical starting point on the way to a clearly developed security program is to define soft targets, active shooters, and workplace violence to ensure that everyone understands the topic. From there, training and awareness

---

[*] Thomas L. Norman, *Risk Analysis and Security Countermeasure Selection* (Boca Raton, FL: CRC Press, 2015).

programs will help create a culture of safety and security within the organization. This culture will become the framework for developing security policies and procedures as part of a security master plan that is in alignment with the mission of the organization.

WHAT ARE SOFT TARGETS?

The *Oxford English Dictionary* defines soft target as "a person or thing that is relatively unprotected or vulnerable, especially to military or terrorist attack."[*]

In her book, *Soft Target Hardening*, Jennifer Hesterman defines soft targets as "civilian-centric places that are not typically fortified." They are vulnerable, unprotected, and undefended privately owned property. They may be resource-constrained, and security may not be a primary mission of the organization. Soft targets may be co-located with or near a hard target.[†]

Soft targets can be identified as churches and other houses of worship; schools, colleges, and universities; hospitals and health care facilities; shopping malls and strip malls; sporting and recreational venues; concerts; hotels, motels, and resorts; office buildings; and critical infrastructure systems. Think about the recent attacks that have occurred in the locations listed above where a soft target was the primary target.

A soft target is any person or thing that is vulnerable to attack but not protected. This could be virtually any location in any environment without sufficient security measures in place to protect its assets. It could also be any person who is not capable of self-protection against attack or is unaware that they are vulnerable to an existing threat. We live, work, and play every day in locations that are identified as soft targets. Many of us are susceptible to attack because of complacency. Some people adopt the viewpoint that "It can't happen here" or "My hometown is different from the town where that incident took place," but the truth of the matter is that we live in a world of soft targets.

A particular location may be considered a soft target simply because it is an environment where we *feel* safe or protected, such as a church or house of worship. The recent shooting in Charleston, South Carolina, at the Emanuel African Methodist Episcopal (AME) Church on June 17, 2015, is a prime example of this. While worshipping, we certainly do not expect an armed gunman to kill eight members of the congregation and the pastor. The mass shooting at Fort Hood in Killeen, Texas, on November 5, 2009, should have been a serious wake-up call to Americans. If such an attack could occur at a military installation, a location that is certainly not perceived as a soft target, and leave thirteen military personnel dead, how vulnerable are we as civilians? Often, killing as many individuals as possible is not the only goal of an attack. Instilling fear is a secondary desired result.

---

[*] *Oxford Dictionaries*, s.v. "soft target," accessed May 18, 2016, http://www.oxforddictionaries.com/us/definition/american_english/soft-target.

[†] Jennifer L. Hesterman, *Soft Target Hardening: Protecting People From Attack* (Boca Raton, FL: CRC Press, 2014).

Solutions

The U.S. Department of Homeland Security was formed in response to the terrorist attacks on September 11, 2001, because we as a nation recognized how vulnerable we were to terrorist attacks. This fear of vulnerability soon diminished, and many Americans returned to "business as usual" because the attacks did not affect them directly and they no longer *felt* vulnerable. As a result of the recent acts of terrorism throughout the world, the American people are beginning to understand once again that this nation is not invincible. Our way of life is being threatened, and we have discovered that our nation is not omnipotent. We had become complacent, and many Americans who now feel as though the government has failed to protect them are losing faith in our system. This certainly compounds the problem. We know that we as a nation could be exposed, but many Americans think that attacks happen to "other people" in "other locations." We do not to live in a heavily guarded fortress with armed security personnel or law enforcement officials on every corner to keep us safe. If that is the direction we take in response to terror, we will have lost not only of our way of life but also what we as a nation hold dear—our freedom.

Along with situational awareness, we need to make our soft targets less attractive to terrorism and terrorism-related crime by "hardening" them with effective physical security measures. Security professionals can conduct a risk assessment to identify concerns and make recommendations that are reasonable, feasible, and affordable to address these vulnerabilities.

Conducting a risk analysis is the step during which a security professional puts all of the information regarding assets, threats, and vulnerabilities together and then considers the potential impact or consequences of a loss event. Patterns and trends are analyzed, but sometimes the only pattern is that there is no pattern. The security professional then makes recommendations to improve individual components in the security system to address identified vulnerabilities. In other words, to effectively secure soft targets, we must take a two-pronged approach: become more vigilant in educating the American people about keeping themselves and their communities safe and implementing effective physical security measures. This approach should be a problem-solving partnership between individuals in their communities and law enforcement. Safety is not just a matter of homeland security, but of hometown security.[*] It is important to recognize that these solutions to protect soft targets and our nation as a whole need to be tempered with reasonable yet effective security measures that are not overly intrusive.

This chapter was written before the San Bernardino, California, shooting that occurred on December 23, 2015. Based on that incident, additional information was added, and we changed the direction of the chapter. We mention this only because it has never happened to us before. For days after the incident, we were writing about the information the media was releasing about the shooting. Larry Fennelly's son, Stephen, called him the night of the shooting and said, "Dad, are you aware that in the last 334 days, there have been 351 mass shootings, so far this year?" The source of his information was an article

---

[*] Janet Napolitano, "Homeland Security Begins with Hometown Security," Department of Homeland Security, August 3, 2010, http://www.dhs.gov/blog/2010/08/03/homeland-security-begins-hometown-security.

in *The Washington Post*, dated November 30, 2015.* Figure 16.1 from the article illustrates how many people have been impacted by the alarming number of mass shootings in this relatively brief period of time.

Active shooters are defined as "individuals or a small group engaged in killing or attempting to kill people in a confined space or populated area."† Attacks on all targets, not just "soft" targets are becoming more commonplace in the United States. It doesn't matter whether these attacks are carried out by domestic or international terrorists, the effect on individuals and our nation as a whole are the same—increased fear. This fear places our freedom and way of life in jeopardy. The most recent example is the following San Bernardino shooting.

SAN BERNARDINO, CALIFORNIA, AT 11:00 AM PST ON DECEMBER 2, 2015

Three people entered an office complex, went into a conference room where a holiday gathering was being held, and opened fire. Fourteen individuals were killed and twenty-two were wounded. The shooters then very calmly left the area. The police were notified immediately and responded in less than 5 minutes. More than 100 officers from the San Bernardino Police Department, the Sheriff's Department, the Bureau of Alcohol, Tobacco, Firearms and Explosives (ATF), and the Federal Bureau of Investigation (FBI) responded to the scene.

The initial description of the shooters that was released stated that three males (later corrected), wearing black, tactical gear and masks and armed with long guns, left the scene driving a dark-colored SUV. As police pursued the shooters, a pipe bomb was thrown out the window of the vehicle. Their guns were reported to be either AK-47s or AR-15s.

As a precaution during this time, the neighborhood surrounding the office complex was put into lockdown. This included schools in the area as well as hospitals. To complicate matters, a bomb threat was called in to a local hospital while law enforcement were searching for the shooters.

It was reported that the scene of the shooting, the Inland Regional Center, was not a secure building. Despite past problems at the location and the fact that typically doors were locked and security personnel were present, there was no security officer on duty the day the shooting took place.

Dr. Hesterman stated, "I have been following the shooting this evening and it sounds like he was a food inspector and worked in the building. The department had some human resources problems lately with supervisors being fired and people quitting. The primary shooter was Syed Rizwan Farook and his wife, Tashfeen Malik (loads of confusion on this). Farook was divorced (strange for a devout Muslim) and had a new baby. He had just returned from paternity leave."

---

* Christopher Ingraham, "There Have Been 334 Days and 351 Mass Shootings So Far This Year," Wonkblog (blog), *The Washington Post*, November 30, 2015, http://www.washingtonpost.com/news/wonk/wp/2015/11/30/there-have-been-334-days-and-351-mass-shootings-so-far-this-year/.

† Joshua Sinai. *Active Shooter: A Handbook on Prevention* (Alexandria, VA: ASIS International, 2013).

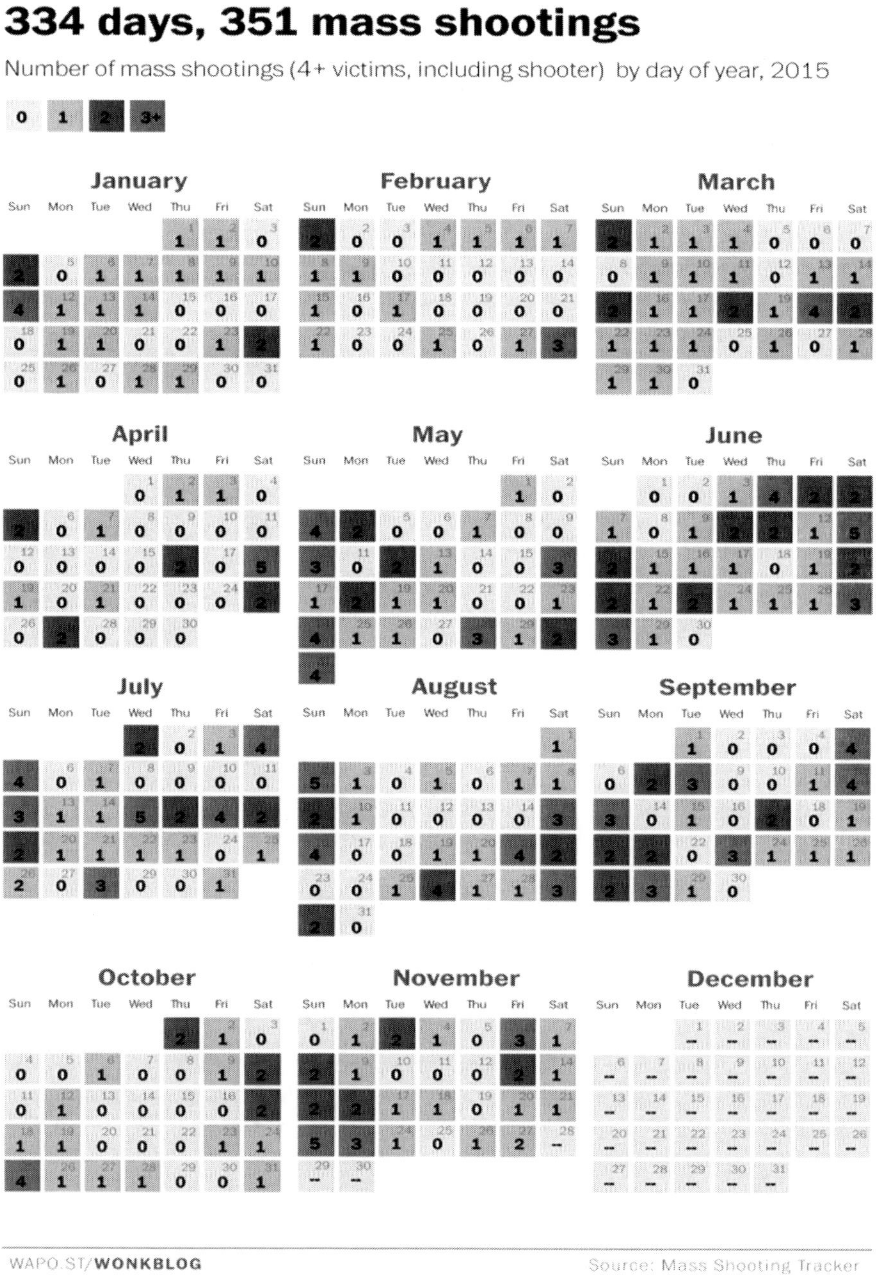

SOFT TARGETS, ACTIVE SHOOTERS, AND WORKPLACE VIOLENCE

334 days, 351 mass shootings
Number of mass shootings (4+ victims, including shooter) by day of year, 2015

Figure 16.1 Calendar figure of mass shootings. (From Shootingtracker.com. With permission.)

An unidentified witness at the scene said Farook came to the holiday party to confirm that his primary target was there (possibly a supervisor), left, and then came back with a weapon. Farook may have also placed some type of crude pipe bomb in the building, and it was speculated that the guns used may have been hidden inside the building prior to the shooting as well. This incident confirms the view of many security professionals—the insider threat is a growing issue.

The news media began referring to the incident as an act of domestic terrorism, a label that was later corrected to an act of internationally inspired terrorism. This shooting ranks as the third worst incident in the United States in terms of fatalities. In the shooting at Sandy Hook Elementary School in Newtown, Connecticut, twenty-six were killed[*] and at Virginia Tech, thirty-two individuals were killed.[†]

A Possible Motive

It was reported that Farook, a Muslim, and another employee at the building had gotten into an argument, possibly over religion. The other employee was a Christian. According to *Fox News* broadcasts, witnesses stated that Farook stormed out of the holiday gathering, and when he came back with his wife, they both began shooting. On social media, the Islamic State took credit for the incident, saying that they were proud of their "three lions." The news that Muslims were involved took about 5 hours to hit national television. Early on, information was being posted on social media, but much of this information was not completely accurate.

Farook's Home

After obtaining a search warrant, the FBI and the ATF carefully entered the home of Farook and Malik and found approximately 5,000 rounds of ammunition, twelve pipe bombs, cell phones, and computers. In the garage, law enforcement discovered hundreds of tools that were used to make the bombs.

WHAT IS WORKPLACE VIOLENCE?

The Occupational Safety and Health Administration (OSHA) defines workplace violence as any act or threat of physical violence, harassment, intimidation, or other threatening disruptive behavior that occurs at the work site. It ranges from threats and verbal abuse to physical assaults and even homicide.[‡]

Annie Le, a 24-year-old doctoral student in the Yale School of Medicine, Department of Pharmacology, was strangled on the campus of Yale University in New Haven, Connecticut,

---

[*] "Sandy Hook Shooting: What Happened?" CNN, accessed May 18, 2016, http://www.cnn.com/interactive/2012/12/us/sandy-hook-timeline/.
[†] "Virginia Tech Shootings Fast Facts," CNN Library, last modified March 30, 2016, http://www.cnn.com/2013/10/31/us/virginia-tech-shootings-fast-facts/.
[‡] "Workplace Violence," U.S. Department of Labor, Occupational Safety & Health Administration, accessed May 18, 2016, http://www.osha.gov/SLTC/workplaceviolence/.

on September 8, 2009. She was last seen in a Yale University research building the day before she was found dead inside. Raymond J. Clark, III, a Yale University laboratory technician who worked in the building pleaded guilty to Le's murder on March 17, 2011, and was sentenced to 44 years of imprisonment.[*]

Le's father said he hoped that as a result of his daughter's death, "there will be greater security provided for all students on campus as well as safer working environments."[†]

According to the U.S. Bureau of Labor Statistics' Census of Fatal Occupational Injuries, of the 4,679 fatal workplace injuries that occurred in the United States in 2014, 403 were workplace homicides. Workplace violence is now a major concern for both employers and employees, and nearly 2 million workers have been victims of workplace violence each year.[‡]

WORK ENVIRONMENT

Over the years, everyone has worked for a boss or a manager who loves drama in the office. We all have had to work under some degree of stress, but what if the stress is deliberately manufactured in order to terminate long-time employees? The results can sometimes be deadly.

In their book, *Risk Analysis and the Security Survey*, Eugene Tucker and James F. Broder stated: "Work environment—what is the relationship between personnel and management? (Loyal? Suspicious?) What are the aggravations of employees? Past labor history? How well do supervisors know employees? What is management's attitude toward employee dishonesty? (Condone? OK within bounds? Dismissal?) How open are lines of communication between employees and supervisors? Supervisors and upper management?"[§]

VIOLENCE AND CRIME IN THE WORKPLACE

Violence and crime in the workplace can take many forms. If an employee's car is broken into in the company parking lot, they feel violated because it is an invasion of his/her personal space. What if an employee is in a car accident on the way to or from work and the shock hits him or her 2 hours later at the office? What if the employee is assaulted on company property and wakes up in the hospital? What if we take the scenario one step further and imagine that the employee is seriously injured while at work in an act of workplace violence but was not the intended target? On the first day back to work, the employee is asked repeatedly how he or she feels. It doesn't stop. They are forced to relive the incident over and over, all day long. Then, the other employees in the organization begin to get

---

[*] *Wikipedia*, s.v. "Murder of Annie Le," last modified January 19, 2016, http://en.wikipedia.org/wiki/Murder_of_Annie_Le.

[†] Randall Beach, "Raymond Clark III on Killing Yale Grad Student Annie Le: 'I Took a Life and Continued to Lie About It' (Video)," *New Haven Register*, June 3, 2011, www.nhregister.com/article/NH/20110603/NEWS/306039942.

[‡] "Workplace Violence."

[§] James F. Broder, *Risk Analysis and the Security Survey*, 3rd ed. (Amsterdam: Butterworth-Heinemann, 2006).

nervous. "What if it happened to me?" This is how the fear of crime originates, and it is usually not addressed. Research suggests that a relatively small number of people can create an atmosphere of fear in which other people find an opportunity to commit more serious crimes.[*] Statistical data indicates that the fear of crime is much greater than actual victimization. This can have a debilitating effect on productivity and the workplace environment. Employers must be cognizant of this and take steps to initiate a response that addresses the specific problem or the cause of the fear, such as increasing the level of lighting in the parking lot, hiring security officers to patrol the area, or offering employees escorts to their vehicles. These approaches, when combined with proactive steps to educate employees about safety and security, both at work and outside of it, will reduce the fear of victimization.

Many times, acts of workplace violence and workplace crimes are committed by disenfranchised individuals who are determined to take out their frustrations on the individual (target) they perceive to be responsible for their unjust or unfair treatment. Experts agree that mental health is certainly a factor in acts of workplace violence. Is it a "societal ill" that has become pervasive in mainstream America, or is it simply copycats looking for their 15 minutes of fame?

The violence that can occur in the workplace can have many possible sources, including issues between employees and management, troubled domestic relationships, stalkers with specific targets, mentally ill individuals with no specific targets, individuals under the influence of drugs or alcohol, and criminals who commit assault or homicide while seeking money or drugs.

All of these sources have to be considered a part of the equation when it comes to looking at solutions.

Solutions

Are there definite solutions to workplace violence that can be used to prevent this type of behavior? What is *the* answer? We believe that the overall solution is much more than just one quick fix, and it may even be a laundry list of items that need to be implemented in order to reduce, and eventually stop, this social problem that is often called domestic terrorism.

Educating employees about the definition of workplace violence is a critical step, along with establishing a zero-tolerance policy. A culture of safety and security must be established so employees feel safe and know how and where to report suspicious or threatening behavior. A workplace violence prevention program has to have buy-in from top management, and all employees should undergo training that teaches them how to identify potentially violent employees and situations, how to recognize behavior indicators that may be a precursor to workplace violence, and how to establish a workplace that has violence prevention techniques built into the business operations. Managers and supervisors should receive training in how to verbally de-escalate potentially violent situations. These elements are part of an effective workplace violence prevention program, as are clear reporting protocols and emergency procedures that are disseminated and practiced.

---

[*] Gil Press, "Bill Bratton on Data and Analytics, Homeland Security and Hometown Security," Forbes, April 27, 2013, http://www.forbes.com/sites/gilpress/2013/04/27/bill-bratton-on-data-and-analytics-homeland-security-and-hometown-security/.

Effective communication skills are necessary within the organization, as well as with outside resources such as law enforcement and counseling centers. OSHA and other governmental agencies have resources that are available for designing and implementing a workplace violence prevention program.

SEVEN THINGS YOU NEED TO KNOW ABOUT SOFT TARGETS

1. Never say, "It can't or will never happen here."
2. Acts of terror, whether domestic or international are usually over within ten minutes.
3. Perceived injustices and stereotypes have a powerful effect on the psyche of individuals. For example, the families of the nine people who were killed at the Emanuel AME Church in Charleston, South Carolina, in June 2015 publicly stated that they forgave the shooter, Dylann Roof, when he appeared in court.
4. Shooters plan and assess how they will commit the act and how they intend to escape. In the above case of the shooting at the Emanuel AME Church, the shooter was at the location for an hour (supposedly praying) before he opened fired on the community. When he was captured, Dylann Roof reportedly said, "I almost backed out because everyone was so nice at the church."[*]
5. Is it possible to harden your target? In some cases, the perpetrators have been insiders or have had help from insiders. In *Soft Target Hardening,* Dr. Hesterman asks the question, "What is the cost of not protecting our people?"[†]
6. Sgt. Glenn French, a 24-year veteran with the Sterling Heights (Michigan) Police Department, offers these six points for consideration as you develop your defense:
   - "The terrorist will first select or identify a vulnerable soft target.
   - The terrorist will determine the method of attack.
   - They will conduct detailed surveillance of the target to measure security forces.
   - They will assess target vulnerability and select the site or move on to another.
   - After site selection, a second round of surveillance will be conducted.
   - Finally, the operation will be scheduled and the attack conducted."[‡]

   "Do not encourage a terrorist attack by becoming a soft target," Sgt. French warns. "The most effective way to avoid indicating as a soft target is to remain in a high level of situational awareness. The sad truth today is that walking down a street on the way to a ballgame in a downtown setting may place you in the middle of a battle space in a split second."

   "Your ability to recognize and perceive potential threats while going about routine activities … will make the difference. Don't deny your gut feelings. As we all know, many good arrests on the streets come from a gut feeling or intuition and

---

[*] http://www.nbcnews.com/storyline/charleston-church-shooting/dylann-roof-almost-didnt-go-through-charleston-church-shooting-n378341.
[†] Hesterman, *Soft Target Hardening.*
[‡] Glenn French, "Terrorist Attacks: Don't Be a Soft Target," PoliceOne, June 5, 2013, http://www.policeone.com/international/articles/6261724-Terrorist-attacks-Dont-be-a-soft-target/.

then most importantly, acting upon that feeling or suspicion. Situational awareness and the apparent readiness of that person indicate whether or not they are a hard target or soft target."*
7. The National Infrastructure Protection Plan (NIPP) risk management framework, and the Commercial Facilities Sector-Specific Plan that builds on it provide an excellent starting point to protecting your assets. We have included guidance from the U.S. Department of Homeland Security (DHS) in the following section.

U.S. Department of Homeland Security Guidance

1. "Set Goals and Objectives. The CF SSA [Commercial Facilities Sector-Specific Agency] uses sector goals that define specific outcomes, conditions, end points, or performance targets as guiding principles to collectively constitute an effective risk management posture.
2. Identify Assets, Systems, and Networks. The identification of assets and facilities is necessary to develop an inventory of assets that can be analyzed further with regard to criticality and national significance. Because of the diverse nature of assets within the CF Sector, the CF SSA works with the DHS Infrastructure Information Collection Division (IICD) to identify and validate these assets. In addition to its collaboration with the IICD, the sector also identifies assets through interaction with trade associations and corporate owners and operators.
3. Assess Risks. The CF Sector approaches risk by evaluating consequence, vulnerability, and threat information with regard to a terrorist attack or other hazard to produce a comprehensive, systematic, and rational assessment.
4. Prioritize. The CF Sector Coordinating Council (SCC) and the CF SSA have found that it is not appropriate to develop a single, overarching prioritized list of assets for the CF Sector. Instead, assets are categorized by using a consequence methodology that allows the CF SSA to drive sector-wide protection efforts.
5. Implement Programs. Given the size and diversity of the CF Sector, there is no universal solution for implementing protective security measures. The owners and operators of CF Sector facilities implement the most effective protective programs based on their own assessments. Protective programs address the physical, cyber, and human dimensions.
6. Measure Progress. The CF SSA has developed relationships with both public and private-sector partners, including trade associations, corporate entities and industry subject-matter experts. Information sharing through these partnerships is used to assist in measuring progress through the creation of sector metrics. By measuring the effectiveness of protective programs and their actions, the CF Sector can continually improve the infrastructure at the facility and subsector levels and improve security overall at the sector level."†

---

* Ibid.
† "Commercial Facilities Sector-Specific Plan: An Annex to the National Infrastructure Protection Plan," U.S. Department of Homeland Security, 2010, http://www.dhs.gov/xlibrary/assets/nipp-ssp-commercial-facilities-2010.pdf.

CONCLUSION

By conducting a risk assessment that will become of part of the security master plan, a security professional can help organizations identify their vulnerabilities and develop countermeasures to address them. These vulnerabilities include the possibility of an active shooter situation as well as workplace violence. The countermeasures will work in conjunction with the culture of safety and security within the organization. Safety and security awareness and training initiatives will become a part of the business operation, and each member of the organization will clearly understand the goals of the security programs and the part they play in keeping everyone in the organization safe. This is taking a proactive stance to address risks. In order for any security program to be effective, it is important that the program is compatible with the mission of the organization.

An excellent program to increase security awareness is the "If You See Something, Say Something" program, which is a government initiative to raise awareness about the indicators of terrorism and terrorism-related crime. The general idea is that "it takes a community to protect a community," and this is accomplished by emphasizing the importance of reporting suspicious activity to law enforcement.[*] The idea behind "If You See Something" does not just apply to reporting terrorism. It is also effective as a crime prevention initiative. During a presentation in 2013 to the CEB TowerGroup Financial Services Technology Conference in Boston, Bill Bratton, the former head of the Boston, New York, and Los Angeles police departments, stated that data or intelligence information is just as critical to preventing terrorism as it is to crime prevention.[†] Homeland security is also hometown security. Just as suspicious behavior or activity within an organization can be reported to the appropriate person in the organization's hierarchy, suspicious behavior or activity in the community can be reported to local law enforcement. According to Bratton, 75 percent of the terror threats that have been detected since 9/11 have been detected by local police. Bratton stated that between 2002 and 2009, when he was the head of the Los Angeles Police Department, many investigations of robberies or stolen vehicles were found to have connections to terrorists.[‡]

We as individuals must become more aware of our surroundings and be "situationally aware."

TEN THINGS YOU NEED TO KNOW

1. Have a security professional conduct assessment of the property to identify potential security risks and vulnerabilities, so recommendations can be made to either eliminate the risk or reduce the impact on the organization if an incident does occur.

---

[*] "If You See Something, Say Something: About the Campaign," U.S. Department of Homeland Security, accessed May 18, 2016, http://www.dhs.gov/see-something-say-something/about-campaign.
[†] "Bill Bratton."
[‡] Ibid.

2. The assessment, along with the mission statement of the organization and a culture of security, should be a part of the overall security master plan.
3. This assessment should also incorporate training and security awareness. The training should include scenarios, drills, and full-scale exercises with local first responders.
4. Review the policies and procedures of the organization and update them every 5 years, or more often if necessary. Organizational policies change less often than organizational procedures. Ensure that emergency procedures are a part of the formal policies and procedures of the organization.
5. Formulate partnerships with local first responders—law enforcement, the fire department, emergency medical services, the state Homeland Security advisor, the Federal Emergency Management Agency (FEMA), and so on. Develop relationships with all resources that are available in the local area. Discuss security concerns and the daily operations of the organization with first responders and involve them in training and drills. The basic idea is to plan, practice, and prepare.
6. Ensure that the organization has a visitor management program in place and that the security officers and employees are trained to identify suspicious behavior and activity. This emphasizes the need to greet visitors or customers and have someone available to answer questions and direct them to the proper location. Many retail establishments, health care facilities, and churches have greeters at their entrances to observe who is entering the facility.
7. If you have concerns with theft at your organization or are concerned about weapons being brought onto the property, consider the implementation of a "no bag" or "clear bag" policy to reduce the risk.
8. Security officers in plain clothes or trained employees are ideal for detecting suspicious behavior and activity.
9. Every organization should have redundant mass notification capabilities. This can be accomplished by an announcement over the PA system, an automatic telephone call to company telephones and/or registered cell phones, an e-mail, a pop-up message on all computers connected to the organization's intranet, a text message to registered cell phones, or an audible siren and visible alarm. More than one of these means of communication can be used for notification. Some means of communication may indicate where in the building or on the property there is an issue and also include a photo. Ensure that there are provisions for those who are physically, visually, or hearing impaired.
10. All security devices and systems should be inspected regularly to ensure they are operational. This can be the responsibility of security officers or a designated company employee, but ensure it is documented. Security devices and systems requiring regular inspection include:
    - Video cameras
    - Keys and key control policies
    - Access control devices and access control card/badge procedures
    - Doors and door hardware, including exit sign illumination
    - Video surveillance systems

- Fire alarm systems and fire extinguishers
- Perimeter of the property and perimeter of the building
- Nonoperational lighting—interior and exterior
- Mass notification systems

As with most things, there is a cost to security or target hardening, but there is also an immediate return on investment—peace of mind and reduced fear.

17

Sport Venue Emergency Planning

Stacey Hall

Contents

Introduction	307
Emergency Management	308
Sport Venue Command Group	308
Preparedness	309
Emergency Response Plan	309
Staff Training and Exercise	310
Establishing a Command Center	311
Response	311
Evacuation Planning	311
Communication and Information Sharing	313
Recovery	314
Mitigation	315
Risk Management	316
Business Continuity	318
Appendix A: General Guidelines Checklist for Emergency Preparedness	319
Appendix B: Evacuation Plan Template for Stadiums	321
References	323

INTRODUCTION

Providing a safe and secure environment is a priority for sport venue operators due to the potential for all-hazard emergencies and subsequent legal implications. The education and training of all key security personnel is critical to ensure that effective security measures are implemented and that an all-hazards approach to emergency planning is employed (Sauter and Carafano, 2005). An effective sporting event security management system requires the involvement and commitment of many agencies and individuals, including professionals, volunteers, public agencies, and outsourced contractors.

EMERGENCY MANAGEMENT

An emergency situation is defined as "any incident, situation, or occurrence that could affect the safety and security of occupants, cause damage to the facility, equipment and its contents, or disrupt activities of the facility" (Center for Venue Management Studies, 2002, p. 1). Examples of emergency situations at sport venues include bomb threats/explosions, medical emergencies, fire hazards, natural disasters/inclement weather, power or equipment failure, crowd control issues, hazardous material release, domestic or international terrorism, and evacuations (partial, full, and shelter in place) (Hall et al., 2012). Emergency planning is imperative for several reasons:

- Sport facility operators must adhere to professional industry standards and expectations to provide a safe environment for patrons.
- Previous incidents at sporting events have demonstrated that planning, training, and implementation of an emergency plan have reduced the potential damage and loss associated with emergencies.
- Sport facility operators must adhere to legal public safety requirements, such as fires safety codes and American Disability Act (ADA) requirements.
- Documentation of an emergency plan will minimize the liability to facility, owners, and operators when an emergency occurs.
- Good planning practices will minimize negative media exposure and enhance a facility's positive image with patrons and local community stakeholders.

Emergency management practices attempt to avoid or reduce potential losses from all hazards, and ensure appropriate assistance to achieve rapid and effective recovery after an emergency (Warfield, 2008). Sport venue managers must focus on all components of this interrelated process (see Figure 17.1). The sport venue command group (CG) must work closely with external agencies involved in the planning, preparedness, response, and recovery operations at venues and events.

SPORT VENUE COMMAND GROUP

A sport venue CG is composed of key personnel from multiple agencies that collaborate to provide a safe and secure sporting environment for spectators, athletes, officials, sponsors, and community stakeholders. The CG is composed of representatives from the following key areas: sport venue management, law enforcement, emergency management, fire/hazardous materials (HAZMAT), and emergency medical services (EMS). Other key entities and individuals that may play a role in security operations include state and federal government security, public health and safety, media/public relations, public utilities, contractors, vendors, and temporary employees (i.e., volunteer groups). The CG serves as the core functional group for all security efforts, including incident management, risk management, staff training, developing and implementing plans and protective measures, conducting exercises, and coordinating response, recovery, and continuity efforts. It is important to establish this multidiscipline team prior to the sport

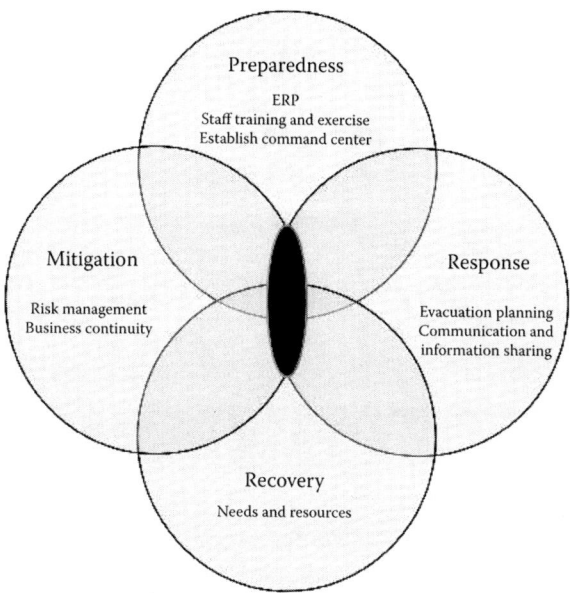

Figure 17.1 The emergency management process.

season (or event day), and it is critical for them to continue to meet on a regular basis (Hall et al., 2012).

Preparedness

Preparedness is a key measure to reduce the impact of an emergency by taking actions before an incident occurs. It is impossible to ensure a risk-free environment; therefore, planning and preparedness activities are critical to build and sustain capabilities to protect against, respond to, and recover from all hazards. Preparedness is a continuous improvement process and involves all key stakeholders in the sport venue CG, as well as local, state, and federal agencies as needed. Preparedness activities may include developing an emergency response plan (ERP), training staff and conducting exercises, and establishing command and control.

Emergency Response Plan

The ERP describes how the sport venue will do business in an emergency situation. The emergency plan highlights the steps and actions to be taken by the sport venue staff and response agencies to minimize or eliminate potential harm to facility patrons and/or damage to the venue (Hall et al., 2012). The ERP's ultimate goal is to reduce risk exposure and potential consequences through prevention, detection, communication, damage control, and recovery efforts (Center for Venue Management Studies, 2002). The sport venue CG becomes the emergency planning team (EPT) responsible for developing the sport venue

ERP in coordination with local and state authorities. The EPT have working knowledge of the venue (including its systems and resources) and are decision makers that have the ability to implement the ERP. The plan should be developed with specific actions and responses to any possible emergency.

Staff Training and Exercise
Once an ERP has been developed, key staff and facility personnel responsible for implementation should be trained and tested. The three primary levels of staff training include the multiagency team (CG), supervisors, and event staff (i.e., ticket takers, ushers, checkers, parking attendants, etc.). The following list highlights possible training techniques (Hall et al., 2012):

- Training sessions (seminars/workshops)
- Practice drills to test the staff's knowledge of the venue ERP
- Pre/Post event briefings
- Hand-held cards provided to event staff containing information pertinent to emergencies and specific roles and responsibilities
- Include external public agencies involved in responding to venue emergencies (i.e., police, fire, public health, media relations) in training sessions
- Produce training videos to orient new employees on the ERP

Once sport venue managers have assessed threats and risks, developed plans, and trained staff members, they should consider testing their operational plans to assess their level of preparedness. The CG should conduct exercises to test plans and promote awareness of roles, responsibilities, and position assignment during an incident scenario. "An exercise is a focused practice activity that places the participants in a simulated situation requiring them to function in the capacity that would be expected of them in a real event" (Federal Emergency Management Agency, 2008, p. 2). Exercises improve readiness by evaluating operational capabilities, reinforcing the concept of teamwork, and addressing identified gaps. Exercises help venue managers to (1) clarify roles and responsibilities, (2) improve interagency coordination and communication, (3) reveal resource gaps, (4) develop individual performance, (5) identify opportunities for improvement, and (6) gain program recognition and support of management (Federal Emergency Management Agency, 2009).

The U.S. Department of Homeland Security Exercise and Evaluation Program (HSEEP) promotes seven different types of exercises, categorized as either discussion based or operation based. Discussion-based exercises familiarize participants with current plans and policies, and may be used to develop new plans, policies, and procedures. Discussion-based exercises include seminars, workshops, tabletop exercises, or game simulations. Operation-based exercises are more complex than discussion-based exercises. Operation-based exercises validate plans and policies, clarify roles and responsibilities, and identify resource gaps in security operations and capabilities. Operation-based exercises normally involve the deployment of resources and personnel; these include drills, functional exercises, and full-scale exercises (HSEEP, 2007).

Establishing a Command Center

Most major sport venues include a command center (command post) with communication capabilities for security forces to monitor activities inside and outside the venue perimeter. The command center controls the security functions for the event and is normally staffed with the facility security director, venue operations manager, police, fire, EMS, private security, and media representatives. Copies of security and facility plans, phone directories, and backup technology systems are located at this site. The command center has reliable communications and the capability to access the venue public announcement system, fire alarm system, voice activation system, turnstile system, and access control systems. Command center capabilities include (Hall et al., 2012, p. 51)

- Coordinate internal response to all minor incidents
- Refer support requests to external agencies for major incidents
- Manage all event communications
- Document venue incidents
- Manage event timeline
- Maintain a safe, orderly environment
- Direct and manage venue evacuation
- Expand or contract based on the incident

The command post is usually established prior to event day and is activated in the event of an incident. The command post signifies the location of the tactical-level, on-scene incident command and management organization. The sport venue CG must decide where the command post will be established, and ensure the location offers enough space for multiagency personnel. The CG must consider what other government officials should be notified and included. The identification and availability of additional emergency resources and amount of time to access these services should be estimated. In addition to possessing adequate resources, certain facilities should be designated for emergency use during a crisis, for example, shelters to house displaced victims, distribution centers for food, water, and emergency supplies, and storage areas for equipment (The Spectrum of Incident Management Actions, 2006). A general guidelines checklist for emergency preparedness is presented in Appendix A.

Response

According to Lindell et al. (2007), emergency response has three distinct goals: (1) protect people, (2) limit damage from primary impact, and (3) minimize damage from secondary impacts. Response efforts begin when an emergency is imminent or immediately after an incident occurs. Response includes tasks and activities that address the direct (short-term) effects of an incident. Response activities may include execution of the ERP and evacuation procedures, increasing security operations, and implementing emergency communication systems.

Evacuation Planning

Planning for an evacuation at sporting events requires coordination, communication, and cooperation by venue operators and the response community (federal, state, local, and private). An evacuation plan should be an essential component of the facility's ERP and should take into consideration all potential hazards for a particular venue. Making the decision to evacuate, shelter in place, or relocate during an incident is complicated and requires input from various entities knowledgeable about the structure of the stadium, the size and distribution of the spectators and participants, the hazard involved, and the anticipated response to that hazard (U.S. Department of Homeland Security, 2008). The incident commander (i.e., police chief/emergency management director) determines whether to conduct a full evacuation, partial evacuation, or shelter in place. A full evacuation may be conducted due to a major structural failure, earthquake, explosion, chemical spill, or severe storm. A partial evacuation may be conducted due to a minor fire, small explosion, bomb threat, minor structural damage, or fan violence. In some instances, it may be necessary to shelter in place by keeping patrons inside the facility, for example, when an event has occurred outside the facility such as inclement weather. A stadium evacuation plan template is provided in Appendix B. Venue managers should consider the following items when developing evacuation procedures (Center for Venue Management Studies, 2002, p. 13):

- Venue staff must direct patrons to a safe area to reduce panic and chaos that may ensue.
- Identify a chain of command, including the person of authority who will make an evacuation decision (normally incident commander).
- Determine alternate points of egress from every point in the facility in advance.
- When an incident occurs and is localized (i.e., fire emergencies), consider staying in place or evacuating to another area of the facility rather than evacuating large crowds to the outside.
- Adequately train supervisory and event staff through pre-event briefings and issuing handheld information cards on specific positions, locations, and evacuation procedures.
- Prepare and display evacuation announcements when an evacuation is deemed necessary.
- Develop a plan on what to do with the patrons once they have been evacuated outside the venue to ensure they are out of danger and the emergency is under control.

If venue management chooses to shelter in place rather than evacuate, the CG must consider several items, such as communication procedures with patrons to provide updates on the emergency situation. This can be achieved through prescripted and prerecorded public address announcements, scoreboard signage, video screens, and verbal commands between event staff and patrons. Patrons should also be allowed the opportunity to contact family and/or friends if communication capabilities exist. Complimentary food and beverages may be offered to patrons if they are required to remain in the venue for an extended period of time. Furthermore, the sport organization should develop partnerships (i.e., mutual aid agreements) prior to the sport season with local emergency response agencies to obtain emergency supplies if needed (Hall et al., 2012).

Communication and Information Sharing

Normal day-to-day communications and emergency communications differ. Emergency information is vitally important and can mean the difference between life and death, or it can provide reassurance to those affected that response and recovery efforts are underway. Unfortunately, people find it difficult to hear messages during an emergency because of stress or change of routine (Federal Emergency Management Agency, 2005). Additionally, information shared with the public should be consistent and relay the same intended message since many parties respond to an emergency. When communicating in a crisis, the speaker should (1) word the message precisely, (2) avoid jargon, codes, or acronyms, (3) use common names for all personnel and facilities, (4) omit unnecessary information, (5) speak in sync with other agencies, and (6) keep messages consistent across all media outlets. The types of communication methods are highlighted below (Hall et al., 2012, p. 35):

- *Emergency alert system*: An established communication system that warns of impending dangers. Individuals should be aware of warning tones, messages across TV screens, cable TV override, and National Oceanic and Atmospheric Administration (NOAA) weather radio.
- *Oral communication*: Phone conversations, briefings, public speeches, TV interviews, radio announcements, and public service announcements.
- *Print communication*: Fax, e-mail, public notice, flier, press release, or feature article.
- *Social media*: Utilization of social media outlets such as Facebook, Twitter, or YouTube.

The most effective communication tool is one that reaches the intended target audience in a timely manner, delivers the message reliably, enhances comprehension of the message content, and can be accessed within resource limitations. The right technology can support and enhance communication capabilities. Most often, a combination of methods is used to deliver a message. The CG should consider (1) how they issue emergency communications, (2) what areas of emergency communications can be improved, (3) what steps need to be taken to improve communications, and (4) whether they should collaborate with other agencies in this effort (Federal Emergency Management Agency, 2005).

The ways of sharing critical information must be agreed upon prior to any incident, for example: What is the chain of command for communication? Does information flow through a preset protocol (during day-to-day activities) or through an all-channel network to preestablished personnel (during an incident response)? Emergency events in the past have highlighted the inability of incident command teams to communicate in a time of crisis because of a lack of basic skills, or incompatible equipment. Venue management should obtain satellite phones to allow continued communication in case hard-wire phone lines into the venue are out of service. All stakeholders involved in response and recovery efforts should focus on alleviating this problem. In times of crisis, all parties must possess high levels of communication skills so that they can send, receive, and understand information (U.S. Department of Homeland Security, 2009).

The Department of Homeland Security (DHS) Office of Intelligence and Analysis has created state and local fusion centers to share information and intelligence within their communities. As of July 2009, there were 72 designated fusion centers in the United States. Sport organizations in the United States should become familiar with their state or local fusion center representatives. This type of support can assist in gathering information pertaining to potential threats or issues at upcoming sporting events. This support is tailored to the unique needs of the locality and serves to (1) help the flow of classified and unclassified information, (2) provide expertise, (3) coordinate with local law enforcement and other agencies, and (4) provide local awareness and access (U.S. Department of Homeland Security, 2009).

Recovery

The goal of recovery is to ensure the sport organizations operations return to normal as soon as possible. Recovery efforts are unique to each incident, dependent on the extent of damage caused and the resources available. This includes (1) identification of needs and resources, (2) addressing long-term care of affected persons, (3) implementing measures for organizational and venue restoration, and (4) identifying lessons learned and incorporating mitigation measures to lessen effects of future incidents (The Spectrum of Incident Management Actions, 2006). Short-term recovery begins immediately post-incident and is an extension of the response efforts to restore basic services and functions. Long-term recovery is the restoration of personal lives and livelihood of the sport organization and surrounding community.

The CG should collect critical data that is needed to recover from a disaster, that is, personnel listings with telephone numbers, inventories of equipment, list of vendors, storage locations, data backup files, and important contracts. The CG should also assess current capabilities by reviewing existing plans and policies, that is, evacuation plan and mutual aid agreements. Additionally, the CG should review the availability of internal assets to respond and recover from an incident, such as emergency medical teams, public relation representatives, fire equipment, communication equipment, warning systems, emergency power, or shelter areas. Consultation with external groups will also provide insight to the availability of external resources available in coordination with agencies such as local emergency management, fire and police departments, ambulance services, public works, hospitals, and volunteer organizations such as the Red Cross. When the CG has decided what measures and procedures will be taken pre-, during, and postincident, they will document their efforts into a comprehensive plan. Supporting documents may also be included, for example, building and site maps indicating utility and shutoff locations, escape routes, location of emergency equipment, and hazardous materials (Hall et al., 2012).

Recovery is primarily the responsibility of local government, but by presidential declaration of a disaster in the United States, a number of assistance programs may be available under the Stafford Act. The Stafford Act defines "emergency" and "major disaster" declarations. An emergency is defined as "any occasion or instance for which, in the determination of the President, Federal assistance is needed to supplement State and local efforts and capabilities to save lives and to protect property and public health and safety, or lessen or avert the threat of a catastrophe in any part of the United States." A presidential declaration of an emergency provides assistance that (1) is beyond state and local capabilities, (2) serves as supplementary emergency assistance, and (3) does not exceed $5 million of federal assistance. The governor of an affected state must request a presidential declaration for an emergency within 5 days of the incident (Fundamentals of Emergency Management, 2010).

A major disaster is defined as "any natural catastrophe or, regardless of cause, any fire, flood, or explosion, in any part of the United States, which in the determination of the President causes damage of sufficient severity and magnitude to warrant major disaster assistance under this chapter to supplement the efforts and available resources of States, local governments, and disaster relief organizations in alleviating the damage, loss, hardship, or suffering caused thereby." Major disasters may be caused by natural events such as floods, hurricanes, and earthquakes. Disasters may include fires, floods, or explosions that the president believes are of sufficient magnitude to warrant federal assistance. Although the types of incidents that may qualify as a major disaster are limited, the federal assistance available for major disasters is broader than that available for emergencies. A presidential disaster declaration provides assistance that (1) is beyond state and local capabilities and (2) supplements available resources of state and local governments, disaster relief organizations, and insurance. The governor of an affected state must request a presidential declaration for a major disaster within 30 days of the incident (Fundamentals of Emergency Management, 2010).

The two major categories of federal aid are public assistance and individual assistance. Public assistance is for repair of infrastructure and public facilities and removal of debris. Individual assistance is for damage to residences and businesses or for personal property losses. Recovery from a disaster is unique to each sport or event business and local community depending on the amount and kind of damage caused by the disaster and the resources that the community has available or has access to (Federal Emergency Management Agency, 2010). The chapter case study provides an example of a sporting business' response and recovery efforts after a natural disaster.

Mitigation

Mitigation activities try to prevent disasters or lessen the damage of unavoidable disasters and emergencies (Hall et al., 2008). Mitigation activities may be implemented prior to, during, or after an incident and are incorporated from lessons learned from prior incidents. Mitigation measures are identified in conjunction with a threat/risk analysis that identifies potential hazards to the sport venue, the probability that an event will occur, and the potential consequences, such as loss of life, destruction of property, disruption of critical services, and economic impact of recovery. The sport or event business' mitigation strategy must consider ways to reduce risks associated with all hazards and potential losses. According to FEMA's Fundamentals of Emergency Management (2010), an effective mitigation strategy is based on several factors:

- *Prevention measures*: To prevent existing risks from becoming worse or implementing new prevention measures in areas that have not been developed or are in an early phase of development, that is, physical protection security features in building design.

- *Property protection measures*: To modify buildings or their surroundings to reduce the risk of damage from a known hazard, that is, raising generators to prevent damage from flooding.
- *Emergency services measures*: To protect people before and after an incident occurs, that is, warning notifications, response tasks, and protective measures for critical facilities.
- *Structural projects:* To protect people and/or property through the construction of man-made structures to control the damage from a known hazard (i.e., bollards to prevent a vehicle-borne improvised explosive device).
- *Public information*: To inform and remind people about potential hazards and measures that should be taken to avoid damage or injury. Public information measures may include outreach projects, technical assistance, or educational programs (i.e., basic security awareness workshops hosted by the sport organization).

The mitigation strategy developed must consider the risks faced, the potential damage or impact, and the overall needs of the organization and venue. The mitigation measures must be consistent with the strategy and considered as part of the larger emergency management cycle.

Risk Management
According to Decker (2001), "risk management is a systematic and analytical process to consider the likelihood that a threat will endanger an asset, individual, or function and to identify actions to reduce the risk and mitigate the consequences of an attack" (p. 1). To determine threats and vulnerabilities, a sport organization must conduct a risk assessment. Input from the CG during the risk assessment process is critical to ensure all information and intelligence sharing about threats and vulnerabilities inside and outside of the sporting venue. Venue management can significantly reduce liability exposure by successfully managing risks and foreseeable actions that lead to injuries (Schwarz et al., 2010). An all-hazard risk management approach is critical for sport organizations to protect physical and human assets against potential threats.

The DHS considers major sport stadia as critical infrastructure in the Commercial Facilities Sector. There are many benefits of conducting an assessment and sport venue operators must embrace risk management processes. The risk assessment will indicate the current security profile of the sport venue and highlight areas in which improved security is required. It will also help to develop a justification of cost-effective countermeasures and increase security awareness by reporting strengths and weaknesses in security processes to all staff members, including management (Broder, 2006).

Sport leagues, teams, and venues must prepare for a wide range of possible threats at their venues (Hurst et al., 2007). "A threat is a product of intention and capability of an adversary, both man-made and natural, to undertake an action which would be detrimental to an asset" (Vulnerability Assessment Report, 2003, p. 11). Sport venue managers can obtain threat information from various local, state, and federal sources. Local sources of threat information are obtained from the venue security director, local law enforcement, and state or regional law enforcement. Many states also have threat and investigation working groups such as the Joint Terrorism Task Force (JTTF) (Biringer et al., 2007).

Timely and accurate information is available from the Federal Bureau of Investigation (FBI) and the DHS. The DHS provides information on current threat levels through the Homeland Security Advisory System (HSAS), including Homeland Security Bulletins. The FBI operates the National Threat Center and maintains the National Security Threat List (NSTL) (2007). Other techniques used to gather information relevant to specific threats include brainstorming and tabletop exercises, modeling and simulation, walk-through of the facility preevent and during the event, knowledge transfer from industry peers, and security surveys (Stevens, 2007). The threat data collected assists the CG to determine who or what the threats are, their capabilities, and the potential and severity of the threat. Sport facility threats include terrorism, natural catastrophes, crowd control, vandalism, theft, fire, fraud, personal assault, traffic incidents, facility intrusion, and technological problems (Stevens, 2007). Terrorism has been cited as one of the most common threats to sport venues, and DHS has issued warnings indicating that sport stadia as critical infrastructure are vulnerable targets. International, domestic, and lone-wolf terrorists have considered stadia as targets since such facilities present open access and opportunity to achieve objectives of mass casualties, economic damage, and social/psychological impact. Scenarios of particular concern to DHS and FBI are explosive devices and the use of aircraft and chemical weapons to attack stadiums and arenas (Office of Intelligence and Analysis, 2009).

Vulnerabilities expose the asset to a threat and eventual loss. The General Security Risk Assessment Guideline (2003) defines vulnerability as "an exploitable capability; an exploitable security weakness or deficiency at a facility, entity, venue, or of a person" (p. 5). Hall et al. (2007) identified vulnerabilities at major sport venues relative to emergency preparedness, perimeter control, physical protection systems, access control, credentialing, training, and communication:

- Lack of emergency response and evacuation plans specific to their facility
- Inadequate searching of the facility prior to an event
- Inadequate lock-down procedures
- Inadequate searches of fans and their belongings
- Unsecure concession areas
- Inadequate signage concerning searches and restricted items
- Lack of closed circuit television (CCTV) coverage of the sport facility or surrounding areas
- Storage of dangerous chemicals inside the sport facility
- Lack of accountability for vendors and their vehicles
- Lack of security notification system for fans, players, staff, and so on
- Inadequate training of staff members
- Inadequate communication capabilities among responding agencies

Countermeasure improvements (also referred to as risk reduction strategies) are recommendations provided to management based on venue assessments to address security weaknesses and enhance emergency planning and recovery efforts. Countermeasures are any actions involving physical, technical, and administrative measures taken to reduce the probability and severity of risks and enhance decision-making abilities (Long and Renfroe, 1999). Possible security measures to reduce the likelihood of an undesired event include increasing security protection system effectiveness through physical upgrades, good personnel practices, information security, staff training, and preventative facility maintenance (Ammon et al., 2010). Physical upgrades include detection, delay, and response strategies; for example, detection methods can include intrusion sensors, identity check access control, alarm communication, and CCTV. Delay mechanisms may include the use of locks and security personnel stationed at restricted access areas of the venue (Biringer et al., 2007). Some measures are designed to be implemented on a permanent basis to serve as a routine protection for a facility; these are referred to as baseline countermeasures. Additional measures can be implemented or increased in their application during times of heightened alert (Department of Homeland Security, 2008). Responding to threat elements requires intelligence and information sharing at the local, state, and federal level. The DHS created an HSAS with corresponding alert levels.

Business Continuity
"Business continuity involves developing measures and safeguards that will allow an organization to continue to produce or deliver goods and services under adverse conditions" (Sauter and Carafano, 2005, p. 333). Continuity planning makes good business sense and is important for numerous reasons, such as reducing the cost of downtime, cost of rebuilding, cost of reconstructing lost critical data, and loss of revenue. The costs associated with the aftermath of a large-scale incident may severely damage the sport organizations operations or even inhibit their ability to fully recover (Broder, 2006). Sport organizations should therefore develop a business continuity plan to ensure operations are maintained. According to Broder (2006), "A business continuity plan is a comprehensive statement of consistent action taken before, during, and after a disaster or outage" (p. 179). The business continuity planning process is also a training exercise for the sport security CG as they must think through contingencies and be familiar with actions required to recover from all types of incidents (Broder, 2006).

Examples of issues to consider include relocation of athletes, use of alternate facilities, enrollment of athletes in other institutions, and/or rescheduling of games or events. This all occurred in the aftermath of Hurricane Katrina and was, in large measure, not planned for. Hurricane Katrina affected many professional and college sports programs in the Gulf Coast region and New Orleans area. Contracts should be in place for immediate restoration and secondary locations identified to hold event bookings in case of an incident. Mutual aid agreements should be included as part of both the ERP and recovery plans. This assures prearranged resources and services are available and provided, if needed. Sauter and Carafano (2005) identified the following key planning steps in developing a business continuity plan:

1. Obtain management commitment
2. Establish a planning committee
3. Perform a risk assessment
4. Establish operational priorities
5. Determine continuity and recovery options

6. Develop a contingency plan
7. Implement the plan

There are several common flaws in continuity plans. First, a one-size-fits-all approach to continuity planning is not effective. Each plan should be customized to a specific venue, as each venue is unique and offers different resources and capabilities. Training venue staff and exercising the plan is important. Not testing your plan could result in failure to notice significant gaps in the plan that may be exposed during a crisis. When a plan has been developed, it should be updated on a regular basis as business regulations or standards may change within the sport industry. A lack of senior management (venue owner/operator) support could prohibit a successful planning process. The CG will need the support of management to develop and test the plan, especially if any financial support needed. In conclusion, the belief that something bad or wrong will never happen to me, or my organization, may result in complacency and a lack of urgency to plan and be prepared for all-hazard emergencies (Hall et al., 2012).

APPENDIX A: GENERAL GUIDELINES CHECKLIST FOR EMERGENCY PREPAREDNESS[*]

Facility Preparedness

- Emergency power to lighting and publics address system, that is, emergency generators.
- An evacuation plan outlining responsibilities and duties of event staff.
- Proper fire equipment available and personnel trained in operating use.
- Adequate signage for exits and general directions are displayed in and around the facility.
- Adequate ingress and egress points and checking emergency exits are accessible in case of an emergency.
- Prepared evacuation messages are activated in event of an emergency.
- A public address system that has the ability to broadcast outside the facility perimeter.
- Cell phone available within facility in case hard-wire phone lines are out of service.
- Copies of local emergency phone numbers are readily available, that is, bomb squad, local utility companies, police, fire, and EMS.

Documentation and Record System

- Predeveloped forms on file to be completed after an incident.
- Facility policy requiring an incident to be reported on every type of situation involving a patron or employee.

---

[*] Adapted from the Center for Venue Management Studies (2002).

- Copies of reports completed by external agencies, that is, police/medical, are filed with the facility's incident report.
- Insure proper documentation and backup information are gathered for legal and insurance purposes.
- Maintain records of all facility emergency preparedness and response efforts, that is, training, policies, standards, and so on.
- Designate an event incident report person to complete all necessary paperwork.
- Use video to record an incident when appropriate.

Emergency Medical

- Ensure appropriate number of emergency medical staff are present at major events.
- Maintain constant communication with EM staff.
- Position EM staff in a predetermined location with adequate supplies/equipment.
- Exercise EM response to certain scenarios.

Bomb Threat

- Have a recording device on phone lines receiving calls during events.
- Individual receiving the call should take note of as much information as possible and keep the caller on the line for as long as possible.
- Take every threat seriously.
- Complete a visual search of the facility.
- Contact local law enforcement, fire department, and facility management.
- Contact event manager about the possibility of evacuating patrons.
- If exploding device is found, immediately evacuate area using security staff and prepared announcements. Refrain from using radio communication as some explosive devices can be triggered by radio waves.
- If nothing is found, the facility manager and police representative should decide whether an evacuation is still necessary.

Fire

- Local fire inspector should conduct periodic inspections of facility.
- Implement facility policies regarding materials that can be taken into the facility.
- A licensed pyrotechnician should be utilized when necessary.
- Contact fire department in event of fire, regardless of size.
- Evacuate facility or threatened area if necessary.
- Fire and smoke alarms should be checked frequently and inspections and certifications should be documented.

APPENDIX B: EVACUATION PLAN TEMPLATE FOR STADIUMS

Introduction

Events at *(Insert Name of Stadium)* are considered premier events hosted in *(Insert Name of State)*. As such *(Insert Name of Stadium)*, needs to be prepared for any eventuality where it may become necessary to evacuate, shelter in place, or relocate spectators, participants, and staff from within the stadium, or to redirect traffic around the stadium. Assessing risk, reducing vulnerabilities, and increasing the level of preparedness will help to minimize potential threats and consequences. It is essential, therefore, that key security personnel at *(Insert Name of Stadium)* are well trained in risk factors, planning an appropriate response, informing the public, and implementing the evacuation plan. This evacuation plan is a supplement to the *(Insert Name of Stadium)* emergency plan (ERP).

Purpose

This evacuation plan provides instructions and guidance to effectively address the safety of all individuals in attendance at *(Insert Name of Stadium)* with regard to evacuation, sheltering in place, or relocation. The emergency plan describes procedures for responding to an emergency or critical incident at the *(Insert Name of Stadium)*. The evacuation plan provides guidance for developing and implementing procedures to evacuate, shelter in place, or relocate in response to an emergency or critical incident.

This evacuation plan was prepared by *(Insert Name)*, *(Insert Name of Stadium)*, Security/Safety Director and *(Insert Name)*, *(Insert Name of County/City)*, Emergency Management Director on X_X_/_X_X_/_X_X_. This document was prepared in coordination and cooperation with the following, and they have signed off with their concurrence:

Chief of Police _____ & Staff _____ Police Department

Fire Chief _____ & Staff _____ Fire & Rescue

Sheriff _____ & Staff _____ Co. Sheriff's Office

Emergency Management Director _____

Emergency Medical Services Director _____

State Highway Patrol Captain _____ & Staff _____

State Bureau of Investigation _____ & Staff _____

FBI Special Agent in Charge _____ & Staff _____

Bureau of Alcohol Tobacco and Firearms _____

Area Substance Abuse Council _____

Federal Aviation Administration, Flight Standard Office _____

Other—if additional or different people, continue to list _____

Relevant Plans

This section provides an overview of the plans, policies, and guidance documents that are applicable to the *(Insert Name of Stadium)*. Plans may be maintained by the county or city where the stadium resides.

A. Owner's security and safety guideline reference manual
B. Emergency action plan
C. Security and safety plan
D. Other (as appropriate)

Command Structure/Response Organization

The command structure/response organization for evacuation, sheltering in place, and relocation activities should mirror the normal command structure, as found in Section *(Insert Section Number)* of the emergency plan.

A. Jurisdiction and liability
   Identify laws, ordinances, and authorities that affect evacuation activities.
   Identify any issues of liability associated with evacuation activities.

B. Evacuation team—Roles and responsibilities
   Define for each entity, designate, and identify key personnel.

C. Direction and control—Roles and responsibilities
   Define for each entity, designate, and identify key personnel.

D. Local, state, and federal assistance—Roles and responsibilities
   Define for each entity, designate, and identify key personnel.

E. Surrounding industry/private-sector assistance—Roles and responsibilities
   Define for each entity, designate, and identify key personnel.

F. Local transportation structure—Roles and responsibilities
   Define for each entity, designate, and identify key personnel.

Pre-Event Planning Considerations

Pre-event planning considerations need to be considered prior to a scheduled event at the *(Insert Name of Stadium)*. This section of the evacuation plan provides further information on the types of potential hazards/scenarios that could occur at the stadium and the number and makeup of the spectators and participants of the stadium.

Potential Hazards/Scenarios

Table 17.1 includes the potential hazards that the *(Insert Name of Stadium)* can expect. The table also illustrates the likelihood of the hazard and whether evacuation, sheltering in place, or relocation, would be the appropriate response for each hazard.

Table 17.1 An Evacuation Planning Guide

Hazard/Scenario	Likelihood of Hazard High/Medium/Low	Evacuation, Shelter in Place or Relocation Decision
(Insert Name of Stadium) Hazards		
Weather		
• Rain		
• Lightning		
• Tornado		
• Heat		
• Severe thunderstorm/heavy rain/flooding		
• High winds		
• Hurricane		
• Heavy snow		
Accidental release (chemical, biological, radiological)		
IED or bomb threat		
Active shooter situation		
Mass casualty event		
Civil disturbance		
Food-borne illnesses		
Fire		
HAZMAT		
Structural collapse		
Terrorism—WMD, explosion, chemical, or biological event, dirty bomb		

Source: From The U.S. Department of Homeland Security, Washington, DC, 2008. Retrieved from http://www.dhs.gov/publication/evacuation-planning-guides

REFERENCES

Ammon, R., Southall, R., and Nagel, M. 2010. *Sport Facility Management: Organizing Events and Mitigating Risks.* Morgantown, WV: Fitness Information Technology.

Biringer, B.E., Matalucci, R.V., and O'Connor, S.L. 2007. *Security Risk Assessment and Management.* Hoboken, NJ: Wiley.

Broder, J.F. 2006. *Risk Analysis and the Security Survey.* Third Edition. Oxford, United Kingdom: Butterworth-Heinemann Business Books.

Center for Venue Management Studies. 2002. *Best Practices Planning Guide Emergency Preparedness.* IAVM Safety and Security Task Force. TX: International Association of Venue Managers.

Decker, R.J. 2001. *Key Elements of a Risk Management Approach.* Washington, DC: U.S. General Accounting Office. Retrieved from www.gao.gov/new.items/d02150t.pdf

Evacuation Planning Guide. 2008. The U.S. Department of Homeland Security, Washington, DC. Retrieved from http://www.dhs.gov/publication/evacuation-planning-guides

Federal Emergency Management Agency, Emergency Management Institute. 2005. Communicating in an emergency. In *Effective Communication: Independent Study.*

Federal Emergency Management Agency, Emergency Management Institute. 2008. Introduction to Exercise Design. In *Exercise Design: IS-139*, Washington, DC.

Federal Emergency Management Agency, Emergency Management Institute. 2009. Exercise basics. In *An Introduction To Exercises: IS-120.A*, Washington, DC.

Federal Emergency Management Agency, Emergency Management Institute. 2010, January 14. *Fundamentals of Emergency Management: IS-230.D*. Retrieved from https://training.fema.gov/is/courseoverview.aspx?code=IS-230.d

General Security Risk Assessment Guideline. 2003. ASIS International. [Online]. Available: http://www.iiiweb.net/files/ASIS_Security_Risk_Assessment_Guidelines.pdf

Hall, S., Cooper, W.E., Marciani, L., and McGee, J.A. 2012. *Security Management for Sports and Special Events—An Interagency Approach*. Champaign, IL: Human Kinetics.

Hall, S., Marciani, L., and Cooper, W.E. 2008. Sport venue security: Planning and preparedness for terrorist-related incidents. *Sport Management and Related Topics Journal*, 4(2), 6–15.

Hall, S., Marciani, Cooper, W.E., and Rolen, R. 2007. Introducing a risk assessment model for sport venues. *The Sport Journal*, 10(2), 1–6.

HSEEP. 2007. Homeland Security Exercise and Evaluation Program. The U.S. Department of Homeland Security.

Hurst, R., Zoubek, P., and Pratsinakis, C. 2007. *American Sports as a Target of Terrorism: The Duty of Care after September 11th*. Retrieved from http://www.mondaq.com/unitedstates/x/19623/Sport/American+Sports+As+A+Target+of+Terrorism+The+Duty+of+Care+After+September+11th

Lindell, M.K., Prater, C., and Perry. R.W. 2007. *Introduction to Emergency Management*. Hoboken, NJ: John Wiley & Sons.

Long, L.E. and Renfroe, N.A. 1999. *A New Automation Tool for Risk Assessment*. 15th Annual NDIA Security Technology Symposium. Session: Risk and Threat Assessment Techniques.

Office of Intelligence and Analysis. 2009, January 26. *Threats to College Sports and Entertainment Venues and Surrounding Areas*. U.S. Department of Homeland Security.

Sauter, M.A. and Carafano, J.J. 2005. *Homeland Security: A Complete Guide to Understanding, Preventing, and Surviving Terrorism*. New York: McGraw-Hill.

Schwarz, E.C., Hall, S., and Shibli, S. 2010. *Sport Facility Operations Management: A Global Perspective*. Oxford, United Kingdom: Butterworth-Heinemann Business Books.

Stevens, A. 2007. *Sports Security and Safety: Evolving Strategies for a Changing World*. London, England: Sport Business Group.

The Spectrum of Incident Management Actions. 2006. In *Principles of Emergency Management: Independent Study*. Emergency Management Institute, Federal Emergency Management Agency (FEMA).

U.S. Department of Homeland Security. 2008. *Evacuation Planning Guide for Stadiums*. Retrieved from http://www.dhs.gov/xlibrary/assets/ip_cikr_stadium_evac_guide.pdf

U.S. Department of Homeland Security. 2009. *State and Local Fusion Centers*. Retrieved from https://www.dhs.gov/state-and-major-urban-area-fusion-centers

Vulnerability Assessment Report. July 2003. Office of Domestic Preparedness, U.S. Department of Homeland Security. Retrieved from https://www.ncjrs.gov/pdffiles1/206046.pdf

Warfield, C. 2008. *The Disaster Management Cycle*. Retrieved from http://gdrc.org/uem/disasters/1-dm_cycle.html

18

Special Events

Patrick J. Jessee

Contents

Special Event Types	326
Intelligence Sharing	328
Personnel	329
Planners	330
Managers	330
Operators	332
Training	332
Planning	334
Volunteers	335
Communications	335
Paperwork	336
Benefits	337

Municipalities and jurisdictions all have different resources. The resources of New York, New York, and Hot Springs, Arkansas, are not necessarily the same. Although some common resources such as law enforcement, fire, EMS, and public works exist, the assets they have are distinctly different. Despite having these clearly different levels of equipment, experience, and training, they have the same common goals of protecting and providing for the civilian population. This includes normal operations throughout the year as well as special gatherings and events.

Special event planning is a unique challenge for emergency managers and public safety professionals. Emergency managers and public safety professionals typically plan for potential events or catastrophic responses. For them, many of the planning phases are for events that may occur, rather than an actual event that will occur. Special events present a known specific event that creates an extra stress on the resources of a municipality beyond normal expectations. Planning for these special events in advance can help alleviate the burden on municipal resources that would occur during a special event.

Should an incident occur during a special event, a lack of planning would create an unpleasant and hazardous environment for those attending. Therefore, it is critical for emergency managers and public safety professionals to analyze potential issues associated with special events to help create a secure venue for these special events. Fiscally responsible emergency managers and public safety professionals may consider the benefits of the following analysis to help plan and prepare for special events, which are applicable to all municipalities—from densely populated urban areas to small and sparsely populated villages throughout the country.

SPECIAL EVENT TYPES

When thinking about special events, the first thought that comes to mind are large-scale festivals such as the Olympics or 4th of July celebrations. These special events require significant planning by emergency managers and public safety professionals. An acceptable definition of a *special event* should be determined. Many different types of events may qualify as a special event, but what constitutes a special event is partially determined by the emergency managers and public safety professionals.

An event may be determined as a special event for a number of reasons: who is in attendance, what the event represents (e.g., a protest rally or march), the significance of the event for the city or community, etc.

Special events may occur in a certain area because of yearly celebrations or festivals, such as the 4th of July celebrations or county fairs. They may also occur because of championship games for sporting events. Another possibility is dignitary visits to an area (e.g., a congressman or foreign ambassador). All of these events fulfill the broad definition of special events, yet they all require significantly different planning considerations.

Photo 18.1 City of Phoenix Food Day, 2015. (From the City of Phoenix, https://www.phoenix.gov/sustainability/foodday)

Emergency managers and public safety professionals can use the same basic considerations for each of these events and then augment the planning process as needed to fulfill the objectives.

Certain events that occur throughout the United States qualify as National Security Special Events (NSSEs) and are deemed of particular importance. These large-scale incidents of national significance bring a large amount of media attention, political importance, and high crowd densities, and present a significant threat for homeland security and counterterrorism experts. When an event is designated as an NSSE, the Secret Service, Federal Bureau of Investigation (FBI), and Federal Emergency Management Agency (FEMA) take lead roles in managing the event. Some examples of previously designated NSSEs are Super Bowl games, political national conventions (both Democratic and Republican), and even the Academy Awards. This section will not discuss the NSSEs because of the federal level involvement in planning for NSSEs and the resources that the federal government brings to these events. Additional open source information on NSSEs can be found online at the DHS or FEMA websites.

Large-scale events require considerable attention and planning by emergency managers and public safety professionals. These events, such as holiday celebrations, may be planned out up to a year in advance (events such as the Olympics may be planned many years in advance). These events may require extra security, fire and rescue, EMS, or first aid tents. Public works may need to contribute with extra sanitation vehicles or roadblocks, to name a few. These agencies all contribute to making a full-scale operation that operates in the background of a public event.

In some special events, emergency managers and public safety professionals only have days to plan as they are last-minute events added into schedules. In order to accommodate this wide array of events and timing, emergency managers and public safety professionals must be flexible and resourceful. The most important factor, however, is to be informed of the event and to share open-source information with all stakeholders. This will allow the rapid synthesis of a plan as all parties involved to efficiently contribute to the development of the response plan.

Smaller municipalities may face more challenges in dealing with a special event because they have fewer resources or personnel. This should not hinder the emergency managers and public safety professionals. Instead, they should think outside the box and use creative problem solving to develop sound and economically feasible solutions. To a smaller municipality, county fairs, art shows, or even parades could be events for which emergency managers and public safety professionals wish to have plans prepared.

Why are special events of particular importance for emergency managers and public safety professionals? These events represent a potential for a mass-casualty event in case of either natural (floods, tornados, fires) or man-made disasters (riots, shootings, terrorism). The sheer number of people gathered in a venue presents a target-rich environment for violent terrorist activities that frequently occur in the presence of media practitioners.

Additionally, many events have limited access or egress from the venue, which creates a bottleneck for both incoming and outgoing crowds. This type of situation can potentially lead to a high number of injuries in case an incident occurs in that location or as a result of trampling as people make a mass exodus because of the incident.

Particular attention should be paid to access points. Establishing multiple access points helps to minimize bottlenecks as people enter the venue. However, it has the drawback of requiring more personnel and equipment to man these posts. Having multiple entrance points also helps to facilitate the exit of people during a potential threat or disaster. Most people, when presented with a threat, will leave by the way in which they entered. A consideration to account for this would be to have many points of egress established throughout the venue that can be opened as needed. Ultimately, the decision to establish multiple points arises from the event planners.

Emergency managers and public safety professionals should give special attention to the layout of the special event venue. A thorough knowledge of the building or topography of the area will help emergency managers and public safety professionals create secondary access/egress points should an event happen. EMS personnel will be able to understand the quickest routes to gain access and remove patients from the area. Law enforcement may be able to understand where potential weaknesses in the boundaries exist. Fire departments will have a knowledge of water resources and building construction. Hazmat will be able to identify inherent chemicals/hazards present within the venue and monitor continuously throughout for unknowns.

Special events also typically capture the attention of those who are not even in attendance. Many of these places have media present also. Local or national news crews may be on-site to report on the events of the festival or gathering. Print media may be present, mingling with the visitors. Television media may have cameras or a stage established to report from. Of course, the ever-present Internet presence is there as people update their status on social networks such as Facebook, Snapchat, or Twitter. A constant stream of reports on the atmosphere of the gathering is inevitable. The atmosphere as well as the public service presence, whether overt or low-profile (intermingled with the crowd), are constantly updated through persistent open-source updates.

This constant presence of media reporting can provide feedback to personnel operating at the scene should it be monitored through the Internet. This constant media presence creates a singular focus for the world should an event, natural or man-made, occur. Any injuries that occur from these incidents turns into an instantaneous news story, one that captivates the audience, especially during the first few hours of an event as it unfolds. This is one primary reason to plan for these events—to quickly respond and rectify any incidents as they occur.

INTELLIGENCE SHARING

It is implicitly understood that intelligence sharing must occur for numerous agencies to work together under one Joint Operations Center (JOC). The decision to plan for a special event generally comes from an elected official. This person may choose what response is appropriate or he may delegate it to his subordinates for the planning sessions. At other times, significant threat information from law enforcement or even a meteorologist may cause the formation of a special event response plan. Regardless of the reason, the information that directed the decision should be shared as much as possible with all parties involved in the planning stages.

Not all information can be shared readily between agencies. On occasion, law enforcement may have credible information regarding a particular threat or have an understanding of protesters who may become unruly. It is the law enforcement agencies responsibility to choose how to handle this information with an understanding that other agencies need to have some degree of justification for the resources that are to be used in the special event. All agencies understand that some of the information is For Official Use Only or Classified. Emergency managers and public safety professionals must be flexible in these considerations when planning for special events.

Numerous technologies can be combined in the sharing of intelligence that is open source, which may help planners share information. Online geographical resources, such as Google Earth, exist to help visualize the operations area. Other software, such as CAMEO (Computer Aided Management of Emergency Operations), allow hazmat TIER 2 reports (quantities of hazardous chemicals onsite), streets, hydrants, gas mains, and other layered information to be presented on top of a map for planners to use. It is ultimately the decisions of the planners as to which technologies to use.

PERSONNEL

In planning for special events, emergency managers and public safety professionals have three levels of "players" that are needed for these events. The emergency managers and public safety professionals need to have planners—those who specialize in moving around resources and accommodating the needs of the emergency managers and public safety professionals. They also need to have managers—those who can manage the operations conducted during the special event. Finally, they need operators—operators are the end product of all the planning and are conducting the "leg work" of the operations. Without the coordination of these three groups, no special event operations can be accomplished.

When establishing a special event responder selection process, a determination needs to occur to establish how to staff. Most special events do not require a full-time complement of personnel. The following issues need to be considered when establishing a special events team:

- Selection of qualified personnel
- Straight pay or overtime rates for those participating
- Compensatory time off
- Contract regulations for unionized special events responders

The selection of the planners, managers, and operators also needs to consider operational security (OPSEC) dependent on the sensitivity of the event. All individuals should be able to pass a standard background check conducted by normal law enforcement agencies as needed. They should understand the importance of not discussing the details of the event with family members and friends, even in a casual setting. Posting event information on their activities via blogs, Twitter, or Facebook may unintentionally compromise the mission of the special events planning.

OPSEC may be more of a concern for higher profile events. Should an NSSE be established, the lead agencies may dictate special needs for the OPSEC. During the creation and recruitment stage of a special events response team, these lead federal agencies should be contacted to determine if they have additional requirements for any NSSE OPSEC.

Planners

Planners are the chiefs, commissioners, and department heads who represent the authority of the department. Once it has been decided that a special event is going to need the support of emergency managers and public safety professionals, the planners should begin establishing what resources are needed and how they can be brought to the event without taxing existing services. A joint meeting with all involved in the special event should be conducted as soon as possible to begin bringing the resources needed for the special event. This will help establish what resources will be needed and who can provide them for the event.

It is important that all stakeholders are present in this meeting so an open discussion of operational objectives can be discussed, and which groups may contribute to fulfilling these objectives. By doing this, the groups can hear what others are bringing and help reduce any unintentional redundancy. If possible, these meetings should be conducted in person rather than through teleconferencing or Internet communications. This helps create an open dialogue between people and organizations and keeps the planners' attention on the task at hand.

Having open communications with the other planners during these sessions is critical. By having an open dialogue, they will be able to obtain information and true operational status by discussing what is needed and what can be brought to the special event. If possible, informal meetings with potential stakeholders for special events should be conducted. These meetings may function as a means to hear what other agencies are doing and what resources they have. These simple "meet and greets" may function to help build open communications between the planners. This degree of familiarity will help the planning sessions for future special events as they occur.

Managers

Managers are the field supervisors who will supervise operations during the special event. These may be fire battalion chiefs, police sergeants or lieutenants, EMS shift supervisors, or foreman of public works. Managers should possess specialized skills or have shown an ability to function well with the public and with other organizations. Unfortunately, the nature of special events planning requires additional training and personnel qualifications that not all managers possess. A 30-year member who has been promoted up internally may not be as qualified as a 20-year employee who has an interest in the field of special events planning or proven competencies with technologies. In order to ensure a smooth operation, it is necessary for a selection process to be conducted by the planners in order to find who has the knowledge, skills, and abilities to accomplish the objectives and be highly efficient during the special event.

Consider possibly using managers for special events who have previously been operators at special events. These managers will understand the dynamics and challenges of working within a special event. Those skills, combined with an understanding of the proper management of a group, would produce a well-qualified manager for a special event. It is understandable, however, that this is not always feasible, especially when using groups at a special event is new. Discretion should then be used in selecting qualified managers from the existing roster.

The managers for the events can function in a variety of positions. Depending on the number or personnel resources that are being used, multiple managers may be needed to conduct the special events operations. The effective span of control should be considered when determining the number of managers. In order to have an effective span of control, no more than seven units should be assigned to an individual manager. The optimal target should be four or five per manager. If strike teams, task forces, or small operations groups are being used for a special event, each group may qualify as a single entity for reporting responsibilities. These suggestions are consistent with the National Incident Management System (NIMS).

Managers may also function over responsibilities other than simply managing personnel resources. They may function as logistic officers handling equipment, food, and perishables during a sustained event. Managers may function as communications officers, dispatching the special events resources to areas that need it. They may also function as liaisons with other organizations or as the Public Information Officer for the event. Obviously, not every manager will have these abilities, so the most qualified person should be placed in these roles.

Managers should have a working knowledge of the organization that they arise from so that they can bring their considerable expertise and experience to help produce a good response for the event. Each manager should operate within their specialty and not cross over into other specialties for the response agencies. Obviously, you do not want a fire supervisor making law enforcement decisions. Ultimately, the responsibilities of managers are flexible with the needs of the event and it is up to the planners to determine what their objectives are.

It is very probable that the managers will be the individuals who are operating within a JOC or a unified Command Post (CP) should either be established during the event. This is dependent on the planners' directions during the planning phases for the special event. Should the managers operate out of the JOC or CP, they must have certain assets at their disposals.

Any person operating out of a JOC/CP should have access to all media outlets (television, radio, Internet). Sometimes these are the quickest ways to find out about something happening within the special event venue. The JOC/CP should have access to maps showing the physical layout of the area. A gridded map would be a much better tool than a simple aerial photograph. A grid layout helps to define boundaries of the operational area and produces a faster response than giving operators directions based on landmarks or approximate locations. This allows the operators a quicker way to determine where they are to respond to an incident or issue demanding their attention.

The JOC/CP should also have access to an Incident Action Plan (IAP). An IAP is a single document that summarizes all pertinent preplanned information for an event. IAPs will be covered in greater detail in the documents section. This document allows a quick reference for all involved with information on the special event planning. IAPs are just the start of the documentation needed for special events.

The JOC/CP should also have access to communications to the planners, managers, and operators of the special event. Communication with the managers allows status reports as needed and adds a fluid dynamic to the special events planning so that it can adapt based on real-time data. Following a strict chain of command is important when operating with a JOC/CP. This keeps all players in the information loop. However, there are times when the JOC/CP may need direct information from the operators. This may occur because of a critical event about to occur or new information that is time sensitive. Giving the JOC/CP the capability to reach directly to the operators is good planning.

Operators

Operators are the end product of the tasks to be completed. The operators at a special event fulfill the tasks assigned to them and are ultimately the public representation of the managers and planners, who are determining the scope, objectives, and logistics, which are supporting the operators. The operators should have a clear understanding of what the group's objectives are and what tasks are beyond the scope of their capability. These objectives must be clearly delineated at pre-deployment meetings.

Selection of operators should begin with examination of the objectives of the group. These operators should be experienced in their respective field so that they may bring a degree of professionalism and ability to the special event. Years of experience should not be the only qualifier for selection for special event operators. The most seasoned operators may be selected for special events or newer members who have special skill sets that can aid in the mission objectives of the special event response. Planners should keep these considerations in mind as they are selecting their operators for an event.

Operators should have the ability to quickly adapt to the situation and overcome obstacles as they occur. To this degree, operators should have a degree of latitude in order to complete tasks but keep their group on target of the mission responsibilities. This is not always an easy task and when staffing small groups, selection of operators should be considered who can facilitate this task.

Special event operators should fulfill the minimum qualifications needed to conduct the operations determined by the various organizations they represent. They should also be able to interact well socially, communicate clearly with others, and have the level of physical fitness required to complete the tasks. Some special event venues are only within a building, whereas some require extensive walking through a number of city blocks. When creating a selection pool for these types of events, the planners and managers should keep these additional considerations in mind and inform potential candidates for special events operations of these requirements. Using letters of reference or interviews may help establish individuals who have the personality types that can fit well into these groups.

TRAINING

Before any type of special event program becomes operational, it is critical for established training protocols to be conducted. While special events utilize basic skills learned during candidacy or new-employee hiring, it also uses skills developed outside of this initial training. The training needed for special events is dependent on the agency but there are several common training objectives that may be covered by all agencies.

Foundational training should include training in NIMS courses. Many of the initial NIMS courses (ICS 100, 200, 700, 800) can be found online through FEMA's online training website (training.fema.gov) as well as numerous other job-specific courses. Some of these brief courses are already required in police and fire departments so that municipalities may apply for homeland security grants. These basic courses provide information on NIMS and the federal preparedness guidelines in place throughout the country. The planners, before establishing special events groups, may wish to use these quick online courses as basis for consideration into the groups capable of responding to special events. They may also wish to require additional online courses that are subject-dependent as prerequisites into the special event response teams. Other more advanced classes in the ICS series help round out this initial training with instruction on the numerous NIMS documents and IAP creation. ICS 300 and 400 are courses taught by various groups. The DHS Training Consortium does teach them on-site at no cost to the attendees at the Center for Domestic Preparedness in Anniston, Alabama.

One type of training to consider before establishing a group is a training program designed to prepare an individual for special events responses. This could be a once-a-year program that brings a presentation from each group involved with special events responses. A brief lecture or PowerPoint session by members of each group (fire, law enforcement, FBI, public health, public works, etc.) should be prepared and presented to those who are in consideration for the groups. This training could be a one-time training requirement for entry into the groups, or it could also be an annual refresher. This training could be limited to just one day in length or it could be longer. It is dependent on what the planners want to cover in the topics and how long they wish the training to be to establish sufficient qualifications.

Establishing this special event training as an annual refresher has several benefits. First, it allows opportunities for new members to be included into the group on a regular basis. This helps bring more qualified people into the pool of members who may work special events. Second, it provides a mechanism to remove people from the list should they no longer be interested. A yearly refresher also allows all levels of people to meet from various agencies, which helps establish the familiarity needed prior to actually working a special event.

Outside training may be required by planners or members wishing to join special events responder groups. Members may be required to attend specialized communications classes, incident command classes, or other classes deemed important for their specialization. DHS hosts numerous courses throughout the United States that cover many of the important homeland security/weapons of mass destruction topics. These courses, under the banner of the Training Consortium, are mostly free for attendees and provide good, consistent information on many topics that special events responders may need. Each state also allows for other specialized training on incident command, hazmat, fire, EMS, tactical law enforcement, or technical rescue. The authority having jurisdiction (AHJ) needs to make the decision of which courses it will recognize as acceptable training courses.

Utilizing these courses as the basis for special events responders helps to separate the special events responders from the normal operational responders and also has the dual benefit of producing individuals who have a variety of skill sets that can be utilized by the planners, managers, and operators. Those working in special events should be chosen from the "rank and file" of the groups they represent as individuals who have skill sets that make them particularly desirable to support these functions.

The more skill sets that each operator has makes that individual more desirable as a special events responder. A law enforcement agent who is also a bomb technician, or a firefighter who is also a paramedic, helps to increase the number of support functions to their operations group while reducing the number of people in the group. This helps to establish a manageable span of control for the operations groups. It also creates special events team members who have a much more comprehensive understanding of the goals and objectives that are to be met by the responders.

PLANNING

Before any planning sessions, a memorandum of understanding (MOU) should be established. The MOUs should contain information from each agency as to the resources that they bring to the special events planning, the duration of the support of the program, and the signatures of the person in charge of each represented organization. The MOUs create a legally binding document between all stakeholders and the limitations of their involvement. All organizations that may be stakeholders in these events may fall under the Emergency Support Functions (ESFs) found within the National Response Framework. ESFs include such positions to be addressed as firefighting, EMS, public works, as well as nearly a dozen other roles. The functions represented in special events planning is entirely dependent on the special event type and the planners' decisions for usage.

The special events responses can be of numerous types. An AHJ may choose to have a visual presence of police, fire, and public works personnel. They may want to establish uniformed officers throughout the venue, a first aid tent staffed by EMS workers, or public works equipment prestaged. The nature of the venue helps dictate how much of a visible presence is needed.

A large visual presence can create an atmosphere of unease, if for example a large law enforcement contingency is present, or there could be a public relations benefit in having a first aid tent or fire apparatus for the attendees to see. Larger venues such as festivals may effectively allow the placement of equipment or first aid tents that can be used to help handle the event. Other smaller venues may, however, not be able to support this type of presence.

Some events, such as sporting events or dignitary events, may dictate a more low-profile presence. The presence of equipment (such as first aid equipment or meters) effectively makes the ability to be covert impractical, yet the lack of uniforms may allow them to be more conspicuous. This low-profile appearance, which is incident-dependent (e.g., suits and ties for a dignitary event, sweatshirts and winter coats for an outdoor sporting event) may make movement throughout the venue more fluid and allow rapid access to areas for treatment, investigation, or intervention. This can help decrease the reflex time to handle a situation and to begin mitigation of an incident. It can also help eliminate any excess response if the incident is unfounded or of a smaller magnitude that can be handled by small teams, which does not require a full-scale response. Ultimately, having an appropriate response helps make the event more successful and the attendees less fearful.

VOLUNTEERS

Volunteer organizations exist as a potential manpower pool to help augment the response to a special event. Volunteers who receive minimal training can function to fulfill some of the low-skill positions that exist. Volunteers can be used to help manage access or egress from an area. They can be a network of ears and eyes for special event planners that can relay pertinent information back to a liaison with the special event managers. Medically trained volunteers can staff first aid shelters.

Although volunteer organizations may not be able to bring in highly skilled, highly trained operators to an event, they definitely bring a vast amount of resources and sheer numbers in volunteers to support the special events plan. They should not be discounted and should be considered depending on the venue and the event response objectives.

COMMUNICATIONS

When planning for special events, communication equipment and abilities is critical. A clear communications plan should be established and tested before the dates of the special events including testing equipment in the special event venue. This allows the planners, managers, and operators to discover where dead zones are and what limitations the equipment has. It may be necessary to build in redundancies for communications by using other technologies.

When selecting the equipment for the special event, the appearance of the operators within the special event venue should be considered. These operators may be directed to function in a more low-profile status than a uniformed response group. For this purpose, smaller, more compact radios with unobtrusive earpieces or throat microphones can be considered. However, if the group is to function in an overt fashion such as fast response fire or EMS response teams, normal radios may be appropriate. The JOC/CP may utilize portable radios or, if possible, use more of a base station–style radio if operating out of a mobile command post or if space is available in the JOC/CP. The use of repeaters may help boost the signals for radios and should be considered.

The frequency channels used for communications should be clearly defined within the IAP for all operators, managers, the JOC/CP, and municipal dispatch. The level of security for the radio encoding should be determined by the emergency managers and public safety professionals The frequencies do not need to be known by all surrounding regular agencies. Should communications need to occur between internal venue response and external responding resources, the JOC/CP may initiate this request. If bandwidth is available, a dedicated special event frequency can be utilized by those operating within the special event venue with the local common frequencies being available to the special event planners, managers, and operators.

At times, normal radio traffic may neither be available nor appropriate. Consider the ambient sound surrounding an operator that is functioning at a rally or concert as examples. When information is passed during loud ambient times, the message may be lost because of the sounds. Using e-mail or SMS (text) messages may function as an alternate source of communicating should the primary option fail. Communication plans may include alternate radio frequencies or cellular phone numbers to use should the primary lines of communications fail. With the prevalence of smartphones in communities, e-mail may be accessible as a form of communication but is not as reliable as direct radio or cellular phone communications. E-mail may be considered for establishing information before an event or having a digital version of the IAP accessible by the operators. If any technologies are to be used by the special events teams (radio, cellular phones, computers), they should all be provided by the agencies and not utilize personnel equipment.

Common terminology should be utilized as much as possible, and use of slang or inaccurate description of events or locations should be discouraged. "Ten codes" have long been used in public safety radio traffic, that were supposedly replaced by "Clear speech" after NIMS was enacted. But, today, many agencies commonly use "Ten codes" such as the word 10-4 which generally means OK, or message received. It helps to deliver a clear, succinct message summarizing the events that are occurring or what the operations groups are to accomplish. Using these communication checkmarks helps to create a professional atmosphere for all involved.

PAPERWORK

IAPs are the cornerstone document upon which operations at a special event should be based. IAPs are a singular document that collect basic information on the details of the event. During the planning sessions, the sections of an IAP should be completed by the appropriate responsible agencies. This information should then be sent back to a single group to compile and produce a single workable document.

After an initial draft is prepared for an IAP, the planners should carefully review the document and consider any changes. The nature of some IAPs cause them to be open documents until nearly the beginning of the event. Some of the information included on the IAPs that may be late additions are as follows: changing itineraries of VIPs for the special event, weather conditions, personnel assigned to various roles, as well as other issues as dictated by the event.

Some information is a bit static that is contained within an IAP. Information such as street closings, radio frequencies, access/egress points, mass casualty collection points, or prestaged resources such as first-aid tents may be planned well in advance. As is clearly evident, some of this information comprises last-minute additions, whereas others are in place significantly in advance of the actual special event.

The IAP should also contain emergency plans for both special event operators and outside personnel responding to the event. This includes preplanned staging areas or rally points should groups be split or become threatened because of a change in the venue status. These may occur if peaceful protests turn into a more mob-like behavior. It may also help if inclement weather appears and disrupts the attendees such as if they were outside and a rush occurs for people to get out of the weather. Having this information readily available, instead of informing after the event, helps responders to move into a response/recovery phase much quicker for all involved.

Other paperwork may be needed beyond the IAP for an event. An IAP only functions as a summary of information. Numerous other documents may be prepared before an event. Attending a FEMA incident command course that contains ICS 100, 200, 300, and 400 may be of particular benefit for managers and planners. Additionally, as resources are available, operators should be minimally trained in ICS 100 and 200, but preferably trained in ICS 300 and 400 as well. This helps all involved understand the documents that they are seeing.

BENEFITS

Planning for a special event is a challenge for even the most seasoned emergency manager or public safety professional. It requires a degree of flexibility while still maintaining a clear view of the objective goals of the response purpose. The development and establishment of many special events at local, state, and federal levels have brought numerous ideas and tactics that have been proven time and time again.

In developing these plans, planners should reach out to those in other agencies that have had experiences, both successes and failures, to see the lessons learned from the events. This can help streamline future planning for special events. This will produce an overall better-orchestrated response that can protect those who are in the special event venue.

The public relations for the agencies and organizations involved with a special event response plan can be beneficial for all involved. The public views a combined group including various agencies that are working well together. This creates a very positive public perception of the organizations involved. It shows government funds being actively used, instead of passively being consumed by sitting in the firehouse or squad car. It helps convey a message of concern for everyone's enjoyment of these special events.

Using special events planning as a foundation for long-term deployment exercises is also an added benefit. The small-scale special events that occur on a regular basis may be used to create foundations of cooperation between agencies. This cooperation may then be beneficial as the experience and lessons learned from the events are applied during a disaster or large-scale event. The overarching success of special events will then help establish a more robust and dependable emergency response community within the municipality that has trained and experienced personnel to help protect the public.

19

Coordinated Terrorist Attacks and the Public Health System

Raymond McPartland and Michael J. Fagel

Contents

Introduction 340
Case Study: Mumbai, India—November 26–29, 2008 340
    The City of Mumbai 341
    Pre-Assault Preparations 341
    Water Incursion and Landing 342
    Armament 342
    Deployment 342
    The Leopold Café and Bar—21:15 Hours 342
    The CST Attack—21:20 Hours 343
    The Taj Mahal Palace Hotel—21:40 and 22:10 Hours 343
    The Trident-Oberoi Hotel—21:50 Hours 343
    Taxi Explosion—22:00 Hours 344
    The Nariman House—22:25 Hours 344
    Taxi Explosion—22:45 Hours 344
Conclusion 344
Swarm Attack Characteristics 345
Terror Medicine 346
    Macro Level: Public Health System Issues When Facing a Coordinated Attack 347
    Micro Level: Untraditional Response Protocols 350
Conclusion 351
References 351

INTRODUCTION

Most areas of the health care system have been preparing for what terrorism experts call "low probability, high consequence" events. These events, such as the detonation of a radiological dispersion device or the dissemination of the plague virus involve the public health system simply by their nature and intent. Public health's preparation has been somewhat centered on events of a scientific nature where the release of some form of contamination would directly involve the response of the medical system and all its functions. A public health outbreak was considered the most dangerous and, at the post-September 11 timeframe, the most likely. But as history revealed, high consequence terrorism has not evolved as predicted. Other than a few isolated incidents that resulted in minimal casualties, the direct involvement of the public health system through an intentional release of biological or chemical contaminate has been hardly evident. With the public health world's lack of involvement followed the difficulty of obtaining a clear and updated operational picture and current threat perspective.

Terrorist groups today still express interest in using weapons of mass destruction whether it is biological, nuclear, radiological, or chemical in nature. However, interests aside, their mainstay to inflict something of a near mass causality incident (MCI) has involved the use of less scientific weapons like explosives and small arms or submachine guns. A central goal of most terrorist groups is the infliction of mass destruction in the form of deaths and injuries but they must weigh the successful deliverance of an attack against their capability and availability of resources. It is much easier to acquire, train, and execute a plan of attack that does not involve any scientific application or complex delivery system. The difficulty in creating and delivering something so dangerous as a biological or chemical weapon, while at the same time not having the ability to accurately target and control its delivery has caused problems for terrorist groups. That has pushed them into a "simpler" yet, equally effective method—*Complex, Coordinated Attacks.*

They have learned the effectiveness of making their attacks more complex by combining various simplistic delivery systems such as the use of firearms, improvised explosives, and multiple assailants; *coordinated attacks carried out using small unit tactics and firearms is the delivery of choice in the twenty-first century.* This was never more evident, nor more effective, than the Mumbai, India attack of 2008. By using ten trained men armed with small arms and explosives, the Lashkar-e-Taiba terrorist organization out of Pakistan was able to hold the city of Mumbai at bay for nearly 3 days.

CASE STUDY: MUMBAI, INDIA—NOVEMBER 26–29, 2008

The attacks in Mumbai, India, beginning November 26, 2008 and ending the afternoon of November 29 can only be described as India's 9/11. After suffering from more than ten coordinated attacks throughout the city and peninsula, Mumbai suffered more than 172 causalities at the hands of ten gunmen armed with small arms, makeshift explosives, and the will to kill.

Directed through the leadership of Lashkar-e-Taiba (LeT), one of Pakistan's most active and militant terrorist organizations, armed gunmen were able to hold a city of more than

13 million at bay and send a message to the world that such a low-tech, inexpensive attack is possible with little way to defend against it.

The City of Mumbai

Formerly named Bombay, Mumbai is the most populated city in India and the second largest city in the world. With a growing populous of over 13 million, projected to rise to 28 million by 2020, Mumbai acts as India's Mecca for entertainment and commercial exploration. Similar to New York City, it acts as a symbol of economic indulgence and cultural power drawing tourists and businessmen alike. Financially, Mumbai generates 5 percent of India's GDP, accounts for 25 percent of their industrial output, 70 percent of maritime trade in India, and 70 percent of capital transactions to India's economy. Essentially, Mumbai acts as India's economic hub. Their seaport handles half of India's cargo traffic.

Along with its financial footprint, Mumbai houses such attractions and soft targets as the City's movie and television industry, famous hotels and tourist attractions, multinational corporations, the stock exchange, and the Reserve Bank of India. Mumbai does not have a single face and all things combined, becomes a melting pot of occupations, attractions, communities, and varying wealth that attract the meagre sightseer to the corporate mogul.

Besides Mumbai's economic output and their cultural footprint, another crucial facet of their existence that needs mention is its reliance on an extremely populated urban transportation system. Statistics show that roughly 88 percent of Mumbai's residents rely on the train and bus system to move about the City on a daily basis. During their average rush hour a new train arrives about every 3 min and within seconds it is filled to three times capacity leaving little room to ride let alone make a mistake; Mumbai loses an average of ten passengers per day to train travel and transportation-involved accidents. This fact of overcrowding is pivotal considering the target selection by the ten men tasked with assaulting the city.

Pre-Assault Preparations

Some of the most notable aspects of the attack on Mumbai were its complexity and pre-planning. Plan building for the attack began mid-2007 leaving substantial time to gather intelligence through digital and personal means. Operatives knew careful pre-planning was crucial to the overall success of a mission involving multiple teams led remotely through what strategists call a "mothership" operation. Assaulters were provided information from other cell members designated solely as surveillance operatives on their various targets. Because they had to maneuver themselves through an unfamiliar city at night while engaging targets, they needed to be as familiar as possible with their environment. Information and digital images were provided via CD by the mission handlers to the attackers in order to accomplish this.

Operators of LeT formed a team of assaulters trained in small arms tactics, demolitions, and waterborne assaults. Narrowed from an initial group of 100 men, ten men were selected once the training was completed.

Water Incursion and Landing

Because Mumbai is a peninsula, surrounded by water on three sides, the choice of a water incursion was considered optimal. The terrorist assault team departed Karachi, Pakistan, on November 22, 2008, at approximately 8:00 am. They departed aboard a Pakistani vessel named *Al Huseenni*, a vessel owned and operated by one of LeT's chief operatives, Zaki-Ur-Rehman Lakhvi. They traveled roughly 30 hours, and then transferred to another boat where their weapons and equipment were waiting. Fearing discovery in Indian waters piloting a Pakistani ship, they ordered the hijacking of a third fishing vessel off the coast and the killing of its crew at around 3:00 pm, November 23. Once aboard the hijacked Kuber shipping vessel, they navigated Indian waters using GPS devices with preprogrammed waypoints. The planners understood a water landing afforded them the ability to bypass security checkpoints on land and air and approach the southern and poorest part of the city under the cover of darkness. Once within striking distance of the shoreline, handlers instructed the team to divide into five teams of two and board two inflatable rafts destined for two separate landing locations. They were to arrive undetected by sea on November 26 as dusk fell and begin their assault at the height of the commuter rush hour.

Armament

Each individual member of the assaulting team was given an AK-47 assault rifle, a 9-mm pistol, hand grenades, a substantial amount of ammunition, and dry fruits or rations. Each team was handed an improvised explosive device (IED) built of RDX explosive and a satellite phone.

Deployment

Once on shore, the assaulters separated into groups based on their preplanned attack schedule. The terrorists divided themselves into four attack teams, one with four men (Team One) and three with two members each (Teams Two, Three, and Four).

- *Team One* boarded taxis headed for the Leopold Café and Taj Mahal Palace Hotel.
- *Team Two* moved on foot to the Nariman House, a commercial-residential complex run by the Jewish Chabad Lubavich movement.
- *Team Three* headed to the Trident-Oberoi Hotel.
- *Team Four* began their assault on a major transportation hub, the Chhatrapati Shivaji Terminus (CST).

Those members of Team One, who took a taxi to their location, armed and secreted a small improvised explosive device consisting of RDX high explosive under the front driver's seat before leaving the vehicle. They were set to detonate nearly an hour later.

The Leopold Café and Bar—21:15 Hours

The Leopold Café and Bar has been in operation since 1871. It is considered a popular tourist site frequented by foreigners and westerners that visited Mumbai. Team One arrived on target via taxi, quickly assessed if the target was viable and open for attack, and began firing at patrons and workers. One grenade was detonated. The assault on the café was over quickly, lasting only 5 minutes. Team One then moved quickly, on foot, toward the Taj Hotel.

The CST Attack—21:20 Hours

The CST Railway Station is the headquarters to the Central Railways of India. More than 3.5 million passengers pass through the station every day. Security was minimal at best, littered with mostly unarmed police, ill prepared for what was to come. Team Four, consisting of two men, Mohammed Ajmal Amir Kasab and Ismail Khan, entered the CST through a newly built entrance and began firing indiscriminately at travelers waiting to depart. They were allowed to roam for nearly 90 min and engage targets at will before being forced to move on due to an increasing police presence. The resulting death toll was 58 dead, 104 injured.

Once the attack on the CST was concluded, Team Four navigated back alleys and entered the Cama and Albless Hospital where they began firing again. It is unclear if this stage of the attack was preplanned or simply a target of opportunity, but nevertheless, the assaulters capitalized on the find of a soft and unprepared target.

As they departed the hospital, they encountered four members of the Indian counterterrorism force arriving via vehicle. They ambushed the officers killing three, and stole their marked police vehicle. Blending easily with a marked police vehicle, they drove slowly firing at bystanders, reporters, and even a movie house.

After switching vehicles once theirs became inoperable, Team Four hijacked yet another vehicle and began heading toward the Trident-Oberoi Hotel. However, police radioed ahead and were able to set up a roadblock capturing one of the terrorists and killing the other. This team of two men was responsible for roughly a third of all deaths during the attack.

The Taj Mahal Palace Hotel—21:40 and 22:10 Hours

The Taj Mahal Palace Hotel was constructed in 1903 and is considered an historical icon and piece of local heritage. It consists of two wings with nearly 300 individually constructed rooms in each.

Team One entered the Taj through the north court entrance, avoiding security in the front of the building. Within the first few minutes, at least twenty patrons were killed. They linked up with the other two members of the team arriving from the Leopold Café attack and together armed their IEDs by the front entrance of the hotel. They traveled to the sixth floor and began engaging civilians and setting fires to slow rescue attempts and create chaos. Team One's members were able to maintain control of the upper floors of the hotel for nearly a full 40 h before Indian Special Forces made entry and neutralized them.

The Trident-Oberoi Hotel—21:50 Hours

The Trident-Oberoi Hotel was a more modern structure consisting of two towers connected by a walkway and formerly operated and owned by the Sheraton and Hilton. It was built with an atrium-like feel to its interior and had the most modern conveniences known in everyday hotels. The total number of rooms amassed 877.

Team Three landed via a raft at an alternate site relatively close to the Trident-Oberoi Hotel and entered the Trident through the front, engaging guests and staff immediately. They moved their way through the hotel and to the upper floors, specifically the 16th and 17th floors, where they were able to hold the location for nearly 42 hours before being killed by responding Indian Special Forces operative. They were able to kill at least thirty-three people.

Taxi Explosion—22:00 Hours

The IED consisting of RDX explodes in the first taxi in the Vile Parle area of the city, north of the attacks areas. The blast kills the driver and passenger and injuries a number of bystanders.

The Nariman House—22:25 Hours

The Nariman House is a contemporary structure of five floors purchased 2 years earlier by the orthodox Jewish organization called Chabad Liberation Movement of Hasidic Jews. Renamed the Chabad House, a Rabbi and family lived there with the task of accommodating Jewish visitors to the city.

Team Two approached the Nariman House on foot and, before making their assault, lobbed grenades into a nearby gas station to cause panic and distraction. They assaulted the front of the house and were able to gain control of the building after taking thirteen hostages, five of whom they killed. Once inside, they fortified their location and commandeered the high ground awaiting the police response. However effective this group was in grabbing media attention and making demands, they only accounted for eight fatalities and held rescuers at bay for more than 30 hours.

The Chabad House was the only location where demands were made and negotiations attempted. Entry coverage into the building was made available to the attackers through live feed television news crews delivered via satellite phone to the attackers from their remote handlers.

Taxi Explosion—22:45 Hours

The second IED explodes killing the driver, passenger, and two bystanders. At this point, the Indian police officials felt they were under attack by more than 100 men and that the entire peninsula was targeted due to the explosions now happening in the North.

CONCLUSION

At the conclusion of the attack there were more than 175 people killed and more than 300 wounded from explosives and small arms fire. This style of attack is known as a "swarm attack" using "Fedayeen," or self-sacrificing, tactics. This is not to be confused with a suicide attack. A Fedayeen raid requires the fighter to fight as long as possible, killing as many as possible, only dying at the hands of the enemy when the mission is complete and/or to avoid capture.

Their goal was the killing of as many people as possible while attracting as much attention from the rest of the world. The entire attack was watched on national television allowing the handlers thousands of miles away the ability to direct and redirect their attackers. Not only were the assaulters using satellite phones to communicate but also any means available taken from tourists and victims. There was no escape plan and very few, if any, negotiations. Police and emergency responders, including the public health system, were overwhelmed by the time the second attack began. Ill prepared logistically from the start, the police force of nearly 30,000 was initially stunned, which allowed the movement of the ten-member attacking team free access.

Outside of direct contact with small arms and explosives, the emergency responders were forced to handle multiple tasks of varying disciplines simultaneously whether it was IED disarmament, fire fighting, medical rescue, and counter-assault tactics. Emergency medical personnel (EMP), whether staged or deployed, were involved in all matters. In order to handle the situation, responders needed to handle each incident from the operational level individually while the command executives managed the larger picture of how all of the events were unfolding. A task more difficult than imagined.

SWARM ATTACK CHARACTERISTICS

The following characteristics of Mumbai-style attacks was compiled by analysts of the New York City Police Department's Intelligence Division and circulated throughout the response community in hopes of aiding with recognition and disruption:[1]

- Use of firearms, mainly military grade, and portable explosive devices
- Extensive intelligence collection and analysis prior to deployment in order to aid in the group's ability to select targets and prepare attacks
- Transportation to and within the target area while minimizing contact with the public and security forces
- Multiple teams attacking several targets simultaneously
- Diversions drawing security forces away from the prime target(s)
- Deployment of IEDs at entrances and exits or in or about target areas eliminating the need for an extra security team
- Tactics such as arson, IEDs, covering fire, and grenades allowing terrorists to reposition themselves
- Attackers reaching targets, seizing hostages, and prolonging the situation as much as possible
- Innovative use of communications technology hindering intelligence collection
- Handlers using open-source media to plan responses against security forces

TERROR MEDICINE

It is important for first responders to assess their calls for service differently than ever before. What was once considered routine must now be scrutinized as atypical. This scrutiny comes from the understanding that the new threat spectrum is considered multitiered and littered with potential hazards not traditionally seen by first responders. Those in the medical profession must now contend with the lone gunman or multiple school shooter, a possible improvised explosive device and possible secondary devices, and even multiple armed men acting in unison using military-style weapons.

Emergency medical responders and public health staff must maintain a heightened sense of awareness while maintaining appropriate and effective medical care. The ideal scenario of treating everyone equally may have to transition to only doing what must be done in order to save those deemed savable. This transition, or application of an *altered standard of care*, will happen after responders evaluate all the necessary avenues using a risk–benefit analysis. The combination of patient criteria and medical response must also weigh in the current threat and operational picture. *Terror Medicine*, a term coined by leading emergency management practitioner and public heath responder Shmuel C. Shapira, may now become the norm more than the rarity.

Terror medicine consists of four main areas; *preparedness, incident management, mechanisms of injuries and responses, and psychological consequences*.[2] All four areas must constantly keep in mind the duality of the threat presented as well as the overall threat or operational picture.

Preparedness can fall into the operational or procedural lane. Training in treatment, establishing protocol, and gathering supplies are all examples that fall under the preparedness umbrella.[3]

Incident Management consists of the system or organizations' ability to manage a crisis. Establishment of an incident command system, even one modified to fit the needs of the public health system, is critical to quicker scene stabilization and recovery.[4]

The practitioner's third area of concern is their ability to effectively understand the nature and *mechanism of injuries* and how best to *respond*. Traditional responders need to familiarize themselves with injuries from blast devices and large-caliber weapons.[5]

Finally, the EMS responder's ability to understand and manage the *psychological consequences* of a terrorist attack becomes paramount. Early psychological intervention is essential. If not appropriately treated during the first 6 months after an incident, patients may suffer irreversible stress disorders.[6]

Question ...

How would a coordinated assault using tactics similar to the 2008 Mumbai attacks affect a major metropolitan area's public health and response system? Public health as a whole is often overlooked due to the immediate need to prepare the law enforcement response instead. Experts agree, however, that without a functioning and effective public health response, a mass casualty incident of this magnitude will surely compound and cost additional lives even after the shooting stops.

To better assess this problem, an incident such as this along with its ramifications needs to be evaluated from both the *micro and macro levels of review*.

- The macro viewpoint is that of the public health system as a whole and all the interconnecting nodes that allow it to function effectively.

- The micro viewpoint focuses primarily on individual responders, a single hospital, or isolated function within the public health system.

Macro Level: Public Health System Issues When Facing a Coordinated Attack

According to the U.S. Department of Human Health and Services or HHS, the Public Health System in the United States is defined as

> ... the complex network of organizations that work towards fulfilling the public health mission of assuring conditions for a healthy population. (HHS, 2011)

Most laypersons see the public health systems as just a single hospital, clinic, or even physician they visit. Their recognition is normally based on their personal experience and interaction but few recognize the magnitude and complexity of how the system operates each and every day. It is a network consisting of various functions including response and transport, treatment, and recovery that rely on the operational capacity of each other. If one node, or connection point, in the system were to deteriorate, the effectiveness of the entire system will be affected and eventually its overall ability to respond to and care for patients in an effective manner will collapse.

Even robust systems like those in major cities like New York, Chicago, and Los Angeles, no matter how prepared, will initially be stressed to the point of failure during such an attack. It may be hours before a proper response and recovery can be put in place. For the sake of discussion, it is those systems we will be referring to. The assumption is that the same issues in response, recovery, and mitigation will be faced even more so and quicker in smaller jurisdictions.

The following is only a partial listing of problems a public health system may encounter:

Lack of immediate medical response due to the severity of the attack and the report of multiple attackers. Emergency medical personnel are normally not equipped with any form of ballistic protection nor are they trained in a tactical response to an event of this magnitude. Most agency protocols require an area to be safe before a non-law enforcement agent can make entry. With the reports of multiple attackers and the possibility the attackers could still be at large, EMS personnel will hold at the outer-perimeter and begin to stage, waiting to receive the victims. This delayed response, although understandable due to available protective equipment and training, will undoubtedly cost additional lives.

Access to the scene needs to be established and secured as quickly as possible in order to begin the triage process. With every second that passes, those injured that were initially viable can become unsavable.

Immediate recognition that the incident they are facing is part of a multitiered coordinated attack. When members of the emergency medical services respond to an incident, their concern is for the care of the patients immediately in front of them. Like many first responders, EMS responders tend to operate in an unintentional vacuum. Understanding that the MCI they are facing may be linked to other incidents around the city may not come immediately. What may become recognizable is the severity of the wounds and similarity to other more serious events in the jurisdiction.

No jurisdiction, even the busiest, is confronted with multiple incidents involving multiple gunshot wounds from high-caliber weapons all within minutes to hours of each other. Couple this with multiple attackers that appear organized and possibly using explosive devices, and the picture becomes a little clearer as long as communication is occurring between responders from different areas within the jurisdictions. Seeing these incidents in a repeated fashion should indicate that the attacks are not random and are more organized allowing the responder to increase their situational awareness. Recognition also lends for better future preparation in transport and receiving as similar incidents begin to unfold.

Triage and Casualty Collection Location. Once the first isolated incident concludes and injury assessments and prioritization of care begins, a *triage location* or *casualty collection point (CCP)* will need to be established by first responders. It is very likely, due to the initial inability to predict another coming incident, that the triage location will begin on site or very near to where the victims were injured. Without proper force protection of the medical responders, treatment will be minimized. As additional events begin to unfold and the bigger picture is revealed, triage locations will become more designated, tighter run, and may be further away in order to have established area security. This will hamper response and slow the treatment process. Security and site evaluation must be done when creating an initial and future CCPs.

Transport and Tracking of Victims. As victims are triaged, the need for immediate transportation will be necessary. Some immediate life-saving treatment may be done on scene but will be temporary and the victim will need to seek hospital care. Taking into account the style of current coordinated attacks, there is a strong possibility of another mobile attacker still in the immediate area or even that the transport vehicle, if unescorted, could enter into another attack area unknowingly. As events progress there are areas of the city that may become unnavigable. Security becomes an issue and requires the immediate establishment of safe routes to designated receiving points. But no matter how important, these safe routes may not be established prior to the areas being declared clear by law enforcement.

Transport of the victims corresponds easily with the tracking of who is being moved and to what location. Tracking becomes critical for three reasons; first, the injured are crime victims who require contact with law enforcement; second, they may have information or actionable intelligence that could further aid the responders and need to be debriefed as soon as possible; third, tracking of who injured are being sent to what hospital will determine whether that area hospital will be at the brink of their surge capacity.

Proper transport and tracking is contingent upon solid communication between the field teams and those receiving. Information such as the number of patients moved and their injury classification, locations being used for casualty collection, and areas in which entry is prohibited is key to keeping the incident manageable.

Lack of Experienced Medical Personnel and Adequate Trauma Hospitals. Many metropolitan areas have various levels of trauma receiving areas or trauma hospitals. These hospitals are medical receiving points designated for the most serious of wounds. They may have their own designated operating rooms and emergency staff trained to deal with high-level emergencies. Given the severity of the attacks, the variety and severity of the injuries, and the overwhelming number of victims, the local trauma hospitals will be overwhelmed in the first hour if not the first few minutes. Once those hospitals are filled to capacity and can no longer receive emergency room patients, incoming victims will be diverted to another nearby location. The staff at this location, although qualified medically, may not be equipped or experienced enough to manage the injuries they are receiving. This is even truer if the incoming victims are arriving en masse.

Public vs. Private Response and Individual Deployment. A majority of the public health emergency response community consists of volunteers and/or private sector–based staff. The level of experience and training for each responder possessing even a common medical baseline, will be different. Many volunteer responders are there for injury management and transport, not critical care under fire. Whether or not these resources will be deployed will be dependent on the incident specifics and command control decisions. Self-deployment of responders without a full understanding of the operational picture is also a dangerous issue. Private resources responding with the best of intentions may find themselves entering the kill-zone complicating matters even further.

Overall Diminished or Altered Standard of Care. Many plans established to respond to and mitigate a mass casualty event normally assume that the standard of care administered will be comparable to what is done in most other emergencies. But an event that unravels this quickly, involves these types of serious and immediate injuries, as well as unimaginable number of victims will compromise even the strongest of systems. Local hospitals are not mirror images of army field hospitals. The standard of care will be different and the ability to operate as current protocols dictate will become evident.

In August 2004, a panel of medical experts in the fields of bioethics, emergency medicine, emergency management, health administration, health law and policy, and public health was convened to discuss matters concerning altered standards of care to an MCI. Their key findings are summarized as follows[8]:

1. The goal of an organized and coordinated response to a mass casualty event should be to maximize the number of lives saved.
2. Changes in the usual standards of health and medical care in the affected locality or region will be required to achieve the goal of saving the most lives in a mass casualty event. Rather than doing everything possible to save every life, it will be necessary to allocate scarce resources in a different manner to save as many lives as possible.
3. Many health system preparedness efforts do not provide sufficient planning and guidance concerning the altered standards of care that would be required to respond to a mass casualty event.
4. The basis for allocating health and medical resources in a mass casualty event must be fair and clinically sound. The process for making these decisions should be transparent and judged by the public to be fair.
5. Protocols for triage (i.e., the sorting of victims into groups according to their need and resources available) need to be flexible enough to change as the size of a mass casualty event grows and will depend on both the nature of the event and the speed with which it occurs.
6. An effective plan for delivering health and medical care in a mass casualty event should take into account factors common to all hazards (e.g., the need to have an adequate supply of qualified providers available), as well as factors that are hazard-specific (e.g., guidelines for making isolation and quarantine decisions to contain an infectious disease).

7. Plans should ensure an adequate supply of qualified providers who are trained specifically for a mass casualty event. This includes providing protection to providers and their families (e.g., personal protective equipment, prophylaxis, staff rotation to prevent burnout, and stress management programs).
8. A number of important non-medical issues that affect the delivery of health and medical care need to be addressed to ensure an effective response to a mass casualty event. They include
   a. The authority to activate or sanction the use of altered standards of care under certain conditions.
   b. Legal issues related to liability, licensing, and intergovernmental or regional mutual aid agreements.
   c. Issues related to effective communication with the public, special needs population and patient transport.

Micro Level: Untraditional Response Protocols

Because the results of a coordinated attack will resemble more a battlefield than a city street scene, it forces the responder to behave more like a soldier than a traditional first responder. In order to do so, lessons need to be taken from past real world battlefield engagements and transplanted into the civilian environment.

The result—the creation of a new set of guidelines that balances the threat, civilian limitations of medical practice, a holistic look at the civilian population, and what they represent as victims and equipment and resource limitations. Based on the military version of combat medical response known as *Tactical Combat Casualty Care or TCCC*, comes the civilianized version, *Tactical Emergency Casualty Care or TECC*.

TECC tries to maintain a realistic look at the medical response to an event reminiscent of a military attack. It provides guidelines on care management of preventable deaths located close to, if not in, the area of attack. This is done while constantly reassessing the scene and allowing the responder to maximize their response potential while minimizing their risk.

Established by responders for responders, TECC is an avant garde look at medical response by civilian responders. Traditionally, medical response and immediate treatment would come to a halt awaiting an all-clear notice by law enforcement. TECC guidelines provide responders with the understanding that in an event of this nature, waiting may take longer than average due to the attack's scope. Treatment needs to be administered once the area is no longer declared a *direct threat area*.

TECC discusses three phases of care defined by the relationship between the responder and the impending threat:

1. Direct Threat Care or Care under Fire
2. Indirect Threat Care or Tactical Field Care
3. Evacuation Care[9]

Direct threat care involves minimal medical actions taken because the threat is still active and immediate. There is an emphasis on responder safety, evacuation of the wounded, and mitigation of any heavy bleeding if feasible.[10]

Indirect threat care involves more complex medical actions and more thorough patient assessment. The threat may be terminated with the possibility of other unknowns or may simply be no longer on scene. The area is not completely secure. Assessment and treatment priorities include major hemorrhage control, airway, breathing/respirations, circulation, head and hypothermia, and anything else that is problematic.[11]

Evacuation Care involves moving the aided to a treatment facility. This phase more closely resembles normal EMS operations.[12]

It is important to note that these phases are fluid. An area that was deemed an indirect threat area allowing EMS personnel to begin their work can suddenly become a direct threat area. This unstable and potentially life-threatening environment combined with the altered standard of care is something for which current first responders may be unprepared.

CONCLUSION

Medical responders need to constantly reassess their current threat perspective. Intelligence analysis is no longer just a law enforcement responsibility; it must be done by all responders, particularly those at the epicenter of the incident. By assessing the current operational picture and allowing the ability to engage in altered standards of care when the need arises, public health personnel will be in a better position to handle a complex, coordinated attack like that of Mumbai, India, 2008.

REFERENCES

1. NYPD Intelligence Division, *Terrorism Awareness Bulletin*, New York City Police Department, 2008.
2. S. C. Shapira, Terror medicine: Birth of a discipline, *Journal of Homeland Security and Emergency Management* 3(2), 2006, Article 9. PDF.
3. Panel on Terror Medicine and Domestic Security, May 2009. Hadassah Hospital, Jerusalem.
4. Environmental Public Health Systems and Services Research, June 2011. Capt. John Sarisky, Senior Environmental Health Officer, Environmental Health Services Branch, Division of Emergency and Environmental Health. http://www.cdc.gov/nceh/ehs/Docs/JEH/2011/June_Sarisky_Gerding.pdf
5. D. W. Callaway, E. R. Smith, M. J. F. Shapiro, and G. Shapiro. The committee for tactical emergency medical care (C-TECC): Evolution and application of TCC guidelines to civilian high threat medicine. *Journal of Special Operations Medicine* Vol. II, Edition 2 (Spring/Summer II). PDF, 2011.
6. M. J. Fagel, (Ed). *Principles of Emergency Management: Hazard Specific Issues and Mitigation Strategies*. CRC Press, Boca Raton, FL, December 2011.
7. World Health Organization (WHO). The elusive definition of pandemic influenza. March 31, 2011. http://www.who.int/bulletin/volumes/89/7/11-086173/en/
8. Altered Standards of Care in Mass Casualty Events. Prepared by Health Systems Research Inc. under Contract No. 290-04-0010. AHRQ Publication No. 05-0043. Rockville, MD: Agency for Healthcare Research and Quality. April 2005.
9. Scientific American. Deadly Pandemic Bird Flu Details Finally Are Made Public. *Scientific American* June 21, 2012. http://www.scientificamerican.com/article.cfm?id=pandemic-bird-flu-studies-public&WT.mc_id=SA_CAT_BS_20120622

10. Brown University. A SMART(er) way to track influenza. June 11, 2012. http://news.brown.edu/pressreleases/2012/06/smart
11. The White House. The National Security Strategy, May 2010. http://www.whitehouse.gov/sites/default/files/rss_viewer/national_security_strategy.pdf
12. CIDRAP. Studies: Antiviral prescribing reflects CDC guidance, flu activity. June 11, 2012. http://www.cidrap.umn.edu/cidrap/content/influenza/swineflu/news/jun1112antivirals.html

20
Hardening Tactics at Global Hotspots

Jennifer Hesterman

Contents

Introduction 353
Soft Target Hardening in the Middle East 355
    Life in a Compound 356
    Schools 358
    Churches 360
    Hospitals 360
    Hotel Security 361
    The Westerner as a Target 367
State Department Engagement 368
United Nations and Soft Targets 370
References 371

We will bankrupt ourselves in the vain search for absolute security.

Dwight D. Eisenhower

INTRODUCTION

At this point in the book, you may be feeling overwhelmed, with knowledge of the vast international and domestic terror threats against soft targets and the unique vulnerabilities at your institution or place of business. How can you possibly protect your facility and its innocent occupants from a variety of threats ranging from anthrax to a car bomb to an active shooter? The good news is that hardening best practices can be harvested from different countries and sectors and then cross-applied to our unique situations.

    Perhaps the best perspective a security expert can have is from afar, as an observer with a broadened perspective. I have been fortunate to live near London twice: during a period of IRA bombing activity in the mid-1990 s and, after 9/11, during al-Qaeda threats and attacks. I recently returned from living and traveling extensively in the Middle East, a region that has experienced years of repelling invaders and whose political unrest with the Arab uprising created a "nursery" for extreme Islam. I have witnessed the loss of freedom and privacy that is the slippery slope of security; however, the threat level is high and being safe is more important than parking right next to an entrance or getting through an airport security line 20 minutes sooner. Living in an Arab culture with a recent history of infiltration, unrest, and violence at the hands of our *mutual* enemy, radical Islam, has given insight into the myriad vulnerabilities we still have at home, and how much risk we assume in the name of convenience or aesthetics.

The 9/11 attacks propelled the United States into two major conflicts carrying exorbitant costs, which likely contributed to the crash of the U.S. economic system. Thousands of young, promising lives were lost on the battlefield, adding to the human toll of the 9/11 attack. Although the successful and tragic strike on the homeland rattled our entire country, the violence was isolated to New York City; one building in Washington, DC; and a field in western Pennsylvania. As much as the attack damaged our psyche and pride, we were back to "business as usual" in no time. Although obviously pained by the events of the day, most of America was immune to seeing the damage firsthand, or smelling the smoke lingering in the hallways of the Pentagon for months. The attack was viewed by many as a "one off" event that could never be replicated, and the attackers as "lucky" to pull it off. However, Americans questioned the government's ability to identify and communicate the threat. They demanded answers and action. The 9/11 Commission did a masterful job in shining a spotlight on our failures and vulnerabilities.

In an attempt to help manage the residual fear, the Department of Homeland Security (DHS) instituted a color-coded Homeland Security Advisory System, which ultimately failed because it barely budged from yellow (elevated risk), despite multiple threats and successful attacks. In a 9-year period, orange (high) was used five times, and red (severe) was used just once, targeted only to inbound aircraft from the United Kingdom. People started ignoring the alert system, which "taught Americans to be scared, not prepared" (CNN 2011). The color-coded system replaced with the National Terrorism Advisory System, or NTAS, was meant to provide more special targeted information regarding threats, yet it has also been met with a general sense of apathy. Months have gone by without alerts despite the unraveling of plot after plot against our country. In December 2015, Secretary Jeh Johnson announced that due to new, complex threats posed by terrorists, as in San Bernardino and Paris, DHS will modify the National Terrorism Advisory System. He stated at a speech given at a Defense One conference that the current alert system, which is triggered by "credible threats" inside the United States, was "becoming impractical in an increasingly fluid and blurred terrorism environment" (Rockwell, 2015). I am not sure the system was ever practical in the first place.

As we move farther from the attacks of 9/11 without the once-dreaded dirty bomb, chemical or biological attack, or other mass casualty event, the citizens of the United States have moved on psychologically. Even foiled plots barely get press coverage and have become routine news. Successful attacks are written off as an "anomaly." Of course, security officials know for every successful, mitigated or failed plot, there could be many others in planning stages but the overarching threat is downplayed in the interest of keeping the public calm.

For example, the first domestic terror attack against the grid, the shooting of a key power substation in California in April 2013, was kept quiet from the public for almost nine months (Hesterman and Kinzer 2014). The way we have chosen to adapt to this asymmetric enemy is to become comfortable with ambiguity and silence, hoping for the best instead of actively and aggressively searching for the "black swans." Except for security professionals, intelligence analysts, law enforcement, and first responders who lie awake at night thinking about the next attack, we have become complacent as a society regarding the threat. There are countries and regions better prepared and engaging in activities to protect their soft targets from physical to personal hardening, many of which we would be prudent to adapt. For instance, traveling in the Middle East, I witnessed firsthand the possibilities of achieving the desired security/convenience/aesthetics balance, and many of these techniques and tactics might be used in the United States as the terrorist threat grows.

SOFT TARGET HARDENING IN THE MIDDLE EAST

Security fatigue and complacency are something the Middle East knows nothing about after decades of conflict along religious and political lines. Often, the threat requires security procedures that would enrage most Westerners. For example, Ben Gurion Airport is one of the most secure in the world. The first ring of security is along the road to the airport, where your driver will stop at a checkpoint and be asked questions about you and your behavior (in Hebrew). He may be asked more questions after pulling up to the terminal. Cab drivers are part of the security fabric in Israel, as are barbers, street corner vendors, bartenders, and anyone else in contact with the public who may hear, see, or sense something out of the ordinary. Relationships between Israeli intelligence and law enforcement and their sources are strong and built over time. The bond may be cemented with occasional gifts and invitations to the home for dinner, activities that are illegal for law enforcement officers and their confidential informants in the United States. The hope is that, one day, the source will make a call and stop a chain of events leading to an attack.

The layers of security inside the airport are impressive, starting with a very direct, eye-to-eye interview about your visit. Every answer leads to another question—prying questions which are meant to intrude and trip you up. Under this intense pressure you will likely start stuttering and sweating, even if guilty of nothing. If you pass the interview, a sticker is placed on the luggage or ticket in a certain color and with a number. Next, you proceed to a screening area, where hand-carried luggage and checked luggage are thoroughly examined. All zipper compartments are opened and every object (including personal care items) is handled and swabbed to detect explosive material. After this intrusive search, the security screener walks you to the check-in counter, to maintain positive control of you and your bags. After check-in, it is time for the security line, which is determined based on the barcode number on the colored sticker. The first number is the key, with 1 being "no risk" to 6 indicating "high risk." In the higher risk lines, you will be subject to a very thorough pat-down, and possibly asked to undress. Once again, the carry-on luggage is unpacked and all electronic items scrutinized, including chargers. Your carry-on bags may even be taken away from you, into a separate room for inspection. USBs, DVDs, CDs, and even computer hard drives may be taken and screened in a separate room, and it is likely your photo and possibly fingerprints will be run through multiple databases. If your computer or phone battery is dead, and you have no way of charging it to prove you can turn it on and off (sign that it is not a bomb), the electronics may not accompany you on the aircraft.

Young men traveling alone are suspicious, as they may be terrorists. Young women traveling alone are suspicious, as they may have a Palestinian boyfriend who asked them to bring something onboard the flight. Checking no luggage is suspicious; it means you may have reason to believe the plane will not be landing. Those with tattoos or piercings or who are sloppily dressed are suspicious: Agents may wonder about their activities in the country in previous days. Anyone who has visited any Arab country buys himself or herself at least an extra hour of discussions. Those with an Iranian stamp in their passport may be detained much longer. Some people are even escorted by military police directly to their seat on the plane, with no bathroom or shopping stops en route. And to think U.S. citizens complain about the backscatter full-body scanners!

The point: there is a continuum of threat assessment and accepted mitigation tactics. The area in between determines the risk. Israel is a small country surrounded by larger ones with terrorist groups with one goal: to annihilate the Israeli nation. Naturally, their security procedures match the threat, and they are willing to forgo the privacy of travelers for security, even at the risk of repelling would-be vacationers and Holy Land pilgrims. The good news for air travelers departing from Ben Gurion Airport is there will be no hijacking, and a plane filled with innocent civilians will not be turned into a missile, as our U.S. aircraft were on 9/11. But security comes with a price.

Life in a Compound

The Middle East has a "compound" mentality. Housing compounds are luxurious, self-contained neighborhoods, often with one thousand or more residents. Western wear, smoking, and alcohol are allowed inside the compound. Our compound had two fabulous pools, a restaurant, an ATM, a beauty salon, a small grocery store, a clubhouse, dry cleaners, a large fitness center, tennis courts, and a daycare. Compound houses are typically constructed with concrete block reinforced with steel rebar, and all external and internal doors have security locks. Compounds are sealed up behind tall concrete walls with one gated entrance and several security guards, pop-up barriers, large sliding gates, and swing-up arms. Often the contract security is augmented by armed local or federal police. Security vehicles are parked outside the compound entrance and just inside, as well. Many compounds have a skewed entrance, to prevent a vehicle (perhaps laden with a bomb or shooters) from running the gate and making it into the housing area or to the clubhouse. Nearly every inch of compound property and exterior wall is covered by CCTV. The guards know both the vehicles and the occupants, and all visitors are stopped and challenged, forfeiting a form of picture identification that security holds until they exit. Outside, in the community, most businesses are set back from the road, also behind concrete walls and with gates and buzzers. There are few or no outdoor cafés. Malls, shopping centers, and grocery stores have robust security inside as well as outside in the parking lot. Most countries have resisted building parking garages and underground parking decks due to security concerns, opting for large parking surfaces despite the searing heat. The security posture is truly impressive.

Interestingly, there is little to no crime in the Middle East. Living under Sharia law means even expatriates are subjected to both punishments meted out in the Koran and given by a judge (sans jury). Third-country national workers value their jobs and do not want to risk deportation. There is basically no theft and no assault. In the city where I resided, there were two murders in 2 years—not bad for a population of a few million. Of course, there is very little alcohol and no drugs, gambling, pornography, prostitution, or weapons, so many of the provocations and instruments driving crime are removed from society. So, why the "compound mentality"?

The answer is terrorism. Since 9/11, al-Qaeda groups have repeatedly struck targets in Muslim countries, lashing out at moderates in the name of extremist Islam. The Arab Spring uprising also opened the door for radical terrorist groups to step in and leverage the chaos to their benefit. Countries such as Jordan and Egypt have experienced thousands of demonstrations since 2011, ranging from a few dozen peaceful participants to more than 20,000 angry protestors. Libya has also been the scene of violent protest, as well as acts of terrorism perpetrated by al-Qaeda loyalists. The Syrian civil war is raging, with spillover into Lebanon, where Hezbollah has stepped up its strikes against soft targets. Stability in Iraq is sliding backward as al-Qaeda splinter ISIS attempts to destabilize new governments and take advantage of U.S. and coalition withdrawal from military and security operations. Yemen is in steep decline, with AQAP battling the Houthi rebels for control of the country. Consider the entire Arabian Peninsula could easily fit inside our country's borders with room to spare; spillover of violence into relatively peaceful countries is a daily threat. In the crosshairs are schools, churches, hospitals, housing compounds, and marketplaces which must be hardened against the potential threat.

The Arabian Peninsula has also been the site of terrorist activity against soft targets. Saudi Arabia was the scene of horrific attacks on luxurious Western housing compounds. On May 12, 2003, three compounds in Riyadh were attacked at night with car bombs and al-Qaeda terrorists with machine guns. In all, 39 people were killed, and over 160 wounded. Proving again that shared religion and ethnicity are not a factor to religious extremists, al-Qaeda terrorists attacked on November 8 of the same year, using a vehicle-borne explosive device against an Arab-occupied housing compound, killing at least seventeen people and wounding 122 others. At 6:45 am on May 29, 2004, seventeen terrorists from the group, "The New Jerusalem," the precursor to al-Qaeda in the Arabian Peninsula, attacked three targets in the city of Khobar: two Arab oil companies and a Western housing compound, the Oasis. During the 25-hour siege at the compound, school children were shot and killed, and Westerners were either beheaded or had their throats slit when they tried to escape at night. The terrorists took eighty-eight hostages, injuring twenty-five and killing twenty-two.

A friend and colleague, Lt. Col. Ed O'Neal (U.S. Air Force, retired) was in the compound and struck five times by ricocheting bullets as he tried to lead others to safety. As soon as he and colleague Lt. Col. James Broome realized the compound was under attack, they ran to the tallest building on the property, a tower. After barricading the front door, they worked their way to the roof, gathering four workers in the building along the way and grabbing a cooler of water. They spent the next 11 hours hiding on the roof in the searing 120° heat, each man only taking a tablespoon of water to conserve supplies. Saudi forces did not arrive at the compound for several hours, while the terrorists kicked in doors, killing Westerners and taking hostages across the compound. They were told via cell phone it was safe to come out, so the two officers made their way down to the back door, removed the barricade, and opened the door to a hail of gunfire from the waiting terrorists. Both were shot by rounds from an AK-47, yet had to wait several more hours for rescue and transport to the hospital (Cobb 2005). Thanks to Ed's story, I now keep a supply of water on the third floor of our compound house and have thought of ways to quickly barricade the front door. Case studies of soft target attacks abroad are very instructional, not only for counterterrorism purposes but also for general safety and security practices in all businesses and homes.

The situation in the Middle East can also change at a moment's notice and are not always along secular lines, as might be expected. When the United States was poised to engage in Syria in August of 2013, the Gulf Cooperation Council (GCC) states were split about military intervention and whether they would give support with their own money and troops, and/or host bases where Americans and allies would launch their assaults against the Assad regime. As the crisis cooled, the GCC resumed discussions about tightening the alliance. However, the good feelings from the summit did not last long; Saudi Arabia, viewed by many in the region as a sleeping giant, unexpectedly rose in March 2014, confronting Qatar about its alliance with the Muslim Brotherhood. The United Arab Emirates and Bahrain joined Saudi Arabia, with the three countries withdrawing their ambassadors from Doha. Many feared the GCC was unraveling and speculated the impact might have on oil prices and relationships with regional wildcard, Iran. Saudi Arabia threatened a blockade on goods coming into Qatar, both on roads and the Arabian Sea. Qatar stockpiled refrigerated trucks filled with juice and other perishables, and locals braced for sanctions. The saber-rattling quieted over time, and rhetoric sounding like war was reduced to the level of a "brotherly fight." Regional turbulence is why the Middle East fortifies soft targets. And there are no color-coded charts or alerts; the threat is consistently high and met with the highest security.

Schools

In October 2014, a post in an ISIS online forum encouraged terrorists to stage lone wolf attacks on American and other international schools as well as teachers in the Middle East. The post generally stated that American teachers are easy prey because they commute without taking security measures. The forum user speculated the best way to attack a teacher is to target them in the bus they take to commute to and from school. He stated this is "something a lone wolf can do, even with using only one weapon and in minutes."

Teachers may seem an unlikely terrorist target, but ISIS considers them influential members of society. Naturally, they detest western education, and the education of girls and women. If boys are in school, they are not as easily available for recruitment as warfighters. Finally, ISIS believes the death of a teacher will cause panic and attacking schools would result in closure and students pushed out of the country. As we harden our schools against attack, know they are definitely targets for ISIS and other jihadist groups, and also comprehend their rationale.

Two months after the call for action against teachers in the Middle East, U.S. teacher Ibolya Ryan was stabbed to death in Abu Dhabi, United Arab Emirates shopping mall. The perpetrator was completely covered in female Islamic dress including gloves, and used a kitchen knife to stab Ms. Ryan in the stall of a bathroom. The suspect waited 90 minutes in the bathroom for the perfect time to strike, and was able to escape the mall. Hours after killing Ryan, the woman placed a makeshift bomb inside a soda can outside the front door of an apartment of an Egyptian-American doctor living in the UAE, but the device was discovered and dismantled. Due to extensive use of CCTV, the killer was caught within days. The killer was radicalized by ISIS and had a pro-ISIS social media page. She was convicted, sentenced to death and executed within months of the attack.

Schools heavily secured in the Middle East. For example, they typically have high, reinforced, blast protective walls and a very low profile. The name of the school is not advertised on signage, and no host country flags are seen from the outside. Students are transported in small buses with both a driver and another adult; the curtains are always drawn. Buses are cleared by security and then are driven into a special secure area obscured from public view, where children walk directly into the school. Students who walk to school, teachers, and visitors access the grounds through entry control points with standoff protection and are manned with at least two guards. All students and visitors pass through a metal detector. Despite its austere look from the outside, once on the property, schools are very open and lush, with gardens, fountains, and lovely buildings. The students don't need to worry about the potential for shootings, stabbings or attacks; they feel safe and likely have an improved learning experience.

Colleges have the same level of security. For example, at the Education City in Doha, Qatar, universities from the United States, including Georgetown, Cornell, MIT, and Texas A&M share a secure campus with stunning architecture and landscaping (Photo 20.1). This would be a different paradigm in the United States but collating schools is something we could consider with new construction projects.

Photo 20.1 U.S. universities in the secured Education City complex, Doha, Qatar.

Photo 20.2 The courtyard in Church City, Doha, Qatar, hosting Catholic, Anglican, and Coptic Egyptian Orthodox churches.

Churches

ISIS is responsible for destroying Christian churches in Iraq and Syria, as well as kidnapping and beheading Christians. In October 2015, ISIS released a video threatening U.S. Christians. The film concludes with this ominous warning: "This is a message to all Christians in the East and West, and to the defender of the Cross – America: Convert to Islam and no harm will come to you. If you obey, you will have to pay the jizya poll tax" (Memri, 2015b). ISIS also released pictures of Christians taken hostage in Syria being forced to comply with a list of rules or face death (Memri, 2015a).

Many countries in the Middle East allow other faiths to openly worship and many with large expatriate populations are allowing construction of megachurches for their Catholic, Anglican, Egyptian Coptic Christian, and Greek Orthodox populations. One county uses the model where Christian churches are grouped together on property outside the city, secured in a compound-like setting with armed security and limited entry points. No crosses are visible from the outside and no signage indicates what is contained within; this posture "lowers the heat" on the religious area, instead of taunting would-be religious attackers (Photo 20.2). I attended services at one of these "church cities" and felt very safe; in fact, my favorite was attending Catholic Mass in Arabic, with Arab worshippers mostly from Lebanon, Jordan, and Syria.

Hospitals

As discussed in previous chapters, hospitals are actively attacked worldwide by groups also threatening the United States; therefore, we should harvest best practices from abroad and use them at home. Hospitals in the Middle East are massive structures with plenty of standoff protection attractively disguised as grassy knolls, decorated stairs, fountains, and sculptures (Photo 20.3). The emergency room, ground zero for many terrorist attacks, is hidden and secured. You may not drive directly to the emergency room; all those who

Photo 20.3 A hospital in Doha, Qatar.

approach the building are pre-screened. Loading docks are only accessed by cleared vehicles and drivers. Security guards are constantly patrolling the grounds and cameras capture every angle. There is a strong security presence inside the hospital, as well.

Hotel Security

Although hotels are not specifically covered in this book, they are a soft target as structures relatively unsecured and filled with vulnerable guests and staff.

International terrorists have targeted hotels repeatedly in the last 10 years, with attacks resulting in thousands of deaths and injuries. Although these attacks seem far from home, remember the groups involved are actively threatening the United States, therefore we should study tactics and prepare accordingly. There were twenty-five hotels attacked by terrorists in 2015 (IntelCenter, 2015). The deadly year started off with an ISIS attack on January 27, 2015, at the 5 star, 28-story Corinthia Hotel in Tripoli, Libya. The attack started when armed men attacked security personnel at the hotel's guardhouse at around 9:00 am, killing three security guards. Two gunmen then shot their way into the lobby of the hotel, killing a desk clerk. While guests were evacuating out the back door of the property, including Tripoli's leadership who live in the hotel, a prepositioned car bomb was detonated, killing several people. The siege inside the hotel lasted 6 hours, as the terrorists kicked in doors and killed guests. As Libyan forces cornered the shooters on the 21st floor, they gathered and blew themselves up by detonating grenades. Ten people were killed, including an American contract and former Marine, David Berry.

In May 2015, ISIS struck two luxury hotels with remotely detonated car bombs in Baghdad, the Cristal Grand Ishtar (formerly the Sheraton) and Babylon Warwick. The front rooms and lobbies of both hotels were severely damaged, with fifteen deaths and more than thirty severely injured. Although the cars were both unable to get right next to the building, they were parked close enough and the blast's shockwaves still created significant damage. Consider the damage done by Timothy' McVeigh's Ryder truck parked 20–30 feet in front of the Murrah Building that killed 168 people and injured over 600. The fuel truck bomb destroying the Khobar Towers military dormitory in Saudi Arabia, killing nineteen and injuring hundreds of U.S. Air Force airmen was 300 feet away from the building. Naturally the size of the weapon is the major factor in the amount of death and destruction caused in a blast, but the proximity of the bomb to the building influences the pressure significantly. For example, a car bomb at 25 feet distance will put 40 percent less pressure on a building than a bomb even a few feet closer. Therefore, maximizing the standoff distance of vehicles is the most efficient deterrence and hardening approach for buildings. When traveling, consider evaluating hotels based on vehicle standoff criteria. How close could a car bomb get to the main building?

In just a 3-month period between November 2015 and January 2016, there were three deadly hotel attacks in Africa with westerner visitors caught in the mayhem. On November 20th, heavily armed gunmen aligned with al-Qaeda in the Islamic Magreb (AQIM) stormed the Radisson Blu Hotel in Mali's capital city, Bamako. The luxury hotel was host to diplomats working on a peace accord and other contractors. In all, nineteen guests were killed, including American Anita Data. In neighboring Burkina Faso, Africa, thirty people were killed when AQIM terrorists, including two women, set fire to the upscale Splendid Hotel and took one hundred hostages on January 15th. Ten people were also trapped in a restaurant in the hotel and killed. The attack spread to a hotel across the street when a gunman was chased inside by responding police. Neither hotel had a strong security posture, the entrances were completely unsecured. On January 26th, al Shabaab gunman stormed the beach at the Beach View Hotel on Lido Beach, shooting at an outdoor restaurant. As panicked guests tried to flee out a gate to the street, a car bomb went off. In all, twenty-five people were killed.

Indeed, hotels provide many security challenges; in the standard construct, vehicles may drive right up to the entrance and are often left unattended during check-in. Baggage is unloaded and brought directly into the lobby, unscreened, where, again, it may be left for periods of time. Most facilities include restaurants and conference/special event facilities that greatly boost the building's population during peak use. Outdoor dining areas, pools and beach access provide other avenues of approach and vulnerabilities for attack.

Due to a recent history of brutal terrorist assaults, hotels in the Middle East are extremely secure. In the fall of 2013, I was fortunate to spend 5 days on the ground (and off the radar) on a personal trip in Jordan with my daughter and a family friend. We first stayed at the Days Inn in Amman, Jordan. My security team was not too excited with this choice, since the hotel was one of three attacked in Amman by Iraqi al-Qaeda operatives on November 9, 2005. On that date, the Days Inn, Radisson SAS (now known as the Landmark), and the Grand Hyatt were simultaneously attacked by suicide bombers. In all, 59 people died and 115 were injured. At the Radisson, a husband/wife team entered a Jordanian wedding reception of nine hundred people and detonated their belts. Fortunately, the female bomber's belt did not work, keeping the casualty count lower than it might have been, but still killing at least thirty-eight. At the Hyatt, the lone bomber went to the hotel coffee shop and ordered an orange juice. After drinking the juice, he left and went to another room, donned his suicide vest, came back, and then detonated the explosive device. At the Days Inn, the suicide bomber walked into the entrance toward the bar/dining area. An alert employee noticed the man and called for security; the man left the building and detonated his vest in the street, killing three members of a Chinese delegation standing nearby.

Of course the Days Inn in Amman now is a fortress. Security officers and vehicles are visible outside the facility; inside, they are undercover, but "present." No vehicles may stop in front of the hotel; the former "pull through" area is now filled with decorative concrete planters and other barricades. All property visitors enter an external building where luggage and handbags are screened for weapons and explosives. Individuals may be wanded (wands are handheld detectors), and questions are asked about country of origin and travel plans inside Jordan and/or neighboring countries. At check-in, passports must be shown and the numbers are recorded in the registry. Yes, this type of security is inconvenient, but what better place to stay than a hotel previously hit and revamped with blast-proof windows, standoff protection, and its own small police detachment? Any would-be attacker will certainly drive by this property in search of an easier target in Amman. Dining in the lobby—steps from the scene of the attack eight years prior—was not even the least bit nerve wracking.

Driving southwest from Amman, we were stopped at several highway checkpoints and asked for our passports and itinerary. The travel agency provided a copy of the itinerary and told me repeatedly not to lose it, and after showing it for the fourth time in a day's travels, I understood. The concept of internal checkpoints is hard to get used to, but in Jordan, home to millions of refugees from Syria, Lebanon, Iraq, and Egypt, they need to know who is moving about the country. The Jordanians are a warm, open people, influenced by the ancient European trade in the southern port city of Aqaba, and they love Americans. Everywhere we were greeted with handshakes and even hugs; however, we were not exempted from the security checks.

Approaching our next destination, the Holiday Inn at the Dead Sea, we had a taste of resort security in the Middle East. At the nearby Red Sea, luxurious beachfront resorts were subjected to horrific radical Islamist attacks against vacationers: in 2005 at Sharm el Sheikh, in 2006 in Dahab, and in Taba, Eliat, and Aqaba, and in January, 2016 in Hurghada—all located at the mouth of the Red Sea where Israel, Egypt, Jordan, and Iraq intersect. As a result, the Jordanian Dead Sea resorts less than 100 miles away are extremely secure. The Holiday Inn had several layers of security, and entering the property was like driving on a very secure military installation. Our vehicle was thoroughly searched—trunk and hood opened, luggage checked. Our hands were swabbed to detect residue. We were asked questions about where we came from and where we were headed, all the while being scrutinized for nonverbals. However, as with most fortified places in the Middle East, the grounds inside were luxurious and the atmosphere extremely relaxed. We felt extremely safe and able to relax and enjoy the special, pristine beauty of the Dead Sea.

Another technique of U.S. hotel chains in certain Middle Eastern countries is using no signage and flying the flag of the host country at the entrance so the property more resembles an office building. I took the picture in Photo 20.4 of the property of a well-known U.S. chain in a major Middle Eastern city; see the multiple concrete barriers in front of the hotel and the addition of blast-resistant glass. The photo was taken from the parking lot, where all guests must park and walk through a gate, clear security, and then enter the lobby. At the center is the flag of the host country. No signage appears on the building.

Of course a terrorist can find a specific hotel if it is the primary target; however, it certainly does not hurt to lower the "heat" and profile of the building for opportunists. Since most travelers nowadays book hotels online and not by line of sight, properties and resorts

SOFT TARGETS AND CRISIS MANAGEMENT

Photo 20.4 A U.S. hotel in a Middle Eastern city.

in major cities might not be adversely affected from a business standpoint by less signage. Again, remember the theory is to lower the heat and make the property look too hard to penetrate so the bad actor passes by.

A few hints received from security professionals when staying in hotels in Middle East hotspots were to stay above the second floor, where blast damage would be less than on the ground floor, but not on the top floors, from which it would be difficult to evacuate. Also, I have learned the importance of asking for a room in the back of the building, away from the front lobby area, typically the scene of an attack. Furthermore, it is prudent not to have a room facing an enclosed courtyard since rescue vehicles and ladders cannot access the area. As in any situation, individual preparedness when staying at high-profile hotels is important, for example, counting the doors to the stairwell so it can be found in the dark and/or thick smoke, and testing emergency stairwells to ensure doors at the bottom are not locked. Travelers may want to reduce their time in public spaces; in the Mumbai attack and others, most people were killed in hotel restaurants and bars. Finally, consider staying in a smaller or boutique hotel in large cities, since these do not attract terrorists looking for high casualty counts.

Staff training is a key part of the hotel security framework. Recognizing the importance, the Department of Homeland Security produced an outstanding film in 2010 called "No Reservations: Suspicious Behavior in Hotels" (http://www.youtube.com/watch?v=ZLCCvjJJZ4w). Going one step further, Safehotels (http://www.safehotels.se/?lang=en), a company based in Sweden, is the first global organization certifying hotel properties based on standard criteria which measures security training and equipment, crisis management response, and fire and evacuation procedures. The certification benefits property owners and travelers alike. U.S. chains with properties certified by Safehotels include the Courtyard by Marriott, Radisson, and Clarion Hotels.

A company in Africa has studied recent terrorist attacks on hotels and designed a new type of secure hotel. The Southern Sunshine Hotel factors the issue of terrorist attack into its design; for example, all of the rooms directly "communicate" with the outside through large openings, making escape and rescue operations easy in case of terror attacks or fire outbreak. The staircase area and the corridor access to all rooms are also open to the outside so there is no cover for terrorists to hide and conduct a siege of the building. The hotel has its own water supply and generator in case of emergency. Finally, the hotels will only have twenty-eight rooms, making them lower profile and less inviting to terrorist attack (Gichuhi 2014).

However, we must understand that as we harden facilities, terrorists will try to exploit other avenues of approach. For instance, the beach attack at the Riu Imperial Marhaba Hotel in Tunisia gives lessons on the importance of extending security to the pool and beach area, considering it an extension of the building to protect visitors and staff.

Case Study: The Marhaba Hotel Beachfront Attack, Tunisia

At approximately 12:05 pm on June 26, 2015, 23-year-old engineering student Seifddine Rezgui opened fire on the tourist-filled Tunisian beach on the property of the Spanish-owned, 5 star Riu Imperial Marhaba Hotel, 10 kilometers north of the city of Sousse. The chaos unleashed by Rezgui would last for 25 minutes before he was eventually gunned down while fleeing the scene. Carrying a single Kalashnikov rifle with four magazines and three grenades, one terrorist killed thirty-eight people, inflicted damage on Tunisian tourism industry and economy, and threatened the stability of the newly elected secular government.

Attack Timeline

- 11:30–12:00: Gunman arrives on the beach dressed in black shorts and a black t-shirt, carrying a beach umbrella hiding his rifle and grenades. Speaks casually to several people. Calls his accomplice on a cell phone to tell him he is starting the attack, and throws phone into the ocean. Phone recovered later and plot evidence gathered by officials.
- 1205: Rezgui begins firing shots at tourists on the beach
- 1210: Continues carnage at the hotel pool, before making his way inside the hotel where most people fled for protection; throws a grenade in the reception area, killing one woman
- 1220: Returning to the beach, shoots some of the injured, then runs north past another beach property (Hotel Riu Bellevue Park) and then runs west on an access road trying to escape, is now being chased by locals
- 1225: Gunman and police exchange fire
- 1230: Gunman shot multiple times and killed

BACKGROUND

In the 2 years before the attack, the government failed to react appropriately to the growing threat to its fledgling democracy. Homegrown jihadists radicalized in Tunisian mosques were receiving militant training in Libya, and Tunisian fighters were rotating to and from the battle in Syria. The authorities knew radicalized youth were crossing the Tunisia/Libya border to receive training at a base affiliated with Ansar al-Sharia (Dearden, 2015). Upon completion of their training, the men were assigned to sleeper cells, which authorities assumed were in the country. Although the group has not completely joined ISIS, Ansar al Sharia's top official, Abu Abdullah al Libi, pledged allegiance to ISIS leader Baghdadi in early 2015 and several group members have defected (Joscelyn, 2015). The gunman who conducted the Marhaba Beach attack was unknown to the government; however, ISIS, which said it was behind the attack, knew Rezgui and referred to him as "our brother in the Caliphate, Abu Yahya al-Qayrawani" and released a photo of him with weapons (Carty, 2015). Officials later acknowledged Rezgui trained with the ISIS terrorists who attacked the Bardo museum (Dearden, 2015).

Before this attack at the Marhaba Imperial, two other tourist venues in Tunisia were targeted by terrorists. On October 30, 2013, a suicide bomber was turned away from entering the Riadh Palm Hotel, located south of the Marhaba Imperial. Instead, he headed to the beach and started running towards hotel guests when his suicide vest exploded (Mendick et al. 2015). On the day Khalil blew himself up, five others were arrested in Sousse on suspicion of planning similar attacks. His attempted attack on the hotel resulted in hundreds of police deploying to the resort area, but their presence dissipated as time went by and there were no further attacks. Just over a year later, on March 18, 2015, foreigners were again targeted by three jihadis at the Bardo Museum, with nineteen tourists killed. As a result of the Bardo tragedy, the President declared Tunisia was in a "war against terrorism." To demonstrate his resolve against terrorist who threatened the stability of his country and to reassure tourist they would be kept safe, he deployed troops to major cities. However, the beach resorts were largely left unprotected. Without significant security measures in place, the Marhaba Beach attack was easily executed.

Finally, in a bold warning that should have evoked a massive increase in tourist protection, the Ajnad al-Khilafa group, affiliated with Ansar al-Sharia in Tunisia, tweeted this in May of 2015 (Spencer, 2015):

> To the Christians planning their summer vacations in Tunisia, we cant accept u in our land while your jets keep killing our Muslim Brothers in Iraq & Sham (sic). But if u insist on coming then beware because we are planning for u something that will make you forget #Bardoattack. @Ajnad al-Khilafa

The tourism industry in Tunisia took a massive hit in the days and weeks after the Marhaba beach attack. Over twenty hotels, including luxurious $500 a night spas closed their doors in the 2 weeks following the attacks as tourist fled home and others cancelled upcoming visits. The Marhaba Imperial closed the hotel due to the "difficult economic situation in Tunisia" and are attempting to reopen in Spring 2016. The parent company of the Marhaba Imperial, the Riu Spanish hotel chain owns ten hotels in Tunisia. Given there were one million fewer tourists in the first 8 months of 2015 compared with 2014, and two million hotel nights expected to be lost in the year following the attack (Leach, 2015), the future of the Marhaba Imperial and Tunisian tourism looks bleak.

Although guests at the resort who witnessed the shooting thought Rezgui seemed to know the layout of the grounds, there is no evidence he conducted prior surveillance. And given the lax 25-minute response time to the scene by authorities, it is clear neither the hotel nor the local authorities were adequately prepared to respond. The Prime Minister's actions after the Marhaba massacre to close eighty mosques where radicalism and violence were preached (including Rezgui's Mosque) and the construction of a wall and sand trench for 100 miles along the border of Libya and Tunisia are steps in the right direction.

On July 10, 2015, in light of the Bardo Museum and Marhaba Hotel massacres, the Ministry of Tourism issued a press release outlining steps to protect tourists. An updated release was issued on October 7, 2015. The measures they directed put in place included the following (Ministry of Tunisia, 2015):

1. Tourist Police Units have been armed and operational inside and outside any tourist area from July 1, 2015.
2. On top of the above, 1,000 security officers are deployed to enhance these measures.
3. Securing airports and border checkpoints.
4. Stationed and patrolled units have been deployed on the beaches and in the surroundings of all the resorts.
5. Security around cultural and archaeological sites has been reinforced
6. Implementing a tracking security system of tours and excursions.
7. Strengthening of the supervisions of car parking areas for tourist vehicles in several tourist areas.
8. Running training courses for hotel security guards.
9. Allocating essential funds to provide hotels with necessary surveillance equipment.

Time will tell if a combination of these initiatives will keep ISIS and AQIM activity away from soft targets in Tunisia. As we protect hotels and other soft targets in the United States against our common enemies, ISIS and al-Qaeda, it's important to consider cross-applying the updated anti-terrorism standards in countries already experiencing terrorist massacres.

(*Source:* Miller (2016). Ms. Miller is a prior Air Force intelligence officer who is learning first-hand about Arab culture while residing in the Middle East.)

The Westerner as a Target

As a Westerner living in the Middle East, I was a possible target or, at the very least, a person of interest. Since I cannot carry a gun, a knife, or even pepper spray for self-defense when traveling in a Middle Eastern country, I have learned to lower my own profile and deter a possible attacker. As a blonde-haired, blue-eyed American, it can be difficult to blend in even wearing a hijab head covering, but there are ways to keep from standing out. For instance, I learned a few Arabic phrases to show willingness to be part of the culture, instead of forcing my language on the host nation populace. I wear loose-fitting, conservative clothing, ensuring my shoulders, knees, and chest are covered, and often add a local accessory such as a scarf or jewelry. I pull my hair back and sometimes cover it with a scarf if shopping or driving in a mostly Arab area. I purchased a vehicle in the country, with Arabic markings and plates. When parking in lots, I was careful not to leave anything in the open that is Western in style or belies my nationality. I have learned how not to make eye contact or start conversations with host nation citizens unless necessary. I use cash at the local stores and give a different first name, if asked. Kidnapping is a rising threat in the region, so great care was taken not to be alone at night on desert back roads, or in shops in the windy, chaotic souqs. Do not misunderstand—Arabs are kind, welcoming people and I enjoy their culture, being in their homes, and having them in mine. I just do not want to be the trigger for a radical Islamist in wait who is having a bad day and looking for a target. This type of opportunistic terror attack has occurred at the entrance of the mosques, museums and in shopping areas, so the best course of action is to lower the "heat" and attempt to blend in, instead of provoking. Interestingly, some Arabs I have spoken with feel the exact same way when visiting the United States, and they attempt to lower their profile by staying in large cities and altering their traditional clothing to look more western, if possible.

STATE DEPARTMENT ENGAGEMENT

All of us want to be able to say truthfully, "I did my very best" when looking into the eyes of grieving survivors and family members.

—Ambassador Prudence Bushnell (U.S. House of Representatives 2005)

Prudence Bushnell made this statement in a congressional hearing on May 10, 2005. She was the ambassador to Kenya during the al-Qaeda attack on the embassy in Nairobi on August 7, 1998, and was knocked unconscious by the bomb blast and cut by flying glass. After receiving basic treatment at a nearby hotel, she oversaw rescue operations. She attended the funerals of twelve embassy staff personnel killed in the bombing, a task no government leader, school principal, or church or business leader ever wants. She was also present at memorial services for the two hundred one local nationals killed in the blast, who were also her embassy employees. Imagine yourself in this scenario as a college president, church leader, hospital administrator or mall owner.

Embassies around the world were fortified in light of the twin bombings that day. At nearly the exact time the vehicle bomb exploded in Nairobi, the embassy in Dar es Salaam, Tanzania, was similarly attacked, killing eleven. However, in 2005, the State Department realized the growing threat to the families of their personnel stationed abroad, stating in a Congressional hearing:

But as embassy and consulate compounds are fortified, US government personnel and their families living and working outside those walls draw the aim of criminals and terrorists looking for the next tier of targets. So hardening official buildings is not enough. The security of soft targets hinges on the harder tasks of building personal awareness and sustaining institutional vigilance. Adding cement to the physical plant is an easy part. Precious lives depend on strengthening protections for America's human capital abroad. (U.S. House of Representatives 2005)

The State Department now has a budget to protect the family members of those serving abroad. Although a specific number is hard to pin down, the State Department receives roughly $15 million a year, including $10 million to increase security at American and international schools abroad. All State Department employees receive basic security training, including crisis response and self-aid and "buddy care" through programs like Simple Triage and Rapid Treatment. Those stationed overseas are given enhanced training on the local threat through the Security Overseas Seminar (SOS) Program. The Security Overseas Seminar, which concentrates on life in overseas environments, is mandatory for all federal employees and recommended for eligible family members. A similar, age-appropriate program, Young SOS, is offered to young family members in grades 2–12. Companies with workers and families abroad might consider using the State Department's curriculum as a best practice and model for their corporate security offerings.

Under the Soft Targets Program, the State Department is spending at least $28 million to improve the protection of U.S. officials and their families at Department-assisted schools from terrorist threats. This multiphase program provides basic security hardware such as shatter-resistant window film, alarms, and radios, and additional protective measures designed based on the threat levels in the country. Also, security walls, bollards, and gate systems were funded for the schools. As a parent of a high school student who attended one such institution in the Middle East, I am eternally gratefully for the funding my government has expended to keep her safe while we served in our military assignment.

In July 2015, the GAO released a report entitled "Diplomatic Security: State Department Should Better Manage Risks to Residences and Other Soft Targets Overseas." The report identified the six key categories of security standards to protect U.S. government employee residences from the threats of political violence, terrorism, and crime according to Overseas Security Policy Board (OSPB) standards:

1. An anti-climb perimeter barrier, such as a wall or a fence, and access control
2. Setback from the perimeter
3. A secure off-street parking area
4. A secure building exterior with substantial doors and grilled windows with shatter-resistant film
5. Alarms
6. A safe space for taking refuge

This criteria and related guidance could certainly be taken into account when protecting schools and home in the United States from crime and terror (GAO, 2015).

Ambassador Bushnell also recognized that in addition to physical improvements, the culture must change to protect the State Department's soft targets. She made three prescient recommendations that are applicable to all sectors:

1. Finding the right balance between living vigilantly and normally. People do not stay on high alert for long periods of time. Scare tactics are ultimately self-defeating, and administrative mandates such as checklists risk becoming rote exercises. To use a metaphor, our challenge is to ensure people are looking both ways before they cross the street, becoming neither paralyzed nor indifferent to the oncoming traffic.
2. Maintaining a consistency of funding and attention to security issues.
3. Changing the ethos, perception and the image of the job. Employees can count on experiencing evacuation, civil unrest, kidnapping, natural disasters, assassination, terrorist attacks, biochemical attacks, and other crises.

Certainly, the State Department's mission overseas is far more vulnerable to terrorist attack than a school in the U.S. heartland. However, the Ambassador's insight as someone who lived through an unexpected and unanticipated terrorist event against a hard target is thought provoking and a departure for further discussion. She ended her testimony with this statement: "My colleagues are fiercely patriotic, willing to put themselves and their families at risk in order to make a difference on behalf of the American people. At the very least, they deserve our best efforts to keep them safe."

UNITED NATIONS AND SOFT TARGETS

The United Nations has also entered the soft target hardening realm around the world. The organization's preferred manner of engagement is through a program called Public–Private Partnerships (PPPs), for the protection of vulnerable targets against terrorist attacks (UN 2009). Through its Counterterrorism Implementation Task Force (CTITF; https://www.un.org/counterterrorism/ctitf/), the UN's PPP task force harvests best practices, leveraging the connections and insights of INTERPOL about threat and response. INTERPOL also ensures timely dissemination of threat and attack information through the PPP construct. The United Nations also runs a Major Events International Academy aimed at those responsible for managing and securing high-attendance events such as the FIFA World Cup, Wimbledon, and multiday stadium concert events which are popular in Europe. Recognizing that event owners must ultimately decide how to secure their venues, the academy provides a good model for staying engaged and being a conduit for information and training, without imposing its will. Since the United Nations must respond during attacks such as the Boko Haram kidnapping of female students in Nigeria and the Westgate Mall massacre, they are investing money and time up front to help prevent soft target attacks around the world.

The State Department's Overseas Security Advisory Council (OSAC) is a PPP to promote security concepts and enhance cooperation between the state and U.S. organizations operating worldwide. OSAC focuses in particular on soft, vulnerable targets and provides a forum for the exchange of best practices and a platform for the regular and timely interchange of information between the private sector and the Department of State. At the city level, one very successful PPP is Project Griffin in London, a joint venture between the City of London and the London Metropolitan Police forces. Its charter is to advise and update the managers, security officers, and employees of large public- and private-sector organizations across the capital on security, counterterrorism, and crime prevention issues. The initiative focuses on protecting the city and the public from terrorist attacks. It brings together and coordinates the resources of the police, emergency services, local authorities, business, and the private-sector security industry, and helps with implementation of counterterrorism and crime prevention policies and procedures. Awareness and reporting of hostile reconnaissance and other suspicious activity have been dramatically increased across London since the implementation of Project Griffin, and this initiative could surely be replicated in our cities.

The Department of Justice (DOJ) has an Office of Community Oriented Policing Services that hosts PPP initiatives called "Building Private Security/Public Policing Partnerships to Prevent and Respond to Terrorism and Public Disorder." Community-oriented policing can help control crime. The private-sector and community members are encouraged to participate actively in the development of prevention strategies including those designed to counter terrorism. The DOJ is helping with the development of knowledge resource products (CDs, guidance documents, and videos) as well as training individuals from the public and private sectors.

The UN's PPP initiative is powerful because it enforces a critical concept related to soft target protection: the role of the private sector should not be limited to involvement in crisis situations. A proactive approach to partner and develop measures to prevent terrorism and enhance overall security can prevent and/or deter attacks. Of course, the type of threat information the government can share with the private sector is an issue and the sector must understand the importance of safeguarding such information. Also, law enforcement should not merely share threat information, but also assist with identifying and mitigating vulnerability. A trusting relationship is the key to successful PPP partnering.

There is much to harvest from soft target hardening efforts around the world, especially through the sharing of best practices. With knowledge of the threat, vulnerability, and worldwide response efforts, we are better prepared to address hardening planning and tactics here at home.

REFERENCES

BBC, Tunisia Attack: What We Know About What Happened. June 30, 2015.
Carty, Peter. Tunisia Hotel Attack: ISIS Releases Picture and Names Killer of the 38 Tourists in Sousse. *International Business Times*, June 27, 2015.
CNN. Color-Coded Threat System to Be Replaced in April. January 26, 2011.
Cobb, Sean E., Air Force Master Sgt. Bronze Star Awarded to Airman for Combat Actions. December 21, 2005.
Dearden, Lizzie. Tunisia attack: Gunman trained at terror camp in Libya with Bardo museum attackers. *The Independent,* June 30, 2015. http://www.independent.co.uk/news/world/africa/tunisia-attack-gunman-was-in-isis-sleeper-cell-and-had-terror-training-in-libya-student-says-10354930.html.
Gichuhi, Francis. Terrorism in Hotels and How to Prevent Attacks through Design. *A4architect,* May 14, 2014.

Hesterman, Jennifer and Sarah Kinzer. PG&E Sabotage. *The Counter Terrorist Magazine*, June 2014: 18–29.

IntelCenter. http://intelcenter.com/reports/charts/hotel-location-country-2015/#gs.rMNFRZM. January 26, 2015.

Joscelyn, Thomas. Ansar al Sharia Libya fights on under new leader. *The Long War Journal*. June 30, 2015.

Leach, Naomi. Tunisia's tourism industry plunges into crisis amid European travel warnings with up to two million hotel nights set to be lost over the next year. *Associated Press*. July 13, 2015.

MEMRI. ISIS in New Video to Christians. 2015a. http://www.memri.org/report/en/print8785.htm.

MEMRI. ISIS Issues Dhimma Contract for Christians to Sign, Orders Them to Pay Jizyah. 2015b. http://www.memrijttm.org/isis-issues-dhimma-contract-for-christians-to-sign-orders-them-to-pay-jizyah.html.

Mendick, Robert, Patrick Sawer, Richard Spencer and Hassan Morajea. Why Weren't We Told of Sousse Suicide Bomber? *The Telegraph*. July 4, 2015. http://www.telegraph.co.uk/news/worldnews/islamic-state/11718614/Why-werent-we-told-of-Sousse-suicide-bomber.html.

Miller, Nicole. Case Study: The Marhaba Hotel Beachfront Attack, Tunisia. January 2016.

Ministry of Tourism, Republic of Tunisia. A Formal Statement from the Government of Tunisia about the Steps Taken to Prevent Further Attacks which UK Tour Companies Can Use to Reassure Their Customers. http://www.tourisme.gov.tn/en/services/news/article/communique-a-lattention-des-voyageurs-et-des-to.html. October 7, 2015.

Rockwell, Mark. DHS chief plans changes to terror alert system. *Federal Computer Week*. December 7, 2015.

Spencer, Richard. ISIL-Linked Terror Group Warned of Tunisia Attack One Month Before. *The Telegraph*. June 30, 2015.

UN (United Nations). Counter-Terrorism Implementation Task Force. https://www.un.org/counterterrorism/ctitf/. 2009.

UN (United Nations). Public–Private Partnerships (PPPs) for the Protection of Vulnerable Targets. Edited by Counterterrorism Implementation Task Force, 2009.

U.S. Government Accountability Office (GAO). Diplomatic Security: State Department Should Better Manage Risks to Residences and Other Soft Targets Overseas. GAO-15-700. July 2015. http://gao.gov/assets/680/671305.pdf.

U.S. House of Representatives. Overseas Security: Hardening Soft Targets, May 10, 2005.

21

Developing Strategies for Emergency Management Programs

S. Shane Stovall

Contents

Introduction 374
Strategy Defined 374
Why Develop a Strategy? 375
    Justification of Program and Projects 375
    Mandated Goals and Objectives 376
    Program Development and Direction 376
    Work Plans and Assignments 376
Elements of the Strategy 376
    Introduction 377
    Mission Statement 377
    Vision Statement 377
    Organizational Values 378
    Organizational Chart 378
    Executive Summary 379
    Standards 379
        State Emergency Management Standards 379
        NIMS Requirements 380
        Emergency Management Accreditation Program 381
    Goal and Initiative Development 382
        Planning Team 382
    Format and Structure of Strategy Document 383
    Developing Strategic Goals 385
    Constructing Strategic Initiatives 387

Defining Challenges 388
Capabilities and Future Needs 388
Summary/Conclusion 388

INTRODUCTION

This book is focused on soft targets, and the prevention, mitigation, and preparedness aspects surrounding soft targets and their exposure to terrorist attacks and other multiple hazards. Soft target planning is often done as part of an overall Emergency Management program or Business Continuity program. This chapter will discuss how strategies can be developed to establish or enhance these programs.

As the profession of Emergency Management continues to advance itself, it is important that Emergency Management agencies and organizations have a clear programmatic direction. In order to accomplish this, it is critical that the Emergency Manager establish a solid mission, set of objectives, and initiatives to carry out these objectives. Each of these elements are components that make up a complete strategy that will provide direction and guidance to the Emergency Manager as they progress with their agency or organization. With the litany of mandates, laws, rules, regulations, ordinances, guidance, and other internal and external influences, managing an Emergency Management program can be daunting at best. No matter what the staffing is or what the resources are for an Emergency Management program, it is critical that a solid departmental or organizational strategy be developed. A strategy can serve as the foundation for an Emergency Management program. It is also considered good business practice. In this chapter, we will look deeper into what strategies are, why departmental and organization strategies are necessary, and the elements of an Emergency Management program strategy.

STRATEGY DEFINED

In order for an Emergency Manager to develop a strategy, it is important to know what one actually is. To achieve this, here are a couple of definitions that can help to clarify the term:

Merriam-Webster's Dictionary: (a) A careful plan or method. (b) The art of devising or using plans of strategems toward a goal.[*]

BusinessDictionary.com: (1) Alternative chosen to make happen a desired future, such as achievement of a goal or solution to a problem. (2) Art and science of planning and marshalling resources for their most efficient and effective use. The term is derived from the Greek word for generalship or leading an army. See also tactics[†]

In these definitions, some common words appear, such as "plans/planning" and "goals." For the purposes of this chapter, a hybrid of these definitions is going to be

---

[*] "strategy." Merriam-Webster Online Dictionary. 2010. Merriam-Webster Online. http://www.merriam-webster.com/dictionary/strategy. Accessed December 14, 2010.
[†] "strategy." BusinessDictionary.com. 2010. Web Finance Inc. http://www.businessdictionary.com/definition/strategy.html. Accessed January 6, 2010.

used—a strategy is a plan or method used to provide efficient and effective direction for use of resources in meeting organizational goals. In this chapter, the purposes for developing an Emergency Management program strategy, and the methodologies that can be used in their development, will be discussed.

WHY DEVELOP A STRATEGY?

There are various reasons for why a strategy should be developed. In this section, reasons for strategy development will be outlined. These reasons can be used to "sell" the idea for developing a strategy to other staff members (strategy development is not a solo project—this will be discussed later in this chapter). Also, whether they realize it or not, Emergency Management departments and organizations are businesses. They can be a division (department) of an overall business (jurisdiction or organization). They usually have a budget (no matter how large or small), and they hopefully are providing a service to their constituents (whether internally or externally). If the Emergency Manager begins to think from a business standpoint, the development of a strategy (business plan) begins to make sense. It does not matter if the Emergency Manager works in the public sector, the private sector, or the volunteer sector—the development of a strategy for the Emergency Management department, program, or organization is a necessity. Let's look at why.

Justification of Program and Projects

As the economy fluctuates, situations arise that may bring people to question the justification for particular programs and functions. Emergency Management is not immune from this. Many have had the unfortunate experience of cutbacks and reductions in funding that have left many Emergency Management programs with minimal staff and resources to deal with an ever-increasing workload. During such economic times, Emergency Managers often find themselves answering the question: "So, what are you working on when there aren't any disasters?" or "What is it that you do?" Emergency Management in its purest sense, as a profession, is largely misunderstood by the public and sometimes even by those that hired them. Emergency Management is a profession that usually is not on the "front lines" in the public eye on a day-to-day basis. Therefore, what the Emergency Manager does is typically not very obvious to the casual observer. For this reason, it is important that Emergency Managers have a written document showing what it is that their department or organization has done, is doing, and plans to do in the future. The strategy document helps to further legitimize programs and efforts put forth by the Emergency Management department or organization.

An Emergency Manager armed with a well-written strategy can provide this document to anyone who may be looking to justify further cuts in personnel, funding, or other resources. A properly written and updated strategy should be able to answer many questions as to what each person (by position) is working on and plans to work on, as well as what current and future programs will require funding. Obviously, a strategy does not prevent or deter potential questions, scrutiny, or even cuts to programs, personnel, and funding. However, it does provide a plan and, if written correctly, a justification for personnel, funding, and other resources that help to support the Emergency Management programs.

Mandated Goals and Objectives

Emergency Managers often find themselves inundated with projects that they have to do in order to meet local, state, or federal requirements. These requirements can come in the form of plan review and acceptance, grant deadlines, and compliance with the National Incident Management System (NIMS), just to name a few. The development of a strategy will assist the Emergency Manager in organizing resources and determining tasks required to meet these mandates. This is critical in order for the Emergency Manager to be proactive rather than reactive with their workload.

Program Development and Direction

Development of a departmental or organizational strategy can give the Emergency Manager a solid idea of what has been accomplished, what is currently being completed, and what needs to be done. The development of a strategy (as will be seen later in this chapter) also allows the Emergency Manager to take a look at overall goals for the department or organization and what steps or tasks are needed to be completed in order to reach that goal. The thought process in developing these tasks and subtasks, or initiatives, is typically very intense and time consuming. However, the end product gives a clear look at what needs to be accomplished in the present and in the future.

Work Plans and Assignments

Once goals and initiatives to meet each goal are set forth, personnel and resources can be assigned to each one of them. The Emergency Manager, at this point, can begin to get a clear picture as to what resources and personnel may or may not be necessary as the organization or department moves forth. This is very important in staffing and budget planning. During this point, a Work Plan can also be developed that outlines each task and subtask within an initiative and assigns personnel and other resources to it. This can be a separate document that can also be used to drive the Performance Objectives of staff (to be discussed later in this chapter).

The preceding should provide the Emergency Manager with ample reasons for developing their own departmental or organization strategy. Not only is it a good business practice, but it also can provide justification and help to outline details of the direction that the department or organization will be taking in the future.

ELEMENTS OF THE STRATEGY

The strategy document should be separated into sections and easy to read. One should feel free to use pictures, charts, and other graphics to illustrate and emphasize any points that need to be made, particularly those that quantify any accomplishments (i.e., amount of grant funds brought in, number of plans written or reviewed, number of exercises held). The end user/reader should certainly be kept in mind as the strategy document is developed. The following outlines sections that the Emergency Manager many want to include in their overall strategy document.

Introduction

This section gives the Emergency Manager an opportunity to explain the departmental or organizational responsibilities. This will be the first page that the reader will see, other than possibly a table of contents. The Emergency Manager must ensure that they say everything that they want to demonstrate about their department on the first page. Some readers may not read past the first page, and this will be the only impression of your department or agency that they get.

Mission Statement

The Mission Statement of any organization is essentially a concise definition of what the Emergency Manager organization or department does. The Mission Statement should be no more than two sentences long. There should be no extraneous wording. Just a simple answer to "What do you do?" Here is an example:

> The City of Plano Department of Emergency Management will serve the citizens of the City of Plano by directing and coordinating emergency management programs to prevent/mitigate, prepare for, respond to, and recover from emergencies and disasters.[*]

This statement is to the point, and clearly defines the role of the organization.

Vision Statement

The vision statement differs from the mission statement in that it gives readers an overall view of the direction of the department or organization. This statement can be a little bit longer than the mission statement, but should be short in length yet encompass the future of the department or organization in general terms. Here is an example:

> The City of Plano Department of Emergency Management shall continue to develop and maintain a leading edge, all-hazards emergency management and homeland security program that encompasses all organizations in the public and private sectors. This will include citizens; government agencies at the city, county, regional, state, and federal levels; school boards, businesses (small and corporate), faith-based, and volunteer agencies. The program will coordinate the comprehensive community planning, training, and exercises needed to ensure maximum efficiency and benefit from hazard prevention/mitigation, preparedness, response, and recovery in order to protect lives and property in the City of Plano. The program will be professional, responsive, and shall strive to serve as a model municipal Emergency Management agency.[†]

---

[*] Stovall, Shane. City of Plano, Texas. City of Plano Department of Emergency Management Strategic Plan 2011–2016, p. 3, Plano: 2011. Print.
[†] Stovall, Shane. City of Plano, Texas. City of Plano Department of Emergency Management Strategic Plan 2011–2016, p. 3, Plano: 2011. Print.

SOFT TARGETS AND CRISIS MANAGEMENT

Table 21.1 The Organizational Qualities of an Emergency Management Agency

Excellence	Teamwork
Dependable	Adaptable
Commitment	Respect
Ethics and Integrity	Loyalty
Empowerment	Safety
Education	Communication

Source: From Stovall, S., City of Plano, Texas. City of Plano Department of Emergency Management Strategic Plan 2011–2016. p. 3, Plano: 2011. Print. With permission.

Organizational Values

This section does not necessarily need to be placed in any particular part of the strategy. However, when stating the department or organization's mission and vision statements, it may make more sense to list the qualities for which the Emergency Management agency is trying to achieve, as shown in Table 21.1.

Following each term, the organization should add a statement such as "We will … ." This helps to demonstrate the departmental or organizational understanding of the term. This section also helps to serve as a reminder of the qualities for which the agency wants to portray.

Organizational Chart

In this section, the Emergency Manager can insert the organizational chart for their department or organization. This will depict the staffing structure for the organization, and hopefully show the areas for which employees are responsible (Figure 21.1).

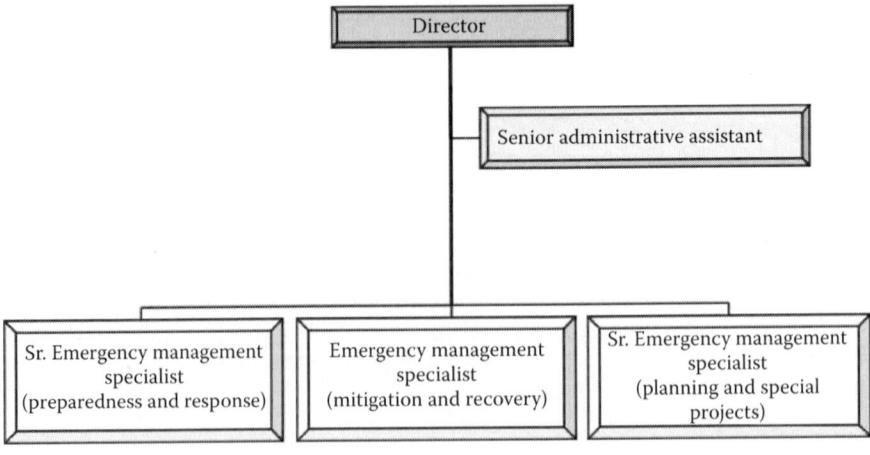

Figure 21.1 A sample organizational chart. (From Stovall, S., City of Plano, Texas. City of Plano Department of Emergency Management Strategic Plan 2011–2016, p. 3, Plano: 2011, Print. With permission.)

The preceding elements of the strategy can be considered preliminary elements that set the stage for the rest of the strategy. The following elements can be considered the body, or core, of the strategy document.

Executive Summary

This section should provide an explanation of the contents and structure of the strategy. The intent of this part of the strategy should also be to provide the readers with an opportunity to understand what they are getting ready to read in the rest of the strategy document. The Executive Summary should contain the following information:

Who is covered in the strategy—what department, organization, or program.
What the department does and what agency has generally accomplished in the past.
Why the strategy is being developed—internal requirement, to provide a foundation for future direction of the agency, other? All of the above?
When the strategy will be reviewed and updated.
How the strategy is laid out structurally in order to give the reader an idea of the layout of the strategy document (section names, chapters, etc.).

In general, the Executive Summary should be no more than one page in length. It is important to keep the word "summary" in mind for this section.

Standards

Emergency Management professionals are faced with many different types of standards that should be figured in when developing a program strategy. This section will take a look at a few of these standards, and why they should be considered when developing an Emergency Management program strategy.

State Emergency Management Standards

Many states put forth requirements that dictate the elements that local Emergency Management programs shall have included in their programs. Oftentimes, these requirements are in the form of requirements for inclusion in a local Emergency Operations Plan (EOP). Some states include training and exercise standards as well.

For instance, the "State of Texas Preparedness Standards" focuses on three activities that contribute to the overall readiness of a community. These three activities are planning, training, and exercises. The Texas Division of Emergency Management (TDEM) uses a series of collective assessments of local emergency preparedness programs to measure their effectiveness and determine where areas of additional emphasis are needed. The "State of Texas Preparedness Standards" set forth criteria for local jurisdictions to meet in order to meet certain levels of preparedness. These levels include Basic, Intermediate, and Advanced levels of preparedness. Each level of preparedness contains a specific set of criteria prescribed by TDEM with respect to planning, training, and exercise programs.

Compliance with these criteria does not imply that a jurisdiction's efforts in emergency management should be limited only to the criteria outlined in the "State of Texas Preparedness Standards." The standards that are set forth do not assess staffing levels, funding for emergency programs, the level of training provided for emergency responders, or the availability of response equipment or emergency facilities. These are seen to be local responsibilities, and are taken into consideration by the local jurisdictions.

Another example can be found with the Michigan Department of State Police Emergency Management Division. They have issued a "Local Emergency Management Standards"[*] document. This listing of standards includes many required and recommended elements that local jurisdictions should include in their Emergency Management programs. The elements in the Michigan document include requirements and recommendations for planning, training, exercise, and administration for the four phases of Emergency Management—mitigation, preparedness, response, and recovery.

Many times these state standards and requirements can be tied to funding streams, and compliance with these standards can affect the eligibility to access these funds. The funds can come in the form of grants, technical assistance, and disaster assistance. Therefore, it is important that the Emergency Manager become familiar with any State Emergency Management standards that may be set forth for their organization, and take them into consideration when constructing their Emergency Management program strategy.

NIMS Requirements

Following the attacks on the World Trade Center and the Pentagon on September 11, 2001, it was determined that a national approach to incident management would further improve the effectiveness of emergency response providers[†] and incident management organizations when dealing with any hazard, whether it be natural, man-made, or technological. This national approach would be applicable to all jurisdictional levels and functional disciplines in order to allow for consistency in emergency and disaster preparedness, response, and recovery efforts.

On February 28, 2003, the President issued Homeland Security Presidential Directive (HSPD)-5, which directed the Secretary of Homeland Security to develop and administer a NIMS. According to HSPD-5:

> This system will provide a consistent nationwide approach for Federal, State, and local governments to work effectively and efficiently together to prepare for, respond to, and recover from domestic incidents, regardless of cause, size, or complexity. To provide for interoperability and compatibility among Federal, State, and local capabilities, the NIMS will include a core set of concepts, principles, terminology, and technologies covering the incident command system; multiagency coordination systems; unified command; training; identification and management of resources (including systems for classifying types of resources); qualifications and certification; and the collection, tracking and reporting of incident information and incident resources.[‡]

---

[*] State of Michigan. Local Emergency Management Standards. Lansing: State of Michigan, 1998. Print.

[†] As defined in the Homeland Security Act of 2002, Section 2(6), "The term 'emergency response providers' includes Federal, State, and local emergency public safety, law enforcement, emergency response, emergency medical (including hospital emergency facilities), and related personnel, agencies, and authorities." 6 U.S.C. 101 (6).

[‡] United States. Homeland Security Presidential Directive 5: Management of Domestic Incidents. Washington, DC: White House, 2003. Print.

To provide the framework for interoperability and compatibility, the NIMS is based on the appropriate balance of flexibility and standardization in order to allow for consistent integration of multiple internal and external agencies during incident management. The major components of NIMS are

Command and Management—Mandates consistent use of the Incident Command System (ICS), Multiagency Coordination Systems, and Public Information Systems.

Preparedness—Requires standardized planning, training, and exercises; consistent methods for qualification and certification of emergency personnel; uniform response and recovery equipment acquisition and certification; and publication management.

Resource Management—Defines uniform mechanisms for inventorying, mobilizing, dispatching, tracking, and recovering resources over the life cycle of an incident.

Communications and Information Management—Identifies the requirement for standardized communications, information management (collection, analysis, and dissemination), and information sharing at all levels of incident management.

Supporting Technologies—Includes identification and acquisition of technology and technological systems that support capabilities that are essential to implementing and continuously refining the NIMS. These include voice and data communications systems, information management systems, and data display systems.

Ongoing Management and Maintenance—Establishes activities to provide strategic direction for oversight of the NIMS. This includes routine review and refinement of the system.

Since the establishment of the NIMS, the U.S. Department of Homeland Security has issued requirements that local, state, and federal governments are to meet annually in order to be compliant with the NIMS for each respective year. These requirements cover each of the elements described above. Many of the requirements placed upon state and local governments typically require many hours of work in order to be fully compliant. In addition, there is a continual refinement of each of the standards in order to clarify the "spirit and intent" of each requirement. Because of these factors, full implementation of the NIMS at the federal, state, and local levels of government is a phased process, and is expected to take several more years for all requirements to be fully mandated.

There has been some debate regarding the life expectancy of the NIMS requirements. However, these requirements and standards should still be considered as the Emergency Management professional constructs their program strategy. This will assist in long-term NIMS compliance for the organization.

Emergency Management Accreditation Program

The Emergency Management Accreditation Program (EMAP) is a voluntary, nongovernmental process of self-assessment, documentation, and independent review designed to evaluate, enhance, and recognize quality in emergency management programs. The accreditation process is intended to improve emergency management program capabilities and increase professionalism at the federal, state, and local levels of government, thus benefiting the communities that these programs serve. This process has been used internationally, and is also being used at colleges and universities. The goal of the accreditation is to evaluate an emergency management program's organization, resources, plans, and capabilities against current standards to increase effectiveness in protecting the lives and properties of residents.

The EMAP has been designed to facilitate compliance with a set of standards called the "EMAP Standard." The *EMAP Standard* is now a stand-alone standard. The *EMAP Standard* was built upon the *NFPA 1600 Standard on Disaster/Emergency Management and Business Continuity Programs* adopted by the National Fire Protection Association (NFPA). The *NFPA 1600* earlier adopted a portion of its program element framework from the Capability Assessment for Readiness created by the Federal Emergency Management Agency (FEMA).

The *EMAP Standard* contains sixty-four standards that are intended to indicate the components that a quality emergency management program should have in place. These standards are often difficult to meet and prove to be challenging for most emergency management agencies. The standards describe "what" a program should accomplish, but not necessarily "how" compliance with a standard should be achieved. This provides flexibility to the local governments in developing emergency management programs based around the *EMAP Standard*.

Although meeting this set of standards is voluntary, it provides a good "bar" or quality level for the Emergency Management professional to strive for. Therefore, strong consideration should be given to using this standard when developing an Emergency Management program strategy.

The standards mentioned above are a few examples that Emergency Management professionals must consider during the development of their overall strategy. The Emergency Manager must stay apprised of local, state, and federal rules, regulations, ordinances, statutes, and laws that may govern the way in which they conduct business. Some standards are voluntary standards, whereas others are mandatory. The Emergency Manager must be able to make a determination as to what is in the best interest of their organization, and include possibly a mixture of standards into their strategy planning.

Goal and Initiative Development

Planning Team
Developing the department or organizational goals for an Emergency Management program strategy should involve all members of the Emergency Management staff, and in some cases, any supervisors of the Emergency Management Coordinator or Director. The Emergency Manager may also decide that it is relevant to invite other Emergency Management stakeholders (i.e., Public Works, Public Information, Schools, etc.) to participate in the strategy building process. Having multiple inputs into what goals and initiatives shall be set forth is critical in gaining holistic understanding and buy-in to the overall strategy. Having multiple people involved in the strategy building process also ensures that the strategy is comprehensive and that all elements in the Emergency Management program are addressed.

Format and Structure of Strategy Document

Before goals can be established, it is wise to develop an outline or format of how the goals and initiative are to be laid out. There are several formats that can be used for structure. The decision as to what format should be chosen is a matter of preference of the Emergency Manager. However, the format should reflect the general foundational structure of the department or organization (if one has been developed). Some examples of formats for Emergency Management program strategies can include

Four Phases Format—As discussed previously in Chapter 1, in this approach the four phases (some now use five to include "prevention") of Emergency Management are used to outlined program areas. The four phases of Emergency Management include prevention/mitigation, preparedness, response, and recovery. Under each area, a list of pertinent goals and initiatives can be developed (Figure 21.2). Discussion of goal and initiative development will be explained later in this chapter.

Incident Command System Format—This approach can be used by Emergency Management agencies that have set themselves up in accordance with the ICS Structure. This format splits goals and initiatives into the four areas defined as General Staff in the ICS structure: Finance and Administration, Logistics, Operations, and Planning. This format is not, contrary to many beliefs, necessarily "consistent" with the NIMS requirements. The reason for this is that ICS is only one component of the overall NIMS structure and mandates. This format is merely a structure that can be used to organize a department, and therefore the strategy (Figure 21.3).

Figure 21.2 The four phases of Emergency Management. (From the Idaho Department of Education.)

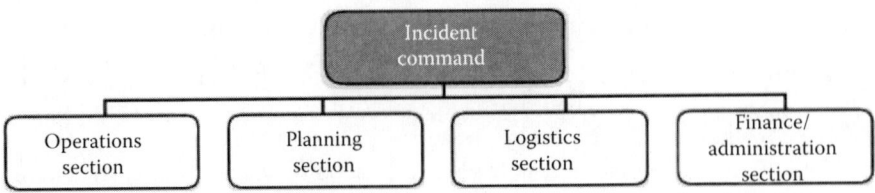

Figure 21.3 The incident command system general staff. (From the Federal Emergency Management Agency [FEMA].)

Emergency Support Function Format—Some Emergency Managers may decide to format their strategy in accordance with the Federal Emergency Support Functions (ESFs) outlined in the Department of Homeland Security's National Response Framework. Some states have also set forth ESF criteria for local Emergency Management Plans that may include additional ESFs. The ESF format is an option, although it can be a labor-intensive format (Table 21.2).

Operational Area Format—Some Emergency Managers may decide to organize their agency, and thus their program, in accordance with specific operational areas. These operational areas can be varied and can include areas listed in previous discussed format, or any hybrid thereof, including

- Administrative
- Natural hazards
- Man-made hazards
- Financial
- Technological hazards
- Public education
- Risk and hazards analysis
- Direction and control
- National Incident Management System

Table 21.2 The Federal Emergency Support Functions—The National Response Framework

ESF-1 Transportation	ESF-9 Search and Rescue
ESF-2 Communications	ESF-10 Oil and Hazardous Materials Response
ESF-3 Public Works and Engineering	ESF-11 Agriculture and Natural Resources
ESF-4 Firefighting	ESF-12 Energy
ESF-5 Emergency Management	ESF-13 Public Safety and Security
ESF-6 Mass Care, Emergency Assistance, Housing, and Human Services	ESF-14 Long-Term Community Recovery
ESF-7 Logistics Management and Resource Support	ESF-15 External Affairs
ESF-8 Public Health and Medical Services	

Source: From the Department of Homeland Security, 2010. With permission.

> **Strategic Goal 4**
>
> By May 2011, the Department of Emergency Management staff (all staff members) shall complete a review of and update the Comprehensive Emergency Management Plan as needed based on lessons learned and new requirements.
>
> > Emergency Management staff, as assigned, shall continue identification of planning gaps and work with emergency management stakeholders to attain information to address gaps. This shall be completed no later than the end of February 2011.
> >
> > By the end of April 2011, Emergency Management staff, as assigned, will complete initial training agencies on CEMP and specific roles outlined in the 22 Emergency Support Functions (ESFs).
> >
> > By the end of March 2011, Emergency Management staff, as assigned, shall perform Tabletop Exercise (TTX) that tests out the new CEMP and make corrections based on After Action Review (AAR) and Improvement Plan.

Figure 21.4 Sample of a goal and objectives. (From S. Shane Stovall.)

- Community Emergency Response Teams
- Public–private partnerships
- Hazard-specific sections
- Evacuation
- Multiagency coordination
- Drills and exercises
- Training

As can be seen, there are a variety of different formats that can be used, as well as sections, to make up an Emergency Management program strategy. There is no "silver bullet" or "one-size-fits-all" strategy format. The Emergency Manager must determine which format best fits their organization, and proceed with the development of their strategy document.

Once the Emergency Manager decides on a format, it is important to lay out the structure within the overall format. Figure 21.4 shows an example of a structure that can be used when laying out the content of the overall strategy. This example can be used in each section under the strategy format to provide for a consistent layout that will make the strategy easier to read and follow.

Once the format and structure of the Emergency Management program strategy document are complete, the Emergency Manager must begin working with the planning team to fill out the format and structure that have been set forth. This is done with the development of goals and initiatives to meet these goals.

Developing Strategic Goals

Before continuing with the development of the agency strategy, the Emergency Manager must understand how to develop a goal. This is one of the most misunderstood elements when building a strategy or a business plan. Many do not develop their goals correctly. As a result, the goals are misunderstood and many times never reached. A "goal" is defined as follows:

Merriam-Webster's Dictionary: "2.: the end toward which effort is directed: (Aim) ..."*

BusinessDictionary.com: "... a goal is an observable and measurable end result having one or more objectives to be achieved within a more or less fixed timeframe."†

As can be seen, a "goal" is considered an end result of some sort of effort or action, and typically is completed within a particular timeframe. Goals that are contained in the Emergency Management program strategy should include both short-term and long-term goals. The strategy is a living document, and the goals will drive the future of the Emergency Management agency.

As an Emergency Manager and their planning team begin constructing the goals for the strategy, there are some things to consider in order to ensure that the goal is written effectively. One of the best tools in developing goals and initiatives is to use the SMART acronym.

Specific—A specific goal is much easier to understand than a general or a vague goal. To set a specific goal, it is important to include answers to as many of the following questions as possible:
- Who is to achieve this goal (can be individuals or a team)?
- What needs to be accomplished?
- Where is this goal to be reached or accomplished?
- Why does this goal need to be accomplished?
- Which way is the goal to be attained, taking into account legal and other requirements?
- When should the goal be completed by (this is also covered under the "T" part of the SMART acronym)?

Measurable—Establish milestones for accomplishing the goal. Determine how the goal can be measured quantitatively so that successful completion can be measured.

Attainable—The planning team has to agree that the goal is attainable. Sometimes it is a good exercise to "reverse engineer," or start with the result and work backward to examine what personnel, funding, and other resources will be needed to attain your goal.

Realistic—Be realistic. Just as with being "attainable," a "realistic" goal is also critical (although many may see how these go hand in hand). A goal that is too lofty or that is not realistic or attainable is not worth developing. Unmet goals can lead to lower morale and unclear direction.

Timely—Time constraints or definitions are probably some of the most difficult attributes of a goal to establish. Emergency Management professionals are often inundated with projects that are influenced by entities external to their departments (i.e., grant deadlines, plan reviews, budgets). However, when possible, it is necessary to tie a timeframe to a goal. Without a timeframe, the goal loses its sense of

---

* "goal." Merriam-Webster Online Dictionary. 2010. Merriam-Webster Online. December 14, 2010 http://www.merriam-webster.com/dictionary/goal.
† "goal." BusinessDictionary.com. 2010. Web Finance Inc. http://www.businessdictionary.com/definition/goal.html. Accessed January 6, 2010.

urgency and may or may not ever be accomplished. Even if it is necessary to move a self-imposed deadline due to unexpected increases in workload or other factors, it is still important to add a time element to goals when possible.

Examples of goals that do not institute the SMART methodology include

- Complete EOP
- Ensure readiness of the Emergency Operations Center
- Establish pre-event contract for debris management

Taking these same non-SMART goals, here are examples of SMART goals:

- Establish a planning team to complete revisions to the EOP no later than July 1 (can add year).
- Develop a schedule for Operations Chief to test and complete operational readiness checks before the beginning of hurricane season each year.
- The director will work with Purchasing personnel to develop specifications and bid package for debris management contractor no later than April 10 (can add year).

Using the SMART methodology helps the Emergency Management strategy planning team by forcing a thought process that makes the goals clearer, manageable, and attainable. SMART goals will also be clearer to whoever reads the strategy document, and will help to develop a clearer direction for the Emergency Management agency.

Constructing Strategic Initiatives

Once the Emergency Management program strategy goals are established, there needs to be initiatives to help meet those goals. In some situations, these initiatives may be called "objectives." However, the word "initiative" implies that an action will be taken. The word "initiative" indicates that a leading action will be taken to meet an end result. For this purpose, we will use the term "initiative" to define those tasks that will help the Emergency Manager and their staff meet a goal.

Initiatives, much like goals, can follow the SMART acronym for their development. Developing initiatives that are specific, measurable, attainable, realistic, and timely can help to clearly outline tasks that are designed to meet an overall goal in the Emergency Management program. When constructing the initiatives, the planning team must think about action words. After all, the tasks required to meet a goal always require some sort of action. The following list suggests some of the action verbs that can be used:

Develop	Evaluate	Speak	Support
Construct	Assess	Execute	Negotiate
Build	Perform	Identify	Hire
Complete	Justify	Evaluate	Identify
Update	Propose	Monitor	Determine
Enhance	Compose	Test	Operate
Review	Generate	Modify	Collect
Change	Operate	Mobilize	
Exercise	Check	Classify	

The list of action verbs is extensive. However, they are necessary in order to show some sort of action taken toward achieving a particular goal. Figure 21.4 shows what a set of goals and initiatives can look like.

Defining Challenges
Following the development of goals and initiatives, it is important that the strategy planning team define any challenges that could arise when working toward completion of the goals and initiatives. Emergency Managers are faced with a litany of challenges that can include apathy, lack of funding, lack of personnel, external deadlines, politics, and having to deal with actual emergencies and disasters. Identification of these challenges allows the Emergency Manager to determine ways that these influences can be controlled (when possible). Once the challenges have been identified, it is important to make sure that the challenges are thoroughly explained. This is important so that readers can understand that there may be certain influences that can affect the timely completion of goals and/or initiatives.

Capabilities and Future Needs
In this section, the Emergency Manager has the ability to describe what their agency's capabilities are in terms of completing the listed initiatives and reaching the strategic goals. Areas that can be covered in this section include budget, facilities, equipment, personnel, and any other area that influences the capabilities of the Emergency Management agency. The Emergency Manager should point out any shortfalls that may exist. Although the Emergency Manager may be hesitant to point out any shortfalls, it is important to indicate them, as they may be obstacles in the completion of initiatives and goals. There is not one Emergency Management agency in existence that has all of their administrative and operational needs fulfilled.

Once shortfalls are identified, it is important that the Emergency Managers describe how they plan to fill any gaps. Solutions to filling gaps can include budget supplements, personnel hiring, acquisition of resources, attainment of grants, further coordination, and development of plans (just to name a few). Once this process is complete, the Emergency Manager should ensure that these action items are included in the Goals and Initiatives set forth in the Emergency Management program strategy.

SUMMARY/CONCLUSION

This section is the wrap-up section where the Emergency Manager can finalize the strategy and make any last statements that they want the reader to know. The Emergency Manager may also want to summarize all strategic goals that have been outlined in the strategy (sans initiatives). The Emergency Manager needs to realize that this section will be the place in the strategy that should leave a final impression about the Emergency Management agency, its current status, and what direction it will be taking in the future.

This chapter has been designed to give Emergency Managers and their planning team steps that can be taken when developing an Emergency Management program strategy. As mentioned previously, a departmental or organizational strategy should be treated much like a business plan for the agency. This strategy should remain a living document that is reviewed and refined over time. Without a solid strategy document, it is difficult for the Emergency Manager to explain what their agency does, and the vision for their agency's future. The strategy document can be a tool used to develop task lists for staff, and to build performance measures for employee evaluation. The preceding steps are not part of an "exact science." As discussed, there are many different formats and structures for Emergency Management program strategies. However, the intent was to help Emergency Manager in examining their program and developing a roadmap of sorts for their agency's successful future.

22

Soft Target Planning

Michael J. Fagel and S. Shane Stovall

Contents

Introduction	391
What Is a Soft Target?	392
Movie Theaters	392
Schools	392
Airports	393
Tailgate Parties and Sporting Events	394
Shopping Malls	394
Other Soft Targets	395
Soft Target Threat Assessment	396
Soft Target Vulnerability Assessment	397
Crisis Management	399
Training	399
Planning	400
Exercises	400
Command Centers	401
Summary	402

INTRODUCTION

In this chapter, we will take a look at the differences between hard and soft targets and discuss a layered planning approach that can be taken in order to protect soft targets from all hazards, to include a terrorist attack. We will also examine the different methodologies for assessing risks and threats affecting soft targets, and how a facility manager can plan and manage these types of risks and threats.

Before looking at protection measures, it is important to understand the difference between hard and soft targets. Hard targets are easier to define because they are typically heavily guarded and have security measures in place. In essence, a terrorist could be detected and met with some level of force as needed. Hard targets include military facilities, government facilities, and other facilities defined as critical infrastructure. Soft targets are often more vulnerable and usually do not possess the level of protection found in hard targets. Soft targets also usually house more vulnerable populations. Soft targets include facilities such as churches, schools, malls, sporting events, and amusement parks. A terrorist could potentially enter and attack these facilities with little, or no, detection or resistance. In this chapter, we will focus primarily on soft targets.

WHAT IS A SOFT TARGET?

In the news, we have seen the phrase "Soft Targets" used recently. What is a soft target?

Movie Theaters

One example of a soft target is a movie theater—a multiscreen movie theater. Think about the scene one may likely see in the theater on a night when a very famous film is going to open. If one goes inside and examines the surroundings, It is not difficult to think of what could be sneaked into this building without much effort.

Schools

Figure 22.1 shows the front of a school building—an example of a soft target. Some thought has been put into the protection of the outer envelope of this building.

Figure 22.1 A school building example of security. (From Michael Fagel.)

SOFT TARGET PLANNING

First, one may notice the concrete bollards that are placed In front of the entrance. This is to protect, to a degree, from vehicle intrusion into the front entrance of the school. Also, if one examines further, one can see that the entrance is has two sets of glass doors. It is almost like an airlock. To the right, once one gets through the first set of doors, there is another glass partition that allows some level of security. These are a couple examples of how the facility managers of this school have attempted to protect their building from an initial attack.

Airports

Another potentially soft target is an airport. Think about the check-in area during a busy time, such as when people are queued up during a flight delay or a holiday. The airport presents itself as a soft target because no one knows who could be walking into this facility with a weapon or a bomb in in his or her bags. If one thinks about the Los Angeles Airport shooting in 2013, where the first TSA officer was killed; the attacker was someone walking in with a duffel bag, a roller bag, a backpack, and a long rifle. They were able to push their way through the security screening point. As one looks at at Figure 22.2, look at the man with the black bag.

From the photograph, one cannot tell if the man in focus is putting the bag on the carousel or off the carousel. Could that bag contain contraband? Could it be loaded with explosives? Did this person take the bag off the carousel that was in a secure site? Did this bag actually come in from the outside? Is he placing it on the carousel to roll into the interior of the airport?

Figure 22.3 is a view of the baggage claim area at Las Vegas. When one looks at the number of people and the depth of the various carousels, it is easy to notice there that are several carousels here. In this section of an airport, there is always commerce and people

Figure 22.2 A potential security threat at an airport baggage claim carousel. (From Michael Fagel.)

SOFT TARGETS AND CRISIS MANAGEMENT

Figure 22.3 Baggage claim at McCarran International Airport in Las Vegas. (From Michael Fagel.)

milling about, coming and going through the doors to the far right and behind the area. All of these elements could be another level of a dangerous operation.

Tailgate Parties and Sporting Events

Another recurring soft target can be found each fall. Tailgating at football games and attending games at stadiums are prime examples of recurring soft targets. In Figure 22.4, look at the stadium in the background.

The people in the foreground of the figure are having the tailgate party. This certainly is a very, very soft target because of the potential of what could be contained in the vans without anyone's knowledge. One would assume that the vehicles have tents and other equipment associated with tailgate parties. However, one can easily think about the possibilities if a terrorist wanted to launch an attack of some sort, or set off a bomb. The proximity of the party to the stadium makes it easy for someone to "slip in" unnoticed. These factors make this venue a very vulnerable soft target because there is no access control or protection of this venue.

Shopping Malls

There have been stories in the news of malls that have been attacked, but stop and think about when would be the absolute worst time to attack them in America—Black Friday. During this time, shoppers line up outside stores to get the good deals and inexpensive items. People are lined up for hours before the stores open to get the bargains.

SOFT TARGET PLANNING

Figure 22.4 Tailgating before a football game can pose a risk. (From Michael Fagel.)

If one reflects on the Nairobi mall attack in 2013, one may remember that some of the perpetrators actually found work inside the mall, and were there for 2 to 3 months prior to the attack. Threats against shopping malls occur worldwide on a somewhat regular basis.

For malls in the United States, the biggest concern is that there are not security checks at the doors. Overseas, some countries station security officers at the entrances to malls. In the United States, it is doubtful, due to our culture, that the public would be open to seeing armed security officers checking everyone as they entered shopping malls. Malls mean different things to people overseas—they are symbols of wealth, symbols of ingenuity and prosperity. In America, profit is a factor that often outweighs security. A mall operator does not want to slow commerce down. They want people visiting their mall to feel comfortable as they spend their time and money at their mall. They do not want people to be inconvenienced by overt security measures. Additionally, they do not believe the cost-benefit exists for physical security measures. The Mall of America in Minneapolis, Minnesota, has over 600 exterior doors. Think of the security aspects surrounding that. Is it worth it considering they are a viable target?

Other Soft Targets

In the fall of 2015, there were brutal attacks on cafés and concert venues in Paris that rocked the world. These elements are extremely soft and very visible. These attacks represent the latest trend in terrorism.

Think of all of those elements that are available to us for mischief and mayhem. As has been discussed above, soft targets come in many shapes and sizes. We use them every day—a school, a movie theater, an airport terminal, a football stadium, a concert, a café.

SOFT TARGETS AND CRISIS MANAGEMENT

Due to concerns with the aesthetics and inconvenience of physical security measures and the perception that such measures are obtrusive, the challenges will continue to remain when crisis managers and planners try to protect the nation's soft targets.

SOFT TARGET THREAT ASSESSMENT

When determining the risks associated with soft targets, one must look at the target, or facility, as a whole. Using the Defense in Depth approach (Figure 22.5), one can examine the elements of a soft target that should be comprehensively examined when determining a risk or a threat.

The layered approach to the Defense and Depth methodology examines the following from outward to inward of the soft target:

- *Policies, procedures, and awareness*—Examination current policies, procedures, rules, regulations, and laws that could either assist or impede protection of soft target against an attack.
- *Physical*—Existing and potential physical protection measures. This could be a fence, a gate, or any other measure that could physically mitigate an attack on the soft target.
- *Perimeter*—Assessment of the outer envelope of the soft target. A perimeter could be another fence, doors and access points where people may not be able to get into the facility easily without being physically checked or without an electronic access device (key card, biometrics, etc.).

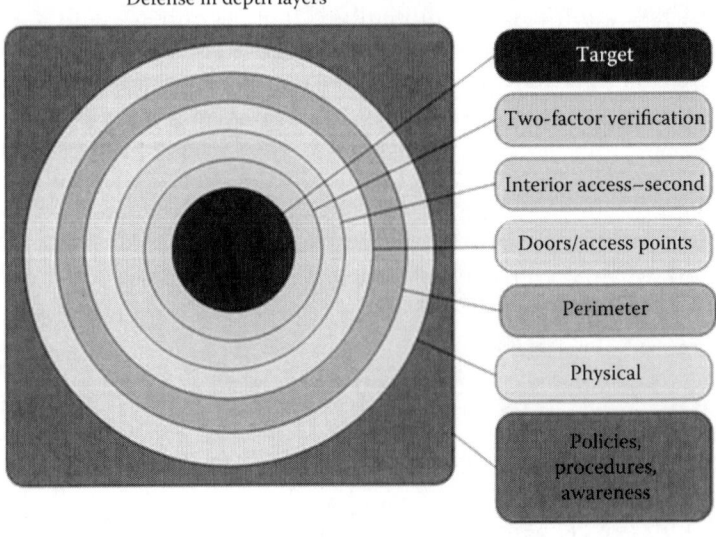

Figure 22.5 Defense in depth layers. (From Michael Fagel.)

- *Doors/Access Points*—Once inside the facility, there may be additional doors, or locks keeping unauthorized personnel away from core or critical components of the soft target.
- *Two-Factor Verification*—Just before one gets to the target, the two-factor verification will be seen. That might be an identification card, it might be a biometric lock, it might be an iPad, or it might just be someone knowing people and checking them in.
- *Target*—Finally, at the center of the circle, is the target—which is what needs to be protected. That target could be a chief executive officer, it could be a data room, it could be anything that is critically important to the organization.

When planning recovery against threats, one important aspect is to make sure that everyone involved understands the rules responsibilities associated with business/operational recovery. Facility stakeholder must be invited to be a part of the threat assessment, preparedness, and recovery planning. This is not a job for one person. These stakeholders must have buy-in to the process. Therefore, gaining their support and input to the process is critical. Without it, any other work that has been done insofar as security is concerned is pretty much wasted.

SOFT TARGET VULNERABILITY ASSESSMENT

Refer back to Figure 22.5, and recall center target area. This section will discuss how to assess vulnerabilities of a soft target. When looking at vulnerabilities, it is Important to look at entry points where a potential terrorist could enter into the facility, or locations where a substance could be introduced into the facility. This could include doors, windows, skylights, vents, roof access, and so on. Questions must be asked:

- Are doors secured 24/7?
- Is their access control on doors?
- Are windows locked, able to be open, alarmed?
- Are vents protected?
- Is the air-handling system filtered to be able to pick up biological or chemical agents?

Additionally, it is important to determine if the building design and the environment around the building lend themselves to security and protective measures.

Refer to the aerial view of a shopping mall In Figure 22.6. This shopping mall has multiple areas of access control, but if one look toward the right side of the larger building one can observe some truck loading docks. There are access doorways, there are skylights, there is a loading dock, there are dumpsters, and, to the right of the picture, one can see the heating ventilation air conditioning systems that may occasionally require access for maintenance. This means that someone would be coming to this facility with a tool pouch with a bottle of refrigerant and could approach this building with relative ease without

SOFT TARGETS AND CRISIS MANAGEMENT

Figure 22.6 An overhead view of a shopping mall. (From the U.S. Department of Homeland Security.)

being questioned. This could be a prime opportunity for a terrorist to access the ventilation system of the building, or enter into the building under the guise of a maintenance worker.

In Figure 22.7, one will see people waiting in line at a convention center—this is interesting because this is for a security convention. Think of who could be standing in that line. Who could be in that facility?

When one assesses vulnerability, one needs to think about what protects the target best and what tools may already exist that provide target protection. As a facility is designed, understand which tool will be better to use. Figure 22.8 shows three fence options that are available to facilities. Moving from left to right the amount of protection provided diminishes due to the type and construct of the fence.

One facility can also present multiple targets and vulnerabilities. In Figure 22.9, on the right of the image, one can see a TV tower, a commercial television station, which is located adjacent to the Department of Homeland Security complex in Washington, DC. One can observe how close the tennis courts and the access roads are to the buildings. Every one of these different may be vulnerability.

Imagine Times Square on New Year's Eve with thousands of people engaging in revels. Located in Times Square, there is an Armed Forces recruiting station, those have been attacked already. Think about the number of people milling about that facility; just think of all those opportunities there. There are hard targets and soft targets. At any point in time, remember to focus on the main question—what is the target? Where is it that needs to be protected the most?

SOFT TARGET PLANNING

Figure 22.7 A convention center entrance. (From the U.S. Department of Homeland Security.)

Figure 22.8 Fencing options for facilities. (From the U.S. Department of Homeland Security.)

CRISIS MANAGEMENT
Training

Crisis Management training is an important part of the layered approach of Defense in Depth. One problem with Crisis Management Training is that there is a thought process from many that crisis training is unnecessary. However, this type of training is critical to the success of a soft target protection team.

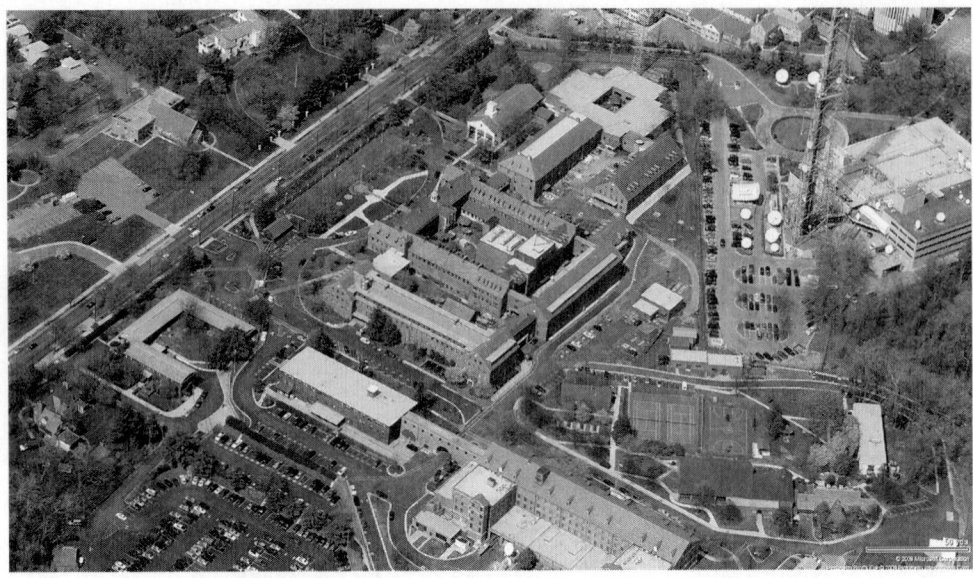

Figure 22.9 Department of Homeland Security complex, Washington, DC. (From the U.S. Department of Homeland Security.)

Planning

In crisis management planning, the key to remember is that it is critical to have a clear purpose, a clear set of objectives, and the input and buy-in of all facility stakeholders, each of who may have their own skill sets to bring to the planning process. This includes everyone from the CEO of the company down to the bottom of the organizational chart.

Another critically important fact to keep in mind is that good managers may not be good crisis managers. A good manager may be very successful on a day-to-day basis. However, in a crisis, that manager may fail. The top leader or manager in the company may not be the best to lead the organization in a crisis. This needs to be recognized and considered during the crisis planning process.

When one considers the word "team," it has to be remembered that there is no "I" in the word "team." The team has to be a group where everyone pulls together. Plan for an absolute worst-case scenario.

Plan for contingencies. There is no such thing as a "perfect plan" due to the variables and the differences with each crisis. Keep this in mind, but do the best with what resources are available.

Remember that recovery begins even before the initial response to an incident is completed.

Exercises

Exercises should never ever be designed for failure. Some people have designed plans for failure or trickery thinking, "Well, let's see how many ways we can make this person or this group fail." However, that kind of thinking is completely unacceptable. Exercises should be well designed to be learning tools and with reasonable objectives. The point of exercises is to allow the organization to progressively improve upon protocols, procedures, and actions before, during, and after a crisis. The objectives of exercises must be reasonable and understandable because the ultimate point of the exercise is to practice a response to a crisis. If an exercise is designed for failure, it can be guaranteed that the organization will fail in an actual crisis.

If one looks at the after action report of most exercises, a common finding is that there are failures in communications. Communications typically fails for a number of reasons, ranging from the failure of a two-way radio to that of basic personal communication failure. Remember that communications is not just a tool or a device—it is a concept.

Simple changes make a very effective difference. Remember that, as an organization first one have to crawl, then walk, and then run. This is how facilities improve their methodologies for hazard prevention, mitigation, preparedness, response, and recovery. Use exercises find opportunities for improvement and build strengths. The culture of protection and preparedness is a continuously evolving process.

COMMAND CENTERS

The following items are recommended for the Command Center and Go Kits (Table 22.1). A good rule to remember is to think of the amount that a crisis management team might need and double it. One can never have too many supplies set aside for an emergency.

Command Centers are often built into the back of an office or a conference room; Go Kits are commonly stored in the back of a vehicle, or as a backpack, or a rolling kit. One thing that is important to understand is that a command center or a "Go Kit" should be three people deep. This is in case someone in the command center needs to trade off with a backup or relief person. All of the kits should have the same things in them. Remember

Table 22.1 Go Kit Supplies for Command Centers

• Laptop	• Charger	• Hot battery
• Cell phone	• Cell phone battery supply	• Portable radio
• Batteries	• Clipboards	• Grease pencils
• Mylar sheets	• Maps	• Blueprints
• Rolodex	• Index cards	• Pencils
• Magic markers	• Tape	• Cellophane
• Masking tape	• Color-coded tape	• Magnets
• Rainproof paper	• Rainproof pens and ink	• Highlighters
• Charts	• Graphs	

Source: From Michael Fagel.

Note: Go kits are a tool that should change and evolve over time. They're are only effective if trained with, and determined what should be in or out of the kit at any given time. They are meant to be portable, accessible and useful. Make them your own!

that the any documents on a thumb drive should be on common platforms—for example, a PDF, excel spreadsheet, or Word document but nothing exotic. Remember that if a thumb drive is plugged into a police car's computer or a fire truck's computer, hey may be able read it. However, if the document on the thumb drive requires specialized software, the document is not going to be able to be opened.

Now, consider rolodexes. Rolodexes are typically considered antiquated items. However, sometimes the old technology works very well. It is important to remember that records must remain up to date. Every 2 or 3 months, change the thumb drive. Continually change these elements out so that everybody who needs to know what is going on has the most common products and the most recent, accurate data.

In all Go Kits, remember to keep charged batteries. This seems simple, but cell phones often run out of battery power at the worst possible time

As one looks at their command post elements and Go Kits, it needs to be made sure that they are functional and that everybody knows what is in them. It is important to know exactly where they are, so they are ready to for use when needed.

SUMMARY

Protection of soft targets, and the planning and training associated with the process, is a team effort. The Crisis Team Manager has to always keep an eye on the target. Remember Figure 22.5, we showed at was shown throughout this chapter—look for failure points. To look at a failure point, one needs use that point to build an opportunity for improvement. Everyone has strengths, yet everyone also has weaknesses. Every element that is proposed may or may not be approved. Regardless, keep the course. Remember that, above all things, protection of a soft target has to become a mindset, it is a culture, and it is a continuous process—it evolves, because threats change on a minute-by-minute basis. An organization's culture and mindset are going to be the tools that can be used most in this continuous evolutionary process as everyone becomes better and safer.

23

Beyond the Response—The July 7 Bombings Inquest
A First-Person Account

Gary Reason

Contents

The Attacks	404
My Role	405
The Coroner's Inquests	406
The 7/7 Team	407
The Challenges (Pre-Inquests)	407
An Absence of Guidance and Procedures	407
Things That Can Conspire against You	407
Access to a Key Witness	408
Information Overload	408
Dealing with the Crate Issue	409
"Hearts and Minds"	410
The Support	410
The Challenges (During the Inquests)	411
Dealing with the Negative Media Coverage	411
Additional Workload Pressures	412
Expect the Unexpected	412
Preparing to Give Evidence	413
Validation	413
The Big Day	413
Things I Learned While Giving Evidence	415
The Challenges (The Verdict)	416
Going Back to the High Court	416

The Verdicts 416
Facing the World's Media 417
Going Back to Work 418
Personal Reflections 418
In Summary 420

Much of this book focuses on the range of preventative measures and preplanning activities that individuals and organizations should consider in order to mitigate and respond to the risks/consequences of a major terrorist attack. While there is no doubt that leading in the aftermath of a catastrophic incident can be one of the toughest challenges that a strategic commander will face during their career, operational command is unlikely to be the only set of difficulties arising from these types of events. This chapter looks at the issues and challenges that strategic managers will often face after the operational response phase of an emergency has concluded, sometimes many years later. These challenges usually arise when an organization's operational response performance and decisions are subjected to some form of external review, such as a public inquiry, coroner's inquest(s), independent review, and so forth. As this type of scrutiny will always occur after the event, it will be undertaken with the benefit of hindsight and informed by a rich source of information that is unlikely to have been available to the operational commanders at the time they were managing the emergency. In addition, people's perceptions of events will have already been shaped and influenced by the plethora of media coverage that is always generated during an emergency and its immediate aftermath, and their views may have been skewed by inaccuracies and/or misinformation. This can lead to unrealistic expectations with respect to what the emergency services' response could or should have achieved. This bias can make actions extremely difficult to explain and defend many years after the event when an organization comes under some form of external scrutiny.

To demonstrate the types of issues and challenges that can arise under these circumstances, this chapter shares my own experiences as the London Fire Brigade's lead officer for the inquests relating to the fifty-two civilians that were killed as a result of the terrorist attacks that occurred in London (United Kingdom) on July 7, 2005. These attacks are often referred to as the 7/7 bombings.

THE ATTACKS

The 7/7 bombings were a series of coordinated suicide bomb attacks in central London that targeted civilians using the public transport system during the morning rush hour.

On the morning of Thursday, July 7, 2005, four suicide bombers separately detonated three bombs in quick succession aboard London Underground trains across the city and, 50 minutes later, a fourth on a double-decker bus in a central London street: Tavistock Square (see Figure 23.1). Fifty-two civilians were killed and over 750 more were injured, some seriously, in the attacks. According to *Wikipedia*, these attacks were the United

Figure 23.1 A double-decker bus following an onboard bomb explosion. (From the U.S. Department of Homeland Security.)

Kingdom's worst terrorist incident since the 1988 Lockerbie Airplane bombing; they were also the country's first ever suicide attack.*

On the morning of Thursday, July 7, 2005, four suicide bombers separately detonated three bombs in quick succession aboard London Underground trains across the city and 50-minutes later, a fourth on a double-decker bus in a central London street: Tavistock Square (see Figure 23.1). Fifty-two civilians were killed and over 750 more were injured, some seriously, in the attacks. London had been warned repeatedly that an attack was inevitable; it was a question of when, not if. The bombings were an act of indiscriminate terror and occurred the day after it was announced that London had won its bid to host the 2012 Olympic/Paralympic Games.†

In total, the London Fire Brigade (LFB) deployed more than fifty front-line fire engines and specialist vehicles to the four bomb scenes, with approximately seventy senior and strategic commanders deployed in incident command and/or support roles. All four bomb scenes were declared major incidents under the London Emergency Services Liaison Panel (LESLP) definition by one or all of the emergency services.

My Role

In early 2010, I was given the lead for overseeing and managing the LFB's preparation for the 7/7 bombing inquests, which, as it turned out, proved to be a key moment in my career. At that time, I was an assistant commissioner (AC), a strategic operational command and management position within the LFB and part of the organization's top management team. I believe I was chosen to take on this task thanks to my experience and the fact that I had detailed knowledge of one of the attacks, having been present at the King's Cross incident in 2005.

---

* *Wikipedia*, s.v. "7 July 2005 London bombings," last modified May 6, 2016, https://en.wikipedia.org/wiki/7_July_2005_London_bombings.
† Ibid.

THE CORONER'S INQUESTS

The purpose of an inquest is to establish the cause of death and to make recommendations regarding any concerns that create a risk of other deaths occurring, or any concerns that will continue to exist in the future. These recommendations, if made, will be directed to a person or organization perceived to have the power to take action. In terms of the July 7 bombings, the scope of the Inquests was defined as follows:

> The purpose of the Inquest is to determine who the deceased were, when and where the deaths occurred, and how the deceased came by his or her death. In relation to those who were not killed immediately by the explosions, the coroner will also consider whether any failings in the emergency response contributed to or were causative of their deaths.
>
> Pre-opening statement of Lady Justice Hallett,
> Inquest held at the Royal Courts of Justice, Central London 11th of October 2010

As can be seen from the second sentence above, the scope of the Inquests clearly brought the response of the emergency services into sharp focus, indicating that all of the operational activities and command decisions made at that time would come under significant scrutiny.

The coroner's Inquests were chaired by Lady Justice Hallett and took place at the Royal Courts of Justice located in central London. The Inquests commenced on October 11, 2010, and ran for a period of nearly 5 months, concluding on March 3, 2011. It has been acknowledged that the Inquests were unprecedented in their scope and detail, bringing to light new information about the bombers and the emergency services' response to the attacks. The coroner's level of scrutiny during the inquests also pushed the boundaries of transparency between the need to maintain national security and the public's right to information.

The coroner chose to hear the evidence without a jury and in chronological order reflecting the sequence in which the bombs had been detonated. As operational firefighters were often the first emergency responders on scene, the LFB's witnesses were typically some of the first witnesses to give evidence for each scene of operation. A number of operational and control room staff gave evidence, and multiple LFB witness statements were read out during the Inquest.

Although I attended the King's Cross bomb scene, I was not called upon to give evidence in relation to my operational command activities. Instead, as the LFB's lead officer for the inquests, I was asked by the coroner to prepare an overarching statement summarizing all of the brigade's evidence and was called to give evidence as the penultimate witness on March 3, 2011, the final day of the Inquests.

The coroner gave the verdicts of her Inquests into the deaths of the fifty-two commuters killed by the bombers on May 6, 2011.

For those wondering why there was such an extended period between the terrorist attacks and the inquests, this was to allow the Metropolitan Police Service (the lead agency for the criminal investigation into the suicide bombings) to reach a point where the evidence that would be a matter of public record during the inquests could not prejudice any potential criminal or civil proceedings.

THE 7/7 TEAM

Being given corporate responsibility for this type of task certainly focuses the mind on how suddenly your role as a strategic manager can change. Given the scale of the 7/7 attacks and the number of fatalities involved, the enormity of the task that lay ahead was fairly daunting, especially given that all of the Inquest-related work had to be undertaken on top of my full-time job. I knew the workload would be enormous and likely to have a significant impact on both my personal and professional resilience. I also knew that I would need help and support just to survive this period, so my first job was to secure a small, dedicated team of three operational officers and two lawyers who, as it turned out, became my saviors in the months that followed. This team, known corporately as the 7/7 team, was tasked with sourcing all the required information, reviewing all the available material submitted to the coroner's office by the other interested parties involved in the Inquests, and dealing with the many welfare and management issues arising from the LFB staff who were likely to be involved. I am hugely indebted to this group of people for the commitment and professionalism shown during a period that lasted the better part of 12 months. There is no doubt that I could not have done what I did, nor could I have overcome some of the personal challenges outlined in the coming paragraphs, without their loyalty and unwavering support.

THE CHALLENGES (PRE-INQUESTS)

The following few paragraphs highlight some of the organizational and personal difficulties I experienced in preparing the LFB's evidence for submission to the coroner's team. I am confident that these issues are not unique to the LFB and demonstrate the types of challenges that a large organization is likely to experience in the run-up to some form of major external inquiry.

An Absence of Guidance and Procedures

The scale and scope of the 7/7 Inquests was something the LFB had never experienced before, and therefore it is not surprising that no specific guidance or policy document existed that I could use to inform and direct my approach. In addition, there were only one or two serving LFB officers who had any inquest experience, so again, there was very limited organizational knowledge available for me to tap into. While I did not particularly worry about this lack of guidance at the start of the journey, it did become a factor as the date of the inquests hearing approached and I started to feel less confident about my preparations.

Things That Can Conspire against You

Like many large organizations, the LFB routinely updates its information technology (IT) and document management systems. One such technological upgrade had occurred in the period after the attacks and prior to the coroner announcing the date of the Inquests. This made accessing some of the key documents more difficult and created additional work pressures. There was also an issue with respect to tracking the version of operational policy/doctrine that had been current at the time of the attacks. I was aware that a number of the LFB's key operational policies had been rewritten and updated in the intervening years since the terrorist attacks, and therefore it was essential that the LFB traced and verified the historical version(s) of each policy in order to submit the correct information to the coroner's team. This proved to be more difficult than it may sound.

The other key issue that emerged in the months leading up to the Inquests was the format and content of some of the LFB's witness statements given by those operational firefighters and officers who undertook key roles on the day of the bombings. In the immediate aftermath of the attack, the Metropolitan Police Service (MPS) took statements from the majority of LFB operational staff who attended the four bomb scenes. These statements were taken under strict conditions, as their content could have potentially been used as evidence in the subsequent criminal investigations. However, when I started to review the statements, it became clear that the MPS's focus in terms of their evidence gathering was, understandably, on the potential criminal/civil cases that may follow. That is, they did not contain a comprehensive record of all the activities and decisions that the respective firefighters and officers undertook during their time at the incident. This made me feel very uncomfortable because I could see gaps in the chronology of the LFB's evidence, along with, in places, no demonstrable rationale recorded for some of the key operational command decisions that were made. Therefore, in order to fill these gaps, I had to arrange for some of the key LFB witnesses to be reinterviewed by my 7/7 team. This proved difficult in some cases, given the time that had elapsed and the fact that a number of staff were reluctant to relive their experiences relating to the day of the attacks.

Access to a Key Witness

One unusual difficulty that I had to overcome was in relation to a firefighter who had been one of the first on scene at the Tavistock Square bus bomb. In the intervening years since the attack, this firefighter had been convicted of a criminal offense and was serving a long prison sentence. As expected, this individual was called to give evidence by the coroner, which meant I had to arrange some sensitive negotiations with both the coroner's office and the Prison Service, which eventually led to an agreement for this particular individual's witness testimony and subsequent cross-examination to be undertaken by video-link from prison.

Information Overload

One of my greatest personal challenges in the lead-up to the Inquests was dealing with the vast amount of 7/7-related information that I needed to absorb and understand. I knew that it was very likely I would be the most senior LFB operational fire officer to be called as a witness, which meant I needed a comprehensive understanding of the fire service's role at each of the bomb scenes. In addition, it soon became clear from looking at the wider sources of evidence material submitted to the coroner by the other interested parties and agencies that I would also need to have a good understanding of how others had portrayed the fire service's response.

In the months leading up to the Inquests, the LFB's legal staff and my own 7/7 team continually delivered large volumes of information to my office, which included policy documents, witness statements, and other material that I needed to read, understand, and remember. These crates of files and paperwork piled up in my office and served as a constant and worrying reminder of the vast amount of information I needed to absorb. At times I felt overwhelmed by the enormity of the task, and due to the work commitments and pressures of my day job, I rarely had any time to devote to the 7/7-related work during my normal office hours. I knew I had to deal with the crate issue in order to make progress.

By this point, my preparations for the Inquests had consumed my life and dominated every spare waking hour. In hindsight, the challenge I had set myself—to understand all of the material that had been submitted to the coroner—was probably unrealistic, but I am stubborn by nature and was determined not to fail myself or the organization. It was around this point that I realized how lonely and detached I had become, particularly from my close family and friends. As soon as I arrived home each day, I would find a quiet space and get my head into the 7/7 material. It is at these times that you test the patience of even the most supportive family members. It also started to dawn on me how professionally exposed I had become. I knew that if the Inquests did not go well for the LFB, my name would be forever associated with failure, and this could ultimately change perceptions of me as a professional fire officer and potentially negatively affect my career.

I can honestly say that the months I spent absorbed in the 7/7 material were some of the hardest and loneliest of my career. I did not feel I had anyone with whom I could discuss my anxieties, and I had no way of benchmarking whether I was making sufficient or satisfactory progress. With no reference points or guiding support, I was on my own, not really knowing whether my preparations were enough to deliver the desired and expected outcomes for the LFB.

Dealing with the Crate Issue

Every morning when I arrived at work, I was faced with the crates that had piled up, which appeared to be getting more overwhelming on a daily basis and were starting to become my nemesis. I therefore knew I had to deal with this very visible and troubling millstone that was only serving to drag me down. I decided that I had to get all the salient 7/7 information into a more user-friendly format that would be portable and easy to review. It was at this point I started the colossal task of creating my own information folder. Unsurprisingly, this folder took a significant amount of time and effort to prepare, but for me it was an instinctive process that was essential to make me feel confident that I had the requisite knowledge and understanding ahead of taking the witness stand.

As it turned out, this information folder fulfilled a number of very important purposes for me. Once completed, it represented an important milestone in my preparations; I felt that I had finally completed my review of all of the available information. The folder acted as a single point of reference for any 7/7-related issue that emerged or needed clarifying in the weeks leading up to and during the Inquests. It also served as an important physical and practical comfort to me, insofar as everything I thought I needed to know and understand was now cross-referenced in a single folder rather than contained in the vast number of crates that had filled my office for many months. I realize that my approach to managing this information challenge may not work for everyone, but I can honestly say that I found it very reassuring to have a single point of reference for such a large volume of material. That said, by the time I took the witness stand on March 3, 2011, I felt that I was familiar enough with the content of my information folder so as not to rely on it in court. It therefore remained in my office, along with all the crates, on the day that I was called to give evidence.

"Hearts and Minds"

For me, what I refer to as the "hearts and minds" issues were some of the most important challenges that I had to deal with, and I was determined to get them right. With respect to the 7/7 Inquests, these issues fell into two categories: (1) responding to the needs of the family and friends of the deceased and (2) supporting all the LFB staff who would be called to give evidence.

In reviewing all of the 7/7 information, it struck me that the LFB's witness testimonies would be incredibly important in helping the victims' families and friends understand what had happened to their loved ones in the immediate aftermath of the explosions. Firefighters had been some of the first emergency responders at the four bomb scenes, and I knew their evidence would shed new light on what actions had been taken to rescue casualties and tend to the seriously injured. It was for this reason that I became a bit obsessed with trying to do the right thing for the families and friends of the deceased, and this definitely influenced my decisions and shaped my entire approach to preparing for the Inquests.

For me, this focus was very much an intuitive and emotional response to the situation I found myself in. I knew that whatever challenges and stresses I was personally struggling with, they all paled into insignificance when measured against the emotional trauma and heartache the victims' families and friends had experienced since the attacks in 2005. I was determined that I would be able to answer any questions directed at me when giving evidence, and I believe this commitment and focus helped to keep me going when I was at my lowest point.

My other overarching focus was on the welfare and support that I wanted to make available for all the LFB staff who would be required to give evidence or have their statement read out in court. I knew from my own feelings of trepidation about what lay ahead that it was likely that the majority of LFB staff would be extremely anxious and worried about the thought of giving evidence. I therefore made it my personal goal to help them through this difficult process.

The Support

As soon as I received the list of LFB witnesses being called by the coroner, I personally wrote to every member of the staff detailing all the support that would be made available before, during, and after the Inquests. I set up a "hotline" to be monitored by the 7/7 team so that the staff had a confidential means of discussing their concerns or requesting additional information. On a practical level, I offered staff the opportunity to visit the Royal Courts of Justice, which as anyone who knows the site will tell you, is a fairly imposing building.

In the lead-up to the Inquests, I ensured that there was face-to-face contact with all the LFB staff who had been called to give evidence. During these meetings, staff were given all the available 7/7 information and encouraged to discuss their concerns. This highlighted a number of unexpected issues that I will share, as they are not necessarily issues that you would immediately think of.

A number of officers expressed concern about how they should respond if questioned in detail about what they had witnessed at the scenes of devastation. This was very much a concern about appearing too insensitive or causing any further distress to the families of the deceased who would be present in court. This issue proved quite a challenge, as I knew that all witnesses would be under oath when on the witness stand and would feel compelled to describe, graphically in some cases, the scene of carnage they had witnessed. As a team, we spent many hours discussing how to describe the four bomb scenes in an accurate but sensitive way. We came up with some delicate phrases and words that we felt would truthfully convey the scale of the devastation without appearing insensitive or risking unnecessary further distress to the families and friends who were present. This information was then shared with all the LFB witnesses.

Once the Inquests started, I encouraged staff to attend the Royal Courts of Justice to watch the proceedings, so that they could get a firsthand feel for what would be involved in giving evidence. I also ensured that the LFB's team of trained counselors would be on standby for any staff seeking more professional support.

Once the LFB staff had completed giving their evidence, I also arranged for every witness to have a formal debriefing. This helped me gauge how they had coped with their ordeal and enabled me to offer/arrange additional welfare or professional counseling support when it was requested or required.

THE CHALLENGES (DURING THE INQUESTS)

Dealing with the Negative Media Coverage

As soon as the Inquests started in October 2010, the world's press and media descended on the Royal Courts of Justice, and it became clear that this would be a very high-profile news event. While this was not a surprise, I had not been fully prepared for the large number of requests for media interviews that the LFB received on a daily basis. The focus on the LFB also intensified on the days when LFB staff had given evidence or witness testimonies had been read aloud. Most days, the issues and perceptions arising from the Inquest proceedings appeared as the lead stories in both national and local media publications (both in print and online).

It was interesting to witness how the media were interpreting the evidence, but unfortunately, in the first few days of the inquests there was some very negative media coverage about the LFB response to the first bomb attack. This was deeply troubling for me, as I realized that this type of media coverage would do little to ease the anxiety of those LFB staff who were scheduled to give evidence later in the Inquests. It was therefore necessary for the LFB to develop an internal and external media strategy that would be in place for the duration of the Inquests.

As a result of this very negative reporting, a decision was made not to respond to any of the specific daily media stories or to allow LFB witnesses to participate in media interviews after they had appeared in court. While this was a sensible approach to adopt from a strategic media management point of view, I had been unprepared for the backlash from the LFB's own staff who were upset by the unfavorable media coverage. Many LFB staff, especially those who had already given evidence, demanded to know why the LFB was not defending its position and participating in media interviews to put the counterarguments in the public domain. Understanding the staff concerns and frustrations, as well as recognizing that there was a risk that this negative media coverage may continue throughout the duration of the inquests, I arranged for regular internal briefings to be published. These internal briefings gave details of the LFB's approach to handling the media and provided regular updates as to how the inquests were progressing. This approach did not appease all of the LFB staff, but I felt it was important that everyone be given the opportunity to understand the rationale for taking this approach to managing the media.

Additional Workload Pressures

By the time that the Inquests had started, I knew for certain that I would be called as the most senior LFB operational officer to give evidence. I also knew that I was likely to be called very near to the end of the proceedings, which meant most of the LFB evidence would be a matter of public record by the time I took the witness stand. This increased my sense of pressure and expectation because I would now need to be poised to deal with any fire service–related clarifications, outstanding issues, or questions that had been generated throughout the Inquests.

Knowing that I was going to be called near the end of the Inquests did give me additional time to prepare, but it also made it more important that I kept abreast of developments, which meant having to read many of the daily court transcripts. These transcripts often spanned several hundred pages, which made my workload challenges even more difficult. However, reviewing the daily transcripts did prove very helpful in highlighting the key lines of inquiry that I was likely to face and gave me further opportunities to improve my knowledge and understanding of certain issues.

Expect the Unexpected

Staff called as witnesses in a traditional criminal case are typically exposed to questioning from two legal teams, once by the prosecution and once by the defense. However, in the 7/7 Inquests, there were more than fourteen interested parties, the majority of which were represented by Queen's Counsel (QC). This made the prospect of giving evidence even more daunting and resulted in many of the LFB witnesses, including myself, being cross-examined by several different QCs, each with a different focus and line of inquiry.

Another issue that arose early in the Inquests was the realization that one of the interested parties, representing a young woman who had sadly lost her life to the Aldgate bomb, was actually the father of the deceased. This was a very unusual situation, seeing as the father had no formal legal experience but had secured agreement from the coroner that he would be allowed to ask the witnesses questions in person. For the operational firefighters that were called as witnesses for the Aldgate bomb, this was a particularly difficult experience. Understandably, the father was desperate to know what specific actions the emergency services had taken in trying to save his daughter's life. His questions to some of the firefighters were among the most heartfelt and challenging cross-examinations that occurred during the Inquests. For example, the father asked firefighters to explain why they had chosen to tend to casualties near the location of his daughter's body before tending to his daughter. He was desperate to uncover whether his daughter could have survived if the emergency services had gotten to her earlier. There is, of course, no answer that anyone can give to these types of questions that will ever help remove the pain and anguish that accompanies them. While this was something that I hadn't foreseen, all of the firefighters who were cross-examined by the father acquitted themselves with great professionalism and sensitivity. Even though they couldn't confirm the details the father was seeking, the father did eventually get the answers he had been searching for when a blast injury medical expert gave evidence that confirmed the injuries sustained by all fifty-two fatalities were not survivable.

Preparing to Give Evidence

In terms of my own preparations, I decided to attend the Inquests on a number of key dates to see how my counterparts from the other emergency services fared on the witness stand. I am not sure if this was a particularly helpful measure, as a number of the most senior officers from the other emergency services had a difficult time on the stand, thereby fueling my concerns about how trying the experience was going be.

Validation

In the weeks leading up to the date of my own appearance as a witness, the LFB engaged the services of a QC to assist with the LFB's legal preparations. By this stage, I had already drafted my overarching witness statement, which had been requested by the coroner—a weighty tome eighty-six pages in length. I first met the LFB's QC approximately 3 weeks before my scheduled appearance in court. During this first meeting, the LFB's QC played devil's advocate with the information detailed in my overarching statement. This was a very challenging session; the QC probed every aspect of my statement and pressed me on all the areas where he felt the LFB could be vulnerable. After a couple of hours of very uncomfortable cross-examination, he quietly said, "That's enough," and calmly announced that he thought I was ready to give evidence. He congratulated me on my "in-depth" knowledge of the evidence and said he thought I was ready to deal with the inevitable "difficult" questions that I would face in court.

This was a pivotal moment for me, as it was the first time in my many months of preparation that my approach and knowledge had been independently assessed and validated. Consequently, it was of some comfort and great relief that the LFB's QC appeared to be broadly satisfied with both my overarching statement and my readiness to present evidence. As the only personal feedback I had received since commencing my preparations, it lifted my spirits and boosted my confidence.

THE BIG DAY

Since it is difficult to know the exact time you will be called to give evidence, witnesses are normally summoned to report to the court on the day the witness before you is scheduled to give their evidence. For me, this was a fellow LFB officer and one who had retired in the period between the bomb attacks and the inquests. Having been summoned, I was able to sit in the courtroom during my former colleague's cross-examination, which, as it turned out, proved to be a difficult session. Again, this made me acutely aware that my time on the witness stand was likely to be very challenging.

The following morning, I arrived very early at the Royal Courts of Justice in an attempt to avoid the media circus. To my surprise, the media had already surrounded the building. This meant I experienced the usual bombardment of questions and requests for interviews, which I politely declined. Once safely inside the court building, I was taken to a waiting area where I sat with the LFB's legal team for a short period. This quiet space was extremely helpful in allowing me to clear my head and focus on the task that lay ahead. In the time I had available, I systematically went through the key LFB issues from each of the four bomb scenes as well as rereading sections of my own overarching statement.

When the LFB's legal team left to take up their position in the courtroom, I was left on my own for a brief period. This was a strangely reflective moment, as I knew that my many months of hard work were about to be tested in the most public and unforgiving way. Remarkably, I did not feel nervous about the impending ordeal, which both surprised me and slightly worried me. While I was used to public speaking and representing the LFB, I almost always experienced a slight pang of nervousness in these types of situations. In fact, I liked the slight feeling of trepidation, as it had always been my experience that having some nervous tension brought out the best in me. I certainly did not feel arrogant in the absence of nerves; it was more like a serene calmness had descended on me in those few solitary moments before the clerk came to collect me. To this day, I cannot explain why I reacted so uncharacteristically to the pressure of that moment, but it did help me to feel more composed and confident when I finally took the witness stand.

When the time came, the clerk of the court collected me from my solitary space and ushered me to the witness stand where I took the oath. Being the final day of the inquests, the courtroom was packed with the various legal teams and all of the relatives of the deceased; there was not a spare seat anywhere. Although there were a few familiar faces in the courtroom, I have to admit to feeling very isolated and alone at that moment. Everyone appeared to be looking at me, waiting for the first of many challenging questions to be asked. As expected, the questions came thick and fast, and while a coroner's hearing is not supposed to be adversarial, it certainly did not feel particularly friendly.

I can honestly say that my time on the witness stand flew by, even though I was cross-examined for several hours. I quickly gathered that my in-depth knowledge relating to all aspects of the LFB's operational response to the attacks was aiding legal counsel to confirm facts, which could ultimately assist the coroner in reaching her conclusions. However, part of the way through my testimony, I sensed that the coroner's QC was attempting to lead me into saying what I thought the LFB's failures had been. Even though I recognized this was happening, there was little I could do to avoid it. At that point, I could feel myself getting frustrated with what I felt was an unhelpful line of questioning, but I knew my frustration would do little to help the families sitting directly in front of me to understand what had happened to their loved ones. Given the promise I had made to myself about supporting the needs of the families, I was brave enough to hold my composure and, where necessary, explained to the QC that a simple "yes" or "no" response would not accurately reflect what had happened. I knew this was a dangerous path to tread, as I would be no match for a QC if he felt I was being argumentative. Fortunately for me, the coroner's QC focused on the evidence from the King's Cross bombing, the incident I had attended. This allowed me answer the majority of the questions in the first person and to explain in great detail what it is actually like for emergency service personnel when they arrive at the scene of a major emergency. I had not planned to do this, but when the opportunities presented themselves, I related in great detail the scenes of chaos and devastation that all emergency service responders had faced when they arrived at the site of the bombings. I went on to explain that these scenes of chaos prevailed for a time even after the emergency responders had arrived. Until I made this point, I truly believe that the people in that courtroom were under the misconception that everything is okay as soon as the first emergency responders arrive on scene. As any emergency service professional reading this chapter will testify, that is not how major emergencies pan out.

THINGS I LEARNED WHILE GIVING EVIDENCE

While I am no expert in giving evidence, nor do I profess to fully understand the nuances of the legal profession, I did learn some things through my experience at the Inquests, which I will share here. First, as a nervous witness under cross-examination, you instinctively feel obliged to fill the awkward silences that the lawyers leave after you have responded to a question. I quickly realized, having watched some of the others witnesses, that these awkward silences are actually used by legal professionals to give themselves additional thinking time to prepare their next question. If you continue to speak or try to add clarification to your answers, you run the risk of saying something that will prompt the lawyer to explore a new or slightly different line of questioning. It is therefore important to be confident in your initial response and only answer the question that has been asked. Be succinct, and stop talking when you are satisfied with your answer; trust me when I say that this is harder to do than you may think.

I also found using the available technology in the courtroom very helpful in checking my responses. I was extremely lucky that the 7/7 courtroom had a number of large screens that showed all my statements. I was therefore able to double check that what I intended to convey in my answers matched what I had actually said. This approach also proved useful in slowing things down while I took the time to read each of my responses. You do, however, have to be very careful to remain focused, as the follow-up questions can come rapidly and it not advisable to give the court the impression that you are not listening.

Finally, as the most senior person giving evidence on behalf of your organization, I believe it is important that you take the opportunity to "tell it how it is." When you take the stand, it is the expectation of the coroner and legal professionals that you will convey information in a way that any layperson could understand. While this is completely reasonable, there are times when it is essential to explain the context of a situation as it would have been viewed and experienced by an emergency responder. It is important to remember that the majority of people in the courtroom do not have any emergency service experience, so it is vital that you try to make them understand the challenges and difficulties the blue light agencies face, especially when responding to a major emergency. Again, this is harder than it may first appear.

The public expectation is that everything will be okay when the emergency services arrive at the scene of an emergency, but as we all know, it takes time to bring a degree of order to the inevitable chaos created by a terrorist act. It is very important that everyone understands this, as it does help to put the emergency services' evidence in the proper context.

THE CHALLENGES (THE VERDICT)
Going Back to the High Court

On the morning of Friday, May 6, 2011, I was nominated to attend the Royal Courts of Justice to receive the verdicts on behalf of the LFB. The court building was again surrounded by the world's media, and I observed the largest group of journalists and TV camera crews/photographers that I had ever seen at the venue.

Inside the Royal Courts of Justice, the courtroom was packed with all the families and close friends of the deceased, and there was a palpable emotional tension in the room, which reflected the significance and importance of the day. I remember feeling more nervous waiting for the verdicts than I had when giving evidence, which seemed bizarre, since I was only required to listen to the coroner give her verdict and collect a copy of the final report.

When the verdicts were read, there was an outpouring of emotion from the gathered family members and friends, which was probably one of the saddest and most poignant moments of my whole 7/7 Inquests experience. The emotion of that moment made the hairs on the back of my neck stand up, and I remember feeling very sad and drained by having shared this incredibly personal and heartbreaking experience with those closest to the deceased.

The Verdicts

At approximately 10:30 am on May 6, 2011, Lady Justice Hallett handed down her verdicts on the deaths of the fifty-two people who were tragically killed on July 7, 2005. The verdicts were given as unlawful killing, with the medical cause of death recorded as "injuries caused by an explosion" with respect to each of the deceased. The coroner also used her powers under the Coroner's Rules 1984 to make nine recommendations that fell under two distinct headings—"Preventability" and "Emergency Response." The full details of the verdicts and the associated recommendations can be found in the coroner's report issued on May 6, 2011.* For the purposes of this chapter, I can confirm that the LFB did not receive any specific recommendations, with the coroner concluding, "The evidence I have heard does not justify the conclusion that any failings on the part of any organization or individual caused or contributed to any of the deaths."

Once the coroner had delivered her verdicts, she went on to explain that each of the interested parties would be given a copy of her final report on the rise of the session. This

---

* Coroner's Inquests into the London Bombings of 7 July 2005, 6 May 2011, NA.

report contained the full details of the verdicts and listed the nine recommendations she had made with respect to the evidence she had heard. The coroner also confirmed that a copy of the report had already been uploaded to the 7/7 website and was now a matter of public record. At this point, it dawned on me that the media were probably already reading the report and therefore would know before I did whether the LFB had been given any recommendations. This development and the knowledge that I was about to face the world's media made me feel extremely uncomfortable.

Facing the World's Media

As I left the courtroom, the clerk of the court handed me a brown envelope containing the sixty-five page coroner's report. This was my first opportunity to establish whether the LFB had been given any specific recommendations. I quickly scanned the document and, at first glance, could not find any recommendations specifically related to the LFB. However, being sixty-five pages in length, I was worried that I may have missed something that could catch me off guard when I faced the media. It had already been agreed that I would deliver a short statement on the front steps of the Royal Courts of Justice once the coroner had concluded the verdict hearing. I had therefore prepared a statement in advance, but I now knew that it would feel too insincere and impersonal for me to read directly from a script in light of the emotion I had just experienced in the courtroom.

When my time came to face the media, I had already decided that I would speak from the heart and try to tap into the emotion of the occasion. While I had a reasonable amount of strategic media experience, nothing could have prepared me for the sight that I was faced with on leaving the sanctity of the courthouse. From my earlier arrival at court, I already knew that there was a large throng of press waiting, but seeing it for the first time directly in front of me and knowing I needed to walk up to my mark and deliver a heartfelt statement was probably the most nerve-racking prospect I faced related to the 7/7 Inquests. Oddly, as I made my way to my mark, the only thought in my mind was trying to avoid stumbling on the large number of steps that led down from the building. Fortunately, I had a very trusted colleague with me from the LFB's communications team who kept whispering in my ear to stay calm and focused. He used the few moments before I reached the bank of microphones and cameras to reassure me, and I remember him saying, "Just be yourself and you will be great." I will always be hugely grateful to my colleague for the trust and faith he showed in me in those seconds before I reached the mark. His final words of advice were also enough to give me the confidence to deviate from the preprepared script and speak from the heart.

When I reached the vast spread of microphones and cameras, I was introduced to the world's media, and then I undertook the obligatory sound check. Seconds later, I took a deep breath, focused on the emotion of the day, and delivered my statement. It rests with others to decide how this went, but I was met with a great sense of relief after getting through this particular ordeal, and I remember feeling grateful that I had been given such a public opportunity to say something that I hope reached out to all the suffering families and LFB staff who had been touched by this tragic event. In my view, the statement I delivered was a fitting end to my involvement with the Inquests, and I remember feeling that a great weight had been lifted.

It had been arranged by the LFB communications team that I would undertake a large amount of media work for the remainder of the day, and I was feeling confident in the knowledge that nothing would be as challenging as what I had just been through. However, as it became clear that the LFB had not received any recommendations from the coroner, the media interest in the LFB quickly dissipated.

Going Back to Work

Once I was made aware that the majority of prearranged media commitments were ebbing away, I decided to return to LFB Headquarters where I was immediately called into a meeting of the LFB's top management team. The meeting acted as a hot debriefing and is where I was formally congratulated on a job well done. It all felt a bit surreal at the time, as I could not quite comprehend the range of emotions I had just experienced. The only thought that kept recurring following the coroner's verdicts was that all the hard work and personal sacrifices had been worth it. I also felt a huge sense of relief that I had fulfilled my commitment to the families of the deceased and done everything in my power to support my LFB colleagues through the Inquests. Achieving these two goals, above all else, gave me the most satisfaction.

However, I am also a realist and acknowledge that the positive outcomes achieved by the LFB on this occasion were influenced by lots of factors outside of my direct control and probably also involved an element of good luck. While I have no doubt that the professionalism of all the LFB witnesses and the hard work put in by the 7/7 team and myself contributed to a positive outcome, I also know that things could have turned out very differently for both the organization and myself. If this had been the case, I am fairly certain that the remainder of my LFB career would have played out very differently.

PERSONAL REFLECTIONS

As a strategic manager, I believe that one should expect to face certain situations and tasks that are both challenging and personally taxing; it goes with the territory. However, when taking on a very unusual and/or unique task that is both high profile and potentially career limiting, strategic managers may find themselves in a position that only relatively few others have experienced. It is through these challenges that you truly find the limits of your own abilities and personal resilience, which can be either extremely rewarding or equally destructive.

Throughout this chapter, I have touched on a range of personal challenges that I faced during my 7/7 Inquests experience. I have tried to express these more personal issues in an open and honest way in the hope that they will assist other strategic managers who may find themselves facing similar challenges in the future. That being said, I am not someone that ever tries to impose my opinions and beliefs on others, and as such, I have only included the following personal reflections to complete my story.

Remember to look after those closest to you. As stated earlier in this chapter, in the months leading up to the Inquests, I managed to secure assistance from a small dedicated team composed of both operational officers and legal professionals. By the time the inquests started, this group of staff had been immersed in all the 7/7 material and evidence for longer than 6 months. This material included some very harrowing survivors' statements, photographic images, and video footage. This same group of staff also sat through the majority of the Inquests, which proved challenging at times when the proceedings were emotionally charged. It was therefore not surprising that all of this exposure had, in some cases, an effect on the welfare of a number of the team. This situation required careful and constant management, and I cannot overemphasise the importance of looking after those closest to you in terms of their support needs and emotional wellbeing.

Who looks after you? When taking on a very high-profile leadership role or responsibility, I believe it is extremely important to ensure you have someone looking out for you. In situations like those described in this chapter, it is highly unlikely you will ever find someone who has all the requisite experience and knowledge to assuage your anxieties and answer your questions. However, having a trusted friend or colleague you can turn to in those periods when you feel very lonely is something I would recommend. I say this because I failed to find this person while I was engaged in the 7/7 work, and, on occasion, this did leave me with a deep sense of isolation. I know from experience that some strong and visible leaders often view asking for help as a weakness, but I see it as a natural human response and a recognition that everyone needs someone to turn to when times get tough. I was extremely lucky to have the support of my family and my close colleagues in the 7/7 team, and this was a tremendous help. If I am ever faced with a similar set of professional and personal challenges in the future, I am fairly certain that I will try to identify the person I can most depend upon long before I ever need them.

In the coroner's final report, she made reference to the following (taken from the case of *Duchess of Argyll v. Beuselinck* 1972):

> "… in this world there are few things that could not have been better done if done with hindsight.'"

I believe that these words are particularly relevant to emergency services when responding to large-scale or major events, since by their very nature these events will always be extremely challenging and commanded without the benefit of having all the information. When you and your organization come under significant scrutiny, often many years after the event, there will always be a high risk of what I call a lose–lose outcome, as things always look different with the benefit of hindsight. Those undertaking the scrutiny will have the luxury of many months, and sometimes years, to prepare their lines of inquiry, while those responsible for managing a major emergency have only seconds to make critical decisions. Therefore, an organization and the person charged with representing the organization always face the possibility of an outcome that could be very damaging to both parties.

Remember that you know more about your profession than the lawyers do—without trying to appear arrogant or overconfident, I believe it is important to maintain a healthy perspective when giving evidence in these types of situations. While it will feel uncomfortable to step into a courtroom and enter the legal profession's domain, it is always worth

---

* Coroner's Inquests into the London Bombings 2005—Report under Rule 43 of the Coroner's Rules 1984, dated May 6, 2011.

remembering that lawyers will know less about your service and organization than you do. If you have comprehensively prepared for the challenge of giving evidence, you should feel confident in your ability to respond effectively and appropriately to any line of inquiry. This will not always guarantee a successful outcome, but it is worth starting the process with a positive attitude regarding your abilities.

IN SUMMARY

I truly believe that while my 7/7 bombing experiences were extremely challenging, they were also professionally very rewarding. I feel very privileged to have been involved in resolving such a major event, and there is no doubt in my mind that the experience shaped me as a person. Having personally witnessed the devastation and carnage reeked by the suicide bomber at the King's Cross site, I can appreciate firsthand the suffering endured by those poor members of the public who happened to be in the wrong place at the wrong time.

Like many emergency service professionals, I have experienced and witnessed a large number of tragic events during my long career, and I understand these incidents change lives forever. But, unlike most incidents where you quickly move on to the next emergency, my involvement in the 7/7 Inquests required me to immerse myself in the wider issues arising from this particular event. This level of immersion changes your perspective and makes the whole experience more real and personal. For my part, I do not believe that anyone can go through these types of events without them having a great impact, and by sharing my own experiences, I sincerely hope to, in some small way, help other strategic managers.

While I am immensely proud of the LFB, both for its response to the bombings and its support during the Inquests 5 years later, I know that this is unlikely to be the last time someone will have to navigate a similar situation. It is for this reason and this reason alone that I agreed to share my experiences in this book.

24

Infrastructure Protection
The Fusion Center's Role

Vincent Noce

Contents

Advocacy for State-Level Intelligence Capability 422
Collaboration in the Practice of Homeland Security 423
Creation of the National Network of Fusion Centers 424
Fusion Centers and the Infrastructure Protection Mission 429
Fusion Process and Infrastructure Protection Overlay 429
Fusion Liaison Officer Programs and Infrastructure Protection 433
Summary and Key Takeaways 433

In tackling information issues, America needs unity of effort.

The 9/11 Commission Report

Fusion Centers are uniquely positioned, both organizationally and geographically, to provide the most timely and accurate information relating to the systems and individual features of critical infrastructure within their area of responsibility (AoR). Further, Fusion Centers have been conceptualized as the touch point for information sharing among and between state, local, and tribal governments, the federal government and the intelligence community writ large. One by-product of this collaboration is the ongoing consensus-building surrounding the evaluation of the most critical threats and most vulnerable critical infrastructure in a Center's AoR, and ostensibly the most effective steps needed to harden its softest targets. This chapter will describe the logic behind why states need such a capability, how Fusion Centers came into existence and some of the most crucial partnerships and programs intended to increase Fusion Centers' effectiveness with relation to their role in identifying and protecting the nation's most critical assets.

Soon after the terrorist attacks of September 11, 2001, policy makers in the United States began to debate and examine ways to improve how the country does the business of intelligence, information sharing, and domestic security. One of the most profound actions that the U.S. government took following the terrorist attacks resulted in the largest reorganization of government in the republic's history. The newly created U.S. Department of Homeland Security (DHS) combined, amalgamated, and merged all or part of twenty-two different departments into a single cabinet-level organization in an effort to form a single, unified, homeland security structure tasked to "improve protection against today's threats and be flexible enough to help meet the unknown threats of the future."* In addition, in a further effort to remedy the well-documented intelligence and information-sharing failures leading up to September 11, 2001, President George W. Bush proposed that:

> The new Department would contain a unit whose sole mission is to assemble, fuse, and analyze relevant intelligence data from government sources, including CIA, NSA, FBI, INS, DEA, DOE, Customs, and DOT, and data gleaned from other organizations and public sources. With this big-picture view, the Department would be more likely to spot trends and would be able to direct resources at a moment's notice to help thwart a terrorist attack.†

Additionally, claiming, "the Federal Bureau of Investigation had failed to merge properly and perform effectively its dual missions of law enforcement and the collection, analysis, and dissemination of foreign intelligence inside the United States,"‡ Senator John Edwards introduced a bill in 2003 titled the *Foreign Intelligence Collection Improvement Act of 2003*. The bill advocated for the establishment of a new member of the United States Intelligence Community called the Homeland Intelligence Agency.§ Although unsuccessful, Edwards' bill recognized the importance of the state and local role in the larger intelligence community.

While each of the efforts mentioned above shows the importance of, and strategic intent to identify, formalize, and, in some cases, create new relationships between existing intelligence practitioners, the United States chose not to pursue the creation of a single domestic intelligence agency. Rather, it chose the path of reform through creating an environment of greater collaboration and engagement with state and local entities, and with existing organizations that were already engaged in the collection and analysis of criminal intelligence.

ADVOCACY FOR STATE-LEVEL INTELLIGENCE CAPABILITY

Since the September 11 terrorist attacks, states have taken a lead role in homeland security, and while it is not practical to duplicate the federal intelligence community in every state, experts have advocated for a single, integrated intelligence enterprise with well-defined lanes-in-the-road.¶ One's first inclination when considering intelligence capability usually

---

* George W. Bush, *The Department of Homeland Security* (Washington, DC: Department of Homeland Security, June 2002).
† Ibid.
‡ *Foreign Intelligence Collection Improvement Act of 2003*.
§ Todd Masse, *Domestic Intelligence in the United Kingdom: Applicability of the MI-5 Model to the United States* (CRS Report No. RL31920) (Washington, DC: Congressional Research Service, 2003).
¶ Dr. James E. Steiner, Needed: State-Level, Integrated Intelligence Enterprises, *Studies in Intelligence*, Vol. 53, No. 3, September 2009, p. 1.

is to focus on the criminal and/or national security mission set with law enforcement as the primary consumer; however, by weaving common intelligence functions throughout a state's office of homeland security's four responses to risk—Prevention, Protection, Response, and Recovery—and by bringing nontraditional partners and customers into the intelligence cycle, more comprehensive and better informed domain awareness products can be delivered. So, whether responding to a pandemic, flood, hurricane, or terrorist attack, all parties with a homeland security mission are involved from the genesis which will presumably result in the effective prosecution of coordinated responses.

In its 2009 state homeland security strategy, excerpted below, New York seems to have seized on this concept:

> OHS [Office of Homeland Security] chairs the Homeland Security Executive Council, which brings together senior officials from the Governor's Office and many State agencies and Public Authorities to enhance communication and coordination on relevant and emerging homeland security issues ... In addition, OHS leads the Homeland Security Strategy Working Group, consisting of program experts from a variety of State agencies that work together to develop, advance and evaluate programs and initiatives necessary to implement the State Strategy.*

And, in the 2010 Governor's Guide to Homeland Security, the idea is perpetuated:

> The size, capability, and jurisdictional reach of the homeland security organization vary considerably among states, but most are charged with uniting their state's preparedness and response capabilities across multiple agencies and jurisdictions. A coordinated state homeland security effort involves many stakeholders.†

One immediate benefit from this model is the way in which the marriage of strategic risk management and operational crisis management can be helpful in handling a live event. Further, it would be foolish not to leverage and coordinate the collective knowledge of subject-matter experts who are probably already working with these issues in the course of their normal duties. Also, well-defined lanes need to be established to eliminate the threat of mission creep, to appropriately manage information collection and production, and to maintain topical focus.

COLLABORATION IN THE PRACTICE OF HOMELAND SECURITY

One of the underlying—and perhaps most critical—missions of the DHS is improving the effectiveness of collaboration and information-sharing among all levels of government within the United States. This nation's weaknesses in these areas was exploited by al-Qaeda in the lead-up to the terrorist attacks of September 11. In addition, considering the findings of the 9/11 Commission Report, collaboration, or joint action, stands to benefit all involved in the homeland security practice by engaging in joint planning, ensuring a unified effort

---

* 2009 New York State Homeland Security Strategy, 6. Available at http://qa.dhses.ny.gov/media/documents/2009_NYS_Homeland_Security_Strategy.pdf.
† Carmen Ferro, David Henry, Thomas MacLellan, NGA Center for Best Practices, Homeland Security & Public Safety Division, *Governor's Guide to Homeland Security*, 7–8.

by identifying definitive leadership, and mitigating the gaps created by critical shortages of experts.* While it will never be possible to know if these activities would have prevented the attacks, it stands to reason that a culture of "need to share" rather than "need to know"[†] prior to September 11 could have increased the odds of seizing on any one of the ten operational opportunities to disrupt the attacks identified by the 9/11 Commission.[‡]

While not related to terrorism, but nonetheless important, the response to Hurricane Katrina again exposed U.S. vulnerabilities within and across organizational and jurisdictional boundaries. Throughout the event, a number of breakdowns in collaboration were evident: a lack of information sharing among agencies, confused interorganizational relationships, competing roles and responsibilities, and shortcomings in leadership.[§]

> The responses to man-made and natural disasters have many things in common, the most impactful of which may be that they are both complex problems that require capabilities of many disciplines that have both aligned and competing interests, and usually function without an overarching command authority.[¶] With the recent designation of the intelligence/investigations function as a section-level stakeholder in the National Incident Management System,[**] and with a substantial number of state and local fusion centers colocated with either state or local emergency operations centers,[††] and supporting all-hazards missions, the collection, analysis, and dissemination of threat information by fusion centers can inform preplanning response activities, as well as operational response and coordination efforts during and after an incident or event.[‡‡]

CREATION OF THE NATIONAL NETWORK OF FUSION CENTERS

The most recent and perhaps most controversial addition to the domestic intelligence enterprise is the National Network of Fusion Centers (NNFC). Composed of seventy-eight individual centers located in forty-nine U.S. states and two territories,[§§] the NNFC is ostensibly the nation's solution for enhancing information-gathering and sharing activities on the state and local levels to inform the national threat picture. Conversely, the NNFC aims to keep state and local entities informed of transnational threats, which may have nexuses to their community.

---

* Thomas H. Kean and Lee Hamilton, *National Commission on Terrorist Attacks upon the United States, The 9/11 Commission Report: Final Report of the National Commission on Terrorist Attacks upon the United States* (Washington, DC: National Commission on Terrorist Attacks Upon the United States, 2004), 401.
† Ibid., 24.
‡ Ibid., 8–9.
§ Susan Page Hocevar, Gail Fann Thomas, and Erik Jansen, Inter-Organizational Collaboration: Addressing the Challenge, *Homeland Security Affairs* 7, The 9/11 Essays (September 2011): 4.
¶ Ibid.
** Department of Homeland Security, *NIMS Intelligence/Investigations Function Field Operations Guide* (Washington, DC: Department of Homeland Security, 2013).
†† Seventy-one percent of state and local fusion centers have an all-hazards mission; 46 percent are co-located with a state or local emergency operations center; and 58 percent assign personnel to emergency management and/or emergency operations centers during events or incidents.
‡‡ Department of Homeland Security, *Fusion Center and Emergency Management Collaboration Meeting After-Action Report* (Washington, DC: Department of Homeland Security, 2014).
§§ The state of Wyoming does not have a designated fusion center; and Guam and Puerto Rico both have centers.

From 2002 to 2012, no fewer than eight reports were published that recommended and guided possible solutions.* Included in these reports was the *9/11 Commission Report*, published in 2002, which recommended that information be shared across a wider audience of stakeholders and across new networks.† The theme of information-sharing for law enforcement was given a concrete framework and course of action in the National Criminal Intelligence Sharing Plan (NISCP), which was published in 2003. The plan, which was not developed solely for fusion centers, arose from the collaboration of the Global Intelligence Working Group, formed following the 2002 International Association of Chiefs of Police Criminal Information Sharing Summit. The vision of the working group was to create:

- A model intelligence sharing plan
- A mechanism to promote intelligence-led policing
- A blueprint for law enforcement administrators to follow when enhancing or building an intelligence system
- A model for intelligence process principles and policies
- A plan that respects and protects individuals' privacy and civil rights
- A technology architecture to provide secure, seamless sharing of information among systems
- A national model for intelligence training
- An outreach plan to promote timely and credible intelligence-sharing
- A plan that leverages existing systems and networks, yet allows flexibility for technology and process enhancements‡

Further defining information-sharing protocols, procedures, and expectations was the 2004 Intelligence Reform and Terrorism Prevention Act (IRTPA). IRTPA called for the creation of an Information Sharing Environment (ISE) and, with relation to the national network of fusion centers, the requirement that the ISE will provide and facilitate "the means for sharing terrorism information among all appropriate Federal, State, local, and tribal entities, and the private sector through the use of policy guidelines and technologies."§

In 2005, the DHS's Homeland Security Advisory Council issued a document titled *Intelligence and Information Sharing Initiative: Homeland Security Intelligence & Information Fusion*. This document asserted that the concept of intelligence/information fusion has emerged as the fundamental process to facilitate the sharing of homeland security-related information at a national level, and, therefore, has become a guiding principle in defining

---

* These reports included, *The 9/11 Commission Report* (2002), *The National Criminal Intelligence Sharing Plan* (2003), *The Intelligence Reform and Terrorism Prevention Act* (2004), *The Intelligence and Information Sharing Initiative: Homeland Security Intelligence and Information Fusion* (2005), *Fusion Center Guidelines: Developing and Sharing Information and Intelligence in a New Era* (2006), *National Strategy for Information Sharing, Success and Challenges in Improving Terrorism-Related Information Sharing* (2007), *Enabling Capabilities for State and Major Urban Area Fusion Centers* (2008), *The National Strategy for Information Sharing and Safeguarding* (2012).
† National Commission on Terrorist Attacks upon the United States, *The 9/11 Commission Report* (Washington, DC: U.S. Government Printing Office), 418.
‡ Department of Justice, *National Criminal Intelligence Sharing Plan* (Washington, DC: Department of Justice, 2003).
§ Quoted from the National Strategy for the National Network of Fusion Centers, Department of Homeland Security, *National Strategy for the National Network of Fusion Centers* (Washington, DC: Department of Homeland Security, 2014).

the ISE (as required by IRTPA).[*] The document identified the priorities of homeland security intelligence/information as efforts to do the following:

- Identify rapidly both immediate and long-term threats.
- Identify persons involved in terrorism-related activities.
- Guide the implementation of information-driven and risk-based prevention, response, and consequence management efforts.[†]

The document further defined the following, which could be considered an early roadmap to fusion center standards:

- How homeland security intelligence/information fusion is the overarching process of managing the flow of information and intelligence across levels and sectors of government and the private sector to support the rapid identification of emerging terrorism-related threats and other circumstances requiring intervention by government and private-sector entities
- The importance of involving every level and discipline of government, the private sector, and the public, and how efforts should be organized in a scalable way and on a geographic basis so that adjustments can be made based on the operating and/or threat environment
- A description of the fusion process that includes the following activities and could arguably be the predecessor to the intelligence cycle adopted by the national network[‡]

A further definition of the operation of fusion centers was contained in a 2006 publication titled *Fusion Center Guidelines: Developing and Sharing Information and Intelligence in a New Era*. The intent of this document was not to dictate the emphasis or priorities of individual centers; rather, it was to provide newly organized fusion centers with standards and guidelines for operation, as well as advice for the effective exchange of information throughout the network. Three phases of development (Phase 1: the Law Enforcement Intelligence Component; Phase 2: the Public Safety Component; and Phase 3: the Private Sector Component) were accompanied by eighteen guidelines that were developed to clarify the standards for the successful operation of a single center and assist with the development of the NNFC. Those guidelines, each with an accompanying instructional document, are organized as follows:

Guideline 1: The NCISP and the Intelligence and Fusion Processes
Guideline 2: Mission Statement and Goals
Guideline 3: Governance
Guideline 4: Collaboration
Guideline 5: Memorandum of Understanding and Nondisclosure Agreement

---

[*] Homeland Security Advisory Council, *Intelligence and Information Sharing Initiative, Homeland Security Intelligence & Information Fusion*, Department of Homeland Security, April 2005, http://www.dhs.gov/xlibrary/assets/HSAC_HSIntelInfoFusion_Apr05.pdf.
[†] Ibid.
[‡] Homeland Security Advisory Council, Intelligence and Information Sharing Initiative. https://www.dhs.gov/xlibrary/assets/HSAC_HSIntelInfoFusion_Apr05.pdf.

Guideline 6:   Database Resources
Guideline 7:   Interconnectivity
Guideline 8:   Privacy and Civil Liberties
Guideline 9:   Security
Guideline 10: Facility, Location, and Physical Infrastructure
Guideline 11: Human Resources
Guideline 12: Training of Center Personnel
Guideline 13: Multidisciplinary Awareness and Education
Guideline 14: Intelligence Services and Products
Guideline 15: Policies and Procedures
Guideline 16: Center Performance Measures
Guideline 17: Funding
Guideline 18: Communications Plan*

In 2007, the *National Strategy for Information Sharing* identified state, local, and tribal entities as "full and trusted partners with the federal government in our nation's efforts to combat terrorism."† The document identified three areas in which federal departments and agencies could better coordinate efforts with their state, local, and tribal counterparts, and also introduced, without naming it specifically, the concept of sharing information in an all-crimes, all-hazards environment. The document mentioned activities that would enhance information sharing, more specifically emphasize the following:

- Foster a culture that recognizes the importance of fusing information regarding all crimes with national security implications, with other security-related information (e.g., criminal investigations, terrorism, public health and safety, and natural hazard emergency responses).
- Support efforts to detect and prevent terrorist attacks by maintaining situational awareness of threats, alerts, and warnings, and develop critical infrastructure protection plans to ensure the security and resilience of infrastructure operations (e.g., electric power, transportation, telecommunications) within a region, state, or locality.
- Develop training, awareness, and exercise programs to ensure that state, local, and tribal personnel are prepared to deal with terrorist strategies, tactics, capabilities, and intentions, and to test plans for preventing, preparing for, mitigating the effects of, and responding to events.‡

In 2008 and 2012, the fusion centers' role was further refined and defined by the publication of *Baseline Capabilities for State and Major Urban Area Fusion Centers*, and the updated *National Strategy for Information Sharing and Safeguarding*. In these documents, the further development of expertise within the NNFC is encouraged and the Network is again identified as a critical partner in the collection and analysis of threat-related information.

---

* Department of Justice, *Fusion Center Guidelines: Developing and Sharing Information and Intelligence in a New Era* (Washington, DC: Department of Justice, 2006).
† The White House, *National Strategy for Information Sharing, Success and Challenges in Improving Terrorism-Related Information Sharing*. https://www.whitehouse.gov/sites/default/files/docs/2012sharingstrategy_1.pdf
‡ Ibid.

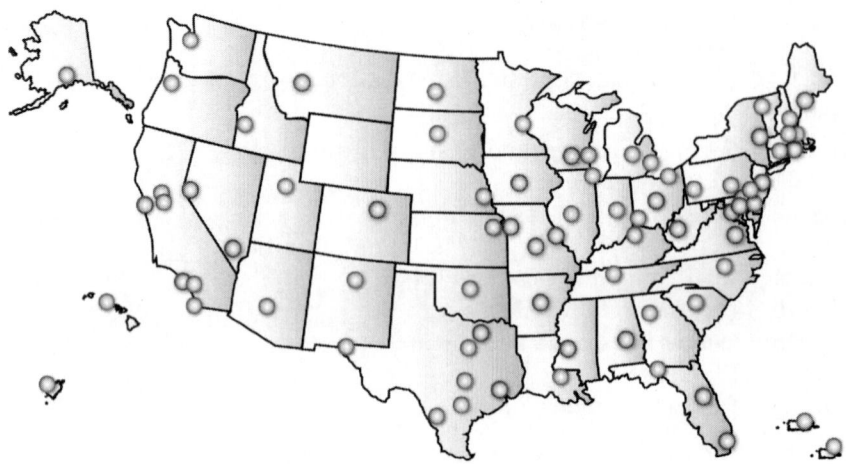

Figure 24.1 A map of the Department of Homeland Security–designated Fusion Centers. (From the Department of Homeland Security, *2013 National Network of Fusion Centers Final Report*.)

With the exception of Wyoming, each state has at least one fusion center, and several states have multiple fusion centers, which may be missioned to support specific local agencies or to provide analytic support on particular issues. While the fusion center concept is not new,[*] centers following the homeland security model began organizing in 2002, growing to twenty-nine in the year 2005, and more than doubling in 2007 to sixty.[†] The number currently stands at seventy-eight, with centers in both Guam and Puerto Rico (see Figure 24.1).

A common theme woven throughout several national-level documents is the idea that the principal activity of a fusion center is to facilitate the sharing of information. Recognizing the unique capabilities of its member centers, and the emergent nature of the NNFC, the mission statement for the national network remains quite general in nature:

> The mission of the National Network is to use the capabilities unique to the NNFC and the state and major urban area fusion centers included in the National Network to receive, analyze, disseminate, and gather threat information and intelligence in support of state, local, tribal, territorial, private sector, and federal efforts to protect the homeland from criminal activities and events, including acts of terrorism.[‡]

In addition, considering that many fusion centers have specific topical focus, it can be accurately stated that their primary goal is to enhance public safety through the timely sharing of relevant information with the most appropriate audience.

---

[*] The New York City Police Department had a system in place in the 1990s to ensure information was shared between the detective, patrol, and organized crime sections.
[†] Department of Homeland Security, *2013 National Network of Fusion Centers Final Report*.
[‡] Department of Homeland Security, *National Strategy for the National Network of Fusion Centers*.

INFRASTRUCTURE PROTECTION

Figure 24.2 The fusion process overview. (From the Baseline Capabilities for State and Major Urban Area Fusion Centers, 2008.)

FUSION CENTERS AND THE INFRASTRUCTURE PROTECTION MISSION

Just as a state fusion center acts as the touchpoint and conduit to inform the national threat picture with information derived from state and local agencies, national-level threat information can be applied to evaluate threats and vulnerabilities to local assets. Fusion Centers also serve a critical role in the risk management process and contribute domain-specific expertise to inform vulnerability reduction activities and can ultimately provide the basis for intelligence-led response and prioritization. Figure 24.2 represents the fusion center's adaptation of the traditional intelligence cycle.

FUSION PROCESS AND INFRASTRUCTURE PROTECTION OVERLAY

Step 1: Planning and Requirements Development—Fusion center leadership and stakeholders develop key intelligence questions and information requirements, and implement strategies and sources to gather the needed data.

Infrastructure Protection Focus—*Set Goals and Objectives*:

- Provide forum for stakeholder identification and engagement.
- Identify stakeholder information requirements.
- Offer access to multidisciplinary subject-matter experts to identify assets.

Step 2: Information-Gathering and Recognition of Indicators and Warnings—Ongoing two-way outreach and strategic communication programs are executed in order to encourage reporting.
Infrastructure Protection Focus—Identify Assets, Systems, and Networks:

- Establish access to databases and public/private agency information.
- Collect site-specific and network information to assist in the creation and/or population of critical infrastructure-related databases.
- Use multidisciplinary subject-matter experts to identify assets.

Step 3: Processing and Collation of Information—Raw information is organized and categorized.
Infrastructure Protection Focus—*Assess Risks:*

- Conduct strategic threat assessments.
- Conduct site- and sector-specific threat assessments.
- Conduct tactical analyses for emerging threats and on-site responses.
- Conduct geographic trend analyses.
- Participate in vulnerability assessments.
- Access tips, leads, and suspicious activities.

Step 4: Analysis and Production—Analytic rigor is applied to information to produce finished intelligence.
Infrastructure Protection Focus—*Prioritize:*

- Analytic capabilities to support decision-making process
- Access to multidisciplinary subject-matter experts to identify assets

Step 5: Dissemination—Finished intelligence is distributed among parties with a need to know in order to answer specific questions and drive further information gathering.
Infrastructure Protection Focus—*Implement Programs:*

- Develop sector-specific products.
- Inform deployment of resources and field personnel, including on-site/incident response, plain-clothes, and undercover teams.
- Provide awareness training for stakeholders.

Step 6: Evaluation—Further sharpening and refinement of intelligence questions, gaps, and effectiveness of efforts.
Infrastructure Protection Focus—*Measure Effectiveness:*

- Refinement of information requirements, collection processes, product development, and dissemination mechanisms
- Continual reevaluation of threats and the impact of regional risk

See the Critical Infrastructure Risk Management Framework at Figure 24.3.

INFRASTRUCTURE PROTECTION

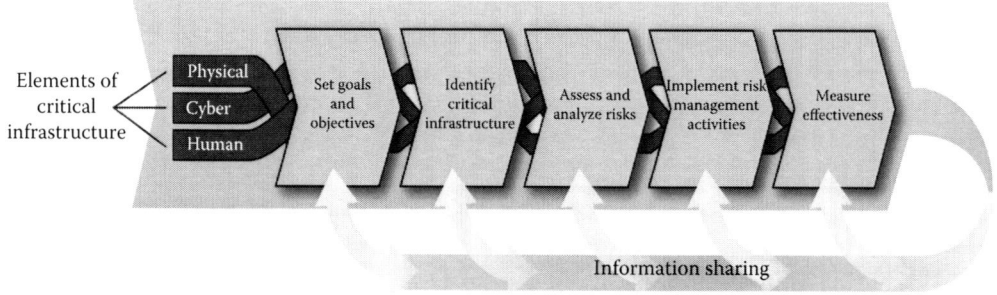

Figure 24.3 The Critical Infrastructure Risk Management Framework. (From the National Infrastructure Protection Plan, 2013.)

Case Study: Silver Shield, Nevada's Critical Infrastructure Protection Program

The concentration of resorts and hotels in Las Vegas provides ample opportunity for bad actors to exploit the inherent vulnerabilities of such facilities. With over 40 million visitors per year, nineteen of the world's largest hotels, and nearly 7,000 private-sector security officers with the unique capability to observe any manner of suspicious activity, the state of Nevada implemented Silver Shield, a critical infrastructure protection program that provides the unifying structure for the integration of existing and future critical infrastructure protection efforts within the state of Nevada. Silver Shield conducts site visits to identify, capture, assess, and catalogue high-priority features of critical infrastructure throughout Nevada. This process consists of making initial contact with the owners/operators or designees of those facilities, systems, or assets that are representative of the area's critical infrastructure, mass population centers, or sites of symbolic importance.[*] Additionally, Silver Shield collects critical infrastructure information and disseminates alerts via a web-based tool known as the Critical Infrastructure Protection System (CIPS), designed for police, fire, and emergency management first responding agencies.[†]

Case Study: Colorado Information Analysis Center, Rubicon Teams

In conjunction with the Colorado Information Analysis Center, the Rubicon Team is charged with the identification, prioritization, and assessment of the state's critical infrastructure assets. To accomplish this mission, Rubicon operations are divided into four distinct segments:

- Full-spectrum, integrated vulnerability assessments
- DHS Constellation/Automated Critical Asset Management System (C/ACAMS)
- Critical infrastructure protective measures templates
- Training and education

---

[*] For more details see http://www.uerlv.com/critical.html.
[†] Fusion Center Spotlight—Silver Shield, Nevada's Critical Infrastructure Program, September 2009.

SOFT TARGETS AND CRISIS MANAGEMENT

The Rubicon Team conducts highly detailed, on-site vulnerability assessments as identified by a number of different methodologies. The end product of these assessments is a series of options for the site operator's consideration that are intended to mitigate threats and reduce potential loss of life, property, and economic damage. Each inspection includes

- Current threat streams provided by the CIAC
- Emergency operations capabilities analysis for response to an incident
- Infrastructure support recommendations for planning and development
- Blast mitigation recommendations for most likely and worst-case scenarios
- Enhancements to the current security profile with low cost/no cost recommendations
- Design suggestions for current elements and planning suggestions for the future
- Information technology recommendations for protection of Internet and internal technologies
- Current specifications directly in line with the Department of Defense standards for all military and defense industry based terrorist threats, domestic threats, and criminal acts
- Detailed out-briefing to site representatives with suggested recommendations, plus a CD-ROM containing the assessment data*

See the CIAC Full-Spectrum Vulnerability Assessment Chart at Figure 24.4.

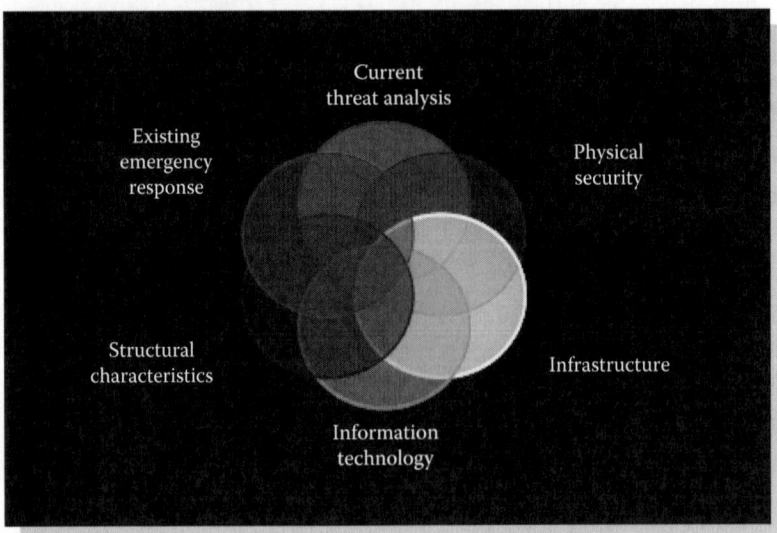

Figure 24.4 The Fusion Center spotlight. (From the Colorado Information Analysis Center Rubicon Teams, February 2009.)

---

* Fusion Center Spotlight—Colorado Information Analysis Center Rubicon Teams, February 2009.

FUSION LIAISON OFFICER PROGRAMS AND INFRASTRUCTURE PROTECTION

One very effective way to complete these information transactions is through an active fusion liaison program. Fusion Liaison Officers, or FLOs, are individuals who work in a variety of disciplines, including Private Sector, Fire Service, Emergency Management, Law Enforcement, and Public Health officials, and who are specially trained by their state, local, or urban fusion center on assessing suspicious activity. FLOs are force multipliers in a fusion center's Area of Responsibility (AOR) and create an additional network of networks within the AOR. FLOs are comprised of individuals with highly specialized knowledge of their communities and the infrastructure features contained therein. Additionally, FLOs add more value and utility to reports of suspicious activity by adding local context and potentially connecting seemingly raw and isolated data points.

In addition to FLO programs, two marquee national programs endeavor to increase the aggregation of suspicious activity reports and sensitize the public to the information needs of the intelligence community. These programs include:

- The Nationwide Suspicious Activity Reporting Initiative: A collaborative effort between the U.S. Department of Homeland Security, the Federal Bureau of Investigation, and the Program Managers for the Information Sharing Environment and the Criminal Intelligence Coordinating Council, this initiative advocates for the responsible gathering, reporting, and analysis of reports of suspicious activity. With a focus on protecting the civil rights and civil liberties of Americans, suspicious activity that meets a functional standard can be shared across the intelligence community to better understand behaviors that may be reasonably considered indicative of terrorist planning.
- Building Communities of Trust: Grounded in the concept of community-oriented policing, this program focuses on developing trust among law enforcement, fusion centers, and their constituent communities by engaging them in two distinct ways; (1) by hosting community roundtables with citizen groups and law enforcement leaders, and (2) by providing materials to support individual engagement with owners of vulnerable critical infrastructure. As of January 2014 these initiatives have been implemented in over fifteen urban areas across the country.

The combination of reports of suspicious activity as a result of fusion center outreach and FLO programs can be indications of inherent vulnerabilities within specific features of critical infrastructure and, when analyzed by trained professionals, the resulting output can be leveraged by decision makers within the AOR to prioritize vulnerability reduction activities.

SUMMARY AND KEY TAKEAWAYS

- The terrorist attacks of September 11, 2001, revealed several weaknesses in how the United States shares information among and between different levels of government.

- One of the key responsibilities of the National Network of Fusion Centers is the analysis of the vulnerabilities of the U.S. critical infrastructure, more specifically those assets that present as "soft targets."
- The intelligence cycle can be leveraged to assist in analyzing soft targets and reducing their vulnerability.
- Fusion center liaison officer programs provide both fusion centers and the intelligence community with highly specialized information and expertise with relation to a specific jurisdiction.

25

Complex Coordinated Attacks

J. Howard Murphy

Contents

Attackers' Tactics, Techniques, and Procedures 439
What Differentiates a CCA from an Active Shooter Incident or a Mass Casualty
Incident (MCI)? 440
    Different Outcomes 440
    Multiple Responses 443
    Indicators and Characteristics 444
    Designed to Overwhelm Responder Resources 445
        Asymmetric: A Smaller Group Attacking a Site Secured by a Much
        Larger Force 446
    Examples of Complex Coordinated Attacks 446
        Nazran, Republic of Ingushetia, Russia (2004) 446
        Beslan, Russia, School Massacre (2004) 448
        Mumbai, India (2008) 451
    Indicators and Characteristics of the Attack 452
        Paris, France (2015) 453
    Indicators and Characteristics of the Attack 455
Conclusion 455

Complex Coordinated Attacks (CCAs) signal a need for a paradigm shift in emergency management and emergency response operations. CCAs are unlike an active shooter incident that involves "an individual actively engaged in killing or attempting to kill people in a confined and populated area."[*] Implicit in the definition of an active shooter incident is that the subject's criminal actions typically involve the use of firearms.[†] CCAs are

---

[*] DOJ/FBI, 2013, p. 44.
[†] U.S. Department of Justice, "A Study of Active Shooter Incidents in the United States between 2000 and 2013," Federal Bureau of Investigation (2013).

characterized as high-risk, coordinated assaults frequently directed against multiple targets or infrastructure complexes (e.g., government or commercial buildings), using mobile groups to circumvent security measures, in which attackers employ at least two distinct classes of weapons systems (i.e., Improved Explosive Devices [IEDs], indirect fire, direct fire, firearms, etc.).* Both active shooter and CCA events will result in mass casualty incidents in which medical resources are overwhelmed by the number and/or severity of injuries. While numbers of casualties may be similar, there are some more obvious differences between an active shooter incident and a CCA provided within this chapter.

Unlike an active shooter/armed assailant or a more typical incident that a conventional emergency response can manage, a CCA is an incident that requires a greater number of resources, specialized resources, and strategic thinking to successfully counter the threat(s). Countering a CCA requires emergency response agencies to effectively and quickly coordinate their response operations. Additionally, this requires developing and maintaining relationships throughout the preparedness cycle of planning, organizing/equipping, training, exercising, and evaluating/improving processes and plans. The greater the effort put forth before the CCA, the more effective the response to counter the CCA will likely be.

While there is not an agreed-upon definition of a CCA, there are recognizable characteristics that allow for definable triggers for levels of emergency response operations. It may become best practice for communities to characterize and define a CCA based upon their respective emergency response capability and capacity, much as they do in defining mass casualty incidents (MCIs).

A CCA differs from an active shooter incident:

1. CCAs involve active shooter incidents, though there will be multiple attackers with preplanning and coordination often only discovered after the attack during the investigation phase.
2. CCAs require responders to maintain situational awareness that focuses on a larger area than the initial or apparent target.
3. CCAs possess a "bigger picture" aspect, as strategy is involved to effectively stage and deploy emergency response assets and capabilities in accordance with the current and forecasted threat(s).

In contrast to CCAs, characteristics of an active shooter incident include

1. Single location
2. Shorter duration than many CCAs
3. Contained
4. Narrow threat
5. Selective goal (criminal instead of terrorist)
6. Not coordinated
7. Typically involves a single assailant at a single location

From emergency management and emergency response management perspectives, a CCA is a dynamic incident with little or no warning mounted by multiple attackers and/or

---

* Jacobson, C., "ISAF Violence Statistics and Analysis Media Brief, Sept. 29, 2011," available at http://www.rs.nato.int/article/isaf-releases/isaf-violence-statistics-and-analysis-media-brief-sept.-29-2011.html.

affecting multiple locations, involving the use of atypical tactics, techniques, procedures, and weaponry. Such incidents require emergency responders to treat the initial incident as a precursor to another attack or attacks.

There currently exist eight generally agreed-upon characteristics of attackers involved in CCAs. Based upon studies of multiple CCAs or swarm attacks—discussed in detail later in this chapter—these eight characteristics include

1. *Number of attackers:* An incident involving more than one attacker should immediately raise concern and awareness among emergency responders and emergency managers, as two or more attackers working together is evidence of planning and a conspiracy. Initial responders should immediately alert each other and communicate to follow-on emergency responders that more than one attacker is involved. While the involvement of two attackers itself does not immediately indicate a CCA, it is one of the strongest indicators of the possibility of a CCA. All responders should adopt a posture appropriate for facing an aggressive, well-armed, and trained attack force whose only objective may be the discriminate or indiscriminate killing of as many people as possible. The involvement of three or more attackers almost certainly constitutes a CCA as coordination is presumed, as is complexity of the attack operation.
2. *Attacker attire:* The attire worn by attackers may provide some insight as to the level of their preparedness and training. Emergency responders should be aware of the following considerations pertaining to attackers' attire:
   - Similar attire
   - Use of camouflage or other tactical colors (e.g., black) and gear
   - Use of hoods or balaclavas
   - Evidence of the use of body armor

   These indicators reflect the level of planning and coordination by the attackers. For example, the more similarly the attackers are dressed, especially if in paramilitary attire, the higher the level of suspicion among emergency responders to a CCA.
3. *Body armor:* The use of body armor is particularly important information. Initial responders should immediately alert each other and communicate to follow-on emergency responders if the attackers are using body armor. Law enforcement responders should consider using weapons and ammunition that are most appropriate for confronting an attacker who is wearing body armor. Until proven otherwise, law enforcement responders should assume that attackers are using the most protective body armor available.
4. *Weapons:* The types of weapons used by attackers are also indicators of a CCA. Attackers using assault weapons should raise the suspicion of a CCA. Emergency responders should consider attackers using multiple weapon types, especially grenades and improvised explosive devices, as extremely dangerous adversaries. Initial responders should immediately alert each other and communicate to follow-on emergency responders that attackers are using assault weapons. To the extent possible, law enforcement responders should arm themselves with assault weapons rather than relying only on their department-issuedsidearms. Emergency responders should emphasize the necessity of working in organized

contact teams, avoid single-officer entries, and attempt to avoid single-officer contact with attackers.
5. *Aggressiveness:* It is difficult to describe the concept of attacker aggressiveness in a meaningful way without stating that emergency responders will recognize it when they see it. Attackers demonstrating no fear or hesitation in confronting responders, moving aggressively, or even using advanced tactics against emergency responders, should be considered extremely dangerous adversaries. Initial responders should immediately alert each other and communicate to follow-on emergency responders that attackers are using aggressive tactics. Emergency responders should emphasize the necessity to work in organized contact teams, avoid single-officer entries, and attempt to avoid single-officer contact with attackers.
6. *Declarative statements:* Attackers' statements made to survivors, media, emergency responders, or others may provide insight into the attack. Declarative statements alone are not a good indicator of a CCA. However, declarative religious, ethnic, or terrorism-related statements in conjunction with other elements previously described strengthen the indications of a CCA. These declarative statements may take the form of yelling religious phrases as they kill, asking victims to identify their religion, or even stating outright to victims or survivors that they are part of a terrorist organization. A key differentiator with declarative statements is insight into the mindset of the attackers. For example, are the attackers likely to take hostages and negotiate a settlement, or is it more likely that hostage-taking would be a stalling tactic with the intent of fortifying positions or continuing the killing of people? Declarative statements may help incident managers in their decision-making process when contemplating entries or counterattacks. Initial responders should communicate declarative statements to on-scene incident commanders.
7. *Tactical movements/behavior:* Observed attacker behavior can also provide strong indications of a CCA. Emergency responders should be alert to tactical behavior on the part of attackers, such as (but not limited to) coordinated movements, use of cover, bounding overwatch (leapfrogging), cover fire, and other combat tactics. Emergency responders observing attackers using tactical behavior should consider them extremely dangerous adversaries. Initial responders should immediately alert each other and communicate to follow-on emergency responders that attackers are using tactical movements. Emergency responders should emphasize the necessity to work in organized contact teams, avoid single-officer entries, and attempt to avoid single-officer contact with attackers.
8. *Remote command and control:* Initial responders should be alert to indications that attackers are communicating with some remote command and control entity directing the attack. Indications may come from direct observation of attackers communicating on cell phones, satellite phones, or by other means, or they may come from witnesses and survivor statements. Remote command and control is a significant indicator of a CCA. Initial responders should immediately alert each other and communicate to follow-on emergency responders that attackers are communicating with each other or other entities. Efforts should be taken to disrupt attackers' communications. Additionally, emergency responders should

emphasize the necessity to work in organized contact teams, avoid single-officer entries, and attempt to avoid single-officer contact with attackers.

Sound tactical practices are appropriate in a CCA response. The tactical differences for consideration lie in recognizing the much higher threat risk that is posed by attackers who are perpetrating a CCA. Emergency responders should assume that the attackers are well trained, skilled in weapons and tactics, and not afraid to die. Perhaps more importantly, responders should assume that attackers will be prepared to not only engage law enforcement, but also to make an effort to inflict heavy damage and even repel counterattacks.

Emergency responders should consider and provide themselves with all possible tactical advantages. These tactical advantages will vary from agency to agency but may include equipment carried in a patrol car, such as Level IV body armor, assault weapons or rifles, shotguns, and sufficient ammunition. Responders should make every effort to form and work in contact teams. Larger teams of four to six officers are preferred over smaller teams of two to three officers. Responders should engage with multiple contact teams and coordinate their counterattacks and movements against the attackers. Maximizing tactical use of the environment for cover and concealment should be a priority. Consideration might be given to containing attackers or limiting their ability to move to a more target-rich environment. Most importantly, clear communication among contact teams about position, movement, and actions is paramount.

Emergency responders must also consider the behavior of the attackers as it applies to mindset and objectives. It is extremely unlikely that initial responders will know the mindset or objectives of the attackers they face. Nevertheless, these factors should be considered, as they may relate to traditional practices and decision-making when facing shootings and hostage situations. Terrorists have repeatedly demonstrated a commitment to and acceptance of death as a result of their actions, and they have repeatedly shown a commitment to killing. Unfortunately, there is little information to guide decision making in CCAs. Incident Commanders must make the best decisions they can based on the information they have at the time.

ATTACKERS' TACTICS, TECHNIQUES, AND PROCEDURES

The CCA tactics, techniques, and procedures (TTPs) allow attackers to perpetrate multiple or mass casualty incidents, obtain news coverage, and inflict considerable damage prior to the neutralization of the attackers.[*] In some cases, attackers have combined multiple attack, with one set of attackers undertaking shooting sprees and hostage-taking, while another group becomes involved in simultaneous barricade siege incidents.[†] CCAs present other challenges to the emergency response community, including

1. Creating a multihazard incident (i.e., direct fire, hazardous materials, IEDs, conflagrations/fires, etc.)

---

[*] Sullivan, J. and Elkus, A., "Preventing Another Mumbai: Building a Police Operational Art," *CTC Sentinel* 2:6 (2009).
[†] Dolnik, A., "Fighting to the Death," *The RUSI Journal* 155:2 (2010): pp. 60–88.

2. Overwhelming available resources such as personnel, equipment, communications systems, and other essential emergency response assets
3. Perpetrating tiered or parallel incident(s) requiring multijurisdictional, multidiscipline responses and coordination
4. Presenting potential cascading system failures within a community (i.e., transportation, medical, power, economic, etc.)

The concept and terminology pertaining to CCAs involves integration of the terms "complex" and "coordinated" in order to develop a "CCA term." One possible definition of a CCA is derived from two separate terms and definitions, provided below[*]:

Complex Attack: An attack conducted by multiple hostile elements that employ at least two distinct classes of weapons against one or more targets.

Coordinated Attack: An attack that exhibits deliberate planning conducted by multiple hostile elements, against one or more targets from multiple locations.

Complex attacks differ from coordinated attacks in that they lack any indication of a long-term planning process or prior preparation. These hybrid operations, such as the attacks in Mumbai in November 2008, the Nazran Raid in June 2004, the September 2013 Westgate Mall attacks in Nairobi, Kenya, and the November 2015 attacks in Paris highlight the devastating impact that trained, mobile groups of attackers can have.[†] These attacks lasted multiple days, generating a considerable amount of news coverage, especially given the 24-hour international news cycle. The Mumbai attackers are known to have had lines of communication to external groups, that appeared to compound the barricade siege situation that followed the initial assault, while groups of attackers in the attacks mentioned above are suspected to have had such lines of communication.[‡]

While an attack such as the one perpetrated by al-Qaeda on September 11, 2001, across a large geographic region of the United States, or one perpetrated within a local jurisdiction like Mumbai or Nazran can certainly be classified as a CCA, the difference between them is one of scale and scope. The economic costs and impacts related to the attacks described above differed significantly, as does the scope related to geographic area (e.g., the entire U.S. airspace and Eastern Seaboard versus the cities of Mumbai or Nazran).

WHAT DIFFERENTIATES A CCA FROM AN ACTIVE SHOOTER INCIDENT OR A MASS CASUALTY INCIDENT (MCI)?

Different Outcomes

CCAs are "asymmetric" operations (often referred to as swarm attacks), in which there is a mismatch existing between the resources and philosophies of the combatants, and in which the emphasis is on bypassing an opposing military or para-military (e.g., law

---

[*] "ISAF Violence Statistics and Analysis Media Brief," September 29, 2011, Brig. Gen. Carsten Jacobson, http://www.isaf.nato.int/article/transcripts/isaf-violence-statistics-and-analysis-media-brief-sept.-29-2011.html.
[†] Sullivan and Elkus, "Preventing Another Mumbai."
[‡] Dolnik, "Fighting to the Death."

enforcement) force and striking directly at cultural, political, or population targets, is a defining characteristic of fourth-generation warfare. CCAs, therefore, may be considered "asymmetric" or fourth-generation warfare. Just as Coalition forces in Afghanistan and Iraq have faced decentralized, non-state actors (perhaps supported by a rogue nation or two) who understand how big an impact attacks on economies/markets, infrastructure, communications, and cultural icons can have on the Western psyche, U.S. law enforcement and other emergency response and emergency management entities are likely to face similar asymmetric TTPs.

Much of the information on CCAs and the asymmetric tactics used within such attacks originates from the study of swarm attacks. Swarm attacks are high-risk, coordinated assaults sometimes directed against multiple targets or building complexes, using mobile groups to circumvent security measures, allowing attackers to inflict casualties, garner news coverage, and, in recent years, to inflict considerable damage prior to neutralization of the attackers. The term "swarm attack" has appeared relatively recently, primarily in a military context.* The CCA, like a swarm attack, is designed to overwhelm the defenses of the target using an attacking force often smaller and with less armament than the opposing force (e.g., law enforcement, security force). Attackers use a decentralized force against their opponent, emphasizing mobility, communications, unit autonomy, and coordination.†

One aspect of a CCA is that it moves away from the traditional model of a rigid chain of command and control.‡ CCAs, by necessity, rely upon agility, focus, and convergence. Agility is the critical capability that organizations need to meet the challenges of complexity and uncertainty. Focus provides the context and defines the purposes of an endeavor, and convergence is the goal-seeking process that guides actions and effects.§

Countering CCAs, like countering other threats, requires doing two things: finding targets, and hitting targets while avoiding being hit. CCAs involve large numbers of relatively small weapons that, with synchronized actions, allow the attackers to react more rapidly than their opponent(s), resulting in defeat—a "successful" CCA. Senior opposition leadership, whether virtually, in person, or through agents, release resources (attackers) to conduct a CCA(s), but do not control them once released. CCAs require autonomous or semiautonomous operating agents, with effective synchronization and communications among them.

CCAs appear to tie in well with the theories of military strategist John Boyd's concept of quick action based on repeated application of the following steps (the OODA loop)¶:

- *Observe*: Make use of the best sensors and other intelligence available.
- *Orient*: Put the new observations into a context with the old.
- *Decide*: Select the next action based on the combined observation and local knowledge.
- *Act*: Carry out the selected action, ideally while the opponent is still observing your last action.

---

* Ibid.
† Edwards, Sean J.A. (2000). Swarming on the Battlefield: Past, Present, and Future. Rand Monograph MR-1100. Rand Corporation. ISBN 0-8330-2779-4.
‡ Alberts, D., "Agility, Focus, and Convergence: The Future of Command and Control," *The International C2 Journal* (Command and Control Research Program) 1 (2007).
§ Ibid.
¶ Hightower, T., "Boyd's O.O.D.A. Loop and How We Use It." Retrieved November 14, 2014, from http://www.tacticalresponse.com/d/node/226.

Emergency managers and emergency responders must prepare themselves for the next threat or incident, as forecasting a threat and preparing for it is the first step to resiliency. This preparation could involve improving the security plans for critical infrastructure and obtaining resources necessary to effectively counter a forecasted threat. Referring to the OODA loop, emergency managers and emergency responders must make decisions quickly using the information they have available from their initial observations, as decisions must be made and actions taken even when information may be unclear, confusing, or absent.

The first 10 to 15 minutes of a CCA—and, often much longer—are going to be volatile, uncertain, complex, and ambiguous (VUCA). Emergency managers and emergency responders need to be prepared to make decisions even in this VUCA environment. The first arriving responder on scene needs to take charge of the scene, regardless of rank, and make the decisions during that first phase of emergency response operations even if those decisions are often being beyond the responder's typical level of decisionmaking.

During a CCA, emergency responders need to:

- Make the best possible decisions from the onset using available information.
- Make decisions using little information and information that may be difficult to prioritize.
- Be prepared to make decisions in a chaotic and stressful environment.
- Be mindful of future implications (second- and third-order effects) of decisions and consider the overall response.

As quickly and as accurately as possible, emergency managers and emergency responders should obtain both situational awareness and domain awareness. The U.S. Coast Guard provides that situational awareness is created by identifying, processing, and comprehending critical elements of information that are relevant to the immediate situation.[*] Of course, emergency responders work to obtain and use situational awareness on a daily basis in their response operations. Domain awareness, however, involves understanding what is occurring throughout all areas of responsibility. For incident or area commanders and their respective staffs, their responsibilities should include taking the big view, the entire jurisdiction, or sector of a jurisdiction considered their area or domain of responsibility.

Situational awareness is a subset of domain awareness, and although there is a relationship between the two, a major distinction exists. While situational awareness pertains to what is happening at a specific scene, domain awareness involves many of the same concerns but on a larger scale. That is, domain awareness focuses its attention not only on the individual situation that responders are handling but also on the common elements among multiple scenes. While line-level responders are managing individual situations, commanders generally are less interested in the particulars of each situation and more interested in the cumulative effect of multiple situations.

The relationship between those responsible for managing situations and those responsible for domain awareness is interdependent; each relies on the other to share information in order to be successful. Emergency responders can develop situational awareness based on information that is passed on to them by those who are responsible for domain

---

[*] U.S. Coast Guard. (2014). Situational Awareness. Team Coordination Training (TCT). Retrieved from http://www.uscg.mil/auxiliary/training/tct/.

awareness, such as command personnel, dispatch, or others. Conversely, domain awareness is developed, in part, by responders in the field sharing information with command personnel through dispatch or direct communication with the command post.

Area Command, the command structure often used in major incidents, relies heavily on information gleaned from situational awareness to establish context, meaning, and patterns; or, in other words, to develop a sense of the bigger picture. Domain awareness is not only created by relaying critical information from the scene; that information must also be assessed and analyzed. The goal of those who provide domain awareness is to not only pass on information but to pass on the right information at the right time. The right information at the right time is actionable information, or what is commonly referred to as intelligence.

Emergency responders in all disciplines rely on accurate and appropriate information to avoid potential dangers at the scene, determine the necessary and appropriate equipment, allocate personnel necessary to manage a scene, and identify appropriate response actions. Although this information is largely gathered based on observations and assessment(s) made by the emergency responder(s) upon arrival at the scene, information is sometimes given to the responder(s) while they are en route to the scene. Emergency responders will generally perform better and more safely when they have information prior to arriving at a scene.

In the context of a CCA, emergency responders need to have awareness of the situation to ensure safety and to determine the best course of action. In addition to the information based on their personal observations and scene assessment, responders also rely on those who have domain awareness to forward information that may be pertinent in effectively managing the situation with which they are dealing.

Once an incident is identified as a CCA, although it may not be immediately apparent, it is critical to ensure that the proper command and management structure exists to coordinate a response. Most emergency responders are familiar with establishing an on-scene command structure using the basic elements of the Incident Command System (ICS) and either a single Incident Commander or a Unified Command. In most incidents, this is a sufficient command structure; however, in the case of a CCA, it is critical to establish a higher level of command that includes agency administrators and senior elected officials. This action ensures a command structure is in place that is able to view the incidents, and their impacts, from a broader perspective, and to respond accordingly.

As a CCA unfolds, no single on-scene command will be able to direct and control the response resources necessary to cope with what is occurring. Furthermore, it is likely that the on-scene command will not have the proper authority to enact and respond to a rapidly escalating incident that may span multiple jurisdictions. Activating a Department Operation Center (DOC), Emergency Operations Center (EOC), or other Multi-Agency Coordination (MAC) facility should be considered immediately. It is also important to ensure that these facilities are staffed and managed with senior agency and elected leaders who have broad authorities to initiate actions based upon statutory or constitutional provisions.

Multiple Responses

CCAs pose considerable problems for law enforcement and other emergency response agencies seeking to develop countermeasures within evolving counterterrorism doctrines,

SOFT TARGETS AND CRISIS MANAGEMENT

tactics, techniques, and procedures. CCAs also pose problems within counterterrorism policymaking circles. The scale of such attacks and the range of targets, which are often attacked simultaneously, means responders must themselves remain mobile, while those negotiating or responding from law enforcement agencies will also have to be highly trained and well-coordinated to manage short-term fluid incident dynamics. The more substantive and permanent effects of such attacks are likely to be the political ramifications that come with acts of terrorism. It is operationally and strategically important, therefore, to develop countering CCA policies, as well as TTPs, as part of the community's emergency preparedness cycle of planning, organizing/equipping, training, exercising, and evaluating/improving (see Figure 25.1).

Indicators and Characteristics

Given the aforementioned descriptions of CCAs, one could convincingly characterize many historical attacks as CCAs, including: the September 11, 2001 attacks; the Mumbai attacks in 2008; the Beslan school siege in 2004; and, by strict interpretation of the definition, the recent *Charlie Hebdo* attack in Paris in January 2015; and the December 2015 San Bernardino attack. The following characteristics are typically present during CCAs and were certainly present at each of these attacks and others described later in this chapter:

- Designed to overwhelm responder resources.
- Various types of weapons used.
- Involved multiple locations (targets) within a jurisdiction or a large facility (e.g., a university, mall, factory, large school, etc.).
- Asymmetric—a smaller group attacking a site secured by a much larger force.

Figure 25.1 The national preparedness cycle. (From the Federal Emergency Management Agency.)

444

- While attack groups may be initially coordinated by a command hierarchy, individual attack elements had autonomy to respond and adapt to changing counterattack contingencies.

As mentioned earlier in this chapter, CCAs such as those mentioned above also present the following challenges to the emergency response community, including

1. Creation of a multihazard incident (e.g., direct fire, hazardous materials, IEDs, conflagrations/fires, etc.)
2. Overwhelming available resources such as personnel, equipment, communications systems, and other essential emergency response assets
3. Involvement of tiered or parallel incident(s) requiring multijurisdictional, multidiscipline responses and coordination
4. Present potential cascading system failures within a community (e.g., transportation, medical, power, economic, etc.)

Designed to Overwhelm Responder Resources

Evolving CCA tactics, techniques, and procedures have increased the capacity of the attacks to overwhelm tactical responses and security measures designed to react to attacks directed against single targets. As documented in the Nazran raid, the attackers used a variety of methods, such as diversionary attacks, roadblocks, attacks on security infrastructure, and targeted assassinations.* Although hostages were taken in Nazran, this was neither a direct aim nor an improvised aspect of the attack, as it appears the attacking units needed to remain mobile to maximize their impact. The Nazran attackers, like other perpetrators of CCAs, overwhelmed law enforcement and other emergency responders resulting in significant damage to infrastructure and numerous casualties—hundreds, in the cases of Nazran, Beslan, Mumbai, and Paris.

CCAs can be directed against multiple targets in coordinated operations, challenging the ability of law enforcement and other response agencies to respond. In rare cases, groups of attackers and networks demonstrate a capacity to learn, tactically and strategically, while engaged in operations, thus requiring continuous evolution of counterattack TTPs within the emergency responder community.

Each of the following attacks, with the exception of the Boston Marathon bombings, involved attackers converging onto or into a large facility. Examples include the following:

1. Nazran (2004): Interior Ministry Building and Police buildings/headquarters
2. Beslan (2004): Multiple areas and floors of a large school building
3. Mumbai (2013): Multiple hotels, a religious center, and transportation centers
4. Nairobi (2013): Multiple areas of the large Westgate Mall
5. Paris (2015): Multiple restaurants and cafes, a stadium, and a concert hall in a metropolitan area

---

* Ibrahaev, S., "Nightly Attack," June 22, 2004, Kavkaz-Center. Retrieved November 18, 2014, from http://www.kavkazcenter.com/eng/content/2004/06/22/2905.shtml.

Asymmetric: A Smaller Group Attacking a Site Secured by a Much Larger Force

Each of the following attacks were designed to overwhelm the defenses of the target using an attacking force smaller and with less armament than the opposing force.* Each of the attacks involved relatively small groups of attackers with large numbers of relatively small weapons that, with synchronized actions, allows the attackers to react more rapidly than their opponents.

While attack groups may have been initially coordinated by a command hierarchy, individual attack elements were given autonomy to respond and adapt to changing counterattack contingencies. Attackers used a decentralized force against their opponents, emphasizing mobility, communications, unit autonomy, and coordination.† Senior opposition leadership, whether virtually, in person, or through agents, released resources (attackers) to conduct the attacks, but did not control them once released.

Examples of Complex Coordinated Attacks

While these examples do provide typical characteristics of a CCA, they cannot possibly provide every possible indicator, characteristic, or type of target of future CCAs. However, they are useful in developing a general understanding of the nature of CCAs to assist course participants in anticipating and recognizing indicators and targets within their respective communities.

Nazran, Republic of Ingushetia, Russia (2004)

The Nazran raid was a large-scale attack (or raid operation) conducted in the Republic of Ingushetia, Russia, on the night of June 21–22, 2004, by a large number of mostly Chechen and Ingush militants (see Figure 25.2). Everything started June 21 at 10:00 pm (local time), when the first rocket-propelled grenade exploded in the building of Ingushetian police administration (Interior Ministry). Simultaneously, battles started in several cities and towns in the Republic.‡

A large unit of the Ingushetian Armed Forces, consisting mainly of Ingushetian fighters (mujahideen), attacked the positions of Russian authorities, police, and members of the Russian Federal Security Service (FSB). The battles took place in Nazran, Karabulak, Troitskaya, Sleptsovskaya, and Orjonikidzevskaya, as well as on the Rostov-Baku arterial highway, which Russian authorities called the Caucasus Motorway ("Kavkaz Motorway"). During the first hours of the battle, Ingushetian attackers took control of the police headquarters of the Republic in the Ingushetian Interior Ministry. The attackers set the building on fire shortly thereafter. Sources in Ingushetia reported that the Ingushetian pro-Russian police chief, Interior Minister Kostoyev, was killed.§

The Ingushetian Armed Forces (the attackers) took control of a pro-Moscow police checkpoint on the Rostov–Baku Highway near the village of Yandare, Nazran District, as well as a number of checkpoints on the approaches to Nazran, and on the road leading toward the city of Magas. Eyewitnesses reported that they spotted armed persons wearing

---

* Edwards, *Swarming on the Battlefield*.
† Ibid.
‡ Ibrahaev, S., "Nightly Attack."
§ Ibid.

COMPLEX COORDINATED ATTACKS

Figure 25.2 Map of the Northern Caucasus of Russia, including the Republic of Ingushetia, Russia. (From U.S. Army Foreign Military Studies Office, http://fmso.leavenworth.army.mil/)

green headbands at the checkpoints. The attackers were reportedly stopping and inspecting vehicles.*

The Ingushetian Armed Forces attacked two police checkpoints in the Sunzha District. Attackers burned district police stations and other facilities belonging to pro-Moscow police and to Russian "invaders" in Orjonikidzevskaya and Karabulak. In Nazran, the attackers destroyed the entire pro-Russian police department with 30 policemen in the barracks. Dozens of special police commandos (Russian OMON) were killed and wounded in Karabulak. Local Ingushetian sources also reported that the attackers partially destroyed or burned a Russian military base in Troitskaya, as well as the base of Russian border guards in Nazran.†

Local sources claimed to have seen dozens of Russian corpses and FSB agents in various districts of the Republic. While a confirmed casualty count does not exist, the Kavkaz Center's source assessed that there may have been hundreds of killed and wounded among the attackers and the pro-Russian forces and police. At some point during the battle, reports came in that the attackers were planning to move toward Magas. However, at about 3:00 am (local time) the attackers' units started withdrawing from the cities and villages in an organized manner and moving toward southern parts of the Republic.‡

While the exact number of attackers who took part in the Nazran Raid is unknown, the Russian forces' command stated that there were approximately 100 attackers. Given the scope and scale (broad geographic coverage) of the attack, it appears the attack was conducted by hundreds of attackers, not dozens.§

---

* Ibid.
† Ibid.
‡ Ibid.
§ Ibid.

Characteristics of the Attack:

- Overwhelmed responder resources.
- Involved multiple locations (targets) within the jurisdiction.
- Asymmetric—a smaller group attacked a site secured by a much larger force.
- Individual attack elements had autonomy to respond and adapt to changing counterattack contingencies.

Beslan, Russia, School Massacre (2004)

On September 1, 2004, a large group of students, teachers, and parents was taken hostage by Islamist terrorists in Beslan, North Ossetia, at a school in an agricultural and industrial community of approximately 40,000 (see Figure 25.3).[*] The number of terrorists was assessed to be at least thirty-two. The situation eventually resulted in a mass murder.

The incident lasted 3 days and eventually culminated in hundreds of deaths and injuries;[†] there were over 370 deaths at Beslan during the 3-day period; 331 were civilians, 317 hostages, including 186 children were killed. Over 700 civilians and more than fifty security forces and military personnel were injured.[‡]

As early as 8 to 10 days prior to the assault, the Russian government had developed some intelligence that an assault might take place in a school somewhere around Chechnya. However, no specific intelligence on where the attack would occur was available.[§] The attack occurred on the first day of school, also known as the "Day of Knowledge," when a large number of family members were present at the school.[¶]

On this day, a group of approximately thirty masked individuals, dressed in camouflage, athletic, and civilian clothing, drove onto the school courtyard. The group was well trained and knew the layout of the school from previous surveillance. The attackers jumped from their vehicles, began barking orders, and then began firing their assault rifles into the crowd.[**] The attackers were well armed with assault rifles, vests laden with ammunition, 40 mm grenade launchers, hand grenades, rocket propelled grenades, and protective masks to counter debilitating gas or chemicals.[††] The attackers acted quickly by isolating the hostages and neutralizing their ability to attempt resistance, escape, or contact with anyone outside the school.[‡‡]

---

[*] Giduck, J. (2005). *Terror at Beslan: A Russian tragedy with lessons for America's schools* (1st ed.). Golden, CO: Archangel Group.
[†] Baker, P., and Glasser, S. B. (2004). Russia school siege ends in carnage: Hundreds die as troops battle hostage takers. *Washington Post*, September 4, 2004.
[‡] United States Army TRADOC. (2007). TRADOC G2: Terror operations: Case studies in terrorism. Fort Leavenworth, KA.
[§] Giduck, *Terror at Beslan*.
[¶] United States Department of State. (2005). Beslan school massacre one year later. August 31, 2005.
[**] Uzzell, L. (2004, September). Officials statements on Beslan: A study in obfuscation. The Jamestown Foundation. Retrieved June 6, 2016, from http://www.jamestown.org/single/?tx_ttnews%5Btt_news%5D=2023#.V1btgML2Zdg.
[††] Borisov, S. (2004, September 6). A vision of hell. Transitions Online: Regional Intelligence. Retrieved November 12, 2014, from http://www.tol.org/client/article/12829-a-vision-of-hell.html. This is a subscription service website, so a username and password is required.
[‡‡] United States Army TRADOC. TRADOC G2: Terror operations.

COMPLEX COORDINATED ATTACKS

Figure 25.3 Beslan is a city located in Russia's North Caucasus republic of North Ossetia-Alania. Beslan is located approximately 100 kilometers from Chechnya. (https://www.lib.utexas.edu/maps/chechen.html)

The school had poor security, with one security guard and a police officer who happened to be in the crowd, equipped only with his sidearm. The security guard and police officer engaged the terrorists and killed one attacker; however, they were outnumbered and outgunned. Both were killed within seconds of the attack's start.* The terrorists searched the school and seized people hiding in some classrooms on the first floor. Most

---
* Giduck, *Terror at Beslan*.

of the hostages were taken to the gymnasium and forced to sit on the floor, while some of the female students were dragged into the gymnasium by their hair where they were brutally raped.* All mobile or cellular telephones were confiscated and the hostages were threatened with death if anyone was found with a phone.†

The attackers then barricaded the school, in a manner that appeared to be rehearsed. The attackers were divided into operational teams with specific duties. Some focused on preparing defenses for an assault from local police and military personnel; some started assembling bombs and tripwires; and a small number of terrorists contained the hostages. Snipers were positioned at key locations in the school building complex.‡ Once inside the gymnasium, explosive devices were set at the doorways to prevent the hostages from escaping and prevent anyone else from entering. The terrorist group had brought IEDs—plastic bottles packed with nails, bolts, and screws as shrapnel and homemade dynamite. It was later determined that the terrorists had 66 pounds of explosives, in addition to hand- and rocket-propelled grenades. The terrorists made it clear that they were prepared to die and that anyone who moved would be killed.§

After the attackers seized the hostages and secured the school, the second phase of the operation began. The terrorists began demanding the release of prisoners captured during the Nazran raid and the complete withdrawal of Russian troops from Chechnya, while local police and other security forces gradually cordoned the school area off by establishing a perimeter.¶

On the third day of the seige, Russian security forces used heavy weapons (tanks) and fired into the school to kill and facilitate the capture of the attackers. Eventually, the Russian forces breached the school walls with small explosives. The Russian FSB initially reported that thirty-two terrorists were involved in the Beslan hostage taking and mass murder incident. Intelligence gleaned from a captured attacker indicates that the Beslan attack was intended to create an expansion of fighting across the Caucasus region, and to incite religious and ethnic hatred based on a compulsion for revenge."** Evidence suggests that additional terrorists were involved beyond those killed or captured and that the group may have been as large as seventy terrorists. Some of the attackers likely escaped, and approximately twenty were killed. The Chechen terrorists used this incident to gain international attention and to seek political concessions from the Russian Federation concerning Chechnya.††

Characteristics of the Event:

- Overwhelmed responder resources.
- Involved multiple locations (targets) within the facility.
- Asymmetric—a smaller group attacked a site with limited security, initially, but created significant destruction and death even when countered by a much larger force.

---

* Ibid.
† Ibid.
‡ United States Army TRADOC. TRADOC G2: Terror operations.
§ Giduck, *Terror at Beslan*.
¶ United States Army TRADOC. TRADOC G2: Terror operations.
** Ibid.
†† Ibid.

COMPLEX COORDINATED ATTACKS

- Individual attack elements had autonomy to respond and adapt to changing counterattack contingencies.

Mumbai, India (2008)

On November 26, 2008, a well-planned and coordinated terrorist attack took place in Mumbai conducted by ten attackers/operatives trained by Lashkar-e-Taiba (LeT) (see Figure 25.4). The attackers killed 172 people and wounded hundreds of others with firearms, improvised explosive devices, and grenades during an attack lasting 3 days.[*] The terrorist group had boarded a small boat in Karachi, Pakistan, at 8 am on November 22, then sailed a short distance before boarding a bigger carrier. From the larger vessel, the ten attackers seized an Indian fishing boat, killing the crewmembers (with the exception of the captain, who was later beheaded as they neared the Mumbai shoreline).[†] They then sailed 550 nautical miles along the Arabian Sea, arriving on the shores of Mumbai on November 26.[‡]

The ten attackers divided into four teams. After arriving by sea, the teams split up and each one attacked separate locations. Team One took a taxi to Chhatrapati Shivaji Terminus,

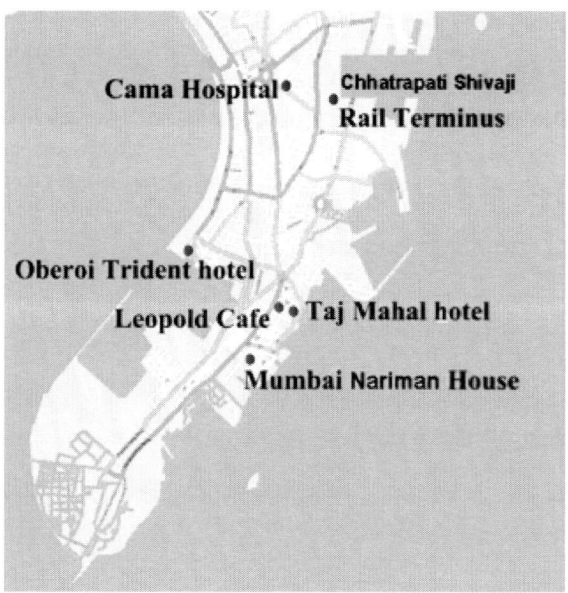

Figure 25.4 Locations of the 26 November 2008 attacks within Mumbai, India. (From Azad, S. and Gupta, A. (2011). A quantitative assessment on 26/11 Mumbai attack using social network analysis. *Journal of Terrorism Research* 2(2). DOI: http://doi.org/10.15664/jtr.187)

---

[*] Jones, S. G. (2012). *Hunting in the shadows*. New York: W. W. Norton and Company.
[†] Gera, Y. K. (2010, February). Mumbai attacks: Confronting transnational terrorism. Retrieved November 8, 2014, from http://fsss.in/agni-volume/2nd/mumbai-terrorattacks-confronting-transnational-terrorism.pdf.
[‡] Sengupta, S. (2009, January 6). "Dossier gives details of Mumbai attacks." *New York Times*. Retrieved November 14, 2014, from http://www.nytimes.com/2009/01/07/world/asia/07india.html.

also known as Victory Terminus, Mumbai's main train station. Each man carried a weapons pack containing an AK-56 rifle, a 9-millimeter pistol, ammunition, hand grenades, a bomb containing a military-grade explosive, and a timer with instructions inscribed in Urdu.[*] Once inside the train station they took out their automatic assault AK-56s, a Chinese version of the AK-47 assault rifle, and opened fire.[†] The attackers began walking through the terminal, killing indiscriminately for 90 minutes before police officers arrived and forced the terrorists to leave. This team then went to the Cama and Albless Hospital, where they again began firing indiscriminately at innocent victims. The terrorists next moved to the Trident-Oberoi Hotel and continued to fire at victims along the way. The attackers had a detailed diagram of the hotel layouts. This team was responsible for approximately 30 percent of the 172 fatalities.[‡]

The second team walked into Nariman House, a commercial-residential complex that was run by the Jewish Chabad–Lubavich organization. The Nariman House attackers killed six of the occupants, including Gavriel Holtzberg and his wife, who was 5 months' pregnant. The third team of attackers went to the Trident-Oberoi Hotel, where they began killing indiscriminately. One of the terrorists was heard involved in a conversation on his cell phone, saying, "… Everything is being recorded by the media. Inflict the maximum damage. Keep fighting. Don't be taken alive … ."[§] The battle in this hotel lasted for 17 hours before the terrorists were killed, and resulted in the deaths of thirty victims.[¶]

The fourth team of attackers entered the Taj Mahal Palace Hotel, after briefly entering the Leopold Café, where they killed ten people with automatic gunfire. Eyewitness accounts from the Taj Mahal Palace Hotel indicated that the terrorists knew their way through the hotel, including the hidden doors and back hallways and, indeed, the terrorists had a detailed diagram of the hotel's layout.[**] Once the terrorists were inside the hotel, they shot at the occupants as they walked from floor to floor. The siege of the Taj Mahal Palace Hotel lasted 60 hours and ended after Indian commandos killed the last four terrorists.[††]

Indicators and Characteristics of the Attack

- Overwhelmed responder resources.
- Involved multiple locations (targets) within the jurisdiction.
- Asymmetric—a smaller group attacked a site secured by a much larger force.
- Individual attack elements had autonomy to respond and adapt to changing counterattack contingencies.
- Handlers monitored live television broadcasts to gather intelligence, which they shared with the attackers.

---

[*] Sengupta, "Dossier gives details of Mumbai attacks."
[†] Gera. Mumbai attacks: Confronting transnational terrorism.
[‡] Rabasa, A., Blackwill, R. D., Chalk, K., Fair, C. C., Jackson, B. A., Jenkins, B. M., Jones, S. G., Shestak, N. and Tellis, A. (2009). *The Lessons of Mumbai*. Santa Monica, CA: RAND Corporation.
[§] Sengupta, "Dossier gives details of Mumbai attacks."
[¶] Rabasa et al., 2009. *The Lessons of Mumbai*.
[**] Gera, Mumbai attacks: Confronting transnational terrorism.
[††] Rabasa et al. 2009. *The Lessons of Mumbai*.

Paris, France (2015)

The attacks in Paris in 2015 were a textbook CCA (see Figure 25.5). Even though Paris's emergency response and counterterrorism capabilities and capacity are robust, as the metropolitan area has extensive CCTV and electronic surveillance, and its emergency responders train and prepare daily for incidents of a significant magnitude, the November 13, 2015, attack initially overwhelmed emergency responders.[*]

A group of nine attackers killed 120 people, wounded nearly 400 others, and inflicted massive damage on a major European city within 3 hours. The attackers combined suicide bombs, diversionary attacks, active shooter tactics, and hostage-taking in such a synergistic fashion that even France's best responders were initially overwhelmed.[†]

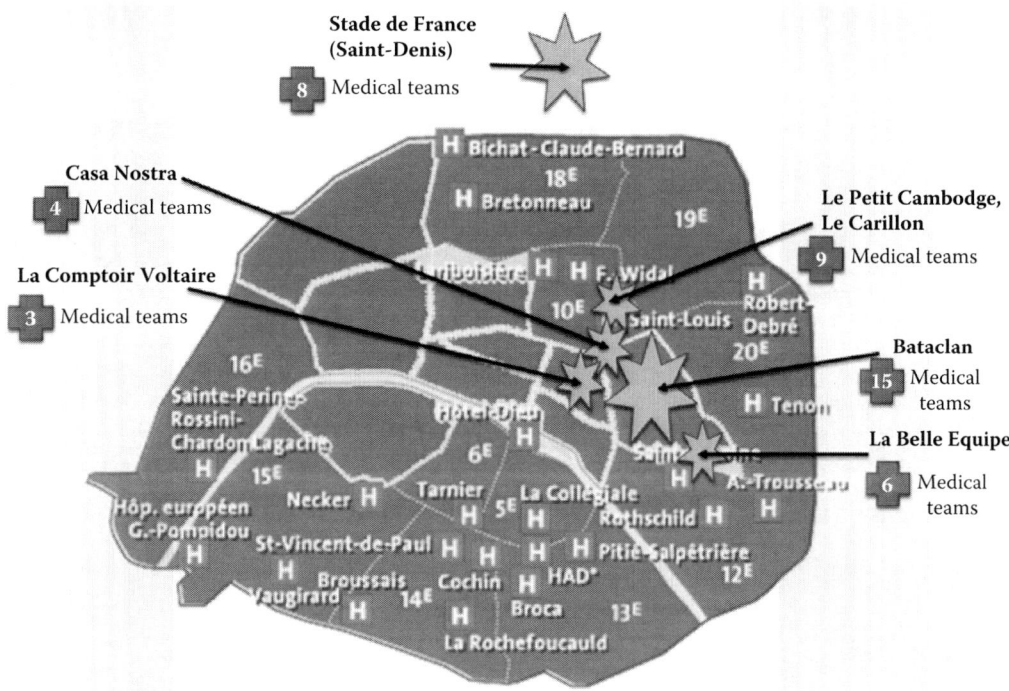

Figure 25.5 Map of Paris attack locations and pre-hospital emergency response. (From Hirsch M, Carli P, Zizard R et al., 2015. The medical response to multisite terrorist terror attacks in Paris. *Lancet* 386(10012): 2535–8. doi: 10.1016/S0140-6736(15)01063-6.)

---

[*] Aisch, Gregor, Wilson Andrews, Larry Buchanan, Jennifer Daniel, Ford Fessenden, Evan Grothian, K.K. Rebecca Lai, Haeyoun Park, Yuliya Parshina-Kottas, Graham Roberts, Julie Shaver, Patrick J. Smith, Tim Wallace, Derek Watkins, Jeremy White, and Karen Yourish, 2015. "Three Hours of Terror in Paris, Moment by Moment." *NY Times.com Europe*. Retrieved from http://www.nytimes.com/interactive/2015/11/13/world /europe/paris-shooting-attacks.html?_r=0.
[†] Ibid.

The attacks began around 9:20 pm, when a suicide bomber detonated an explosive belt outside the Stade de France's Gate D during a soccer match. The explosion took the life of only one victim, but numerous emergency responders were brought in to respond to the blast and investigate the damage. Spectators inside the stadium were told nothing and assumed the blasts were fireworks. The French president, François Hollande, and the German foreign minister, Frank-Walter Steinmeier, were among the crowd.*

At approximately 9:25 pm, shooting attacks occurred at two restaurants in the city center, in which fifteen people were killed and another ten were injured. The attackers were reported to be mobile, driving a black SUV and carrying assault rifles, with one shouting "Allahu Akbar" when firing. Witnesses on the scene reported that the assailants used short, controlled bursts from their weapons and fled quickly but not in a panic, signifying calm and/or practiced attackers. At 9:30 pm, there was another explosion outside Gate H at the Stade de France, killing one person. Emergency responders already on the scene responded, but concerns about another planned blast began to rise.†

Around the same time (and lasting for another 10 minutes), men with machine guns emerged from black SUVs and attacked a series of restaurants with small arms fire. These attackers killed twenty-four people and wounded seventeen. On average, the attacks lasted 3 to 5 minutes each and targeted a busy Friday-night-dinner location. As emergency responders attempted to stop the shooters and render aid to the victims, a suicide bomber detonated an IED at another café. Reports of the gunmen began to come in from the different locations but also sounded somewhat similar. One witness at the last restaurant attacked described one shooter:

> He was standing in a shooting position. He had his right leg forward and he was standing with his left leg back. He was holding up to his left shoulder a long automatic machine gun—I saw it had a magazine beneath it. Everything he was wearing was tight, either boots or shoes and the trousers were tight, the jumper he was wearing was tight, no zippers or collars. Everything was toned black. If you think of what a combat soldier looks like, that is it—just without the webbing. Just a man in military uniform, black jumper, black trousers, black shoes or boots, and a machine gun.‡

Reports began circulating about an attack on the Bataclan Concert Hall where an American band was playing to a sold-out crowd of nearly 1,500 people. The Bataclan attack began when three gunmen stormed the front door and spread out across the floor of the viewing area. They fired into the crowd while also throwing grenades to create more casualties. Patrons not directly affected by the initial attack did not know what was happening until the entire crowd began to move, running for alternate exits and even climbing to the top floors and roof of the Concert Hall and out of windows.§

---

* Ibid.
† Aisch, Gregor, Wilson Andrews, Larry Buchanan, Jennifer Daniel, Ford Fessenden, Evan Grothian, K.K. Rebecca Lai, Haeyoun Park, Yuliya Parshina-Kottas, Graham Roberts, Julie Shaver, Patrick J. Smith, Tim Wallace, Derek Watkins, Jeremy White, and Karen Yourish, 2015. "Three Hours of Terror in Paris, Moment by Moment." *NY Times.com Europe*. Retrieved from http://www.nytimes.com/interactive/2015/11/13/world/europe/paris-shooting-attacks.html?_r=0.
‡ Boffey, Daniel and Henry Zeffman. 2015. "How the Terror Attacks in Paris Unfolded." *The Guardian*. Retrieved from http://www.theguardian.com/world/2015/nov/14/paris-attacks-timeline-of-terror.
§ Aisch et al., "Three Hours of Terror in Paris."

The Bataclan attackers also made statements about Iraq and Syria and shouted "Allahu Akbar." Eventually, they switched from the active shooter tactic to a hostage siege. For the next two hours, police attempted to contain the attackers inside the hall but realized, through social media contact from inside, that the attackers continued to kill innocent civilians, forcing the police to storm the location. When challenged, the attackers detonated their suicide vests. Within approximately 20 minutes, forty people were killed and twenty-eight were injured.[*]

The Paris attacks involved four suicide bombings and shootings at four locations within the span of 3 hours. The perpetrators were mobile and somewhat fluid in their attack, allowing them access to a series of soft targets. In total, 130 people were killed and nearly 400 were injured. Most of those wounded were in serious condition because of the types of weapons used (high-caliber ammunition and explosives).

The medical services of Paris activated their White Plan, which involves the total recall of all medical personnel and releasing of hospital beds to prepare for the surge of casualties.[†]

Indicators and Characteristics of the Attack

- Overwhelmed responder resources (initially).
- Involved multiple locations (targets) within the jurisdiction.
- Asymmetric—a smaller group attacked a site with limited security, initially, but created significant destruction and death even when countered by a much larger force.
- Individual attack elements had autonomy to respond and adapt to changing counterattack contingencies.

CONCLUSION

Emergency management, law enforcement, and other emergency response agencies must continue to learn from the CCAs occurring around the world in order to improve their respective TTPs for effectively deterring, preventing, and responding to CCAs. The accumulated evidence gathered thus far clearly indicates that law enforcement personnel must be prepared to engage attackers rapidly and effectively, lest the attackers cause a significant loss of life in a short period of time. Law enforcement officers and other emergency responders should anticipate preemptive diversionary attacks prior to the main attack, and also expect secondary and tertiary attacks during CCAs.

A way to prepare for these low probability–high consequence incidents (or, rarely occurring incidents with disastrous consequences) is to analyze similar past incidents in order to identify patterns inherent in CCA incidents, along with identifying the mistakes that were made. By gaining a better understanding of these incidents, and by learning from tactical deficiencies and mistakes of the past, the likelihood of repeating the same mistakes and suffering consequences of those mistakes will be lessened.[‡]

---

[*] Ibid.
[†] Ibid.
[‡] Borsch, R. (2007). Stopwatch of death: Part 1. PoliceOne.Com. Retrieved November 14, 2014, from http://www.policeone.com/police-products/police-technology/Emergency-Response/articles/1349058-The-Stopwatch-of-Death/.

The CCAs described earlier in this chapter demonstrate the challenges faced by law enforcement when attacks are well organized and launched by trained and dedicated attackers. While it is almost impossible to completely stop every CCA with zero casualties, speedy and effective law enforcement responses integrated with other emergency response agencies (e.g., EMS and firefighting) is critical to reducing the number of fatalities and injuries.

26
Violent Attacks and Soft Targets

Rick C. Mathews

Contents

Introduction 457
Threat/Vulnerability, Frequency, and Consequence: The Core of Risk Management 458
    Mitigation Measure 1 459
    Mitigation Measure 2 459
        Soft versus Hard Targets 460
Summary and Application 464

INTRODUCTION

On Friday, November 13, 2015, over a period of about 4 hours, approximately 130 people were killed and over 350 injured, many critically, from a set of coordinated attacks against multiple soft targets in Paris, France, by several terrorist teams.* The targets included a concert theater, an outdoor stadium, and a number of restaurants. The Islamic State claimed responsibility. An internal "situation report" produced by the National Center for Security & Preparedness (NCSP) at the University at Albany, SUNY, provides a detailed timeline as well as a well-documented summary of the attacks.† It is clear that the attacks were coordinated and complex, meaning the terrorists used multiple methods of attack, including automatic weapons, assault rifles, improvised explosives, and suicide vests. The attacks were reminiscent of the siege against Mumbai, India, in 2008. In both attacks, the locations were primarily soft targets.

    On December 2, 2015, two terrorists carried out a targeted attack against the Inland Regional Center in San Bernardino, California—a soft target.‡ In this attack, fourteen people were killed and over twenty injured. According to the National Consortium for

---

* http://www.bbc.com/news/world-europe-34823938; http://www.cbsnews.com/news/injuries-from-paris-attacks-will-take-long-to-heal/.
† http://www.bbc.com/news/world-europe-34823938; http://www.telegraph.co.uk/news/worldnews/europe/france/11995246/Paris-shooting-What-we-know-so-far.html.
‡ http://www.latimes.com/local/california/la-me-san-bernardino-shooting-terror-investigation-htmlstory.html.

the Study of Terrorism and Responses to Terrorism (START) Global Terrorism Database, approximately thirty-nine terrorist attacks against soft targets occurred in the United States between 2010 and 2014.* Factoring in violence events across the globe, both terrorist instigated and non–terrorist instigated, hardly a week goes by without one or more attacks against some type of soft target.

Facilities, structures, businesses, social establishments, educational facilities, and others types of locations have all been targeted by terrorists or other perpetrators. In addition to the use of some form of violence in the attacks, the commonality is that each target was considered "soft." When discussing targets for potential attacks, a key element is the degree to which the target is considered soft as opposed to hard. This is primarily a vulnerability assessment according to various types of potential threats. Generally, a "harder" target is one that has been deemed more difficult to assail given a specific type of attack or threat. A facility that has few public entrances is harder to attack than one that has many points of entry. Locking entrances and limiting the number of keys or swipe cards to a building is an example of mitigating the location's security vulnerability. Adding a security guard is an increased "hardening" measure, and so on.

This chapter will provide a brief discussion of types of vulnerabilities that can render a specific location "softer" with respect to a potential terrorist-instigated attack. The discussion will expand to include examples of vulnerability mitigation strategies. This discussion is not intended to be exhaustive, but rather will provide an overview of the topic.

THREAT/VULNERABILITY, FREQUENCY, AND CONSEQUENCE: THE CORE OF RISK MANAGEMENT

Any discussion of target vulnerability assessment in the context of specific threats should be prefaced by a basic understanding of risk assessment and management. Risk has been defined by the U.S. Department of Homeland Security (DHS) as the "potential for an unwanted outcome resulting from an incident, event, or occurrence, as determined by its likelihood and the associated consequences."† The DHS lexicon defines a threat as a "natural or man-made occurrence, individual, entity, or action that has or indicates the potential to harm life, information, operations, the environment and/or property."‡ Consequence is defined as the effect of an event, incident, or occurrence, including the number of deaths, injuries, and other human health impacts, along with economic impacts (both direct and indirect) and other negative outcomes to society, while vulnerability has been defined as a "physical feature or operational attribute that renders an entity open to exploitation or susceptible to a given hazard."§ An assessment of any particular potential target must be made with these terms in mind.

---

* The National Consortium for the Study of Terrorism and Responses to Terrorism. Global Terrorism Database. https://www.start.umd.edu/gtd/search/Results.aspx?page=2&casualties_type=b&casualties_max=&start_yearonly=2010&end_yearonly=2014&dtp2=all&country=217&expanded=no&charttype=line&chart=target&ob=GTDID&od=desc#results-table.
† Risk Steering Committee, U.S. Department of Homeland Security. DHS Risk Lexicon. https://www.dhs.gov/dhs-risk-lexicon.
‡ Ibid.
§ Op. cit. note 4.

From a practical perspective, managing risks is the process of balancing the expenditure of resources in terms of time, money, and policy with the potential likelihood of an attack occurring and the potential consequences of the attack, should it happen. Given the various and often competing priorities and demands upon an organization, to what degree should it expend resources or create policy in order to achieve the highest level of mitigation? From the other direction, to what level can a potential threat be mitigated at the cost of certain finite resources or the imposition of policy? Simply stated: Does the benefit outweigh the cost?

Mitigation Measure 1

For an arguably drastic example, consider a moderately sized department store. After conducting a threat and vulnerability assessment, the store's management decides to do the following:

1. Remove and seal over all but two doors leading to the outside.
2. Erect a 12-foot-high chain-link fence topped with barbed wire that encircles the building with a single gate that is now secured by two armed guards who check the identification of everyone seeking entrance and inspect any and all packages, purses, backpacks, and so on, that are brought through the gate.

Question 1: Would this make the store more secure (harden the facility)?
Answer: Yes, it would certainly be more secure.
Question 2: Would these measures have any impact on the store's day-to-day business and purpose for existence?
Answer: Yes, these measures would almost certainly result in a significant decline in business, most likely to a point that would cause the store to close.

In Mitigation Measure 1—an extreme action—the goal of hardening the facility so as to mitigate any direct assault on it by terrorists would likely be achieved, but at what cost? Assuming the desire for the business to continue to operate profitably, these mitigation measures would likely be deemed too extreme and unsuitable for the business and facility. The cost–benefit analysis would show that the measures would be too costly to be a desirable strategy. Frankly, almost any threat mitigation strategy that would negatively impact business would likely be discarded as being unsuitable. But what type of measure might be implemented? In Mitigation Measure 2, an approach is taken that would likely not have a negative impact on the store's business and would likely mitigate some of the vulnerability of the store to an attack.

Mitigation Measure 2

Consider a moderately sized department store. After conducting a threat and vulnerability assessment, the store's management decides to do the following:

1. Station security guards at all doors and train them to present a positive/helpful demeanor and to be observant for concerning behavior.

Question 1: Would this make the store more secure (harden the facility)?
Answer: Yes, it would increase the security of the store and likely mitigate some of the assessed vulnerability.

Question 2: Would these measures have any impact on the store's day-to-day business and purpose for existence?
Answer 1: These measures would, if properly executed, likely not have any negative impact on the store's business.
Answer 2: If these measures were implemented in a manner that did not appear to be friendly, the impact on business could be negative.

Both mitigation measures can reduce the vulnerability of the store to an attack, but at very different costs. Which is the better choice? To help make this comparison more practical, an additional piece of information could be beneficial. This information would be the frequency of attacks experienced in the past and/or the likelihood of attacks in the future. If it is determined that violent attacks against the store or in the store's vicinity are likely to be frequent, then more stringent hardening measures would be more palatable. The data point of consequence is also important to consider. Even if an attack is determined to be very unlikely, the consequence of a successful attack could be very high, again supporting the desire for more stringent measures. Managing risks is basically assessing the cost–benefit balance and then determining the best course of action to take in context of the likely frequency and the potential consequences of attacks. Most would argue that the consequences are the more important part of the equation.

Soft versus Hard Targets

There is no clear, widely accepted definition for a "soft target." Generally, the terms "soft target" and "hard target" are used to differentiate the degree to which vulnerabilities have been mitigated or "hardened." The more vulnerable a particular target is, the softer it is said to be, and vice versa. A nuclear power plant is arguably one of the hardest civilian targets. They all have the following:

- Armed, highly trained security teams
- Strict access controls and physical barriers in place
- Redundancies in place for critical electronic information flow and control systems, mitigating the possibility of single-point control failure of critical processes

A restaurant, on the other hand, typically has relatively open access during normal business hours and most establishments typically have few, if any, security guards in place. Of the two, the restaurant is clearly a softer target. To determine the degree to which a particular building, facility, or organization is "soft" requires a vulnerability assessment of the location (physical) and potentially an internal assessment with respect to processes and individuals (see Figure 26.1). DHS has defined a vulnerability assessment as a "process for identifying physical features or operational attributes that render an entity, asset, system, network, or geographic area susceptible or exposed to hazards."[*]

There are a number of tools that are used to conduct assessments. George Baker described the vulnerability assessment process in a paper entitled "A Vulnerability

---
[*] Ibid.

VIOLENT ATTACKS AND SOFT TARGETS

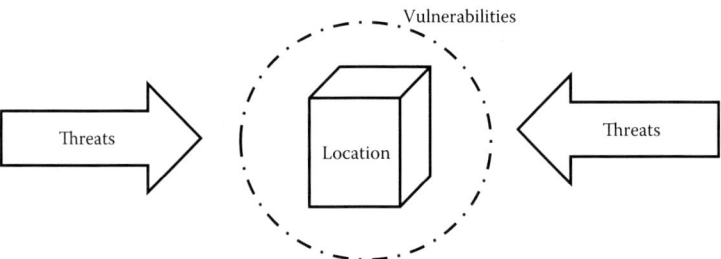

Figure 26.1 *Step 1*: Identify threats and expose potential vulnerabilities. (From Rick Mathews.)

Assessment Methodology for Critical Infrastructure Facilities."[*] The first step in the process is to develop a complete understanding of the facility, business, or organization's purpose and operational and strategic priorities. What is the purpose or mission of this organization or business? Is it to make a profit selling food and drink or is to build widgets? Next, the critical systems supporting the operations must be identified. Given a restaurant, one could argue that it must be able to prepare and serve food consistent with its published menu and in a style or setting acceptable to its clientele. It must be able to keep cold foods cold and hot foods hot, and the chef needs to have access to the tools and systems needed to prepare the food. The dining area should be safe and comfortable. The waitstaff must be able to move around the patrons efficiently, process orders, clean up, and prepare checks and process payments. Patrons must be able to locate the restaurant, park their vehicles, enter the place without undue hardship, and in general enjoy their dining experience. Assessments can be conducted based on physical security, information security, intellectual property security, and continuity of business operations and business practices, among other factors. For the purposes of this discussion, the primary focus will be physical security against violent threats.

When starting the assessment, it is essential to identify the key processes, systems, and "people flow" that make up the business of the company or organization. Examples of these include:

- Means for ingress and egress. This takes into account traditional points of entry (doors) as well as windows, connective tunnels, overhead walkways, etc.
- Human load times, such as the normal business day, weekends, evenings, late night, etc. Does the load and flow vary during these times? Are these times predictable or random?
- Access controls. This includes keys, swipe cards, guards, receptionists, etc.
- Active surveillance. Included in this area are those operations that actively scan for potential issues and threats. This would typically not include surveillance cameras and videos that simply provide evidentiary documentation.
- Lighting and alarm systems. How are they controlled? What threats are they intended to detect? How secure are they?

---

[*] Baker, G.H. 2005. A vulnerability assessment methodology for critical infrastructure facilities. Institute for Infrastructure and Information Assurance, James Madison University. Available at http://www.jmu.edu/iiia/wm_library/Vulnerability_Facility_Assessment_05-07.pdf.

SOFT TARGETS AND CRISIS MANAGEMENT

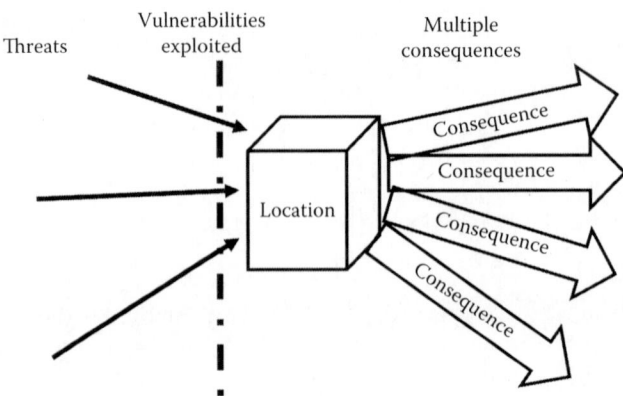

Figure 26.2 *Step 2*: Determine the potential consequences of threats, should they come to fruition. (From Rick Mathews.)

After identifying the key systems and processes as discussed in the list above, the next step is to assess each of these for weaknesses, voids, and ways to defeat them (see Figure 26.2). The red team approach is a good way to do this. A "red team" is typically a team of subject-matter experts that assume the role of one or more types of adversaries with a mission of successfully attacking a particular facility or operation. They look for threats and develop tactics to exploit them. In most cases the red team conducts its operations virtually, identifies the vulnerabilities, and develops plans to conduct the attacks. The most extreme level would involve the red team actually carrying out a penetration of the location. Regardless of the approach taken, the goal is to clearly identify any and all vulnerabilities of the facility or organization. Concurrently, a second team would be tasked with determining the likely consequences of the vulnerabilities being exploited. In doing this, each point of vulnerability is assessed separately with a subsequent review conducted to look at combined vulnerability exploitations. Consequences of violent attacks can include:

- Injuries
- Death
- Temporary loss of business
- Permanent loss of business
- Physical damage to facility
- Loss of revenue
- Damage to reputation and/or good name

A vulnerability assessment, regardless of how robust and thorough it is, will not in and of itself provide greater security for a facility or an organization. To be more secure requires action—action that results in the hardening of the facility. The information compiled in the vulnerability assessment provides the to-do list, so to speak. Examining this list in concert with the identified potential consequences will help prioritize items on the list. The next step in the hardening process is to determine direct and indirect costs, tactics, and timelines required to mitigate each vulnerability (see Figure 26.3). In many cases

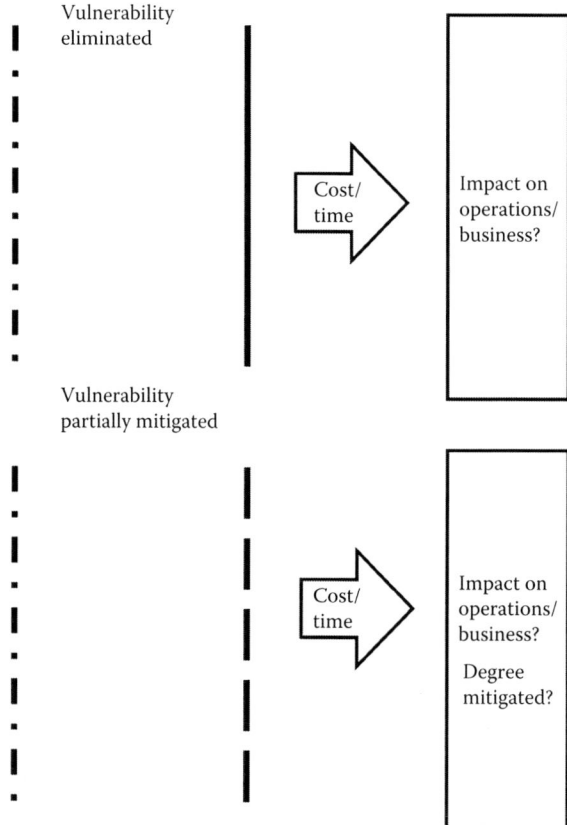

Figure 26.3 *Step 3*: Determine alternatives for the elimination or mitigation of the vulnerabilities. (From Rick Mathews.)

it will not be feasible to completely eliminate a particular vulnerability; a partial fix may be the best option. In other cases, a series of measures will be taken, with each step building upon the previous one and resulting in either total elimination of vulnerabilities or a significant mitigation. As in all business decisions, an informed cost–benefit analysis will usually rule the day (see Figure 26.4).

The process of conducting a vulnerability assessment and formulating a mitigation plan will, upon implementation, harden the target, making the facility and/or organization more secure (see Figure 26.5). The result is that the site will usually present a more

Figure 26.4 *Step 4*: Cost–benefit analysis. (From Rick Mathews.)

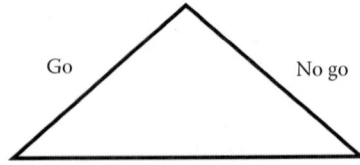

A. Make decision
B. Develop implementation time, budget, timeline

Figure 26.5 *Step 5*: Make a decision and determine an implementation plan. (From Rick Matthews.) *Step 6*: Execute the plan. *Step 7*: Reassess: The effort to mitigate and/or eliminate vulnerabilities must be accomplished by using a systematic approach and incorporating relevant and current information. It will take time and will require the expenditure of resources. Alternatively, the consequences of a violent attack will have more than just financial costs. Part of the mitigation strategy should be to train staff on how to recognize concerning behavior that could provide warning of an impending attack. Staff should also be trained on how to respond to an attack and how to administer immediate care to the injured. These measures can reduce the number of deaths resulting from an attack, which is arguably the most important and costly consequence.

difficult target to terrorists or others desiring to carry out a violent attack, hopefully deterring the attack. In most cases a blend of physical hardening and systems hardening will work best. Physical hardening includes measures that help deny physical access to unauthorized individuals and especially to tactics that would involve breaking into the facility, driving a vehicle into the facility, or deploying explosives or weapons of mass destruction into or against the facility. Systems hardening encompasses such actions as issuing access control cards or keys, active surveillance monitoring, training personnel in concerning behavior detection, and the addition or enhancement of security guards, among others.

SUMMARY AND APPLICATION

In general, most facilities, businesses, offices, and organizations are relatively soft targets, as they are not heavily fortified or protected by a team of well-trained, armed security operators. Measures can and should be considered to harden the location within the certain parameters. It wouldn't be productive to heavily fortify and secure a location that is intended to be an open, public facility, such as a restaurant, store, or theater, to the extent that it is extremely secure but nonfunctional. A comprehensive risk management approach should be useful in determining the extent of hardening that is possible and cost-effective, and that has minimal impact on the actual business or operations of the facility.

27

Soft Target Cybersecurity
The Human Interface

Michael J. Fagel, Erin Mersch, and Greg Benson

Contents

Introduction	465
Basic Device Security: How Many Devices Do You Have?	466
Home Hardware	467
Removable Data Storage	467
"Remember Me" Checkboxes	467
Free Wi-Fi	467
Cell Phones	467
Free Apps	468
Access Cards	468
Credit Cards	469
Passwords	469
Social Media Dangers	470
Built-In Cameras	471
Disposal of Devices	471
Keyboards as Weapons	472

INTRODUCTION

In this book, there has been a lot of focus on identifying soft targets and the prevention, mitigation, and planning methodologies that can be used in protecting these vulnerable targets. However, one soft target that is often overlooked remains to be discussed: the human interface with technology. People can visualize gates, bollards, and access barriers to a structure, as we have shown. However, they may not realize that the same principles *must* apply to their own world of technology. Data and digital footprints exist in almost all

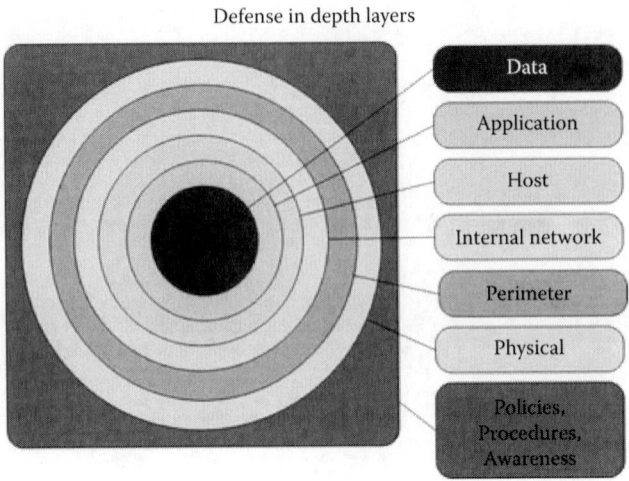

Figure 27.1 Defense in depth layers. (From Michael Fagel, College of DuPage Soft Target Seminar 2016.)

of our digital interactions. These become attractive targets for acquisition and can be used in a variety of damaging ways. Cyber hygiene is crucial in today's culture of the Internet of Things, interconnectivity, and the myriad of data streams everywhere. With the increasing use of the Internet, social media, and electronic devices in personal and business applications, among a litany of other exposures, everyone can consider themselves a soft target. The pace of technological innovation will continue to increase, with new products brought to market at an increasingly rapid pace. In some cases, the expectations for a product will never come to fruition and it will largely fade into oblivion. In every instance, vulnerability exists and can be exploited. The explosive use of personal fitness devices illustrates this trend, with some manufacturers producing their second-generation device while the success of others has begun to wane. In this chapter, we will discuss this vulnerability and briefly explore how individuals can protect themselves from cyberterrorism.

As mentioned previously in this book, soft target planning can be thought of as a set of concentric circles that protect the target (see Figure 27.1). In the case of people and technology, the core target of a cyberattack is data. The human interface threat seeks to capture personal data. Targets can include banking, medical, and legal information. Protection of data is left to a third party in many cases.

Basic Device Security: How Many Devices Do You Have?

Stop and think about how many connections people carry with them on a daily basis. Think of all of the devices that people use to keep themselves "connected." People often carry phones, a tablet, a laptop or two, a camera that can download photographs directly to a device with an interconnected card, a Wi-Fi hotspot, and a key fob for a car and/or office. Personal fitness devices provide an additional level of connectedness. The average person has seven interconnected devices at most times. This includes battery-powered devices

SOFT TARGET CYBERSECURITY

that are used in both work and personal environments, along with other devices that are readily addressable and accessible. Keeping these items secure requires vigilance.

Home Hardware
Routers used in homes will come with default account names and passwords. Changing the account name and password after the Internet service provider has installed the devices is highly recommended. Failing to do so could allow neighbors to steal access to the Wi-Fi. This could also place people at risk if the neighbors are engaged in illegal activity. Powering devices down or unplugging Internet cables when not in use will create air gaps that will prevent unauthorized access.

Removable Data Storage
Removable USB thumb drives and external hard drives are commonly used. Thumb drives and hard drives that require a password to gain access are available for purchase. Document folders or individual documents can also be password protected.

"Remember Me" Checkboxes
Many e-mail services and websites requiring accounts have an option that allows customers to store their usernames. This makes it easier to sign in during future visits, but it can also give other people access to usernames if they use the computer subsequently. Some sites even have an option to store passwords. Make sure that options to remember usernames or passwords aren't checked.

Free Wi-Fi
Everyone needs to be wary of free Wi-Fi hotspots (see Figure 27.2). At times, these may be traps that are disguised to look legitimate; when people log in, they are at risk of being monitored and having passwords and log-in data stolen.

Cell Phones
Cell phones may be "bumped" by near field communication (NFC), which allows others to potentially download contacts and data. NFC is convenient because it allows a person's phone to become their virtual wallet, but it also makes cell phones more vulnerable to

Figure 27.2 The Wi-Fi hotspot symbol. (From the Washington, DC Office of the Chief Technology Officer, octo.dc.gov)

SOFT TARGETS AND CRISIS MANAGEMENT

surveillance and data collection. People should exercise caution when inputing banking information in cell phone apps. Syncing cell phones with rental cars via Bluetooth is also risky, as this requires the car to download contact data. Also, with some makes and models of cars, that information will still be available even after the vehicle is returned.

Free Apps

There are many credible free cell phone apps. At the same time, apps can be used to install backdoors or Trojans on phones, which allow access to otherwise private data such as pictures, videos, text messages, and e-mails. Why would a flashlight or game app request access to location services? Apps from merchants may send coupons or offers when they detect cell phones in proximity to one of their locations.* Being aware of services and hardware ports that are being accessed is an important means of protecting personal data.

Access Cards

Access cards, when kept on the same lanyard as identification cards, are easy prey if lost. Always keep access cards in a separate pocket away from identification cards (see Figure 27.3). This will make it less likely that they will get lost or compromised.

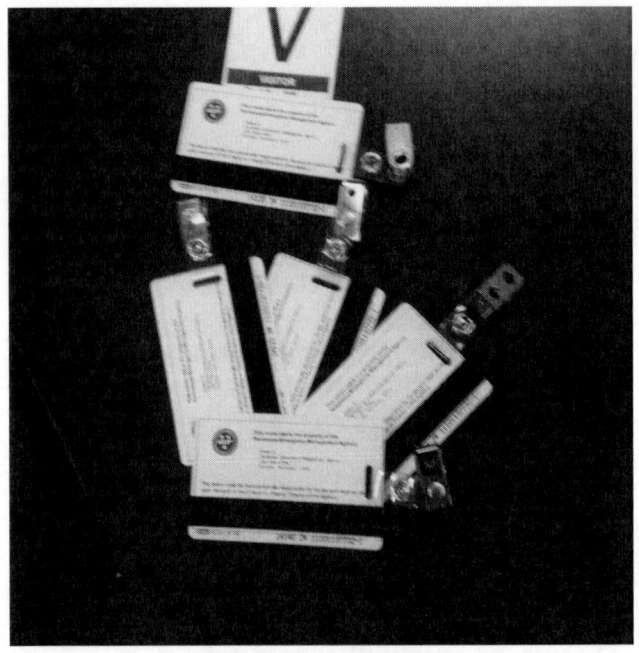

Figure 27.3 Assorted identification cards. (From Michael Fagel.)

---

* SnoopWall. Threat assessment report: Summarized privacy and risk analysis of top 10 Android flashlight apps by SnoopWall mobile security experts and the Privacy App scanner. http://www.snoopwall.com/wp-content/uploads/2015/02/Flashlight-Spyware-Report-2014.pdf (accessed February 27, 2016).

Figure 27.4 Assorted credit cards. (From mycreditunion.gov)

Credit Cards

Credit cards are being developed that use chip technology, and there are many of these types of cards used by banking and credit card institutions now (see Figure 27.4).[*] The visible chip can easily be read by certain nefarious, readily available devices in order to skim data and compromise the card's security. While the chip technology helps ward against point-of-sale breaches in security with the new way they process transactions, many of them are also equipped with NFC contactless reading abilities, which could leave information available to the public. The best defense is a protective sleeve or wallet that has RFID-blocking technology. In addition to credit cards, new passports have a chip inside that can be scanned by authorities—as well as by those who wish to cause harm and capture data.

There have been numerous criminal activities that have involved false skimmers or readers in ATMs and on gas station pumps. In these instances, the victim swipes their card and nothing happens, so the victim moves on to a different machine. Often, that first swipe has captured the victim's card data via the reader for later illegal usage. "Shoulder surfing" can occur when someone peers over the shoulder of a person entering a PIN.

To protect our softest target, the human, and harden our own defenses, a definite cultural shift needs to take place. Personal awareness is key. Vulnerability increases with each connected device one has. Taking proactive steps to protect yourself and your data is critical.

Passwords

Simple password strings should be at least twelve characters and contain uppercase and lowercase letters (no dictionary words), symbols, and numbers. The following steps describe a good way to build a secure password:

---

[*] Kossman, S. 8 FAQs about EMV credit cards. CreditCards.com. http://www.creditcards.com/credit-card-news/emv-faq-chip-cards-answers-1264.php (accessed March 7, 2016).

1. Think of a phrase.
   - For example, use the phrase "The quick brown fox jumped over the lazy dog."
   - If one reverses the phrase and takes the first letter from each word, one will get dltojfbqt, from "Dog lazy the over jumped fox brown quick the."
2. Capitalize a letter and add a space and a few numbers to the string. This will strengthen the password.
3. Make a mental note that the chosen password will always have a four- or six-digit number string and that the third letter is capitalized.
   - This may be a methodology for even-numbered years, and then for odd-numbered years the string can be modified to further increase the strength of the password.
4. Having the discipline and diligence to follow through with this is key. It is recommended that, until this habit is formed, people set reminders to change their passwords in their calendar programs.

Password derivation is a method recommended by the National Institute of Standards and Technology.* In this method, a user-chosen base password phrase is altered for each system. The alteration can add letters or special characters based upon the system name or some other pattern detail. Utilizing this system may encourage the use of strong passwords due to the fact that only the base and pattern need to be remembered, not the individual password.

Banking, medical, work, school, and personal accounts should *never* have the same password strings or even similar characters. If someone gets a password, whether legitimately or not, this vulnerability would allow them to wipe out and collect all sorts of security data. For security, passwords must be interspersed with uppercase letters, symbols, and numbers; these added characters make passwords more difficult to crack with brute-force devices. Everyone needs to remember to change passwords every 30 days and to never, ever share passwords with anyone.

Social Media Dangers

Many social media sites are in use today (see Figure 27.5 for some examples). There are many more to pick from as well. Posting travel plans or family photographs with geotagging are just a few vulnerabilities that many people do not think about when they share their details. Avoid taking photographs with tags that may be exploited later to find the vulnerability in family members' whereabouts.

Social media is a very useful tool for communication. Many emergency notifications get to consumers faster through these readily available platforms. However, they must be used wisely. Practice cyber hygiene on a daily basis with social media applications as well. There have been many people who have been exploited by others monitoring their social media accounts. For example, if someone posts information stating that they are out of town enjoying their vacation, this information can be used by burglars as an invitation to break into that person's house while they are gone. Other vulnerabilities may include

---

* Scarfone, K., and M. Souppaya, April, 2009. Guide to Enterprise Password Management (Draft) (Special Publication 800-118). National Institute of Standards and Technology, U.S. Department of Commerce. http://csrc.nist.gov/publications/drafts/800-118/draft-sp800-118.pdf (accessed March 12, 2016).

SOFT TARGET CYBERSECURITY

Figure 27.5 Social media logos. (From the College of DuPage soft Target Seminar March 2016.)

people learning details about family members that could be used for a variety of nefarious purposes. Photographs can inadvertently reveal security question answers that can be exploited to gain access to password reset operations.

False social media sites can also be set up to trick people into providing personal data or engaging in social engineering to gain more in-depth information. Criminal or civil liabilities can result, but first the creators of the fake sites have to be found.

Built-In Cameras

Most laptops and some desktop computers come with built-in cameras. These are used with programs such as Apple's FaceTime or Microsoft's Skype. However, there are malware programs available that can turn on that tiny camera lens on a tablet, laptop, or monitor without anyone's knowledge or consent. For protection from unwanted observation, cover the camera (see Figure 27.6).

There are inherent vulnerabilities in some applications as well. Keystroke monitoring and key logging are often used by cybercriminals to capture important data. Many organizations offer a warning on the log-in or sign-on screen that tells users that they "may be monitored" and that "the user has no right to expect privacy" on the organization's device. Many fail to heed that warning and wind up doing illegal or otherwise embarrassing and exploitable things on the device. Exploitation can come in many shapes and sizes, and people's digital footprints may leave them very susceptible to that sort of compromise.

Disposal of Devices

Selling old cell phones or computers can provide a treasure trove of personal data for the buyer. Deleting items does not mean they cannot be retrieved, placing pictures, videos,

SOFT TARGETS AND CRISIS MANAGEMENT

Figure 27.6 Be cautious of who can see your keyboard, forms, keystrokes remotely. Cover your camera with a sticky note so you can not be spied upon. (Courtesy of IIT Cyber Security PA 597.)

e-mails, text messages, and other personal details at risk. The same applies to computers and scanners. Hard drives, memory cards, and SIM cards should be removed from devices to minimize data loss and unintentional access.

Keyboards as Weapons

The connectivity that we have today is amazing, and future generations will grow up being even more connected than we are. Most young people today can run circles around most adults in the cyber world. Gaming, social networking, and even schoolwork are much more available and "real-time" than they were just a few years ago.

Figure 27.7 Using a keyboard as a weapon. (From fbi.gov/library)

Our interconnectivity has also been used to hack, maliciously infect systems, and hold systems hostage for ransom. Bullying on social media has led to the death of several people. Digital resources are great tools, but like any great tool they have the potential to become dangerous weapons and must be used properly. The image of the keyboard in Figure 27.7 is a cautious reminder that with keystrokes, people can wreak havoc on people, systems, and infrastructure. The Department of Homeland Security has information available on its website to help people protect themselves from cyberattacks; these tips include paying attention to website URLs in e-mails and not downloading attachments that look suspicious.* Remember that we are all soft targets. Be aware of the amount and type of personal data that is vulnerable. Take steps to minimize your footprint. Change passwords frequently to increase the difficulty of illegally accessing your accounts. Learning and utilizing these concepts to harden your own cyber world can increase your security in a connected world. Good cybersecurity occurs in layers and is driven by a proactive, not reactive, system that recognizes interdependencies. A good cybersecurity plan that is utilized daily is better than a perfect plan never implemented.

---

* U.S. Department of Homeland Security. Protect myself from cyber attacks. https://www.dhs.gov/how-do-i/protect-myself-cyber-attacks (accessed March 7, 2016).

AFTERWORD

Roland Calia

Dr. Michael Fagel is an internationally recognized expert in emergency planning, response, and preparedness, with more than four decades of public service. He has worked in fire services, emergency medical services, public health, law enforcement, emergency management, and corporate safety and security. For more than a decade, he has actively assisted with Homeland Security operations, and has also served in roles with the Federal Emergency Management Agency (FEMA), the Department of Justice, the Department of Defense, and the Department of Energy. He has served in leadership capacities with the International Association of Emergency Managers (IAEM), and is the author of several books and collections on emergency management, crisis management, and emergency planning.

I have been fortunate to work with Dr. Fagel over the past 5 years through the Master of Public Administration program at the Illinois Institute of Technology Stuart School of Business—one of his many classroom appearances throughout his career, and yet another demonstration of his unwavering commitment to prepare tomorrow's emergency responders for the challenges they may face. In my interactions with Dr. Fagel and in observing his approach to instruction, I am continually impressed by both his wealth of knowledge and firsthand experiences in these topics, and also his ability to stay at the forefront of these issues.

In this book, Dr. Fagel has assembled a team of outstanding experts who have extensive knowledge on the threat of soft targets, tactics for hardening soft targets, and strategies for emergency management programs that incorporate and address the unique elements of soft target attacks.

Numerous government agencies have spoken on the growing and unique risk that soft targets present. In testimony on the global war on terror before the Senate Select Committee on Intelligence of the U.S. Senate in 2003, then-FBI Director Robert S. Mueller, III, listed soft targets as one of several likely threats from al-Qaeda: "Multiple small-scale attacks against soft targets—such as banks, shopping malls, supermarkets, apartment buildings, schools and universities, churches, and places of recreation and entertainment—would be easier to execute and would minimize the need to communicate with the central leadership, lowering the risks of detection" (Mueller, 2003).

The transformation of the places where civilians work, socialize, study, and live into targets is not a new phenomenon, and the work to address this threat continues to evolve. These targets are countless, widely distributed, and with various levels of organizational control, making the process of hardening soft targets uniquely complex. Dr. Fagel's book recognizes that complexity, offering perspectives and strategies at all levels, from the psychology of soft targeting, to the impact on businesses and management teams, to assessing the threat levels at specific identified soft targets, to issues to consider during an emergency response. As the list of potential targets broadens, so does the importance of preparing potential responders.

At a hearing before the Subcommittee on National Security, Emerging Threats, and International Relations of the Committee on Government Reform in the House of Representatives in 2005, it was noted that even people, referencing ambassadors, government officials and their families, are now being labeled as soft targets. Representative Shays commented: "Defeating the myths of invulnerability and inevitability requires teaching government employees and their families how to recognize threats, how to take reasonable precautions, and how to handle themselves appropriately in menacing situations. Those lessons need to be reinforced regularly as part of a strategic focus that links embassy security and personnel safety to harden today's soft targets against the very real threats waiting outside" (Shays, 2005).

For more than a decade, these conversations have been occurring, and in recent history, the need for instruction on hardening soft targets has become all too apparent. In November 2008, ten gunmen attacked a railway station, cafe, two hotels, a movie theater, a community center, and a hospital, leaving 164 dead in Mumbai (CNN Library, 2015). On Friday, November 13, 2015, gunmen and suicide bombers attacked numerous locations throughout Paris in a coordinated attack, including a concert hall, athletic stadium, and several bars and restaurants, leaving 130 dead and hundreds wounded (BBC News, 2015). In December 2015, two shooters attacked an office holiday party at the Inland Regional Center in San Bernardino, California, in an act of terrorism, killing fourteen and wounding twenty-two (Winton and Queally, 2016).

An article in the *Washington Post* on the Paris attacks quoted Peter Bergen, vice president at the think tank New America, who observed that "In the United States, school shooters study other school shooters, in particular Columbine. This is true of terrorists as well. They study tactics that have worked before" (Achenbach, 2015).

It is critical, now more than ever, that public safety professionals, security personnel, and yes, even civilians, are equipped with the knowledge to recognize, assess, and harden soft targets; to deter and mitigate potential attacks; and to respond to emerging threats.

In a 2003 public hearing of the National Commission on Terrorist Attacks Upon the United States, international terrorism expert Rohan Gunaratna noted that "Hardening of government targets will also displace the threat to softer targets making civilians prone to terrorist attack," and that "The reality is that government countermeasures have increased the vulnerability of population centers and economic targets" (Gunaratna, 2003). The strategies to address these threats require greater awareness and empowerment among civilian organizations, and the authors who are contributing to this book have an array of practical and critical tools to support these efforts.

This is a conversation that strikes at the heart of our personal feelings of safety and preparedness. It is also a situation where greater and more effective knowledge plays a critical role in preventing, minimizing, and responding to these threats. Dr. Fagel and his colleagues have assembled a wealth of relevant information, practical tools, and real-life scenarios to advance this conversation. This book is an essential resource for those who manage the spaces where people live, work, and play.

REFERENCES

Achenbach, Joel. "Experts: Terrorists learning from one another and going after soft targets." *Washington Post.* November 15, 2015. Accessed https://www.washingtonpost.com/national/health-science/experts-terrorists-learning-from-one-another-and-going-after-soft-targets/2015/11/15/68405564-8bb2-11e5-acff-673ae92ddd2b_story.html, March 17, 2016.

BBC. "Paris attacks: What happened on the night." December 7, 2015. Accessed http://www.bbc.com/news/world-europe-34818994.

CNN Library. "Mumbai Terror Attacks Fast Facts." November 4, 2015. Accessed http://www.cnn.com/2013/09/18/world/asia/mumbai-terror-attacks/index.html.

Gunaratna, Rohan. "The Rise and Decline of Al Qaeda." Third public hearing of the National Commission on Terrorist Attacks Upon the United States, Statement of Rohan Gunaratna to the National Commission on Terrorist Attacks Upon the United States, July 9, 2003. Accessed http://www.9-11commission.gov/hearings/hearing3/witness_gunaratna.htm, March 17, 2016.

Mueller III, Robert S. "Testimony before the Senate Select Committee on Intelligence of the United States Senate." The FBI Federal Bureau of Investigation, February 11, 2003. Accessed https://www.fbi.gov/news/testimony/war-on-terrorism, March 17, 2016.

Shays, Christopher. "Overseas Security: Hardening Soft Targets." Hearing before the Subcommittee on National Security, Emerging Threats, and International Relations of the Committee on Government Reform, House of Representatives, May 10, 2005. Accessed U.S. Government Printing Office: https://www.gpo.gov/fdsys/pkg/CHRG-109hhrg22704/html/CHRG-109hhrg22704.htm, March 17, 2016.

Winton, Richard, and Queally, James. "FBI is now convinced that couple tried to detonate bomb in San Bernardino terror attack." *Los Angeles Times*, January 15, 2016. Accessed http://www.latimes.com/local/lanow/la-me-ln-fbi-san-bernardino-bombs-20160115-story.html.

INDEX

A

AAC, *see* After action conference
AAR, *see* After action report
AAR/IP, *see* After action report/improvement plan
Abdulmutallab, Umar Farouk, 28
AC, *see* Assistant commissioner
Access cards, 468; *see also* Soft target cybersecurity
ACLU, *see* American Civil Liberties Union
Acronyms, 155–156
Active shooters, 296; *see also* Security organization; Soft target; Workplace violence
   calendar figure of mass shootings, 297
   Farook's home, 298
   possible motive, 298
   San Bernardino, California, shooting, 296, 297
Active shooter threat, 143; *see also* Exercise
   continued response/evacuation and recovery, 153–155
   exercise guidelines, 148
   exercise objectives, 147
   exercise schedule, 145
   exercise structure, 146
   handling instructions, 144–145
   instructions, 145–146
   notification and initial response, 151–153
   participants, 148
   preface, 144
   purpose, 147
   scope, 147–148
   shootings at Sandy Hook Elementary School, 144
   TTX facilitator's guide, 144
   warning, 148–151
ADA, *see* Americans with Disabilities Act
AEDs, *see* Automated external defibrillators
African Methodist Episcopal (AME), 294
African Union force (AMISOM), 243

After action conference (AAC), 143, 155
After action report (AAR), 132, 155
After action report/improvement plan (AAR/IP), 142, 143, 155
AHJ, *see* Authority having jurisdiction
Airports, 393–394
Alarms, 198–200; *see also* Security
Alcohol, Tobacco, Firearms and Explosives (ATF), 296
Aleph, *see* Aum Shinryko
Alfred P. Murrah building, 15
Allied Barton Security Services, 259
Alpha, bravo, and zulu plans, 205; *see also* Emergency preparations
Al-Qaeda, 40
Al-Qaeda in the Arabian Peninsula (AQAP), 245
Al-Qaeda in the Islamic Maghreb (AQIM), 41, 362
AME, *see* African Methodist Episcopal
AMEF, *see* Ansar al-Mujahideen English forum
American Civil Liberties Union (ACLU), 241
Americans with Disabilities Act (ADA), 125, 308; *see also* Disaster preparedness and law
AMISOM, *see* African Union force
Anger, 45
Ansar al-Mujahideen English forum (AMEF), 234
AoR, *see* Area of responsibility
AQAP, *see* Al-Qaeda in the Arabian Peninsula
AQIM, *see* Al-Qaeda in the Islamic Maghreb
Area of responsibility (AoR), 421
Assistant commissioner (AC), 405
Assorted credit cards, 469; *see also* Soft target cybersecurity
Assorted identification cards, 468; *see also* Soft target cybersecurity
ATF, *see* Alcohol, Tobacco, Firearms and Explosives
Aum Shinryko, 42
Authority having jurisdiction (AHJ), 333
Automated external defibrillators (AEDs), 206

INDEX

B

BCM, *see* Business continuity–management
BCMM, *see* Business Continuity Maturity Model
Behavioral profiling, 43
Bergendorff, Roger, 41
BIA, *see* Business impact analysis
Biological agents, 74; *see also* Weapon of mass destruction
  indicators of possible use of, 75
Biological warfare, 41; *see also* Weapon of mass destruction
Bioterrorism incident, 90; *see also* Planning for terrorism
Black swan; *see also* Churches; Soft target
  axioms of, 17
  phenomenon, 242
  theory, 16
Bloggers, 175
Bomb threat, 205–206; *see also* Emergency preparations
Boston Marathon bombing, 22
Built-in cameras, 471; *see also* Soft target cybersecurity
Business continuity, 283, 318–319; *see also* Hospital business continuity; Sport venue command group
  management, 288
  planning, 284–285
  planning tool advantages, 288–290
  program and focus, 285
  program design checklist, 285–286
  tasks and processes associated with, 286–287
Business Continuity Maturity Model (BCMM), 288
Business impact analysis (BIA), 285, 286, 290; *see also* Business continuity; Hospital business continuity

C

CAMEO, *see* Computer Aided Management of Emergency Operations
Casualty collection point (CCP), 348
CAUSE, *see* College and University Security Effort
CBR, *see* Clinical and business recovery
CBRNE, *see* Chemical, biological, radiological, nuclear, and explosive
CCAs, *see* Complex Coordinated Attacks
CCP, *see* Casualty collection point
CCTV, *see* Closed circuit television
C/E handbook, *see* Controller/evaluator handbook
Cell phones, 467–468; *see also* Soft target cybersecurity
CEM, *see* Comprehensive Emergency Management
Central Intelligence Agency (CIA), 241
CEOs, *see* Chief elected officials
CF SSA, *see* Commercial Facilities Sector-Specific Agency
CG, *see* Command group
CHAOS, Operation, 241
Chemical agents, 73; *see also* Weapon of mass destruction
  indicators of possible use of, 74
Chemical, biological, radiological, nuclear, and explosive (CBRNE), 40
Chhatrapati Shivaji Terminus (CST), 342
Chief elected officials (CEOs), 52
  background information, 60–61
  checklist, 59, 65, 66, 68
  community's EOP, 57
  federal expectations, 53–54
  goals, 54
  IEMS and, 65, 67–68
  immediate actions, 61–62
  legal issues, 63
  personal actions, 62–63
  political issues, 63–64
  public information, 64–65
  responsibilities and public's expectations, 52
  risk reduction, 54, 69–70
  self-assessment, 55–56
  survival kit, 57–58
Chlorine, 41; *see also* Weapon of mass destruction
Churches, 211–212, 242; *see also* Deterring and mitigating attack; Soft target threat assessment
  black swan phenomenon, 242
  courtyard in Church City, 360
  megachurches, 243
  in Middle East, 360
  proliferation of attacks, 243–245
  religious bias, 243
  U.S. churches as targets, 245–246
CIA, *see* Central Intelligence Agency

CI/KR, *see* Critical infrastructure/key resource
CISDs, *see* Critical Incident Stress Debriefings
CISM, *see* Critical incident stress management
Clinical and business recovery (CBR), 284, 290
Closed circuit television (CCTV), 200–201; *see also* Security
COINTELPRO, *see* Counterintelligence Program
College and University Security Effort (CAUSE), 209
Command centers, 401–402
Command group (CG), 308; *see also* Sport venue command group
Command Post (CP), 331
Command post exercise (CPX), 136, 155
Commercial Facilities Sector-Specific Agency (CF SSA), 302
Complex attacks, 440
Complex Coordinated Attacks (CCAs), 435, 455–456; *see also* Swarm attack
　vs. active shooter incident, 436, 440
　asymmetric, 446
　attacker characteristics, 437–439
　emergency responders, 439, 442
　examples of, 446
　indicators and characteristics, 444–445, 452, 455
　vs. MCI, 440
　multiple responses, 443–444
　Mumbai, India, 451–452
　national preparedness cycle, 444
　Nazran, Republic of Ingushetia, Russia, 446–448
　outcomes, 440–443
　to overwhelm responder resources, 445
　Paris, France, 453–455
　school massacre, Beslan, Russia, 448–451
　situational awareness, 442
　TTPs, 439–440
Compound houses, 356
Comprehensive Emergency Management (CEM), 65; *see also* Chief elected officials
Computer Aided Management of Emergency Operations (CAMEO), 329
Consequence, 458; *see also* Violent attacks and soft targets

Contemporary terrorism, 2
Controller/evaluator handbook (C/E handbook), 138, 155
Conventional explosives and secondary devices, 76
Coordinated attacks, 340, 351, 440; *see also* Mumbai, attacks in
　public health system issues, 347–350
　swarm attack, 344–345
　terror medicine, 346–347
　untraditional response protocols, 350–351
Counterintelligence Program (COINTELPRO), 240; *see also* Schools
Countermeasure improvements, 317
Counterterrorism Implementation Task Force (CTITF), 370
CP, *see* Command Post
CPX, *see* Command post exercise
Credit cards, 469; *see also* Soft target cybersecurity
Crisis leadership, 30; *see also* Soft target hardening
Crisis management, 399; *see also* Soft target planning
　exercises, 400–401
　planning, 400
　training, 399
Critical Incident Stress Debriefings (CISDs), 272
Critical incident stress management (CISM), 153
Critical infrastructure/key resource (CI/KR), 246; *see also* Hospitals
CR Pilot, *see* Resilience Pilot Program
CST, *see* Chhatrapati Shivaji Terminus
CTITF, *see* Counterterrorism Implementation Task Force
Cybercriminals, 236
Cyber hygiene, 466; *see also* Soft target cybersecurity
Cyberterrorism, 77–78; *see also* Planning for terrorism

D

Deadly force incidents (DFIs), 242
Deep packet inspection (DPI), 268
Delphi methodology, 33
Department of Health and Human Services Hospital Preparedness Program (DHHS-HPP), 281

INDEX

Department of Homeland Security Exercise and Evaluation Program (DHS-HSEEP), 144
Department of Justice (DOJ), 371
Department Operation Center (DOC), 443
Deterring and mitigating attack, 190, 217; *see also* Emergency preparations; Hardening college campus
Device; *see also* Soft target cybersecurity
    disposal, 471–472
    security, 466
DFIs, *see* Deadly force incidents
DHHS-HPP, *see* Department of Health and Human Services Hospital Preparedness Program
DHS, *see* U.S. Department of Homeland Security
DHS-HSEEP, *see* Department of Homeland Security Exercise and Evaluation Program
Dining halls, 202–203; *see also* Deterring and mitigating attack
Dirty bomber, *see* Padilla, Jose
Disaster, major, 315
Disaster preparedness and law, 124; *see also* Emergency Operations Plan
    attorney as team member, 124
    emergency management phases, 127–129
    federal laws, 125
    local laws, 126
    negligence, 126
    potentially applicable laws, 125–126
    state laws, 125–126
Disasters, 315
DOC, *see* Department Operation Center
DOJ, *see* Department of Justice
Domestic terrorists, 212
Double bomb tactic, 39–40
DPI, *see* Deep packet inspection

E

EAP, *see* Emergency action plan
EBH, *see* Effects-based hardening
EBO, *see* Effects-based operations
EEGs, *see* Exercise evaluation guides
Effects-based hardening (EBH), 191–192; *see also* Deterring and mitigating attack
    matrix, 193
Effects-based operations (EBO), 191

EMAP, *see* Emergency Management Accreditation Program
Emergency, 314; *see also* Sport venue command group
Emergency action plan (EAP), 132
Emergency Management; *see also* Emergency Management program strategy
    accreditation program, 381–382
    phases of, 383
    stakeholders, 382
    standards, 379–380
Emergency Management Accreditation Program (EMAP), 381–382; *see also* Emergency Management program strategy
Emergency management and media, 171
    dealing with media in crisis, 176–177
    do's and don'ts, 177
    formats, 173
    inverted pyramid, 172
    joint information system/joint information center, 180–186
    newspapers, 172
    nonverbal behavioral clusters, 176
    print organizations and way of news, 173
    public information officer, 178–180
    radio, 172–173
    social network sites, 175
    television, 174
    Weblog, 175
    World Wide Web, 174
Emergency Management Institute (EMI), 170
Emergency Management program strategy, 374, 379, 388–389; *see also* Emergency Management
    capabilities and future needs, 388
    challenges, 388
    elements of, 376
    Federal Emergency Support Functions, 384
    format and structure of strategy document, 383–385
    goal, 376, 382, 385–387
    HSPD-5, 380
    incident command system staff, 384
    justification of program and projects, 375–376
    mission statement, 377
    NIMS requirements, 380–381
    organizational chart, 378–379
    organizational values, 378
    planning team, 382

program development and direction, 376
qualities of emergency management
    agency, 378
standards, 379
strategic initiative, 387–388
strategy, 374–375
vision statement, 377
work plans and assignments, 376
Emergency medical services (EMS), 73, 120, 308
Emergency operation center management,
    167, 170
    hazard and vulnerability analysis review, 168
    incorporating terror analysis, 168–169
    local EOC, 167
    policies and procedures, 169–170
    preparation for terrorist incident, 168
    revisit each EOP annex, 169
    test, train, and exercise, 170
Emergency operation centers (EOCs), 60–61,
    84, 136, 157, 165, 443; *see also* Chief
    elected officials; Emergency operation
    center management; Planning for
    terrorism
    communication SOP development, 160–161
    communications policies and procedures, 160
    document retention, 164–165
    emergency management EOC duties, 158
    emergency operations plan, 159
    emergency phones, 161
    EOP and hazard analysis to design, 159–160
    foundations for establishing, 158
    hazard and vulnerability analysis, 159
    life support policies and procedures, 161
    maintenance contracts, 163
    management and operations, 158
    mutual aid agreements, 164
    operating equipment and supplies, 162–163
    policy and procedures development, 160
    primary, 85
    scheduling system, 163
    setup, 159
    terrorism-related hazards, 159
    tracking goods and services, 161–162
Emergency Operations Plan (EOP), 52, 72, 103,
    112, 129–130, 155, 158, 282; *see also*
    Chief elected officials; Disaster
    preparedness and law; Emergency
    operation centers; Emergency
    planning process; Emergency
    planning team; Hazard-specific
    appendices; Planning for terrorism

administration and logistics section, 113
annexes, 114–115
assumptions, 83
authorities and references section, 114
basic, 89, 91
basic plan, 111
components of, 111
concept of operations, 83
concept of operations section, 113
jurisdiction, 104
needs for, 104–105
organization and assignment of
    responsibilities section, 113
plan development and maintenance section,
    113–114
purpose statement, 112
situational report elements, 78–79
situations and assumptions section, 112
Emergency planning process, 105; *see also*
    Emergency Operations Plan;
    Hazard; Sport venue emergency
    planning
    factors in community profile, 109
    hierarchy for response prioritization
        profile, 110
    jurisdiction profile creation, 107
    risk analysis completion, 107–110
    scenario creation, 111
    steps in, 129
Emergency planning team, 119, 309; *see also*
    Emergency Operations Plan
    active participation, 121–122
    agencies for planning teams, 120–121
    members, 119
    team characteristics, 123–124
    team formation stages, 123
    team operation, 122–123
    team roles, 123
Emergency preparations, 203; *see also* Deterring
    and mitigating attack; Hazard
    alpha, bravo, and zulu plans, 205
    bomb threat, 205–206
    emergency response team and command
        center, 203
    guideline checklist, 319–320
    hold room, 203
    medical program, 206–207
    sealing the room, 207
    shelter in place, 207–208
Emergency public information (EPI), 105, 114
Emergency response plan (ERP), 133, 309

INDEX

Emergency situation, 308; *see also* Sport venue emergency planning
Emergency Support Functions (ESFs), 334, 384; *see also* Emergency Management program strategy
EMI, *see* Emergency Management Institute
Employee Retirement Income Security Act (ERISA), 125; *see also* Disaster preparedness and law
EMS, *see* Emergency medical services
End of the exercise (ENDEX), 143, 155
Environmental Protection Agency (EPA), 93
EOCs, *see* Emergency operation centers
EOP, *see* Emergency Operations Plan
EPA, *see* Environmental Protection Agency
EPI, *see* Emergency public information
ERISA, *see* Employee Retirement Income Security Act
ERP, *see* Emergency response plan
ERT, *see* Evidence Response Team
Evacuation plan template for stadiums, 321–323; *see also* Sport venue emergency planning
Evidence Response Team (ERT), 258
Exercise, 132; *see also* Active shooter threat
  acronyms, 155
  after action report, 141–142
  corrective actions, 143
  design and development of, 134–136
  discussion-based exercises, 134–135
  document types, 140
  evaluation and improvement planning, 142–143
  exercise conduct, 136, 138–141
  foundation to plans, 133–134
  HSEEP, 134
  key personnel involved in, 136, 141
  operation-based, 136
  planning process, 134
  planning team members for, 139
  statement of purpose, 140
  testing plans and capabilities, 132
  types of, 135, 137–138, 310
Exercise evaluation guides (EEGs), 138, 155
Exercise plan (ExPlan), 138, 155
Expressive targets, 16

F

Face of rage, 45–48; *see also* Soft target hardening
Faces of threat, 45
Facial expression and lie detection, 47; *see also* Soft target hardening
Fair Labor Standards Act (FLSA), 125; *see also* Disaster preparedness and law
Family and Medical Leave Act (FMLA), 125; *see also* Disaster preparedness and law
Farook, Syed, 10
FBI, *see* Federal Bureau of Investigation
Fedayeen raid, 345; *see also* Mumbai, attacks in
Federal Bureau of Investigation (FBI), 79, 296, 317
Federal Emergency Management Agency (FEMA), 67, 242
Federal laws, 125; *see also* Disaster preparedness and law
Federal Security Service (FSB), 267, 446
FEMA, *see* Federal Emergency Management Agency
FEs, *see* Functional exercises
Field training exercises (FTXs), 136, 155
Final planning conference (FPC), 143, 155
FISs, *see* Foreign intelligence services
FLSA, *see* Fair Labor Standards Act
FMLA, *see* Family and Medical Leave Act
Foreign intelligence services (FISs), 235
For official use only (FO UO), 145
FPC, *see* Final planning conference
Free apps, 468; *see also* Soft target cybersecurity
Free Wi-Fi, 467; *see also* Soft target cybersecurity
FSB, *see* Federal Security Service
FSE, *see* Full scale exercise
FTXs, *see* Field training exercises
Full scale exercise (FSE), 132, 155
Functional exercises (FEs), 136, 155
Functional targets, 16
Fusion Center, 421; *see also* Infrastructure protection
  fusion process overview, 429
  and infrastructure protection mission, 429
  map of, 428
  national network creation, 424–428
  operation of, 426
  role, 427
  spotlight, 432

G

GCC, *see* Gulf Cooperation Council
Geneva Conventions, 21, 36–37
GOAL, *see* Grassroots Ontario Animal Liberation
Goal, 385
Go Kit Supplies, 401; *see also* Soft target planning
Grassroots Ontario Animal Liberation (GOAL), 255
Gulf Cooperation Council (GCC), 358

H

Hardened targets, 1; *see also* Soft target
Hardening college campus, 209; *see also* Deterring and mitigating attack
    building relationships, 209–210
    exercising due diligence, 211
    students and staff as force multipliers, 210
Hardening tactics, 353; *see also* Soft target hardening in Middle East
    9/11 attacks, 354
    OSPB standards, 369
    State Department engagement, 368–370
    United Nations and soft targets, 370–371
    UN's PPP initiative, 371
Harder target, 458; *see also* Violent attacks and soft targets
Hard targets, 22, 391–392; *see also* Soft target planning
Hazard
    analysis, 105–107
    data types, 110
    profile development, 107
    profile worksheet, 108
    severity rating, 110
Hazardous materials (HazMat), 73, 97, 308
Hazard-specific appendices, 115; *see also* Emergency Operations Plan
    checklists, 118
    forms, 118
    implementing instructions, 119
    information cards, 118
    instruction implementation, 115–116
    job aids, 117–118
    maps, 118–119
    standard operating procedures, 117
    topics for functional annex, 116
HazMat, *see* Hazardous materials
Health and Human Services (HHS), 89
Heating, ventilation, and air-conditioning (HVAC), 162, 288
Hexamethylene triperoxide diamine (HMTD), 244
HHS, *see* Health and Human Services; Human Health and Services
HICS, *see* Hospital Incident Command System
HMTD, *see* Hexamethylene triperoxide diamine
Hold room, 203; *see also* Emergency preparations
Home hardware, 467; *see also* Soft target cybersecurity
Homeland Security Academic Advisory Council (HSAAC), 242
Homeland Security Advisory System (HSAS), 317
Homeland Security Exercise and Evaluation Program (HSEEP), 132, 155, 310
Homeland Security Presidential Decision Directive (HSPD), 180, 380
Hospital business continuity, 281, 283–286; *see also* Business continuity
    business impact analysis, 286–287
    continuity definitions, 290–292
    historical perspective, 281
    hospital disaster plan, 282
    hospital EOP, 282, 284
    hospital preparedness, 281–283
    Joint Commission Accreditation Requirements, 282, 283
    risk assessment process, 287–288
    steps used to prepare contingency and continuity plans, 291
Hospital Incident Command System (HICS), 281
Hospitals, 212–213, 246; *see also* Deterring and mitigating attack; Soft target threat assessment
    CI/KRs, 246
    DHS advisory, 249
    in Middle East, 360–361
    nefarious use of ambulances, 248–250
    violent crimes in, 246–247
Hotel security, 361
    case study, 365–367
    Days Inn, 362–363
    security challenges, 362
    Southern Sunshine Hotel, 365
    staff training, 364
    U.S. hotel in Middle Eastern city, 364

INDEX

HSAAC, *see* Homeland Security Academic Advisory Council
HSAS, *see* Homeland Security Advisory System
HSEEP, *see* Homeland Security Exercise and Evaluation Program
HSPD, *see* Homeland Security Presidential Decision Directive
Human Health and Services (HHS), 347
Human shields tactics, 36–37; *see also* Soft target hardening
HVAC, *see* Heating, ventilation, and air-conditioning

I

IAEM, *see* International Association of Emergency Managers
IAP, *see* Incident Action Plan
ICP, *see* Incident command post
ICS, *see* Incident Command System
ICSC, *see* International Council of Shopping Centers
IED, *see* Improvised explosive device
IEMS, *see* Integrated emergency management system
If You See Something, Say Something program, 303
IICD, *see* Infrastructure Information Collection Division
IM, *see* Incident management
Improvement plan (IP), 132
Improvised explosive device (IED), 76, 228
Incident Action Plan (IAP), 331, 336; *see also* Special event
Incident commander, 312; *see also* Sport venue command group
Incident command post (ICP), 147
Incident Command System (ICS), 53, 84 133, 155, 443
Incident management (IM), 290
Information cards, 118; *see also* Hazard-specific appendices
Information Sharing Environment (ISE), 425
Information technology (IT), 283, 407
Infrared (IR), 201
Infrastructure attacks, 77; *see also* Planning for terrorism
Infrastructure Information Collection Division (IICD), 302

Infrastructure protection, 421, 433–434; *see also* Fusion Center
    case study, 431
    collaboration in practice of homeland security, 423–424
    critical infrastructure risk management framework, 431
    fusion centers and, 429
    fusion liaison officer programs and, 433
    fusion process and, 429–432
    International Association of Chiefs of Police Criminal Information Sharing Summit, 425
    marquee national programs, 433
    on-site vulnerability assessments, 432
    2009 state homeland security strategy, 423
    state-level intelligence capability, 422–423
Initial planning conference (IPC), 155
Inquests, 406
Insider threat, 31–32; *see also* Soft target hardening
Integrated emergency management system (IEMS), 67–68; *see also* Chief elected officials
Intellectual property theft, 236
Intelligence Reform and Terrorism Prevention Act (IRTPA), 425
International Association of Chiefs of Police Criminal Information Sharing Summit, 425
International Association of Emergency Managers (IAEM), 475
International Council of Shopping Centers (ICSC), 214
Internet, 18
Internet service providers (ISPs), 268
Intuition, 30–31; *see also* Soft target hardening
Inverted pyramid, 172
IP, *see* Improvement plan
IPC, *see* Initial planning conference
IR, *see* Infrared
IRTPA, *see* Intelligence Reform and Terrorism Prevention Act
ISE, *see* Information Sharing Environment
ISIS, *see* Islamic State in Iraq and Syria
Islamic State in Iraq and Syria (ISIS), 2, 253, 255; *see also* Soft target threat assessment
    Orlando nightclub shooting, 271–274
ISPs, *see* Internet service providers
IT, *see* Information technology

J

J-1 visa, 236
Jamiyyat ul-Islam Is-Saheeh (JIS), 243
JIC, *see* Joint Information Center
Jihadists, radical, 28
JIS, *see* Jamiyyat ul-Islam Is-Saheeh; Joint Information System
Job aids, 117; *see also* Hazard-specific appendices
JOC, *see* Joint Operation Center
Joint Information Center (JIC), 178; *see also* Chief elected officials; Emergency management and media
 noteworthy details on, 181
 readiness assessment form, 183–186
 types of, 182
Joint Information System (JIS), 178; *see also* Emergency management and media; Joint Information Center
 noteworthy details on, 181
Joint Operation Center (JOC), 85, 328
Joint Terrorism Task Force (JTTF), 28, 316

K

Kenyan Defense Forces (KDF), 257
Keyboards, 472–473; *see also* Soft target cybersecurity

L

Lady al-Qaeda, *see* Siddiqui, Aafia
Lashkar-e-Taiba (LeT), 246, 340; *see also* Mumbai, attacks in
LEOP, *see* Local Emergency Operations Plan
LEPCs, *see* Local Emergency Planning Committees
LESLP, *see* London Emergency Services Liaison Panel
LeT, *see* Lashkar-e-Taiba
LFB, *see* London Fire Brigade
Liberation Tigers of Tamil Eelam (LTTE), 37, 263
Life space, 13
Local Emergency Operations Plan (LEOP), 181
Local Emergency Planning Committees (LEPCs), 93
Locks, 197–198; *see also* Security
Logistical targets, 16
London Emergency Services Liaison Panel (LESLP), 405
London Fire Brigade (LFB), 405
Low-technology devices and delivery, 77; *see also* Planning for terrorism
LTTE, *see* Liberation Tigers of Tamil Eelam

M

MAC, *see* Multi-Agency Coordination
MAD, *see* Marineland Animal Defense
Malik, Tashfeen, 10
Mall of America in Bloomington, 254; *see also* Shopping centers
Malls, 213–214; *see also* Deterring and mitigating attack
Marineland Animal Defense (MAD), 255
Mass casualty incidents (MCIs), 4, 436; *see also* Complex Coordinated Attacks
Mass shootings on military installations, 22
Master scenario event list (MSEL), 138, 155
McCain, Douglas McAuthur, 13
Medical program, 206–207; *see also* Emergency preparations
Megachurches, 243; *see also* Churches
Memorandum of understanding (MOU), 90, 164, 334; *see also* Emergency operation centers
Metropolitan Police Service (MPS), 408
Midterm planning conference (MPC), 143, 155
Millennium attack, 41
Modern black swan events, 17
Mole hunters, 34
MOU, *see* Memorandum of understanding
Movie theaters, 392
MPC, *see* Midterm planning conference
MPS, *see* Metropolitan Police Service
MSEL, *see* Master scenario event list
Multi-Agency Coordination (MAC), 443
Mumbai, attacks in, 340, 344; *see also* Complex Coordinated Attacks
 armament, 342
 CST attack, 343
 deployment, 342
 Leopold Café and Bar, 342–343
 Mumbai City, 341
 Nariman House, 344
 pre-assault preparations, 341
 Taj Mahal Palace Hotel, 343
 taxi explosion, 344

Mumbai, attacks in (*Continued*)
    Trident-Oberoi Hotel, 343–344
    water incursion and landing, 342
Mumbai, India, 451–452; *see also* Complex Coordinated Attacks

N

National Center for Security & Preparedness (NCSP), 457
National Counterintelligence Working Group (NCIWG), 209
National Criminal Intelligence Sharing Plan (NISCP), 425
National Employment Law Project (NELP), 33
National Fire Protection Association (NFPA), 382
National Incident Management System (NIMS), 53, 133, 155 169, 180, 232, 376, 331; *see also* Chief elected officials; Emergency Management program strategy
    components of, 381
    initiatives promulgated by, 53–54
National Infrastructure Protection Plan (NIPP), 302
National Network of Fusion Centers (NNFC), 424
National Oceanic and Atmospheric Administration (NOAA), 313
National Security Higher Education Advisory Board (NSEAB), 209, 242
National Security Special Events (NSSEs), 327
National Security Threat List (NSTL), 317
National Terrorism Advisory System (NTAS), 87
National Weather Service (NWS), 106
Nazran, Republic of Ingushetia, Russia, 446–448; *see also* Complex Coordinated Attacks
NCIWG, *see* National Counterintelligence Working Group
NCSP, *see* National Center for Security & Preparedness
Near field communication (NFC), 467
NELP, *see* National Employment Law Project
Newspapers, 172; *see also* Emergency management and media
NFC, *see* Near field communication
NFPA, *see* National Fire Protection Association
NGOs, *see* Nongovernmental organizations
Nigeria's Boko Haram terrorist group, 36
NIMO, *see* Not in my organization
9/11 attacks; *see also* Hardening tactics
    psychological trauma after, 23
    U.S. economic system, 354
NIPP, *see* National Infrastructure Protection Plan
NISCP, *see* National Criminal Intelligence Sharing Plan
NNFC, *see* National Network of Fusion Centers
NOAA, *see* National Oceanic and Atmospheric Administration
No dark corners approach, 34–35
Non-firearm weaponry, 226
Nongovernmental organizations (NGOs), 67
Nonprofit Security Grant Program (NSGP), 213
Nonverbal behavioral clusters, 176
Not in my organization (NIMO), 35
NRC, *see* Nuclear Regulatory Commission
NSEAB, *see* National Security Higher Education Advisory Board
NSGP, *see* Nonprofit Security Grant Program
NSSEs, *see* National Security Special Events
NSTL, *see* National Security Threat List
NTAS, *see* National Terrorism Advisory System
Nuclear/radiological agents, 75–76; *see also* Weapon of mass destruction
Nuclear Regulatory Commission (NRC), 93, 132
NWS, *see* National Weather Service

O

Occupational Safety & Health Administration (OSHA), 94, 125, 132, 298; *see also* Disaster preparedness and law
Office of Homeland Security (OHS), 423
Online geographical resources, 329
On scene commander (OSC), 147
Open-air market, 261; *see also* Shopping centers
Operational security (OPSEC), 329
Operation CHAOS, 241

OPSEC, *see* Operational security
Organization for local response management, 91; *see also* Planning for terrorism
  emergency responders, 92, 93
  interjurisdictional responsibilities, 92
  medical service providers, 93
  state and local public health authorities, 92–93
OSAC, *see* Overseas Security Advisory Council
OSC, *see* On scene commander
OSHA, *see* Occupational Safety & Health Administration
OSPB, *see* Overseas Security Policy Board
Overseas Security Advisory Council (OSAC), 370
Overseas Security Policy Board (OSPB), 369

P

Padilla, Jose, 14
Pan, tilt, and zoom (PTZ), 201
Paris, France, 453–455; *see also* Complex Coordinated Attacks
Parking lots, 194; *see also* Physical security
Passwords, 469–470; *see also* Soft target cybersecurity
Peripherally inserted central catheter (PICC), 207
Personal protective equipment (PPE), 82, 134, 155
Physical profiling, 28
Physical security, 193; *see also* Deterring and mitigating attack
  alarms, 194
  all-hazards approach security plan, 190
  churches, 211–212
  dining halls, 202–203
  EBO approach, 191
  effectiveness of burglary deterrents, 193
  effects-based hardening, 191–192
  hospitals, 212–213
  malls, 213–214
  parking lots, 194
  pre-positioned vehicles, 194–196
  red teaming soft targets, 216–217
  SAFE Washington program, 212
  schools, 208–209
  soft target facilities, 196
  sports and recreation venues, 214–216
  SUV repurposed as mock security vehicle, 195
  threat assessment model, 208
  traffic duty, 196
PICC, *see* Peripherally inserted central catheter
PIO, *see* Public Information Officer
Place vulnerability index (PVI), 221; *see also* Soft target threat assessment
Planning for terrorism, 71; *see also* Emergency operation centers; Emergency Operations Plan; Organization for local response management; Weapon of mass destruction
  administration and logistics, 93–95
  bioterrorism incident, 90
  combined hazards, 76
  communications, 86
  conventional explosives and secondary devices, 76
  cyberterrorism, 77–78
  direction and control, 83–86
  emergency public information, 87–88
  hazard containment, 82
  health and medical, 89–90
  incident command from federal level down, 85
  infrastructure attacks, 77
  initial detection, 81–82
  low-technology devices and delivery, 77
  mass care, 88–89
  planning process, 72–73
  potential targets, 79–81
  pre-event readiness, 87
  protective actions, 88
  recovery, 91
  release area, 82
  resources management, 90–91
  terrorism hazards, 73, 76–77
  urban search and rescue, 91
  vulnerability areas, 79, 80–81
  warning, 86–87
Planning team development, 97, 100–101
  active participation of players, 98–99
  effective team characteristics, 100
  emergencies and response capabilities, 97
  emergency planning team, 97
  potential planning team members, 98
  team formation stages, 99–100
  team operation, 99
  team roles, 100

Point of contact (POC), 143, 155
PPE, see Personal protective equipment
PPPs, see Public–Private Partnerships
Pre-positioned vehicles, 194–196; see also Physical security
Profiling, 28, 29; see also Soft target hardening
Psy ops (Psychological operations), 14
PTZ, see Pan, tilt, and zoom
Public address systems, 201–202; see also Security
Public Information Officer (PIO), 64, 178–180; see also Chief elected officials; Emergency management and media
Public–Private Partnerships (PPPs), 370
PVI, see Place vulnerability index

Q

Queen's Counsel (QC), 412

R

R&D (Research and Development), 210
Radio, 172–173; see also Emergency management and media
Radiological dispersal device (RDD), 76
Rage, 45, 46
Rajneesh, Bhagwan Shree, 42
RCIWG, see Regional Counterintelligence Working Group
RDD, see Radiological dispersal device
Recovery point objective (RPO), 290
Recovery time objective (RTO), 285, 290
Red teaming, 216; see also Deterring and mitigating attack
Regional Counterintelligence Working Group (RCIWG), 209
Reid, Richard, 28, 29
Religious terrorism, 234
Remember me checkboxes, 467; see also Soft target cybersecurity
Removable data storage, 467; see also Soft target cybersecurity
Rescue Task Force (RTF), 272
Resilience Pilot Program (CR Pilot), 242
Resource-constrained operation, 30
Ressum, Ahmed, 41
Risk, 107
    management, 316

Rostelecom, 268
Routine emergency, 94
Rozin, M., 43
RPO, see Recovery point objective
RTF, see Rescue Task Force
RTO, see Recovery time objective
Rudolph, E., 39

S

SAFE Washington program, 212; see also Deterring and mitigating attack
Sandy Hook Elementary School, 197
SARA, see Superfund Amendment and Reauthorization Act of 1986
Sarin gas attack, 42
SARs, see Suspicious activity reports
SCC, see Sector Coordinating Council
School Massacre, Beslan, Russia, 448–451; see also Complex Coordinated Attacks
School resource officers (SROs), 233
Schools, 208–209, 223, 392–393; see also Deterring and mitigating attack; Soft target threat assessment
    attacks on college campuses, 237–239
    bad blood, 240–242
    campuses already under attack, 235–236
    case study, 227–232, 239–240
    CHAOS, 241
    COINTELPRO, 240
    colleges, 234–235
    CR Pilot, 242
    cybercriminals, 236
    intellectual property theft, 236
    K–12 in crosshairs, 226–227, 232
    K–12 vulnerability, 232–234
    knife attacks, 225
    in Middle East, 358
    MS-13 attack, 225
    religious terrorism, 234
    ripple effect of attacks, 224
    stabbings, 224–226
    U.S. universities in secured complex, 359
Sealing the room, 207; see also Emergency preparations
SEAR, see Special event assignment rating
Sector Coordinating Council (SCC), 302

Security, 196; *see also* Deterring and mitigating attack
  alarms, 198–200
  in Ben Gurion Airport, 355
  closed circuit television, 200–201
  code systems, 201–202
  fatigue, 24–25
  hotel, 361–367
  locks, 197–198
  public address systems, 201–202
  Sandy Hook Elementary School, 197
  ventilation systems, 202
  visitor access and badges, 200
Security organization, 294; *see also* Active shooters; Soft target; Workplace violence
  need to know, 303–305
  program to increase security awareness, 303
  security devices and systems, 304
Security Overseas Seminar (SOS), 369
SERCs, *see* State Emergency Response Commissions
7/7 bombing and experiences, 404, 420
  absence of guidance and procedures, 407
  access to key witness, 408
  attacks, 404
  challenges, 407
  coroner's inquests, 406
  crate issue, 409–410
  dealing with negative media coverage, 411–412
  double-decker bus following onboard bomb explosion, 405
  expect the unexpected, 412–413
  facing world's media, 417–418
  giving evidence, 413–416
  going back to work, 418
  hearts and minds issues, 410
  information overload, 408–409
  MPS's focus, 408
  personal reflections, 418–420
  7/7 team, 407
  support, 410–411
  technological upgrade, 407
  validation, 413
  verdicts, 416–417
  workload pressures, 412
Shekau, Abubakar, 18
Shelter in place, 207–208; *see also* Emergency preparations
Shoe bomber, *see* Reid, Richard

Shopping centers, 254, 394–395, 398; *see also* Soft target threat assessment
  Allied Barton Security Services, 259
  al-Shabaab video, 260
  case study, 256–259
  ISIS, 255
  mall attacks, 259
  Mall of America in Bloomington, 254
  mall violence, 255
  terrorist threats against U.S. malls, 259–261
  vulnerability, 254, 261
Siddiqui, Aafia, 41
Situation manual (SitMan), 138, 155
SMEs, *see* Subject-matter experts
Snowden, Edward, 32
SNS, *see* Strategic National Stockpile
Social media, 470–471; *see also* Soft target cybersecurity
Social network, 175, 328; *see also* Emergency management and media
Soft civilian-centric targets, 2
Soft target, 1, 16, 294, 303; *see also* Active shooters; Security organization; Terrorism; Terrorist; Workplace violence
  effect of attacks against, 3
  black swan theory, 16–18
  Boston Marathon bombing, 22
  challenges in protecting, 3
  fear underlying terrorist attacks, 23
  1949 Geneva Conventions, 21
  goal of attack, 294
  Internet as tool, 18–19
  measured response to terror event, 24–25
  profit-making, 4
  psychological trauma, 23
  psychology, 10
  radicalization of Americans, 6
  security fatigue, 24–25
  security training and resources, 4
  shooting in San Bernardino, 10
  soft targeting motivations, 19–21
  solutions, 295–296
  takeaways, 48–49
  things need to know about, 301–302
  threat, vulnerability, and risk model, 6–8
  U.S. Department of Homeland Security guidance, 302
  unique vulnerability, 21–22
  violence, 16–18
  Westgate Mall, 5

Soft target cybersecurity, 465
  access cards, 468
  assorted credit cards, 469
  assorted identification cards, 468
  basic device security, 466
  built-in cameras, 471
  cell phones, 467–468
  credit cards, 469
  cyber hygiene, 466
  defense in depth layers, 466
  disposal of devices, 471–472
  free apps, 468
  free Wi-Fi, 467
  home hardware, 467
  keyboards as weapons, 472–473
  passwords, 469–470
  remember me checkboxes, 467
  removable data storage, 467
  social media dangers, 470–471
Soft target hardening, 27; *see also* Deterring and mitigating attack
  background checks, 33–36
  behavioral profiling, 43
  cafeteria option, 30
  crises, 38
  Delphi methodology, 33
  double bomb tactic, 39–40
  face of rage, 45–48
  facial expression and lie detection, 47
  fighting the assumptions, 35
  first response and threat of secondary attacks, 39–40
  Geneva Conventions, 36–37
  goal of, 190
  good hiring practices, 32
  human shields tactics, 36
  insider threat, 31–32
  intuition, 30–31
  kidnapping and human shields, 36–37
  mole hunters, 34
  Nigeria's Boko Haram terrorist group, 36
  no dark corners approach, 34–35
  profiling, 28, 29
  resource-constrained operation, 30
  SIRA method, 43
  Snowden, Edward, 32
  soft target takeaways, 48–49
  steady-state and crisis leadership, 29–30
  suicide terrorism, 43–45
  weapons of mass destruction, 40–43
Soft target hardening in Middle East, 355; *see also* Hardening tactics
  churches, 360
  compound mentality, 356–358
  crime rate, 357
  hospitals, 360–361
  hotel security, 361–367
  schools, 358–359
  security in Ben Gurion Airport, 355
  Westerners as targets, 367–368
Soft target planning, 391, 402
  airports, 393–394
  command centers, 401–402
  convention center entrance, 399
  crisis management, 399–401
  defense in depth layers, 396
  DHS complex, 400
  fencing options for facilities, 399
  Go Kit supplies, 401
  movie theaters, 392
  schools, 392–393
  shopping malls, 394–395, 398
  soft target, 392, 395–396
  soft target threat assessment, 396–397
  soft target vulnerability assessment, 397–398
  tailgate parties and sporting events, 394, 395
Soft Targets Awareness Course, 190
Soft target threat assessment, 220; *see also* Churches; Hospitals; Schools; Shopping centers; Sports venues
  actors and target selection, 222–223
  ISIS, 253
  place vulnerability index, 221
  psychological vulnerabilities, 220
  small cities and rural areas in crosshairs, 221
  soft targets, 223
  tourist sites, 275–278
  vulnerability in United States, 220–221
SOPs, *see* Standard operating procedures
SORM, *see* System of operative-investigative measures
SOS, *see* Security Overseas Seminar
Special event, 325
  benefits, 337
  City of Phoenix Food Day, 326
  communications, 335–336
  IAPs, 336
  intelligence sharing, 328–329
  managers, 330–332
  operators 332
  paperwork, 336–337

personnel, 329–330
planning, 330, 325, 334–335
training 332–334
types, 326–328
volunteers, 335
Special event assignment rating (SEAR), 263
Sports and recreation venues, 214–216; *see also* Deterring and mitigating attack
Sports venues, 261; *see also* Soft target threat assessment; Sport venue command group
    attack on Boston Marathon, 264
    attack on Olympic Games, 263
    attacks in United States, 264
    case study, 265–269, 270–271
    college stadium vulnerabilities, 262
    football stadiums, 262
    ISIS-inspired Orlando nightclub shooting, 271–274
    managers, 310, 312
    performing arts and recreation venues, 269, 271
    stadiums and arenas, 261
Sport venue command group, 308; *see also* Sport venue emergency planning
    areas, 308
    business continuity, 318–319
    command center capabilities, 311
    communication and information sharing, 313–314
    countermeasure improvements, 317
    emergency response plan, 309–310
    establishing command center, 311
    evacuation planning, 311–312
    exercise types, 310
    incident commander, 312
    mitigation, 315
    physical upgrades, 318
    preparedness, 309
    recovery, 314–315
    response, 311
    risk management 316–318
    sport venue managers, 310, 312
    staff training and exercise, 310
    threat, 316
    vulnerability, 317
Sport venue emergency planning, 307, 308; *see also* Sport venue command group
    checklist for emergency preparedness, 319–320
    emergency management, 308, 309
    emergency planning, 308
    evacuation plan template for stadiums, 321–323
SROs, *see* School resource officers
Standard operating procedures (SOPs), 111, 117, 155, 160
START, *see* Study of Terrorism and Responses to Terrorism
State Emergency Response Commissions (SERCs), 93
State laws, 125–126; *see also* Disaster preparedness and law
Steady-state operations, 29; *see also* Soft target hardening
Strategic National Stockpile (SNS), 90
Strategy, 374; *see also* Emergency Management program strategy
Study of Terrorism and Responses to Terrorism (START), 458
Subject-matter experts (SMEs), 141, 217, 155
Suicide bombings, 43; *see also* Soft target hardening
    signs of suicide terrorist activity, 44–45
Superfund Amendment and Reauthorization Act of 1986 (SARA), 93
Suspicion Indicators Recognition and Assessment (SIRA), 43
Suspicious activity reports (SARs), 28
Swarm attack, 344–345; *see also* Complex Coordinated Attacks; Mumbai, attacks in
Symbolic targets, 16
System of operative-investigative measures (SORM), 267

T

Tabletop exercises (TTXs), 135, 156
    facilitator's guide, 144; *see also* Active shooter threat
Tactical Combat Casualty Care (TCCC), 350; *see also* Complex Coordinated Attacks
Tactical Emergency Casualty Care (TECC), 350
Tactics, techniques, and procedures (TTPs), 439
Tailgate parties and sporting events, 394, 395
TARA, *see* Toronto Aquarium Resistance Alliance
Target capabilities list (TCL), 136, 156
TATP, *see* Triacetone triperoxide
TCCC, *see* Tactical Combat Casualty Care

INDEX

TCL, *see* Target capabilities list
TCT, *see* Team Coordination Training
TDEM, *see* Texas Division of Emergency Management
Team Coordination Training (TCT), 442
TECC, *see* Tactical Emergency Casualty Care
Technology recovery (TR), 290
Tehrik-i-Taliban Pakistan (TTP), 247
Television, 174; *see also* Emergency management and media
Terrorism, 10–11, 28; *see also* Planning for terrorism; Soft target
    Alfred P. Murrah building, 15
    amoral societies, 14
    case of Jose Padilla, 14
    effectiveness, 13–16
    enhanced law enforcement activity, 15
    hazards, 76–77
    life space, 13
    measured response to terror event, 24–25
    physical profiling, 28
    product of marketing campaign, 18
    profiling by age, 28
    radical jihadists, 28
    rise of modern, 15
    scientific approach to, 15
    target, 16
    triplet of opacity, 24
Terrorism Risk Insurance Act (TRIA), 222
Terrorist, 11
    behavior, 12–13
    goal of, 340
    incidents, 167
    information from operations, 22
    Internet usage, 19
    motivations, 11–12
    operations against civilians, 2
    rage, 47
    targets, 16
Terror medicine, 346–347; *see also* Complex Coordinated Attacks
Terror multiplier effect (TME), 21
Texas Division of Emergency Management (TDEM), 379
The Joint Commission (TJC), 291
Threat, 316, 458; *see also* Violent attacks and soft targets
Threat assessment model, 208; *see also* Deterring and mitigating attack
TJC, *see* The Joint Commission

TME, *see* Terror multiplier effect
Toronto Aquarium Resistance Alliance (TARA), 255
Tourist sites, 275; *see also* Soft target threat assessment
    case study, 275–278
TR, *see* Technology recovery
Traffic duty, 196; *see also* Physical security
Transportation Security Administration (TSA), 194
TRIA, *see* Terrorism Risk Insurance Act
Triacetone triperoxide (TATP), 264
Triplet of opacity, 24
TSA, *see* Transportation Security Administration
TTP, *see* Tehrik-i-Taliban Pakistan
TTPs, *see* Tactics, techniques, and procedures
TTXs, *see* Tabletop exercises

U

UCC, *see* Umpqua Community College
UCS, *see* Unified Command System
Umpqua Community College (UCC), 239; *see also* Schools
Underwear bomber, *see* Abdulmutallab, Umar Farouk
Unified Command System (UCS), 133, 156
United States in post-9/11 world, 1
Universal task list (UTL), 142, 156
UN's PPP Initiative, 371; *see also* Hardening tactics
Urban Search and Rescue (US&R), 91
USA PATRIOT Act, 15
U.S. Department of Homeland Security (DHS), 23, 53, 194, 302, 354
    complex, 400
U.S. Universities in secured complex, 359
UTL, *see* Universal task list

V

VA, *see* Veterans Affairs
VB, *see* Vehicle-borne
VBIED, *see* Vehicle-borne IED
Vehicle-borne (VB), 248
Vehicle-borne IED (VBIED), 41
Ventilation systems, 202; *see also* Security
Veterans Affairs (VA), 89
Victim rescue unit (VRU), 207

Violent attacks and soft targets, 457
    application, 464
    consequence, 458
    consequences of violent attacks, 462
    mitigation measure, 459–460
    risk management, 458–459
    soft *vs.* hard targets, 460–464
    threat, 458
    vulnerability, 458
    vulnerability assessment process, 460–461
ViSAT, *see* Vulnerability self-assessment tool
Visitor access and badges, 200; *see also* Security
Volatile, uncertain, complex, and ambiguous (VUCA), 442
VRU, *see* Victim rescue unit
VUCA, *see* Volatile, uncertain, complex, and ambiguous
Vulnerability, 317, 458; *see also* Violent attacks and soft targets
Vulnerability self-assessment tool (ViSAT), 216

W

Weapon of mass destruction (WMD), 40, 73, 202; *see also* Planning for terrorism
    al-Qaeda, 40
    Bergendorff, Roger, 41
    biological agents, 74–75
    biological warfare, 41
    chemical agents, 73–74
    chlorine, 41
    combined hazards, 76
    detection of attack, 81
    hazard agents, 73
    incident, 91
    nuclear/radiological agents, 75–76
    Salmonella contamination, 42
    sarin gas attack, 42
    synthetic manufacturing, 42
Weblog, 175
Westgate Mall, 5
Workplace violence, 298; *see also* Active shooters; Security organization; Soft target
    solutions, 300–301
    violence and crime in workplace, 299–300
    work environment, 299
World Wide Web, 174; *see also* Emergency management and media